Supernova 1987A in the Large Magellanic Cloud

Supernova 1987A in the Large Magellanic Cloud

Proceedings of the fourth George Mason Astrophysics Workshop held at the George Mason University, Fairfax, Virginia, 12–14 October, 1987

Edited by

Minas Kafatos
George Mason University,
Fairfax, Virginia

and

Andrew G. Michalitsianos
NASA – Goddard Space Flight Center

CAMBRIDGE UNIVERSITY PRESS

Cambridge

New York New Rochelle

Melbourne Sydney

CAMBRIDGE UNIVERSITY PRESS
Cambridge, New York, Melbourne, Madrid, Cape Town, Singapore, São Paulo

Cambridge University Press
The Edinburgh Building, Cambridge CB2 2RU, UK

Published in the United States of America by Cambridge University Press, New York

www.cambridge.org
Information on this title: www.cambridge.org/9780521355759

First published 1988
This digitally printed first paperback version 2006

A catalogue record for this publication is available from the British Library

ISBN-13 978-0-521-35575-9 hardback
ISBN-10 0-521-35575-3 hardback

ISBN-13 978-0-521-03161-5 paperback
ISBN-10 0-521-03161-3 paperback

Contents

FOREWORD

This volume includes all of the invited and contributed scientific papers presented at the Fourth George Mason Fall Workshop in Astrophysics "Supernova 1987A in the Large Magellanic Cloud". The Workshop was held at George Mason University, Fairfax, Virginia, on 12 - 14 October 1987. The papers provide a comprehensive review of the major observational and theoretical results of SN 1987A, which were obtained prior to the meeting. Highlights of the Workshop included contributions from the three major neutrino experiments involved in the detection of the neutrino burst; optical, infrared and radio observations from Southern hemisphere observatories; ultraviolet spectroscopy from the International Ultraviolet Explorer (IUE), and the detection of hard and soft x-ray emisssion from the KVANT Experiment onboard the Soviet MIR Space Station, as well as x-ray observations from the Japanese Ginga x-ray satellite. Finally, a large number of theoretical papers addressed the physics of the outburst and core collapse, modeling the emission from the ejected shell, and the interaction of the ejecta with the environment. The uniqueness of this supernova event, the first naked-eye supernova since the time of Kepler, underscores the importance of assembling experts from a variety of scientific disciplines in a common forum. The interdisciplinary interest in this supernova, for which members of the nuclear physics research community were present, testifies to the great opportunities afforded the scientific community by this singular event. The Workshop was particularly successful in achieving cross disciplinary interaction.

The Editors
December, 1987

ACKNOWLEDGEMENTS

We would like to thank other members of the local Organizing Committee: Drs. R. Ellsworth, L.W. Fredrick, Y. Kondo, S.P. Maran, C.F. McKee, S. Mitton, B.J. Teegarden, V.L. Trimble, R. Viotti and N.R. Walborn for their assistance for planning this Workshop. We would also like to extend our thanks for generous support: to the Dean of the College of Arts and Sciences at George Mason University Dr. G.P. Lewis, to the Dean of the Graduate School at George Mason University, Dr. C.K. Rowley, and to the Chairman of the Physics Department, Dr. R.E. Ehrlich; Cambridge University Press; the Laboratory for Astronomy and Solar Physics at NASA-Goddard Space Flight Center; finally the National Aeronautics and Space Administration, Headquarters-High Energy Astrophysics Branch for the generous grant in support of this Workshop, NAGW-1229. We would also like to extend our thanks to Ms. C. Dion and the physics staff at George Mason for their assistance in organizing the meeting; to Mrs. D.F. Smith for typing; and to the staff of the Office of Community Services at George Mason University for their assistance in organizing this Workshop.

The Editors
December, 1987

WORKSHOP PARTICIPANTS

W. D. Arnett
J. N. Bahcall
N. Bartel
R. Bates
A. Bazzano
A. K. Bhatia
L. Bildsten
R. Bloomer
S. Bludman
E. Boldt
K. Brecher
U. Briel
J. C. Brown
F. Bruhweiler
A. Bunner
A. S. Burrows
A. A. Chalabaev
L. Chase
D. Chernoff
D. Cherry
R. Chevalier
T. L. Cline
A. Clocchiatti
S. Colgate
W. J. Couch
C. Crannell

A. Crotts
J. J. D'Amario
A. C. Danks
I. J. Danziger
A. Dar
R. J. Drachman
E. Dwek
R. Eastman
T. Ebisuzanki
R. Ehrlich
J. Elliott
D. Ellison
R. Ellsworth
R. Fahey
M. W. Feast
J. E. Felten
A. V. Filippenko
G. Fishman
J. Franklin
C. Fransson
L. W. Fredrick
G. Frye
I. Fusjiki
P. Galeotti
G. Garmire
M. Gaskell

N. Gehrels

B. Geldzahler

D. Goldsmith

J. A. Graham

T. R. Gull

H. Gursky

A. K. Harding

F. R. Harnden, Jr.

M. Harris

M. Hauser

H. Heaton

D. J. Helfand

M. Henricksen

P. Hertz

S. Holt

P. C. Joss

G. V. Jung

M. Kafatos

L. Kaluzienski

M. Karovska

D. Kazanas

R. Kinzer

R. P. Kirshner

O. Kjeldseth-Moe

Y. Kondo

M. Koshiba

J. Kristian

H. Kunieda

G. S. Kutter

D. Lamb

H. P. Larson

F. Lasche

B. M. Lasker

J. M. Lattimer

M. D. Leising

L. Lucy

G. Malinie

P. Mannheim

S. P. Maran

L. A. Marschall

F. Marshall

J. Matthews

S. M. Matz

A. G. Michalitsianos

J. Miller

K. Miller

E. S. Myra

G. Nakano

P. Nisenson

K. Nomoto

R. Novick

W. Paciesas

F. Pacini

N. Panagia
J. Pantaleone
C. Papaliolios
C. Pennypacker
L. Peterson
R. Petre
M. Phillips
G. Pizzella
P. Podsiadlowski
T. A. Prince
G. R. Riegler
S. Rodano
A. Rossiter, Jr.
W. Sandie
B. Schaefer
P. J. Schinder
E. M. Schlegel
M. Schmitz
R. A. Schorn
P. J. Serlemitsos
M. Shapiro
M. Shull
R. Silberberg
G. Skinner
G. Sonneborn
D. Spicer
R. Stachnik
T. Stanev
R. Starr
T. Stecher
B. Stiller
M. Strickman
N. Suntzeff
A. Szentgyorgyi
R. Talcott
Y. Tanaka
B. Teegarden
D. Teplitz
V. Teplitz
D. Thompson
E. Thomsen
V. Trimble
J. Trumper
J. W. Truran
R. E. Turner
P. Ubertini
C. Vaganos
S. Van den Bergh
D. Voss
N. R. Walborn

M. Waldrop

J. C. Wang

T. A. Weaver

D. Wentzel

F. West

J. Wheatley

J. C. Wheeler

G. Wilmot

G. W. Wolf

M. T. Wolff

S. Woosley

IMAGES AND SPECTROGRAMS OF SANDULEAK -69°202, THE SN 1987A PROGENITOR

Nolan R. Walborn
Space Telescope Science Institute, Baltimore, MD 21218, USA

Abstract. The resolution of the image of Sk -69°202 on CTIO 4-meter plates into three components is described, and the various suggestions concerning a fourth star in the system are reviewed. A preliminary description of the ESO spectrograms of the progenitor is also presented.

INTRODUCTION

Among the unprecedented attributes of Supernova 1987A in the Large Magellanic Cloud, it is the first such event for which direct observational information about the progenitor star is available. Hence it is of interest to extract the full content of these precious preoutburst data. It is also important to understand their limitations, to avoid overinterpretation. Sanduleak -69°202 is but one of 1272 objects in the objective-prism OB survey of the LMC to about twelfth magnitude whose name it bears. In the approximately one hundred years since relevant southern-hemisphere surveys began, it did not distinguish itself by prominent light variations or spectral peculiarities, which might have attracted more detailed attention. This (non)observation may be significant in itself, but as a result only data of either serendipitous or survey character exist for the SN 1987A progenitor.

DIRECT IMAGES

Walborn, Lasker, Laidler, & Chu (1987) have scanned the image of Sk -69°202 on eight (of 32 available) blue through near-infrared photographic plates of the 30 Doradus region, obtained at the prime focus of the Cerro Tololo Inter-American Observatory 4-meter telescope during 1974-1983. Both intensity syntheses of the image and density differences (Figure 1) were derived by means of reference stars from the same plates, including the similar nearby object Sk -69° 203. The analysis shows that the 12^{m} blue supergiant in Sk -69°202 (Star 1) has two companions with V magnitudes, position angles, and separations $15^{m}\!.3$, 315°, 3" (Star 2) and $15^{m}\!.7$, 115°, $1''\!.5$ (Star 3), respectively. Both companions appear to be early-type stars, and there is no evidence for a bright red star in the system.

Astrometric measurements by others have shown that the position of the SN image coincides with that of Star 1. As discussed by G. Sonneborn in these proceedings, the two companions are responsible for the stellar spectra observed by the International Ultraviolet Explorer following the decline of the SN in the far UV, so that Star 1 has disappeared and was most probably the progenitor.

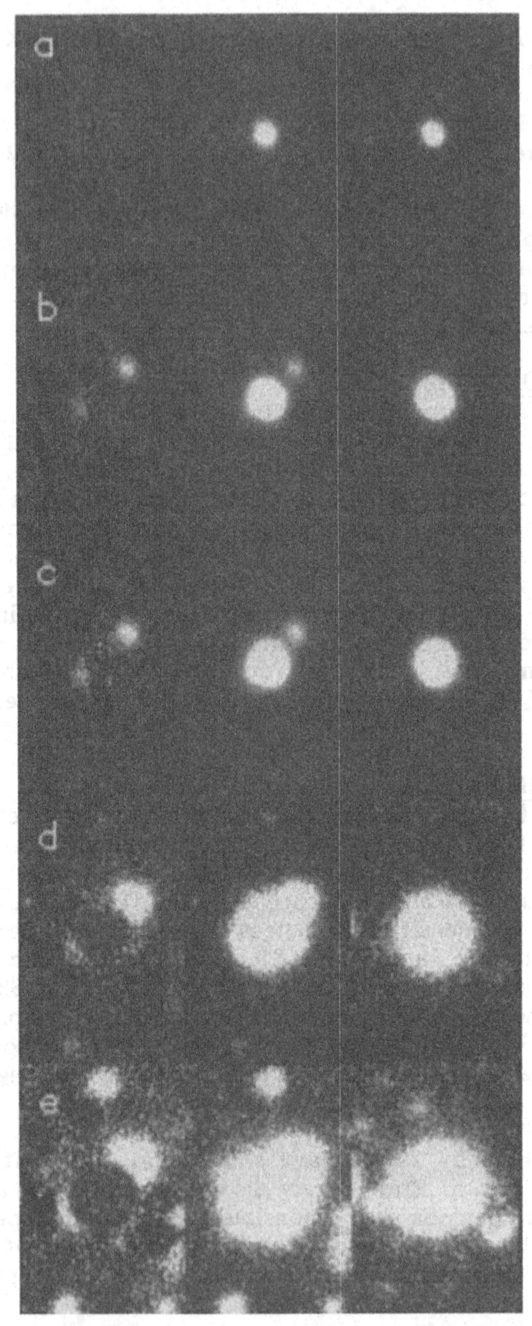

Figure 1 - Density differences from 5 plates, Sk -69°202 minus Sk -69°203. The left panel of each row shows the image subtraction from a given plate, while the center panel is the original image of Sk -69°202 and the right panel is that of Sk -69°203, from the same plate. North is up and east to the left in each case. Star 2 is at 3" in PA 315° and Star 3 is at 1".5 in PA 115°. (a) Plate No. 345 (4765 Å, 2 min--the apparent "nebulosity" is spurious), (b) No. 5973 (4765 Å, 10 min), (c) No. 5976 (5000 Å, 10 min), (d) No. 719 (6725 Å, 30 min), (e) No. 4858 (IV-N, 90 min).

STAR 4?

The existence or otherwise of a fourth star in the Sk -69°202 system has become a subject of confusion for two basic reasons, namely that the term has been applied to four distinct proposed objects, and none of them is definitely established at the present time.

(1) Testor & Lortet (1987) suggested that Star 1 was a 0".4 north-south double from a preoutburst [O III] image. Heap & Lindler (1987) found possible evidence for a similar structure in one of the Walborn *et al.* [O III] images, although with a separation twice as great and uncertainty due to image saturation. Neither component could have been very red and neither is now visible in the UV, although a companion at that distance could not yet have been engulfed by the SN ejecta. The reality of this structure is subject to verification by further analysis of additional available images.

(2) Djorgovski (1987) reported a 10^m companion to the SN at 235°, 1".8 in post-outburst [O II] images. However, this result is now understood as an artifact due to the combination of a red leak in the [O II] filter, the very red color of the SN, and atmospheric differential refraction, and it has been withdrawn. Curiously, Heap & Lindler found a very faint [O III] image at essentially the same position, which therefore must have a different origin.

(3) The idea that the SN progenitor might have been a fourth, unobserved star positionally coincident with, and possibly a binary companion to, Star 1 originated in the initial surprise at a blue supergiant progenitor, before the characteristics of the explosion itself had been shown to be consistent with the latter and plausible evolutionary paths were discussed (e.g. Woosley, Pinto, & Ensman 1987; Maeder 1987). A specific model of this type is developed by P. Joss *et al.* in these proceedings. In this hypothesis, Star 1 has merely been engulfed and will reappear when the debris become optically thin; however, it now appears neither necessary nor likely.

(4) Nisenson *et al.* (1987) have reported a very bright companion 0".06 from the SN in speckle interferometric images, which must be an effect of the outburst if real. There appears to be some doubt as to whether or not the companion was redetected with a slightly greater separation at a later epoch (C. Papaliolios, these proceedings).

SPECTROGRAMS

So far as is known, the available preoutburst spectrograms of Sk -69°202 are all objective-prism photographs, and those of highest information content were obtained with two different instruments at the European Southern Observatory.

W. Wamsteker obtained a plate of dispersion 460 Å/mm at $H\gamma$ with the ESO Schmidt in 1977, a preliminary discussion of which has been given by González *et al.* (1987). In comparison with the spectrum of the normal LMC B3 Ia star Sk -67°78 (Fitzpatrick 1987), that of Sk -69°202 indicates stronger He I lines; a feature which at this resolution is a blend of N II λ 3995, He I λ 4009, and He I λ 4026 is particularly striking. However, the resolution is insufficient to determine independently whether these features correspond to a chemical anomaly, or to a temperature/luminosity difference.

L. and M. L. Prévot have prepared high-quality tracings of the 110 Å/mm (at H_γ) plates taken with the ESO astrograph in 1972-1973, upon which the original B3 I classification of Sk -69°202 is based (Rousseau *et al.* 1978). These spectrograms were limited to λλ 3850-4150 by an interference filter, to alleviate overlapping in crowded fields; hence the useful spectral features are delimited by H_ζ and H_δ. A number of appropriate comparison stars for which optimum digital data are available (Fitzpatrick 1987) are also being extracted from the objective-prism plates, to enable a reliable differential description of the Sk -69°202 spectrum. Preliminary inspection indicates that it is He I λ 4026 rather than the N II line which is primarily responsible for the enhanced blend seen at the lower resolution. A full presentation of the independent information provided by these unique spectrograms will be made in a subsequent joint publication, once the analysis has been completed.

The Space Telescope Science Institute is operated by the Association of Universities for Research in Astronomy, Inc. under contract with the National Aeronautics and Space Administration.

REFERENCES

Djorgovski, S. G. (1987). *IAU Circ.* No. 4376; preprint.

Fitzpatrick, E. L. (1987). IAU Symposium No. 132.

González, R., Wamsteker, W., Gilmozzi, R., Walborn, N., & Lauberts, A. (1987). ESO SN 1987A Workshop.

Heap, S. R. & Lindler, D. J. (1987). *Astr. Ap.*, in press.

Maeder, A. (1987). ESO SN 1987A Workshop.

Nisenson, P., Papaliolios, C., Karovska, M., & Noyes, R. (1987). *Ap. J. Letters*, **320**, L15.

Rousseau, J., Martin, N., Prévot, L., Rebeirot, E., Robin, A., & Brunet, J. P. (1978). *Astr. Ap. Suppl.*, **31**, 243.

Testor, G. & Lortet, M.-C. (1987). *IAU Circ.* No. 4352; preprint.

Walborn, N. R., Lasker, B. M., Laidler, V. G., & Chu, Y.-H. (1987). *Ap. J. Letters*, **321**, L41.

Woosley, S. E., Pinto, P. A., & Ensman, L. (1987). *Ap. J.*, in press.

THE PROGENITOR OF SN 1987A

G. Sonneborn
IUE Observatory, Goddard Space Flight Center,
Greenbelt, MD 20771 and
Astronomy Programs, Computer Sciences Corporation,
Beltsville, MD 20705

INTRODUCTION

Ultraviolet spectra of SN 1987a obtained with the International Ultraviolet Explorer (IUE) satellite within 15 hours of its discovery showed that the supernova's far ultraviolet flux (λ <2000A) was rapidly fading, even as the optical brightness continued to rise (Kirshner et al. 1987). Within several days the flux decreased by three orders of magnitude, presenting a unique opportunity for observing the ultraviolet stellar surroundings of the supernova even while the optical glare obscured its immediate neighbors from ground-based measurement. By 1987 February 28 the far ultraviolet flux had dropped to a level where a constant, stellar background was detected shortward of 1500A.

Examination of preoutburst plates revealed an excellent positional coincidence (<0".1) between the supernova and the 12th mag B3 I star Sk -69° 202 (White and Malin 1987), the presence of a second star 3" to the NW (Lasker 1987), and some evidence for a third star about 1" SE of Sk -69° 202 (Walborn et al. 1987a; Chu 1987). Similarities between the IUE spectrum of the ultraviolet source at the position of the supernova and that of a luminous early-type star led to a suggestion that Sk -69° 202 might have survived the explosion (Cassatella et al. 1987, Sonneborn and Kirshner 1987). In addition, the IUE spectra were broadened perpendicular to the dispersion direction shortward of 1500A, indicating the presence of more than one point source in the 10"X20" aperture during the supernova exposures (Sonneborn and Kirshner 1987).

The detection of two early-type ultraviolet stellar spectra at the site of the fading supernova and the identification of two early-type stars in the Sk -69° 202 system on preoutburst plates led to a suggestion that the spectra came from the two stars: Sk -69° 202 (Star 1) and Star 2 (Sonneborn and Kirshner 1987b). Because Chu (1987) estimated the visual magnitude of Star 3 to be near 17.5, this did not seem a plausible source for the light detected in our IUE exposures.

Subsequently, high-precision measurements and image syntheses of the Sk -69° 202 field (Walborn et al. 1987b, West et al. 1987a,b) provided more accurate relative positions and magnitudes for the three stars. These investigations showed that Star 3 was about two magnitudes brighter than previously thought, at m=15.5, about 1".5 SE of

Star 1, and blue in color, prompting further analysis of the available IUE spectra of SN 1987a, as reported by Sonneborn, Altner, and Kirshner 1987). Their results are consistent with an independent analysis by Gilmozzi et al. (1987). Here I review the evidence which demonstrates that the objects detected near the supernova were Stars 2 and 3 and that Sk -69° 202 was not present in our ultraviolet spectra of SN 1987a.

OBSERVATIONAL MATERIAL

Questions concerning the identity and characteristics of the stars within several arc seconds of the supernova, and detected in the IUE observations, are best addressed by analysis of SWP spectra (1150-2000A) taken after 1987 March. During this period the far UV flux level of the supernova was below that of the other stars in the aperture and, as discussed below, the projected separation between the stars was sufficiently large to allow deconvolution of their spectra. The observations were made from the IUE Science Operations Center at Goddard Space Flight Center. All exposures were taken in the low-dispersion mode (resolution ~ 6-7A) under conditions of optimum telescope focus. The supernova's optical center of light, as measured by the IUE Fine Error Sensor, was centered in the large aperture for each exposure.

The analysis in subsequent sections is based on IUE spatially-resolved spectra. The IUESIPS spectral extraction procedures which preserve the spatial information in low-dispersion IUE images are described by Turnrose and Thompson (1984) and updated by Munoz Piero (1985) for the so-called "extended line-by-line" (ELBL) enhancement. The ELBL spectral file is a two-dimensional array of IUE Flux Numbers with about 800 wavelength points (1150-2000A) and 110 lines in the spatial direction at each wavelength. The spatial separation between adjacent lines is $\sqrt{2}/2$ diagonal pixels. Adopting the spatial scale of 1".51±0".05 per diagonal pixel (Panek 1982), this corresponds to 1".07±0".04 per line. The separation between the spectra of two point sources in the ELBL data is the projection (perpendicular to the dispersion) of their true angular separation on the plane of the sky.

The positions of Stars 1, 2, and 3 as measured by West et al. (1987b) and by Walborn et al. (1987b, hereafter WLLC) are in very good agreement within their estimated errors. We adopt the measurements of WLLC for comparison with our data due to their large-scale plate material and smaller estimated errors. The orientation of the three stars and the IUE large aperture for SWP 30408, 30512, and 30592 are shown to scale in Figure 1. Since the aperture is fixed within the IUE scientific instrument, the aperture's orientation on the sky is determined by spacecraft solar array orientation constraints. Because the LMC is located near the south ecliptic pole, the IUE satellite's roll orientation changes by about 1° per day. It was fortuitous that the stars within arc seconds of SN 1987a were nearly aligned with the spatial direction of the spectrograph when the supernova's ultraviolet flux faded.

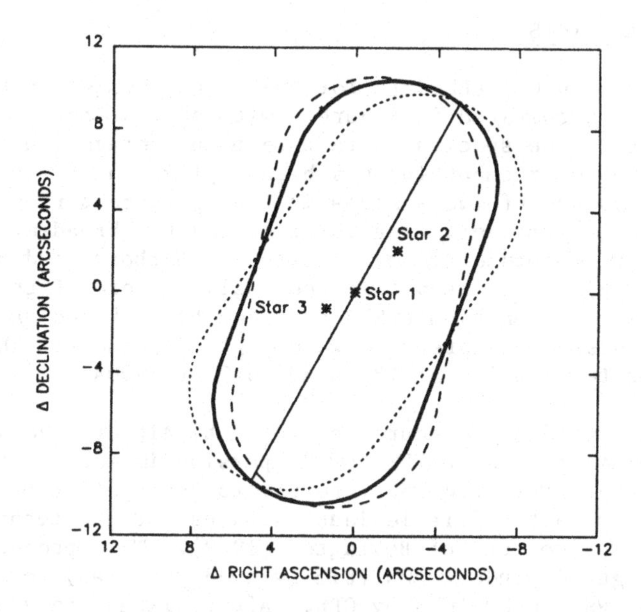

Figure 1 - Changing IUE aperture orientation with respect to the SN 1987a field. The WLLC positions of Stars 2 and 3 are shown relative to Star 1. The outline of the aperture is shown for three images, SWP 30408 (small dash), 30512 (solid), and 30592 (large dash), taken over a three week period (see Table 1). The straight line is the direction perpendicular to the dispersion for SWP 30512.

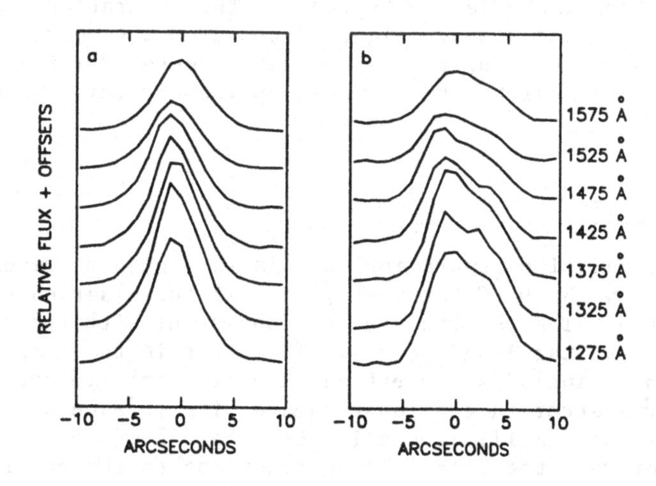

Figure 2 - Low-dispersion SWP spatially-resolved spectra. a) The ELBL data for a single point source, Sk -70 50 (SWP 28765), is shown at 50A intervals from 1275 to 1575A. b) The ELBL data for one of the supernova images (SWP 30512) shows the presence of two sources close to the position of SN 1987a.

DATA ANALYSIS

The 1250-1600A ELBL data for one of our images (SWP 30512, 1987 March 13.60) are compared in Figure 2 with similar data for a single point source. The spectral data have been averaged in 50A intervals. No binning or smoothing has been applied in the spatial direction. The presence of two sources in the aperture during the supernova exposure is apparent. The IUE instrumental broadening function has been well documented by Cassatella, Barbero, and Benvenuti (1985, hereafter CBB). They show that the analytic form of the cross-dispersion point-spread function (PSF) is given by a skewed gaussian, where the width and skewness are both wavelength-dependent. The FWHM in SWP spectra varies between 4".6 at 1350A and 6".0 at 1900A.

We have used a procedure developed by Altner (1987) to separate overlapping spectra in IUE low-dispersion images of the cores of globular clusters. Multiple PSFs are fit to the spatial profile of the ELBL data with a multi-variable least-squares fitting technique based on the CURFIT algorithm of Bevington (1969). This procedure adopts the wavelength-dependent analytic (skewed gaussian) form of the IUE low-dispersion PSF determined by CBB. After fixing the gaussian width and skewness, the two other parameters which describe each stellar component, the position along the spatial axis and the peak flux, are determined by the least-squares analysis.

This procedure was used to fit two point sources in the 1250-1550A region of ten well-exposed line-by-line spectra taken at the position of SN 1987a. The separation and peak flux ratio of the PSF gaussians were computed at 25A intervals for each image. The mean separation for each image is the average of the separations measured in the individual 25A bandpasses. Figure 3 compares the measured separations as a function of time with that expected for Stars 1 and 2 and for Stars 2 and 3 along the direction perpendicular to the dispersion.

RESULTS

The gaussian separations are in very good agreement with those predicted by the WLLC astrometry only if the observed stars are Star 2 and Star 3. The results are not consistent with the expected separation between Star 1 and Star 2. The error in the component separations, which includes the estimated uncertainty in the IUE spatial scale and the rms error in the least-square fits to each section of the spatial profile, are small, generally less than 5% of the gaussian FWHM. We do not know the relative contributions to the small systematic differences between measured and predicted separations from measurement error in the WLLC positions and in the IUE analysis and spatial scale. However, these difference can be accounted for by a four degree rotation of the Star 2 - Star 3 position angle, as shown in Figure 3.

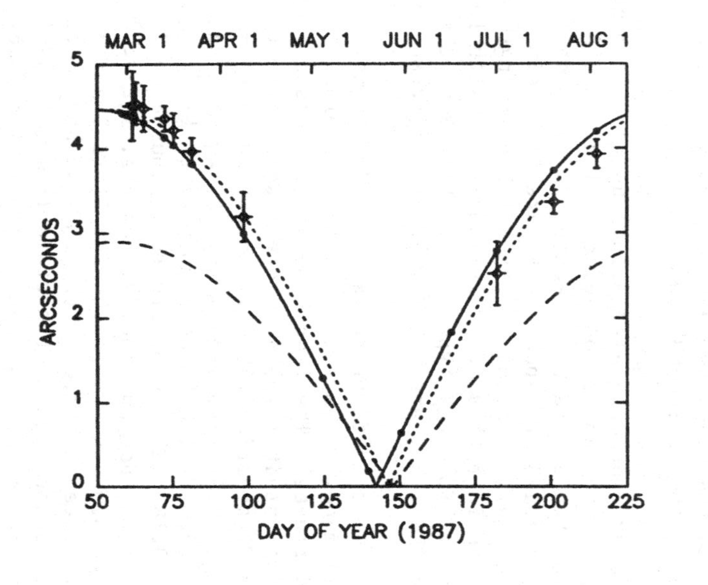

Figure 3 - The measured (diamonds) and predicted (solid dots and line) separation between Stars 2 and 3 in the IUE ELBL spectra are shown as a function of time. The expected separation between Stars 1 and 2 is also shown (large dashed line). The small dashed line represents a 4° change in the Star 2 - Star 3 position angle, within the stated errors of WLLC.

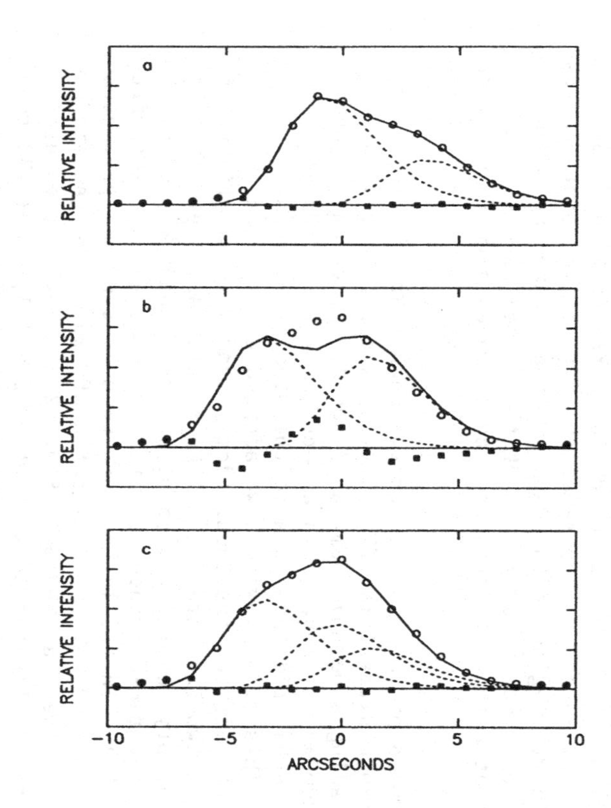

Figure 4 - Least-square fits of IUE PSFs to spatial profiles of the 1350-1375A region of ELBL data. a) Two components are fit to SWP 30512. The computed fit (solid line), observations (open circles), individual PSFs (dashed lines), and residuals (solid squares) are also shown. The left- and right-hand components correspond to Stars 2 and 3, respectively. b) A two-component fit to SWP 30408. The component separation was 4".50. c) A three-component fit to the same data as in Figure 4b is in excellent agreement with the observations (see text).

Figure 4a shows the two-component fit to the ELBL data for the 1350-1375A interval of SWP 30512, which is representative of our images taken after 1987 March 4. The excellent fit of two point sources to this short-wavelength region of the data and the general agreement with the astrometry of preoutburst plates of the Sk -69° 202 system makes it unlikely that a third point source contributes significantly to the IUE spectra.

There was a period of a day or two when Star 2, Star 3, and SN 1987a had comparable ultraviolet flux levels in the 1230-1450 A region, while the fading supernova dominated the SWP spectrum at longer wavelengths (1500-2000A). We have taken a section of SWP 30408 (1987 February 27.54) and subjected this data to the same two-component fitting procedure. This time, however, we adopted a projected separation consistent with our previously determined values. Figure 4b shows the result when two PSFs are fit to the spatial profile. This is clearly an unsatisfactory result; two point sources cannot reproduce the observations. When a three-component fit is computed, an excellent fit is obtained (Figure 4c). In the latter case we used the same projected separation as Figure 4b and set the flux ratio of two of the gaussians to that found in later spectra when only Stars 2 and 3 were present. The third point source is located between Stars 2 and 3, $2''.89 \pm 0''.30$ from Star 2 (the left-most component in Figure 4c), in excellent agreement with the position of Sk -69° 202/SN 1987a with respect to Stars 2 and 3. This procedure demonstrates that the spectrum of SN 1987a, when present, was positioned between the other two ultraviolet sources in the field, consistent with the position of Sk -69° 202.

The gaussian fitting procedure also allows us to determine the relative contribution of Stars 2 and 3 to the total flux in 25A intervals (e.g. Figure 4a) for the entire SWP spectral range. The deconvolved spectra of Stars 2 and 3 are shown in Figure 5. Their spectra have the appearance of early-B spectral types. The strength of the numerous interstellar lines in both spectra is consistent with IUE low-dispersion spectra of other LMC early-type stars. Both Star 2 and Star 3 must be located in the LMC, and are not foreground objects.

We have compared the spectra of Stars 2 and 3 with the Kurucz (1979 and unpublished) model atmosphere grid to determine the best fit to their dereddened fluxes. The reddening was estimated from the colors of Sk -69° 202 (Isserstedt 1975) and recent work on intrinsic colors of LMC supergiants by Fitzpatrick (1987). We assumed a two-component extinction: $E(B-V)_{GAL}=0.08$ from the galaxy (Savage and Mathis 1979) and $E(B-V)_{LMC}=0.1$ from the LMC (30 Doradus curve from Fitzpatrick 1985), for a total E(B-V) of 0.18. The one-third solar abundance Kurucz models which give the best fit to the continuum slopes are: T_{eff}=20000K and 25000K (±2000K) and log g= 3.0 and 4.5 (±0.5) for Stars 2 and 3, respectively. Assuming a distance of 50kpc to the LMC, we have derived estimates for the photometric properties of the stars from the model fluxes. Our derived V magnitudes for Stars 2 and 3 are 14.4 and 16.0 (±0.4), respectively, in agreement with the magnitude estimates of West et al. (1987b) and WLLC.

In summary, only two of the original three stars comprising the Sk -69° 202 system are detected in spatially-resolved ultraviolet spectra taken at the position of SN 1987a. The temporal variation of the projected separation of the spectra are in good agreement with the positions of Stars 2 and 3 measured on preoutburst plates. Futhermore, in late 1987 February the position of the supernova spectrum shortward of 1500A, relative to Stars 2 and 3, was in excellent agreement with the expected position of Star 1. We identify the stellar spectra with Stars 2 and 3. Sk -69° 202 is absent from the field and is plausibly identified as the star which exploded. The known characteristics of Sk -69° 202 are consistent with the interpretation that the progenitor of SN 1987a was a relatively compact star (Woosley et al. 1987), having a high-velocity, low-density stellar wind prior to the outburst (Chevalier and Fransson 1987). Our recent SWP spectra of SN 1987a (1987 July) show no evidence that Sk -69° 202 still exists inside the expanding shell.

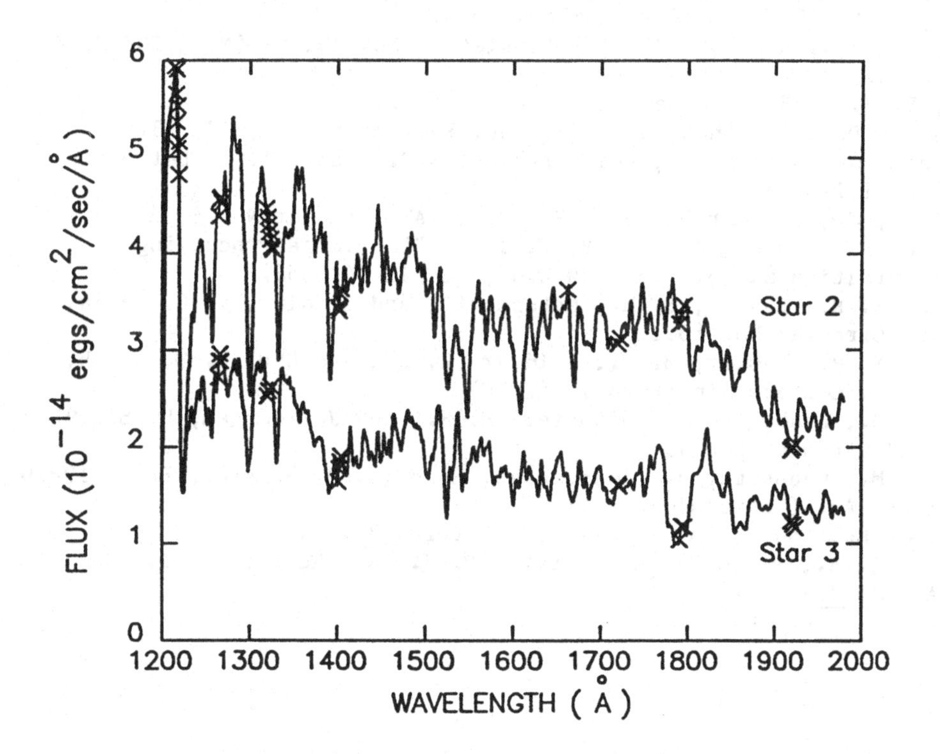

Figure 5 - The deconvolved spectra of Stars 2 and 3. The strong interstellar lines imply both stars are in the LMC. Xs mark portions of the spectrum effected by camera reseaux.

REFERENCES

Altner, B. 1987, PhD Dissertation, Rutgers University (in preparation)

Bevington, P. R. 1969, Data Reduction and Error Analysis for the Physical Sciences, McGraw-Hill, New York.

Cassatella, A., Barbero, J., and Benvenuti, P. 1985, Astr. Ap., 144, 335. (CBB)

Cassatella, A., Wamsteker, W., Sanz, L., and Gry, C. 1987, IAU Circular No. 4330.

Chevalier, R. A., and Fransson, C. 1987, Nature, 328, 44.

Chu, Y.-H. 1987, IAU Circular No. 4322.

Fitzpatrick, E. L. 1985, Ap. J. 299, 219.

________________. 1987, in preparation.

Gilmozzi, R., Cassatella, A., Clavel, J., Fransson, C., Gonzalez, R., Gry, C., Panagia, N., Talavera, A., and Wamsteker, W. 1987, Nature 328, 318.

Isserstedt, J. 1975, Astr. Ap. Suppl. 19, 259.

Kirshner, R. P., Sonneborn, G., Crenshaw, D. M., and Nassiopolous, G. 1987, Ap.J. 320, 602.

Kurucz, R. L., 1979, Ap. J. Suppl., 39, 1.

Lasker, B. M. 1987, IAU Circular No. 4318.

Munoz Piero, J. R. 1985, ESA IUE Newsletter No. 23, 58 (NASA IUE Newsl. No. 27, 27).

Panek, R. J., 1982, IUE Newsletter No. 18, 68.

Savage, B. D. and Mathis, J. 1979, Ann. Rev. Astr. Ap., 17, 73.

Sonneborn, G., Altner, B., and Kirshner, R.P. 1987, Ap. J. Letters (Dec 1 issue)

Sonneborn, G., and Kirshner, R. P. 1987, IAU Circular No. 4333.

Turnrose, B. E. and Thompson, R. W. 1984, "IUE Image Processing Information Manual, Version 2.0", CSC/TM-84/6058.

Walborn, N. R., Lasker, B. M., McLean, B., and Laidler, V. G. 1987a, IAU Circular No. 4321.

Walborn, N. R., Lasker, B. M., Laidler, V. G., and Chu, Y.-H. 1987b, Ap. J. Letters (in press). (WLLC)

West, R. M., Lauberts, A., Schuster, H.-E., and Joergensen, H. E. 1987a, IAU Circular No. 4356.

West, R. M., Lauberts, A., Joergensen, H. E., and Schuster, H.-E. 1987b, Astr. Ap. (Letters) 177, L1.

White, G. L., and Malin, D. F. 1987, Nature, 327, 36.

Woosley, S. E., Pinto, P. A., Martin, P. J., and Weaver, T. A. 1987, Ap. J. 318, 664.

ANOTHER SUPERNOVA WITH A BLUE PROGENITOR

C. Martin Gaskell
Astronomy Department, Dennison Building, University of Michigan, Ann Arbor, Michigan 48109-1090, USA.

William C. Keel
Department of Physics & Astronomy, University of Alabama, P.O. Box 1921, Tuscaloosa, Alabama 35486, USA.

Abstract SN 1984e in NGC 3169 showed strong Hα emission which appeared between 2.2 and 23 years before maximum light. We suggest that this was produced by sudden mass loss from a blue supergiant. From the intense circumstellar Hα emission at maximum light we can estimate that at least 0.4 solar masses of gas had been suddenly expelled.

INTRODUCTION

Although the LMC supernova SN 1987a is perhaps not as unique as was first thought (see Schmitz and Gaskell 1987), it has helped emphasize that, in their light curves and spectral properties, type II supernovae form a very inhomogeneous class. In this paper we want to draw attention to another unusual and (so far) unique supernova, 1984e. It is now clear that the progenitor of the LMC supernova, SN 1987a, was the blue supergiant Sanduleak $-69^\circ 202$ (see Walborn 1988). There is also growing evidence for relatively recent significant mass loss from the progenitor (e.g. Chevalier 1987). We shall see that the evidence points to a blue supergiant progenitor for 1984e (although probably a more massive star than Sanduleak $-69^\circ 202$) and that in this case there was clearly mass loss right before the explosion.

OBSERVATIONS

SN 1984e in NGC 3169 was first reported by Rev. Robert O. Evans (MacLean, NSW, Australia) on March 29, 1984 (Evans 1984). It was also independently discovered on photographs taken a few days earlier by N. Metlova (Sternberg Inst., Crimea, USSR) and Kiyoni Okazaki (Japan). The published light curve (Metlova et al 1985) shows that maximum light occurred on April 1, 1984 (±2d). Spectroscopy by one of us (C.M.G.) with the McDonald observatory 82" on March 30 led to the discovery of remarkable strong narrow Balmer emission superimposed on what otherwise appeared to be a normal type II supernova at maximum light (Gaskell 1984a). Hα had an equivalent width of 66A and a total luminosity (uncorrected for reddening) of about 2.0×10^{40} ergs s^{-1} (assuming H_0 = 50 km s^{-1} Mpc^{-1}). This is easily the largest Hα luminosity ever seen from a single star and is in fact comparable to the luminosity of an entire Seyfert galaxy narrow line region!

At the suggestion of J. Craig Wheeler (Texas) spectra were obtained at higher resolution (on April 4). These show that Hα is resolved with an FWHM of about 250 - 340 km s^{-1}. This shows conclusively that the narrow Balmer emission is not arising from a surrounding normal H II region which would instead show an FWHM of about 10 km s^{-1}. The width of this Hα "spike" is, on the other hand, much too narrow for the supernova ejecta (which has a FWHM of 4000 km s^{-1} in this case; Henry & Branch 1987). The narrow component is quite distinct from the blueshifted P-Cygni profile broad supernova Hα emission (our high resolution data do not confirm the P-Cygni absorption in the narrow component suggested by Dopita et al 1984). Gaskell (1984b) suggested that the narrow spike was the result of the outer layers of a star being blown off some years before core collapse. Dopita et al (1984) suggested instead that the gas was left there by a powerful stellar wind before the supernova explosion (i.e. something continuous rather than a single event). Gaskell (1984b) argued from the absence of forbidden lines that up to several solar masses of gas could have been expelled some years before the final explosion.

This prediction led us to inspect an Hα image (reproduced in Keel 1987) of NGC 3169 which had fortuitously been taken as part of an Hα imaging survey of galactic nuclei (Keel 1983) just a few years earlier (December 19, 1981) with the Video camera (ISIT) on the Kitt Peak 84". We found a point source corresponding to a luminosity of 2.3×10^{37} ergs s^{-1} ($H_\circ$ = 50 km s^{-1} Mpc^{-1}) at the precise position where the supernova would explode. Another Hα image taken in May 1985 (i.e. a year after the supernova) showed that the earlier point source had vanished and only the nearby H II region remained. We also inspected a Lick Observatory 120" prime focus plate taken in 1961 by R. Fish and failed to detect the point source on it even though the resolved comparably bright nearby H II region was clearly visible.

DISCUSSION

From the evidence of the pre-discovery imaging we conclude that the surrounding shell was the product not of a wind, but of a discrete "event" that took place sometime between 1961 and 1981 (i.e. between 2.2 and 23 years before maximum light). At the observed expansion velocity the shell would have travelled between about 130 and 1400 AU. Such a shell is clearly going to be destroyed by the supernova blast wave fairly quickly. The light curve of 1984e is that of a "linear" (i.e. fast) type II so the explosion probably began 20 days before maximum light. To be seen in our spectroscopy the shell had to have survived 25 days at least. Since we find that the probable absorption troughs in H I and He I extend to about 15,000 km s^{-1} this would imply a lower limit to the distance from the star of about 200 AU. An upper limit to the radius can be set by the observation of Henry and Branch (1987) that the shell had been destroyed by the blast wave by May 6, 1984 (i.e. 55 days after the initial explosion). These limits agree well with those set by the constraints of the prediscovery observations.

Knowing the approximate radius of the shell ($5*10^{15}$ cm, say) and the Hα luminosity at maximum light (2×10^{40} ergs s^{-1}) we can estimate the density and hence find the mass of gas assuming that Hα is produced by recombinations. This comes out to be about 6×10^{8} cm^{-3} [R_{300}^{-3}], where R_{300} = R/300 AU. This gives about 0.2 [$R_{300}^{3/2}$] solar masses of ionized hydrogen in the shell. If our estimate of the radius of the shell is too small then the mass of ionized gas will be greater and, of course, there could be a significant amount of gas which is not ionized.

The Hα luminosity in 1981 is at the upper limit of the Hα luminosity of the brightest Hubble-Sandage variables in the local group. We therefore speculate that the progenitor might have been such a star. Although nothing quite like what happened to SN 1984e has been observed before the idea of a star undergoing the expulsion of its outer layers a few years before core collapse is not new. Woosley, Weaver and Taam (1980) found that one of their models expelled its outer layers at just the velocity we observe 5 years before core collapse. This was however only a 10 solar mass model but perhaps something related happened in SN 1984e.

ACKNOWLEDGEMENTS

The observations reported here were obtained at the McDonald Observatory of the University of Texas and at Kitt Peak National Observatory. C.M.G. is grateful to Craig Wheeler for providing real-time theoretical input during the spectroscopy and to Stan Woosley for discussions. We also wish to thank George Herbig for making available the Lick 120" archive plate.

REFERENCES

Chevalier, R. A. (1987), this volume.
Dopita, M. A., Evans, R. O., Cohen, M. & Schwartz, R. D. (1984), Ap.J. Letts., 287, L69.
Evans, R. O. (1984), I.A.U. Circular No. 3931.
Gaskell, C. M. (1984a), I.A.U. Circular No. 3936.
Gaskell, C. M. (1984b), P.A.S.P., 96, 789.
Henry, R. B. C. & Branch, D. (1987), P.A.S.P., 99, 112.
Keel, W. C. (1983), Ap.J., 268, 632.
Keel, W. C. (1987) Sky & Telescope 74, 234.
Metlova, N. V., Metlov, V. G. & Tsetkov, D. Yu. (1985), Inf. Bull. Var. Stars, No. 2780.
Schmitz, M. F. & Gaskell, C. M. 1987, this volume.
Walborn, N. R. (1987), this volume.
Woosley, S. E., Weaver, T. A. and Taam, R. E. (1980) in Type I Supernovae, ed. J. C. Wheeler, p96. Austin: Texas Univ.

OPTICAL AND INFRARED OBSERVATIONS OF SN 1987A
FROM CERRO TOLOLO INTER-AMERICAN OBSERVATORY

M. M. Phillips
Cerro Tololo Inter-American Observatory,
National Optical Astronomy Observatories,
Casilla 603, La Serena, Chile

Abstract. Results from optical and infrared observations of SN 1987A obtained at Cerro Tololo Inter-American Observatory over the first seven months since core collapse are reviewed. Around 130 days after outburst, the bolometric light curve began to smoothly decline at a rate of ~0.01 mag day^{-1}, providing dramatic confirmation of the prediction that radioactivity had powered the optical display after the first month. The peculiar color changes and kinks observed beginning on the 25th day probably signaled the initial release of trapped energy from mass 56 material. The bolometric luminosity of SN 1987A was unusually low at first, but reached a value more typical of other type II supernovae by the time that the final exponential decline had begun. Over much of the period covered by these observations, the optical and infrared spectra were characterized by strong absorption lines of Ba II and Sr II. Comparison with the spectra of other type II supernovae at similar stages of evolution supports the suggestion that s-processed elements were enriched in the hydrogen envelope of the progenitor, Sanduleak -69°202. Radial velocity measurements of the Bγ line indicate that the slowest moving hydrogen in SN 1987A was traveling no faster than ~2100 km s^{-1}. Strong He I absorption and emission at 1.0830 μm is tentatively identified in infrared spectra obtained between the 75th and 135th days. Symmetric bumps in the blue and red wings of the hydrogen emission line profiles were observed for a two month period beginning 25 days after outburst, suggesting that a significant asymmetry had developed in the explosion.

1 INTRODUCTION

Supernova 1987A in the Large Magellanic Cloud has provided an unprecedented opportunity to study in detail the evolution of a type II supernova event. Taking into consideration the available instrumentation and excellent photometric characteristics of Cerro Tololo Inter-American Observatory, we elected to concentrate our efforts in three major observational areas: (1) UBVRI photometry, (2) low-resolution optical spectrophotometry from 3800-7000 Å, and (3) low-resolution infrared spectroscopy from 0.9-4.1 μm. In this paper, I will review the major results from this work covering the first seven months since core collapse.

2 UBVRI PHOTOMETRY

Photoelectric $UBV(RI)_C$ photometry of SN 1987A was begun with the CTIO 0.4m telescope on 25 February 1987 (the first night after Shelton's discovery). Over the first seven months, more than 400 independent observations were carried out on a total of 125 nights. The supernova observations were always bracketed by measurements of the nearby standard star δ Dor. Absolute photometry was carried out on several photometric nights to derive transformation coefficients to the $UBV(RI)_C$ system and to independently determine standard magnitudes for δ Dor.

In order to minimize errors due to possible variations of the atmospheric extinction during the night, the supernova observations were reduced differentially with respect to those of δ Dor. Mean extinction coefficients determined from the photometric nights were assumed. However, since the difference in airmass between the supernova and δ Dor was never larger than 0.05, the final results are fairly insensitive to variations in the actual extinction coefficients. Further details concerning the observations and data reduction are given by Hamuy et al. (1987).

The combined errors due to photon statistics, deadtime corrections, extinction, and color terms for the averaged BVRI data are less than 0.01 mag; for the U band the errors are more like 0.02 mag. Nevertheless, comparison with the photometry carried out at SAAO (Menzies et al. 1987; Catchpole et al. 1987) and ESO (Cristiani et al. 1987) shows systematic differences at the 5% level in the UBVR measurements, and up to 10% in the I band. Such disagreement is not surprising, and almost certainly has its roots in the very different spectral shapes of the supernova and the photometric standards. The Cousins I filter is particularly vulnerable, since it has no red cutoff and therefore relies on the red cutoff of the S-20 photomultiplier. A change in the latter of only a few hundred Ångstroms can lead to a significant difference in the observed magnitude. This is a fundamental shortcoming of broad band photometric systems, and ultimately limits the precision of the SN 1987A photometry (absolute or relative) to no better than 5%.

The individual light curves obtained at CTIO are shown in the upper half of Figure 1. The explosion is assumed to have occurred on J.D. 2446849.82, which was the time of the neutrino burst observed independently by the IMB and Kamiokande II experiments (Bionta et al. 1987; Hirata et al. 1987). The first twenty days of the light curves are amplified in the lower half of Figure 1. In the latter plot, we have included (a) all of the published (and some unpublished) photoelectric data available to us, corrected to the zero point of the CTIO measurements, and (b) the pre-discovery photographic visual magnitudes reported by McNaught (1987).

Figure 1. $UBV(RI)_C$ light curves for SN 1987 plotted as a function of time since outburst. Measurements for the first 20 days are shown on an expanded time scale below.

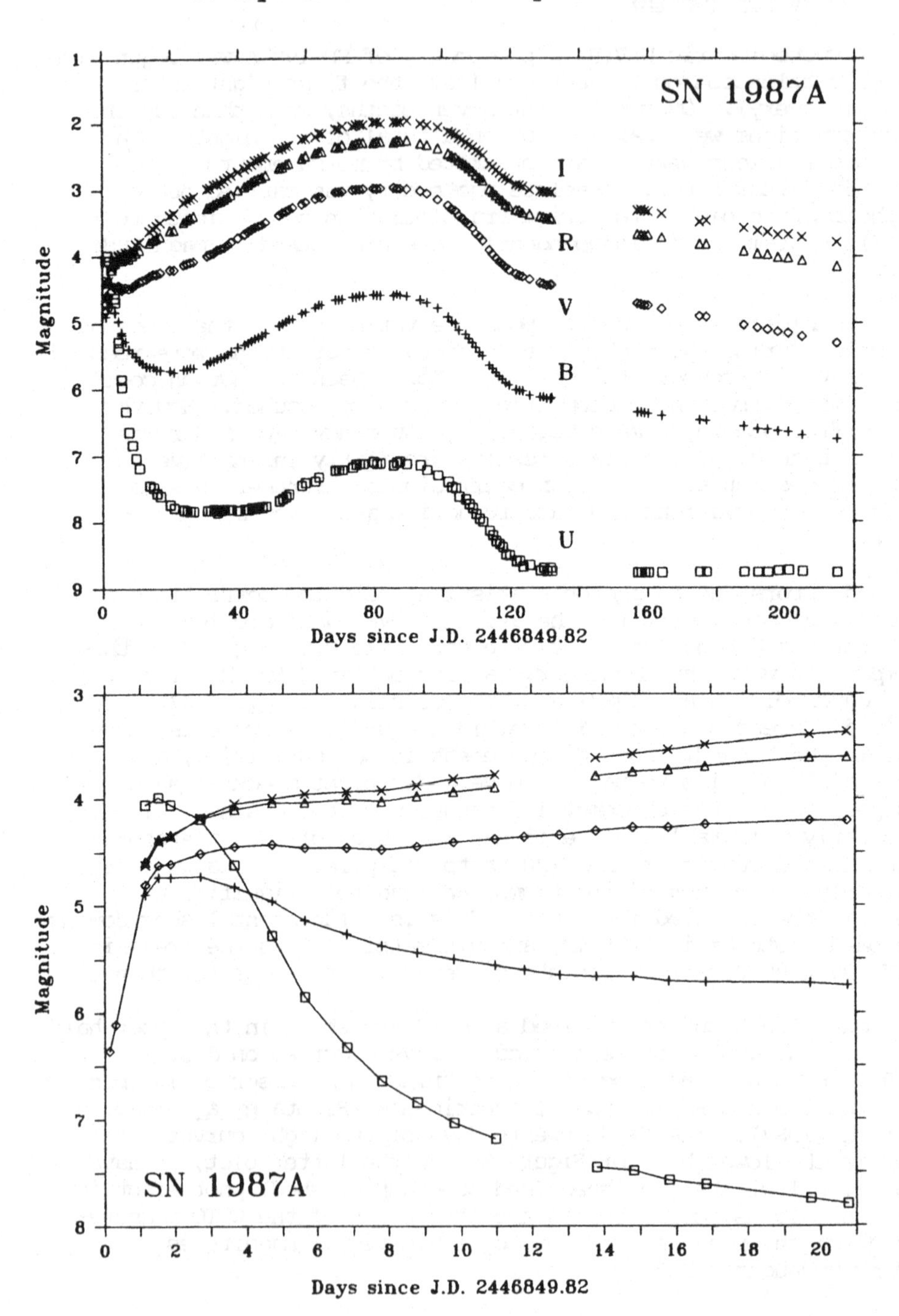

The color evolution of SN 1987A is illustrated in Figure 2. During the first twenty days, all of the colors quickly reddened as the supernova envelope expanded and cooled. The extreme color change observed in U-B over this period is due in major part to the onset of severe line blanketing in the ultraviolet (see next section). Around the 25th day, the R-I color evolution abruptly halted in its reddening trend, and actually began to move slightly to bluer values again. Similar kinks in the V-R, B-V, and U-B colors took place on roughly the 33rd, 45th, and 53rd days, respectively. The trend to bluer colors continued until the time of visual maximum (on approximately the 85th day), at which point the colors began once again to redden slightly. A final blueward inflection occurred in the U-B, B-V, and R-I curves around the 125th day, as the bolometric light curve settled into an exponential decline (see below). The V-R curve showed a similar kink at this same time, but still continued to redden in absolute terms (most likely as a result of the increasing equivalent width of the Hα emission at this epoch).

<u>Figure 2</u>. Optical color curves for SN 1987A plotted as a function of time since outburst.

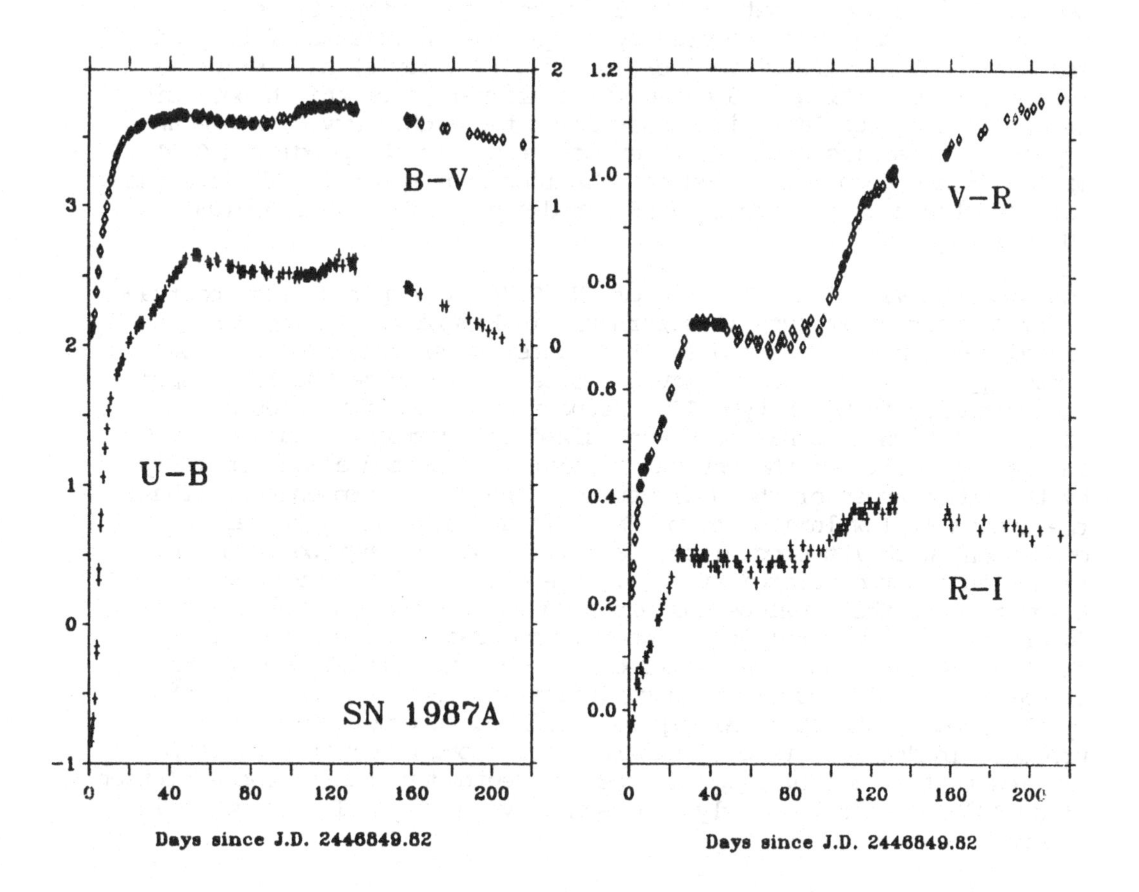

The optical photometry described above was combined with published ultraviolet fluxes (Kirshner et al. 1987) and infrared photometry (Menzies et al. 1987; Catchpole et al. 1987; Bouchet et al. 1987) to derive bolometric luminosities. For these calculations, we assumed a true distance modulus for the LMC of 18.5, and a total reddening of the supernova of E(B-V) = 0.15 (Fitzpatrick 1987). The optical and infrared photometry had to first be converted to equivalent monochromatic fluxes at the effective wavelength of each filter. A logarithmic interpolation was then employed to obtain the final bolometric luminosities, which are plotted in Figure 3. Temperature estimates derived from black body curve fits to the monochromatic fluxes are shown in the same figure. The reader is referred to Hamuy et al. (1987) for more details concerning the these steps, as well as a discussion of the various sources of uncertainty.

By now, the peculiar nature of the bolometric light curve of SN 1987A has been widely discussed. The initial expansion of the supernova envelope accounts for the rapid decline in luminosity and temperature observed over the first few days. A phase of slowly increasing luminosity followed, with the temperature holding nearly constant at ~5500 K. This nearly linear rise in luminosity, which culminated in a second maximum approximately three months after outburst, is now thought to be have been powered by the diffusive release of trapped energy from the radioactive decay of ^{56}Co (Woosley et al. 1987b; Woosley, this meeting). Spectacular confirmation of this hypothesis is provided by the late time behavior of the light curves, where an exponential decline rate almost exactly equal to the predicted 0.010 mag day^{-1} has been maintained for the last four months. (The precise value of the decline rate in V measured since mid-July is 0.0105 ±0.0001 mag day^{-1}).

The absolute visual light curve of SN 1987A is compared with observations of four other type II supernovae in Figure 4. The reddenings and distance moduli assumed for the latter supernovae are given by Hamuy et al. (1987). This figure serves to emphasize the large spread in luminosity found in type II supernovae over the first 100 days of their evolution. As has been emphasized by a number of authors, the initial intensity of the optical display is dictated almost entirely by the compactness of the hydrogen envelope of the progenitor. Thus, the unusually low luminosity of SN 1987A at these early stages is consistent with its progenitor, Sanduleak -69°202, having been a blue rather than a red supergiant (e.g., see Woosley et al. 1987a). However, once the envelope has expanded and recombined, the luminosity is derived from the release of energy from radioactive material. By the time that SN 1987A had settled into the exponential decline phase, it was not at all under-luminous, indicating that the amount of ^{56}Ni synthesized in the original explosion was quite comparable to that produced in "normal" type II supernovae. Among other things, this implies that from approximately the hundredth day onward, the evolution of SN 1987A should be fairly representative of type II supernovae in general.

One of the more interesting properties of the broad band photometry is the abrupt blueward changes in slope of the color curves observed approximately one month after outburst. Such discontinuities could conceivably be produced by rapid changes in strong absorption or emission features, but there is no evidence in the optical spectra for this (see next section). Also, the fact that the kinks occurred in the same sense in all four colors argues persuasively that a change in

Figure 3. Black body temperature fits and bolometric luminosities derived from IUE spectra and optical/infrared photometry. The data are plotted as a function of time since outburst.

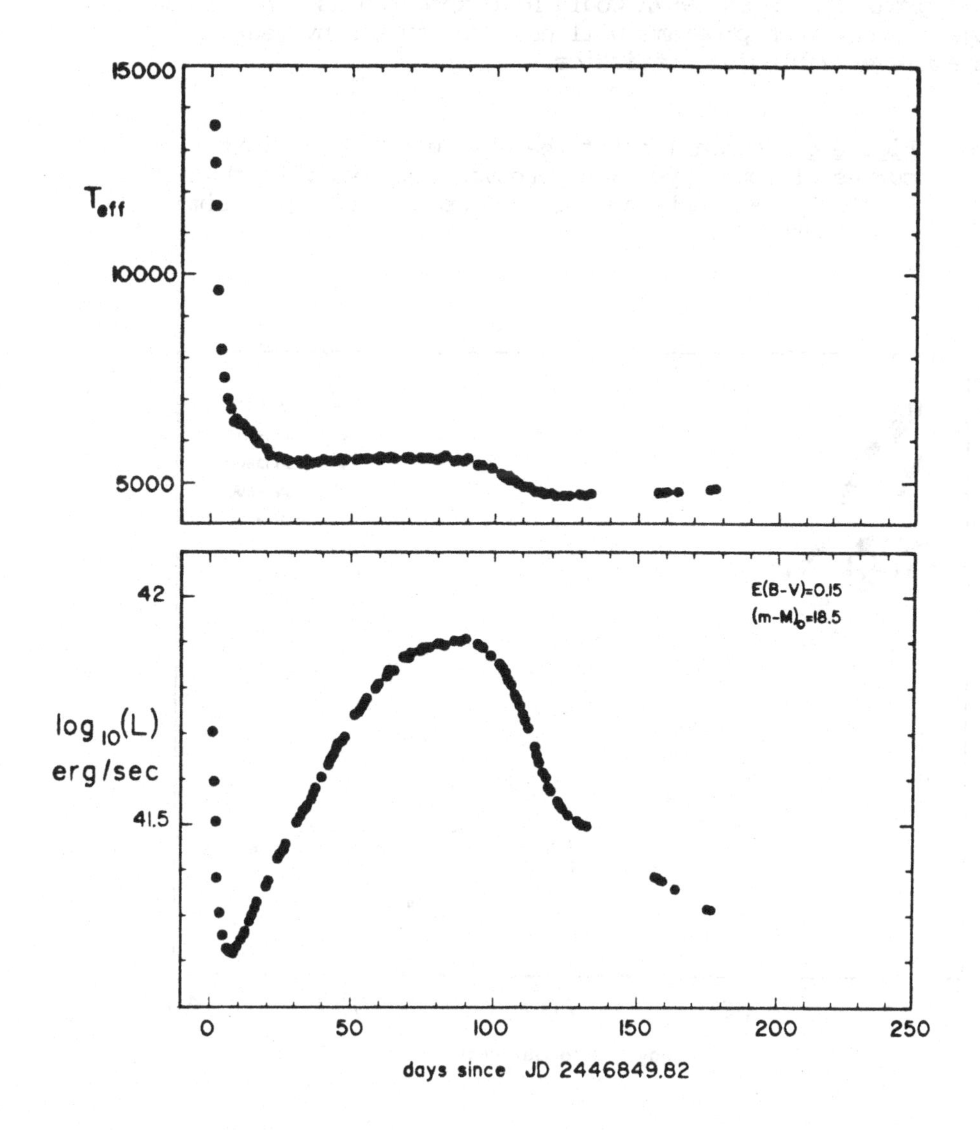

the continuum radiation rather than spectral lines was responsible. As noted earlier, the slight blueward trend which followed the initial kinks came to a halt almost precisely at the time of the second maximum in the bolometric luminosity curve. This strongly suggests that the first blueward kinks marked the moment of emergence of the trapped energy produced by the radioactive decay of ^{56}Ni and ^{56}Co.

It is difficult to know if supernovae like 1987A are common events. As pointed out by Blanco et al. (1987), the rather bright limiting magnitude of the Rev. Robert Evans' survey makes it unlikely that he would have turned up a low-luminosity supernova such as 1987A, even if they were as common as the more luminous type II events. This question can perhaps be answered through monitoring of nearby clusters such as Virgo and Fornax (where SN 1987A would have appeared as a magnitude 15-17 object), but such programs will have to cover many years in order to achieve meaningful statistics.

Figure 4. Comparison of the absolute visual light curves of four type II supernovae compared with that of SN 1987A. All data are plotted as a function of time since outburst.

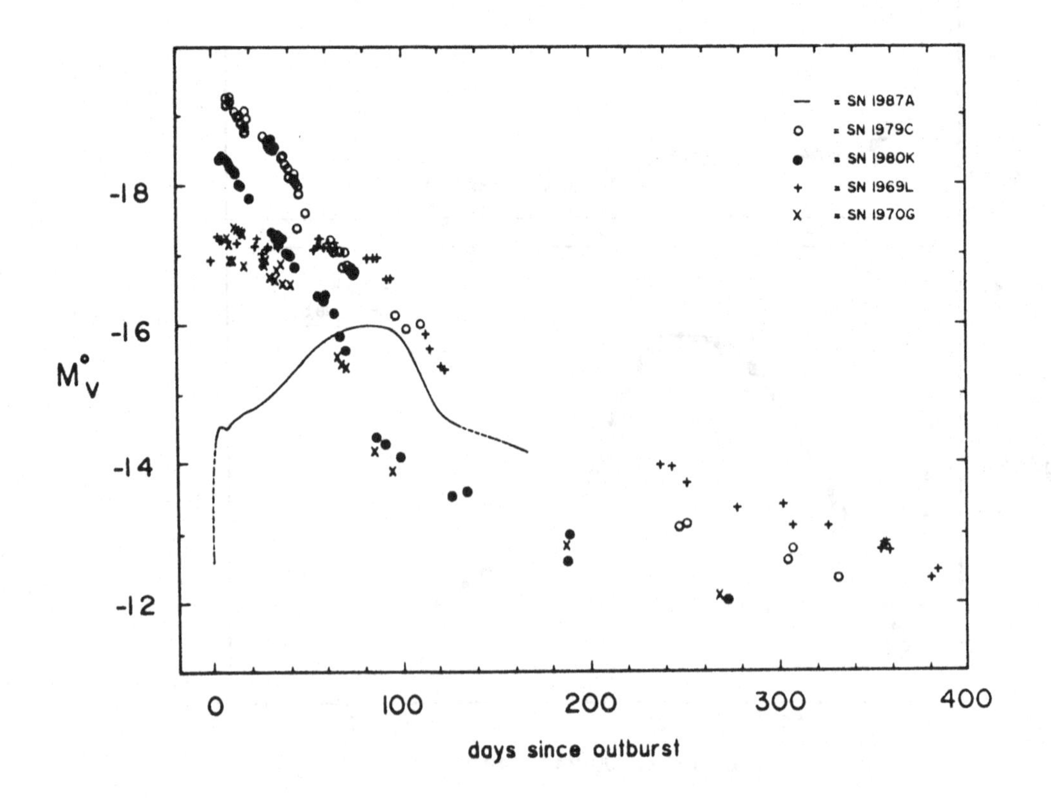

3 OPTICAL SPECTROPHOTOMETRY

Low-resolution optical spectrophotometry of SN 1987A has been intensively pursued at CTIO since the first night of observation. Except for a few spectra taken with CCDs, most of the data have been obtained with the 1m telescope spectrograph and "2D-Frutti" photon counting detector. These observations cover the spectral range 3700-7100 Å at a resolution of ~5 Å. Nightly spectra were obtained for the first two months after outburst. During the next three months, a rate of one spectrum every other night was maintained (weather permitting). Since August, we have tried to get a least one observation per week. As in the case of the UBVRI photometry, δ Dor was observed as a comparison star and the spectrophotometric reductions carried out differentially. The first set of 2D-Frutti spectra were recently submitted for publication (Phillips et al. 1987), along with details of the calibration process. Errors in the absolute spectrophotometric are at the ±0.05 mag level, while the relative flux calibration of any individual spectrum is good to ±0.03 mag.

The optical spectral evolution during the first four months is illustrated in Figures 5a and b. On the first night of observation (February 25th), the only identifiable lines were due to H I and He I. These all showed broad and symmetric P Cygni profiles. The wings of the Hα line in this spectrum extend to approximately ±30,000 km s^{-1}. This extraordinarily large value is yet further evidence of the relatively compact nature of the progenitor (see Woosley et al. 1987a). Within five days of outburst, the He I lines disappeared and were soon replaced by absorption lines of Na I, Ca II, Fe II, Sc II, Ba II, and Sr II. (See below for a further discussion of the Ba and Sr identifications). At the same time, heavy line blanketing set in at blue and ultraviolet wavelengths. The absorption components of the H I Balmer lines reached their maximum depths between the 10th and 20th days, and then began to decline again with, Hβ and Hγ nearly disappearing. Around the 60th day (late-April), however, the Balmer line absorption began to dramatically increase once again. At the same time, the Na I D absorption broadened and deepened. Following the second maximum in the bolometric light curve (late-May), the equivalent width of the Hα emission began a marked increase, as the spectrum slowly evolved from a stellar to a nebular phase.

The evolution of the optical spectrum through the 130th day (early-July) is displayed in a somewhat different fashion in Figure 6. Here we have combined the data to form a two-dimensional image where, as in a trailed spectrum, wavelength and time are the axes. [This figure was inspired by Minkowski's similar diagram for type I supernovae (see Zwicky 1965)]. At the bottom, the bolometric light curve of SN 1987A is reproduced for reference. Except for the disappearance of the He I lines (which occurred on too rapid a time scale), the spectral changes described in the previous paragraph are all illustrated in this figure. The rapid initial decrease in radial velocities of the blueshifted absorption lines as the photosphere ate its way through

Figure 5a. Optical spectral evolution of SN 1987A from 25 February to 28 March 1987. All spectra are plotted on the same F_λ scale, with zero levels shown as tick marks.

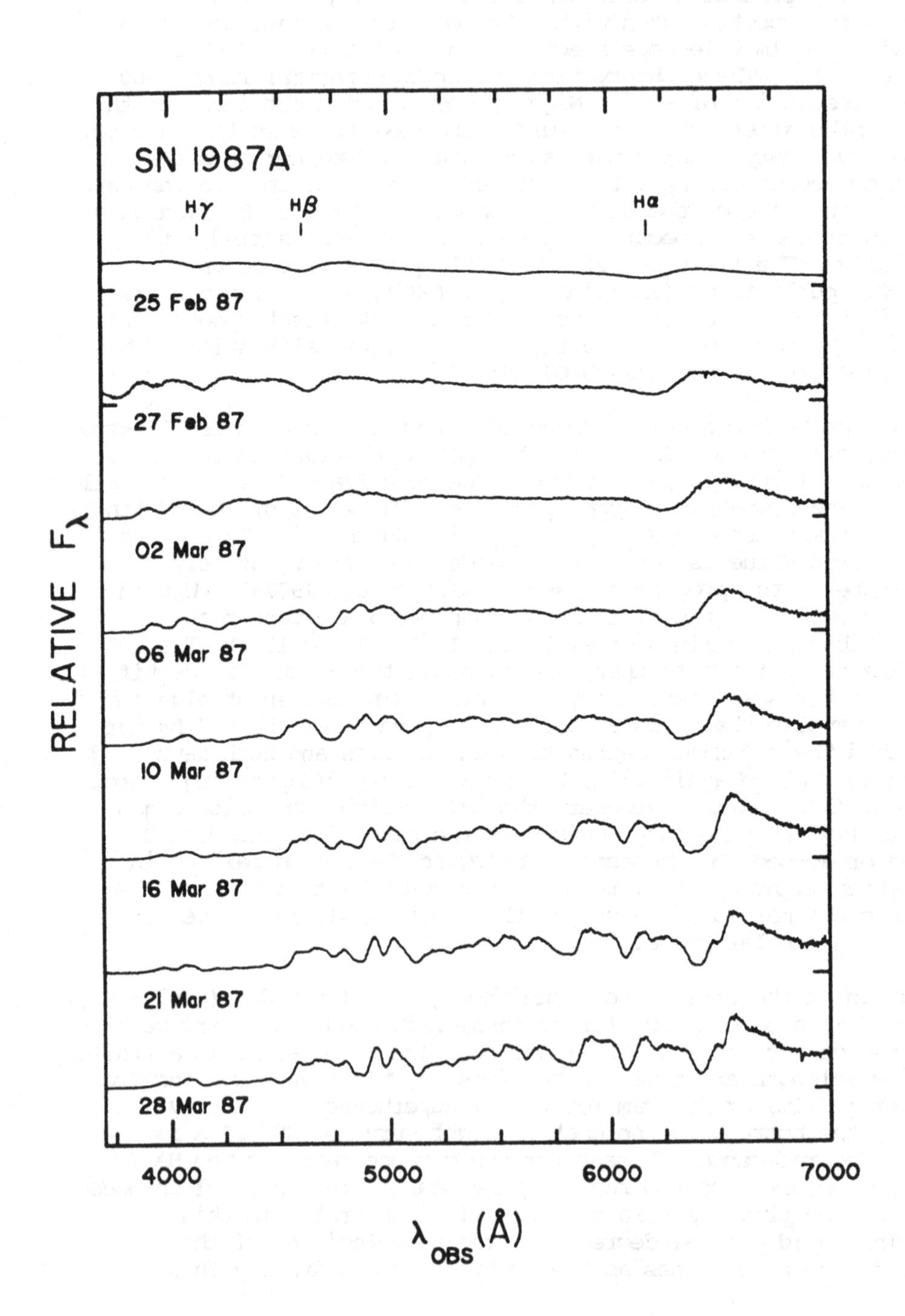

<u>Figure 5b</u>. Optical spectral evolution of SN 1987A from 02 April to 04 July 1987. All spectra are plotted on the same F_λ scale, with zero levels shown as tick marks.

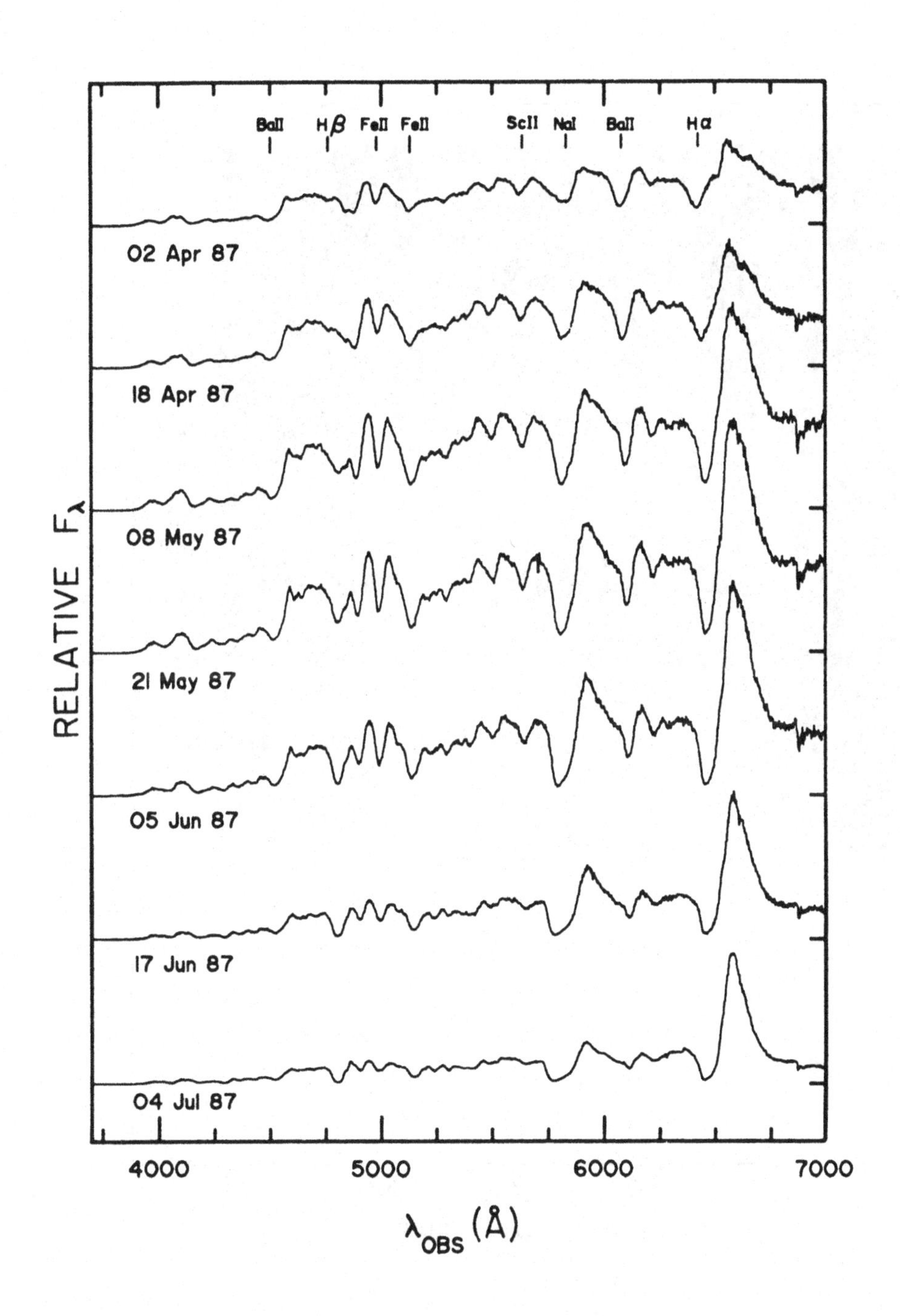

Figure 6. 2-D representation of the optical spectral evolution of SN 1987A. The bolometric luminosity curve is plotted along the bottom for reference.

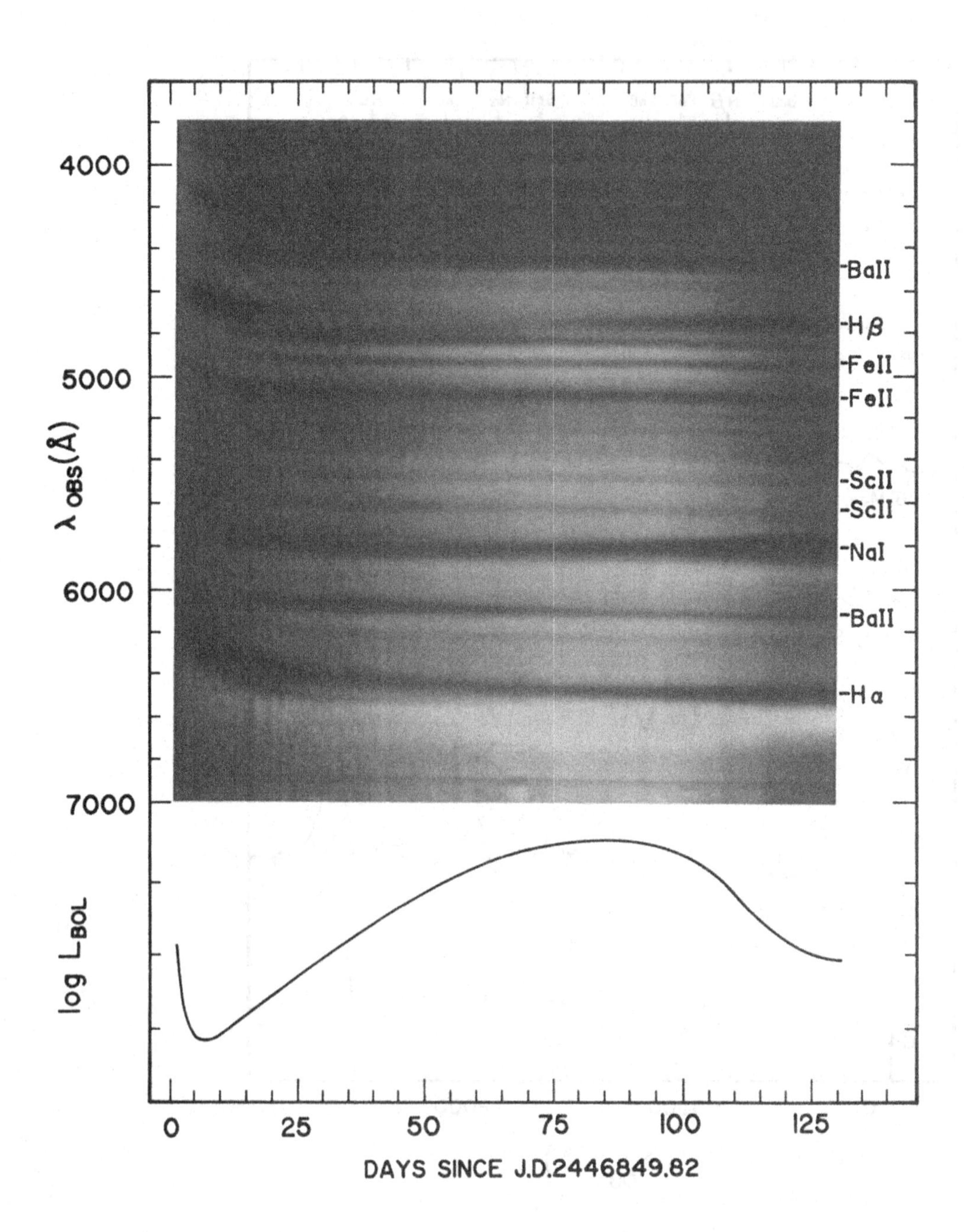

the outer envelope is also clearly visible. However, the real power of this sort of display is in revealing much more subtle spectral variations. An outstanding example is the blueshifted emission components which appeared in the Hα and Hβ line profiles around the 25th day. Attention was called to this feature in the Hα profile by several groups (e.g., see Blanco et al. 1987; Hanuschik & Dachs 1987). However, in Figure 6 it is immediately obvious that a similar emission component was present at Hβ, removing all doubt from the identification with hydrogen. The presence of this emission accounts for some, but not all, of the apparent weakening of the Balmer line absorption which began around the same date. The blueshifted emission component persisted until approximately the 80th day after outburst, with the radial velocity of the peak perhaps decreasing only slightly from an initial value of approximately -4000 km s^{-1}.

Interestingly, the emergence of this blueshifted emission was accompanied only a few days later by the appearance of a redshifted component, at almost exactly the same radial velocity (with the opposite sign, of course). This event is faintly visible in Figure 6. As pointed out by Hanuschik & Dachs (1987) and Lucy (this meeting), such structure in the hydrogen line profiles implies the development of an asymmetry in the exploding material of the supernova. We note that the appearance of the double emission apparently closely coincided with the emergence of trapped energy due to the radioactive decay of mass 56 material (see previous section), which was also the approximate time that the photosphere reached the boundary between the hydrogen and helium zones of the expanding envelope (Woosley, this meeting).

The renewed deepening of the H I Balmer line absorption around the 60th day is somewhat puzzling, but probably owes its explanation to the recombination of the hydrogen envelope. Evidence in support of this is found is the dramatic increase in the Na I D absorption that begins around the same date. The deepening of the Na I absorption also seems to be accompanied by an increase in the emission component of the line profile, as might be expected under recombination conditions.

Radial velocity measurements of the absorption minima of the most prominent optical metal lines are plotted in Figure 7. Similar measurements for the H I Balmer absorption minima are shown in Figure 8. The accuracy of these measurements for the lines with symmetric profiles is estimated to be ~35 km s^{-1}. [See Phillips et al. (1987) for further details.] The velocity evolution of the Fe II and Sc II lines is essentially identical, and probably closely tracks the photosphere after the first few days. The two Ba II lines show significantly higher velocities between the 15th and 100th days. This is most likely due to the lower ionization potential of Ba II (10 eV vs. 16 and 13 eV for Fe II and Sc II, respectively). Note that after the 10th day, the velocities measurements for the Ba II λ4554 line are probably biased to larger values as a result of the severe line blanketing which depresses the continuum just blueward of line center. The hydrogen Balmer line velocities show the effects of high optical

Figure 7. Radial velocity measurements for the absorption minima of selected Fe II, Na I, Sc II, and Ba II lines plotted as a function of time since outburst.

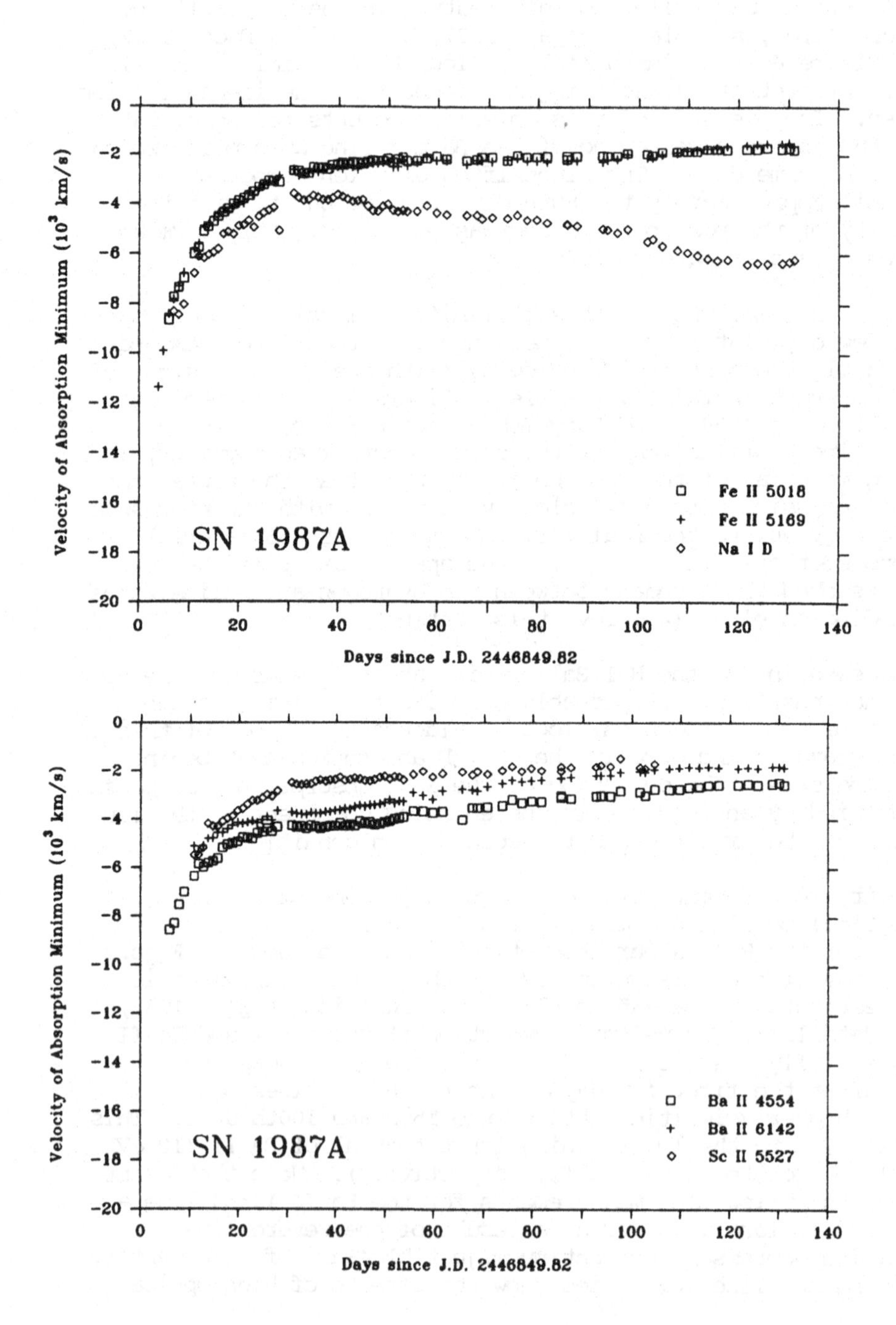

depth, and are also influenced by the blueshifted emission component which appeared between the 25th and 80th days (see above). The Na I D velocities are somewhat greater than the Ba II λ6142 values early on (as might be expected from the lower ionization potential of neutral Na), but begin to deviate to higher values after the 50th day. These measurements are strongly affected by changing asymmetries of the Na I line profile (see Figures 5a and b) and, hence, should be treated with considerable caution.

A CCD spectrum of SN 1987A obtained on Sept. 9th (197 days after outburst) is shown in Figure 9. The late time spectrum is now clearly evolving into a nebular phase, with strong emission lines of H I, Na I, Ca II, and O I beginning to dominate. As the outer layers of the supernova envelope grow more and more optically thin, the heavy elements synthesized in the explosion should make their appearance. The narrower profiles of the O I and Ca II lines may reflect the slower velocities of this core material. Continued monitoring of the late time spectrum is obviously of great importance, and should provide important new insights in the study of explosive nucleosynthesis.

Figure 8. Radial velocity measurements for the absorption minima of Hα, Hβ, Hγ plotted as a function of time since outburst.

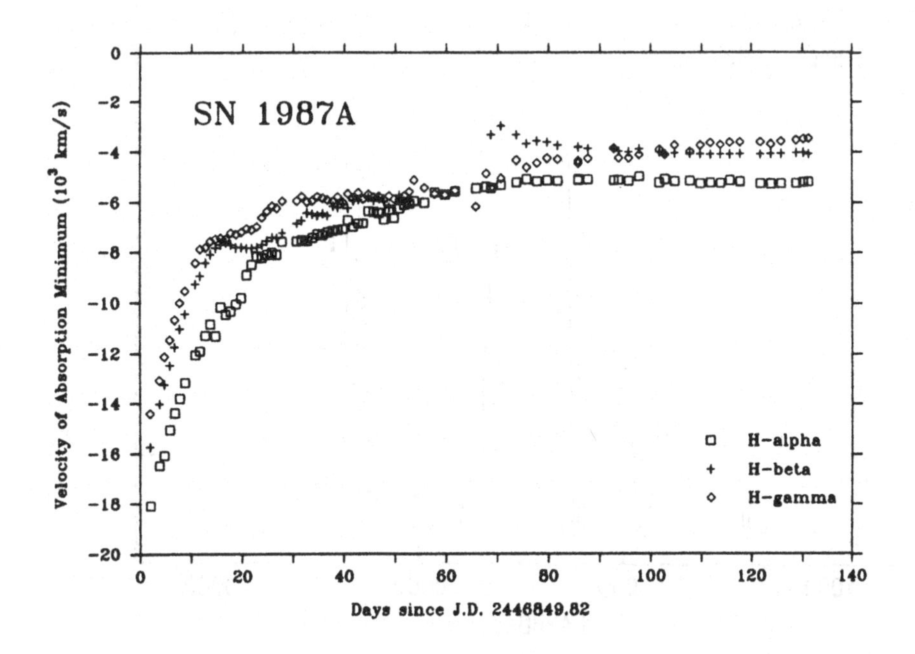

4 INFRARED SPECTROSCOPY

Low-resolution ($\lambda/\Delta\lambda \sim 1000$) spectra of SN 1987A covering the wavelength range 0.9-4.1 μm have been obtained at roughly monthly intervals at CTIO with the Infrared Spectrometer (IRS) on the 4m and 1.5m telescopes. The IRS employs an array of eight InSb detectors, so that full wavelength coverage requires the grating to be stepped many times. A monitor channel which receives approximately 10% of the light was used to calibrate variations in the signal due to poor seeing, guiding errors, or thin clouds. Further details of the observations and data reduction are given by Elias et al. (1987).

Representative IRS spectra obtained through the first 134 days (July 7th) after outburst are plotted in Figure 10. With few exceptions, the strongest spectral features in these data are due to hydrogen. Note that the profiles of the H I lines in the April and May spectra show clear evidence of the redshifted emission component observed at Hα over the same period. An enlargement of the 0.9-1.4 μm wavelength

Figure 9. CCD spectrum of SN 1987A taken on 09 September 1987 on the CTIO 1.5m telescope.

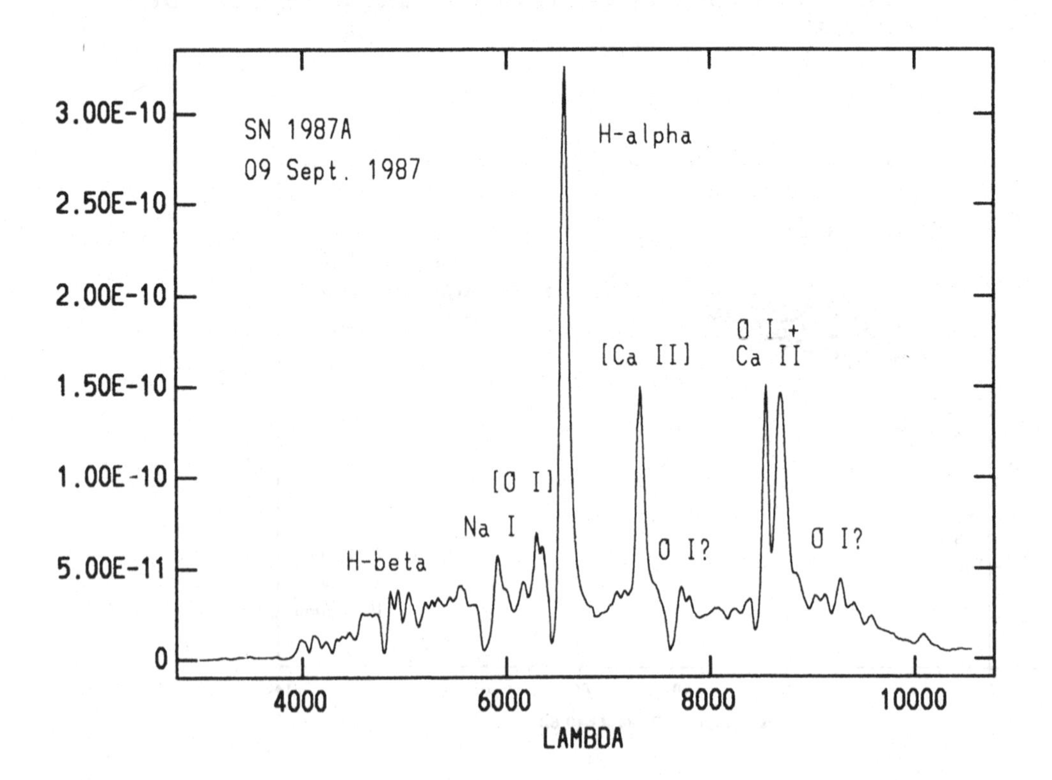

region is shown for the May, June, and July spectra in Figure 11. The strong absorption line observed at ~1.025 μm is almost certainly due to the 1.0327 μm transition of Sr II multiplet 2. Unfortunately, the other two members of this multiplet at 1.0037 and 1.0915 μm are blended with Pγ and Pδ. However, an earlier infrared spectrum published in the AAO Newsletter (No. 41) shows clear evidence for the presence of the 1.0915 μm transition. The strength of the infrared Sr triplet lends credence to the identification by Williams (1987a) of two absorption features in the optical spectra with Sr II $\lambda\lambda$4078,4215. The significance of strong Sr lines in SN 1987A is considered in the next section.

Aside from the obvious Paschen and Brackett series lines, the identifications of the remaining spectral features in the 0.9-1.4 μm wavelength region are not so obvious. It appears likely that there is weak

Figure 10. Sample IRS spectra covering the time period from 09 March 1987 to 07 July 1987. The data are plotted on the same relative flux scale.

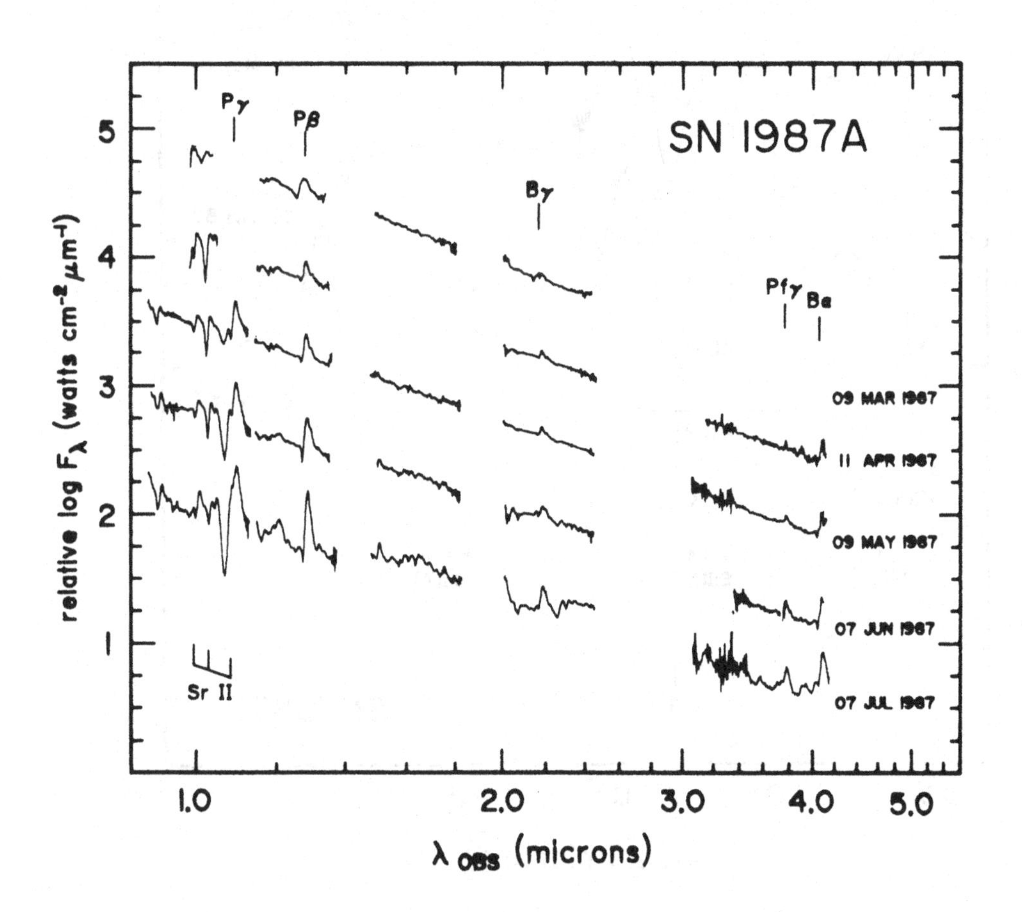

absorption due to multiplets 1 and 3 of S I (see Figure 8). Like the other spectral features identified with neutral elements, the strength of the S I lines shows a gradual increase over this time period. The identity of the strong, broad absorption feature at ~1.07 μm is even more of a puzzle. On the basis of the large increase in strength displayed between May and July, it would seem very probable that this feature is also due to a neutral element. An obvious candidate is

Figure 11. Enlargement of the wavelength region 0.9–1.4 μm from the spectra for 09 May, 07 June, and 07 July 1987. Possible line identifications are shown below the spectra.

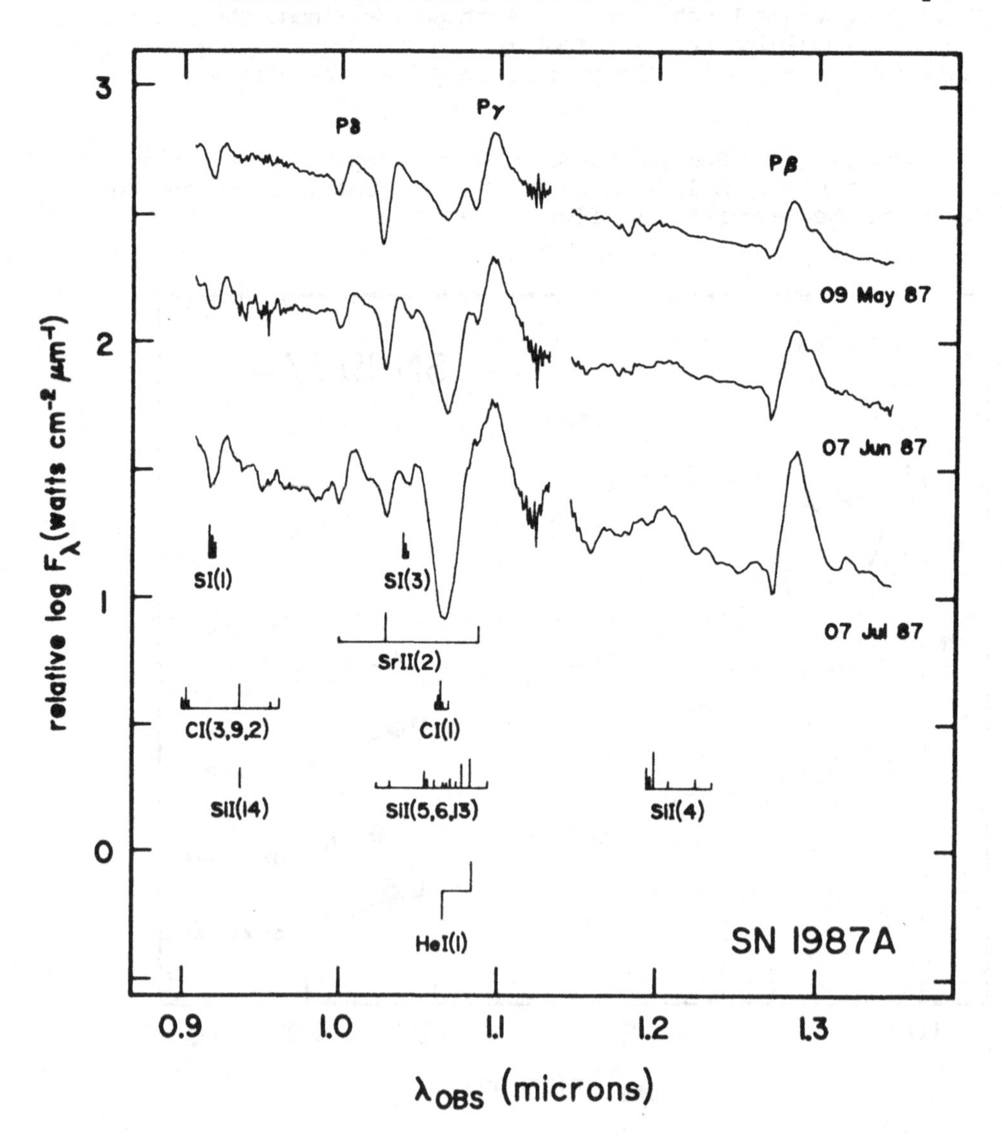

Si I, which accounts for the strongest lines in the solar spectrum observed in this spectral region. However, as Figure 8 shows, this identification can be rejected on the grounds of (a) the poor fit of the relevant multiplets to the observed absorption, and (b) the absence of absorption from other Si I transitions such as multiplet 4. A second possibility, multiplet 1 of C I, provides a much better fit to the data. However, once again we would expect to see comparable absorption features due to other C I multiplets such as 3 and 9 which are clearly lacking.

Barring an explanation in terms of some exotic element, it seems most likely that the strong absorption at 1.07 μm is due to He I 1.0830 μm. This line has a large oscillator strength and, more importantly, arises from the highly metastable $2s^3S$ level. As the helium atoms recombine in the outer envelope of the supernova, many of the electrons will eventually funnel down into the $2s^3S$ level. For $N_e < 4 \times 10^3$ cm^{-3}, collisional de-excitations to lower levels cease to dominate the depopulation of the $2s^3S$ level (see Osterbrock 1964). Such low density conditions may apply in the outer portion of the hydrogen envelope, giving rise to strong absorption and emission in the 1.0830 μm. Indeed, there is considerable evidence in Figure 8 for the growing presence of an emission feature in the blue wing of Pβ at almost exactly the rest wavelength of the 1.0830 μm line. The wavelength measured for the minimum of the strong absorption in the July spectrum implies a radial velocity of approximately -5000 km s^{-1}, which is consistent with an origin in the outer envelope of the supernova. However, self-consistent modelling of the supernova spectrum would be highly desirable to confirm this identification.

Radial velocity measurements of the absorption minima of three of the infrared hydrogen lines (Pβ, Bα, and Bγ) and Sr II 1.0327 μm are plotted in Figure 12. The Sr II values agree quite well with those of the optical Fe II lines (shown schematically in the same figure). The Bγ velocities are also very near the Fe II values, but the Pβ and Bα measurements are slightly higher throughout (presumably due to larger optical depths). These measurements would seem to indicate that the velocity of the slowest moving hydrogen in SN 1987A was no greater than ~2100 km s^{-1}.

5 ENHANCEMENT OF S-PROCESSED ELEMENTS

Within twenty days of the outburst of SN 1987A, a strong absorption feature had developed at a wavelength of ~6100 Å (see Figures 5a and b). This line was identified in a few early papers with one of the transitions of Fe II multiplet 74. However, Williams (1987a) correctly pointed out that, if this were the case, at least two other members of multiplet 74 should be observed with comparable strengths. This led him to suggest Ba II λ6142 as an alternative identification, and to postulate that s-processed elements were unusually enhanced in the envelope of the progenitor, Sanduleak -69°202 (Williams 1987a;

1987b). The identification of strong absorption in the infrared due to another s-processed element, Sr, is particularly interesting in this context. As Williams has pointed out in his papers, the presence of s-processed element enhancements could provide an invaluable clue to the evolutionary status of Sanduleak -69°202 just before it exploded.

The radial velocity data I have presented in the previous sections are relevant to this topic. In his second paper, Williams (1987b) carried out a simple analysis comparing the abundances of Ba, Sr, and Sc with that of Fe. An underlying assumption in this calculation was that the Ba II, Sr II, Sc II, and Fe II lines occupied the same zones in the expanding hydrogen envelope. The different behavior of the Ba II and Fe II radial velocities seen in Figure 7 would indicate that this assumption is not entirely correct. It is difficult to predict how

<u>Figure 12</u>. Radial velocity measurements for the absorption minima of $P\beta$, $B\alpha$, $B\gamma$, and Sr II 1.0327 μm plotted as a function of time since outburst. Mean velocities for the optical Fe II lines are indicated.

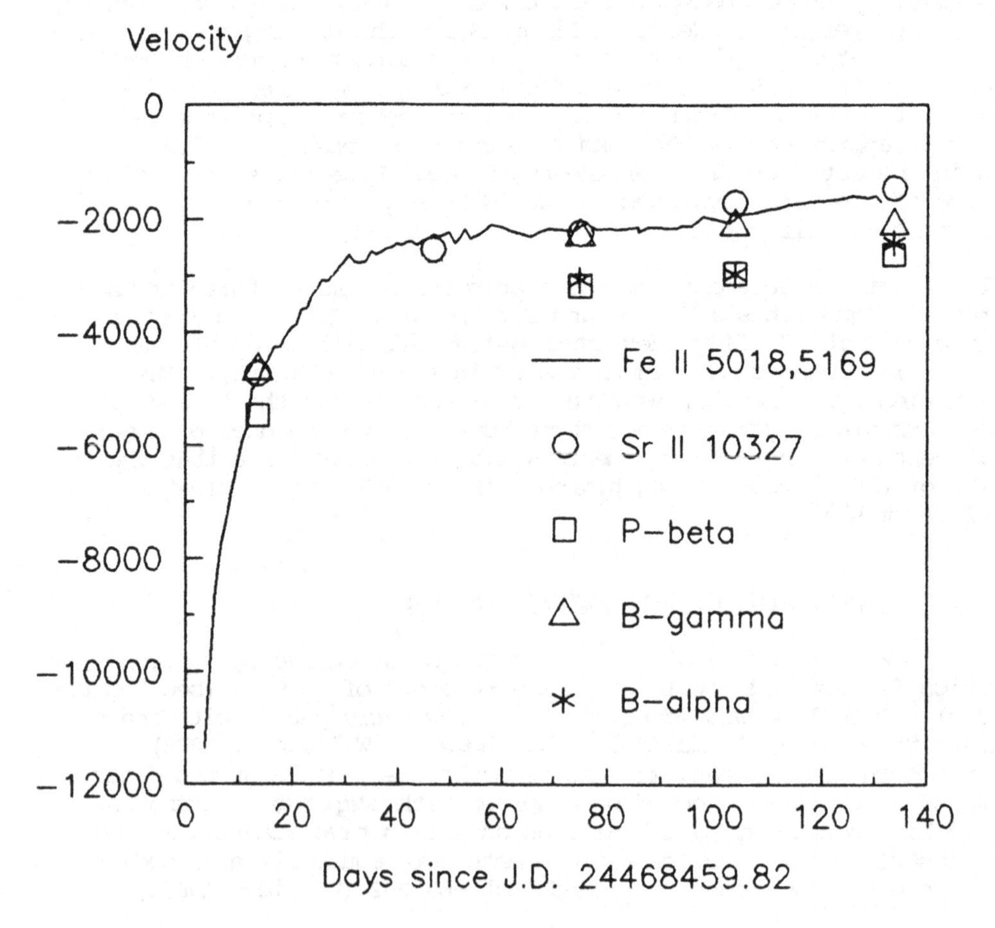

this fact might change the relative Ba and Fe abundances calculated by Williams. On the other hand, it should be pointed out that Williams also found Sr and Sc to be enhanced with respect to Fe. The radial velocities of the Sr II and Sc II lines agree much better with those of Fe II (see Figure 12). Hence, the effective zones occupied by these ions are probably not significantly different.

Perhaps the most persuasive evidence in support of s-process element enhancements in the envelope of SN 1987A is found in simply comparing the spectrum with those of other type II supernovae at similar states of evolution. We have done this for the plateau supernovae 1985P in NGC 1433 (Chalabaev & Cristiani 1987) and 1986I in NGC 4254 (Foltz, private communication). Spectra of these two supernovae taken approximately 2-3 months after outburst are amazingly similar to that of SN 1987A at roughly the same age. The Sc II and Fe II absorption lines in all three objects appear to have nearly the same relative strengths. However, Ba II λ6142 is many times stronger in the spectrum of SN 1987A. The same also appears to be true for the Sr II λ4215 line, but this is much more difficult to ascertain. Unfortunately, there are no infrared spectra of SNe 1985P and 1986I to allow comparison of the infrared Sr II line strengths. Nevertheless, if the physical conditions in the envelopes of the three supernovae are fairly similar (as the overall close resemblance of the spectra would indicate), the evidence for Ba and Sr enrichment in the envelope of SN 1987A is rather convincing.

REFERENCES

Bionta, R. M. et al. (1987). Phys. Rev. Lett., 58, 1494-1496.

Blanco, V. M., Gregory, B., Hamuy, M., Heathcote, S. R., Phillips, M. M., Suntzeff, N. B., Terndrup, D. M., Walker, A. R., Williams, R. E., Pastoriza, M. G., Storchi-Bergmann, T. & Matthews, J. (1987). Astrophys. J., 320, 589-596.

Bouchet, P., Stanga, R., Le Bertre, T., Epchtein, N., Hamann, W. R. & Lorenzetti, D. (1987). Astron. Astrophys., 177, L9-L12.

Catchpole, R. M., Menzies, J. W., Monk, A. S., Wargau, W. F., Pollacco, D., Carter, B. S., Whitelock, P. A., Marang, F., Laney, C. D., Balona, L. A., Feast, M. W., Lloyd Evans, T. H. H., Sekiguchi, K., Laing, J. D., Kilkenny, D. M., Spencer Jones, J., Roberts, G., Cousins, A. W. J., van Vuuren, G., Winkler, H. (1987). Mon. Not. R. Astron. Soc., in press.

Chalabaev, A. A. & Cristiani, S. (1987). In E.S.O. Workshop on SN 1987A, in press.

Cristiani, S., Babel, J., Barwig, H., Clausen, J. V., Gouiffes, C., Günter, T., Helt, B. E., Heyndericks, D., Loyola, P., Magnusson, P., Monderen, P., Rabattu, X., Sauvageot, J. L., Schoembs, R., Schwarz, H. & Steeman, F. (1987). Astron. Astrophys., 177, L5-L8.

Elias, J. E. et al. (1987). To be submitted.

Fitzpatrick, E. L. (1987). To be submitted.

Hamuy, M., Suntzeff, N. B., Gonzalez, R. & Martin, G. (1987). Astron. J., 94, in press.
Hanuschik, R. W. & Dachs, J. (1987). Astron. Astrophys., 182, L29-L30.
Hiriata et al. (1987). Phys. Rev. Lett., 58, 1490-1493.
Kirshner, R. P., Nassiopoulos, G. E., Sonneborn, G. & Crenshaw, D. M. (1987). Astrophys. J., 320, 602-608.
McNaught, R. H. (1987). I.A.U. Circ., No. 4389.
Menzies, J. W., Catchpole, R. M., van Vuuren, G., Winkler, H., Laney, C. D., Whitelock, P. A., Cousins, A. W. J., Carter, B. S., Marang, F., Lloyd Evans, T. H. H., Roberts, G., Kilkenny, D., Spencer Jones, J., Sekiguchi, K., Fairall, A. P. & Wolstencroft, R. D. (1987). Mon. Not. R. Astron. Soc., 227, 39P-49P.
Osterbrock, D. E. (1964). Ann. Rev. Astron. Astrophys., 2, 95-120.
Phillips, M. M., Heathcote, S. R., Hamuy, M. & Navarrete, M. (1987). Astron. J., submitted.
Williams, R. E. (1987a). Astrophys. J. Lett., 320, L117-L120.
Williams, R. E. (1987b). In I.A.U. Colloquium No. 108: Atmospheric Diagnostics of Stellar Evolution, in press.
Woosley, S. E., Pinto, P. A., Martin, P. G. & Weaver, T. A. (1987a). Astrophys. J., 318, 664-673.
Woosley, S. E., Pinto, P. A. & Ensman, L. (1987b). Astrophys. J., in press.
Zwicky, F. (1965). In Stellar Structure, ed. L. H. Aller & D. B. McLaughlin, pp. 367-423. Chicago: The University of Chicago Press.

SN 1987A: OBSERVATIONAL RESULTS OBTAINED AT ESO

I.J. Danziger[1], P. Bouchet[1], R.A.E. Fosbury[2,5], C. Gouiffes[1], L.B. Lucy[1], A.F.M. Moorwood[1], E. Oliva[3], F. Rufener[4]

[1]European Southern Observatory, Karl-Schwarzschild-Str. 2, D-8046 Garching bei München, FRG

[2]Space Telescope - European Coordinating Facility, European Southern Observatory, Karl-Schwarzschild-Str. 2, D-8046 Garching bei München, FRG

[3]Osservatorio Astrofisico di Arcetri, Largo E. Fermi, 5, I-50125 Firenze, Italy

[4]Observatoire de Genève, CH-1290 Sauverny, Switzerland

[5]Affiliated to the Astrophysics Division, Space Science Department, European Space Agency

Abstract. The visual and IR photometry at La Silla has demonstrated the existence of the exponential decay after 140 days. A rise in the UV luminosity during the same period is noted. We have mapped the temporal behaviour of the IR excess. The early optical spectra are used to discuss the barium abundance and the identification of other absorption features. Later optical spectra show emission lines including forbidden ones, and differences in the velocities are noted among these as well as those in the IR. Recent IR spectra show lines from H, HeI, OI, NaI, KI, MgI, SiI, FeII and resolve the band heads in the first overtone band of CO. The temporal behaviour of CO first detected in June is briefly discussed. An upper limit to the Li/H ratio in the LMC is given.

INTRODUCTION

Since the time of outburst of SN 1987A there has been continuous monitoring of its progress at ESO La Silla by staff astronomers and visitors. The main emphasis has been on regular photometry and low resolution spectroscopy at optical and IR wavelengths, although at the very earliest times many valuable and unique high resolution spectral observations were made revealing details of the interstellar medium not hitherto known. This high resolution work has now been published in the journals and in the ESO Workshop on SN 1987A. Similarly the early photometry and spectroscopy has been published and so it is intended here to place more emphasis on the later stages.

OPTICAL AND INFRA-RED PHOTOMETRY

UBV photometry extending 203 days beyond outburst has been published by Cristiani et al. (1987). Bolometric luminosities have not yet been computed beyond 130 days, but the V magnitude (which seems from other published work to follow the bolometric luminosity rather closely at this later stage) has an e-folding time of ~ 112 days without significant deviation. It may be noted here that the standard U magnitude remains almost constant in the interval 130-203 days.

Figure 1 shows the results of Geneva system photometry (Rufener and Maeder 1971) in the U and V bands for the complete interval 1-261 days obtained at the Swiss Telescope on La Silla. During the period 140-261 days the V magnitudes follow an exponential curve with an e-folding time of 114 days with no deviation from this even at late stages. During the same period the Geneva system U magnitude has steadily brightened by 0.21 magnitudes although there have been no other signs of an increase in temperature of the photosphere. A similar brightening has been reported for the IUE ultra-violet by Kirshner (1987). Why the difference in behaviour between the Geneva system U magnitudes and the normal U magnitudes? The Geneva system U filter is of intermediate width centred at 3456 Å while the normal U magnitude is defined by a broad band filter with an effective wavelength at ~ 3800 Å for conventional late-type stars. This suggests that we may be seeing the result of an increasing rate of hydrogen recombination in the envelope resulting in a greater contribution to the continuum shortward of the Balmer jump than longward. If this were the case, significant line-blocking might still be occurring as the absolute UV flux seems too low for the absolute flux of Hα if it were all resulting from recombination. Of course only a fraction of Hα may be produced by recombination. The temporal behaviour of Brackett and Pfund recombination lines has not yet provided a clear indication whether there has been an increasing rate of recombination over this whole period. Another possible explanation for this UV increase is a systematic decrease of opacity in this region as the line blocking varies as a result of contributions from ions in lower stages of ionization at the expense of contributions from the higher stages, and a decrease in overlapping of lines as the photosphere retreats to low velocity matter. This can only adequately be tested with detailed models. In the wavelength interval covered by the Geneva U band there are not any expected emission lines whose increasing strength might result in the observed brightening.

Concerning derivation of bolometric magnitudes and temperatures, it is worth noting that, particularly during these later stages (130-250 days) the fitting of optically thick black body curves to fluxes derived from broad band photometry can be misleading in principle because of the strong emission line features that have developed. For example Hα must affect the R magnitudes and [CaII]7291,7323 and the CaII triplet (8500 Å) must affect the I magnitudes. At ESO it is intended in the future to use calibrated spectroscopy in conjunction with broad-band photometry to separate out the effects of strong emission lines.

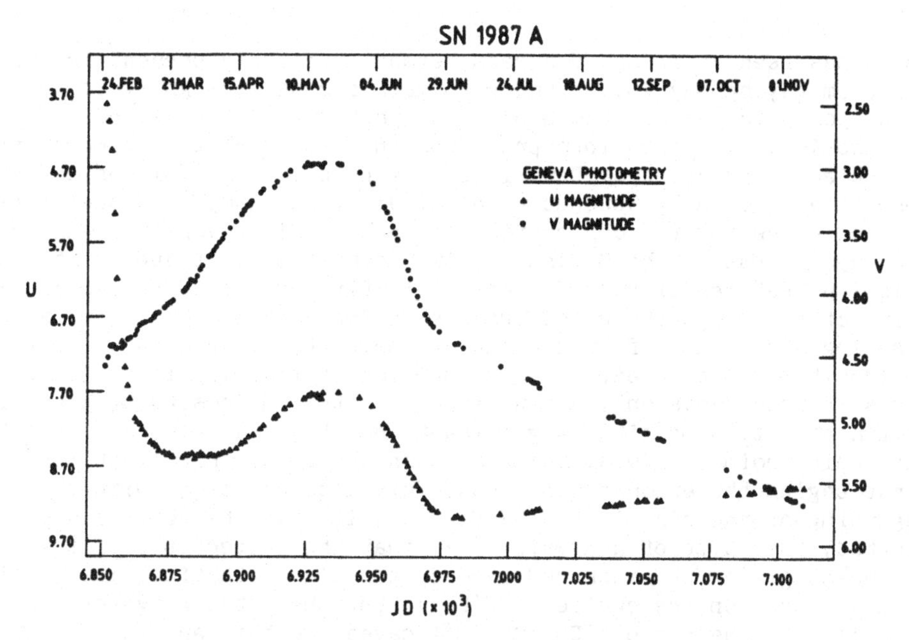

Figure 1. Geneva system U and V magnitudes as a function of time.

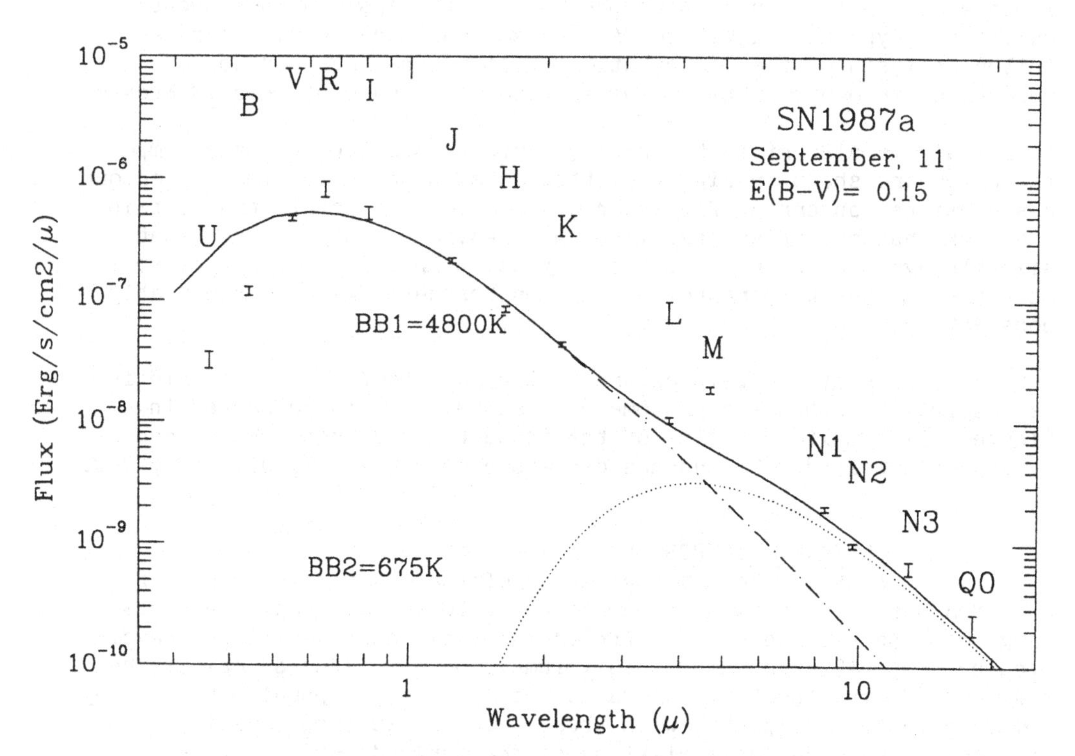

Figure 2. The decomposition of the Sep. 11 photometry into 2 BB components.

There has been an infrared excess beyond 2μ clearly present in the spectrum beyond day 40. In order to gain further insight into the nature of this excess, the broad band photometry, visual and infrared (out to 18μ), has been interpreted as the result of the superposition of 2 black body spectra. Details of the procedure as well as early results are given by Bouchet et al. (1987). An example of such a fitting procedure for the observations of Sep. 11 is given in Figure 2. The huge excess in the M band is now understood to be due to strong emission features discussed below. This fitting procedure leads to a value of the temperature and luminosity for each of the 2 components. Assuming a distance of 50 kpc and reddening E(B-V) = 0.15 we plot the luminosity of the IR excess as a function of time in Figure 3. At all times it represents only a few percent of the luminosity of the main component. It is possibly significant that the peak occurs near 130 days which coincides with the time when the bolometric light curve first begins the exponential decay. This also coincides with the beginning of the rise in the UV fluxes. Although the details are not plotted, this type of analysis shows that the temperature of the component giving the infrared excess decreases with time. The precise values depend on the choice of E(B-V), but the total range of variation lies in the range 1200-700 K. A caveat to this analysis ought to be stated. It is by no means certain that a single black-body is producing this IR excess. This first order analysis is intended to isolate a parameter such as temperature that helps one to conceptualize the type and magnitude of the temporal behaviour. Again, a fuller analysis of the contribution of emission lines has to be undertaken in order better to understand the nature of this IR excess.

This analysis also gives black-body temperatures for the main component which are shown in Fig. 4 plotted as a function of time. Given our reservations concerning fitting black body curves to spectra containing large numbers of strong emission lines, one must conclude that sizeable systematic uncertainties may exist in this temperature scale. Nevertheless the temperature of the "photosphere" remains remarkably constant over the interval 110-250 days.

With a luminosity and temperature of the main component available it is possible to compute a photometric radius. This is presented in Figure 5. The maximum radius of the instantaneous photosphere occurred at approximately day 95 and has decreased monotonically until day 250.

OPTICAL SPECTROSCOPY

At ESO, the interpretation of the evolving spectrum of SN 1987A has been continuing by reference to the spectral synthesis program using the models of differentially expanding envelopes developed by Lucy (1987a,b, also this volume). Here we address mainly the question of line identifications, abundances and temporal behaviour by comparing observed spectra with those from models incorporating density structures based on variations of the early density profile computed for hydrodynamical models by Arnett (1987). Earlier attempts have been presented by Danziger et al. (1987) and Fosbury et al.

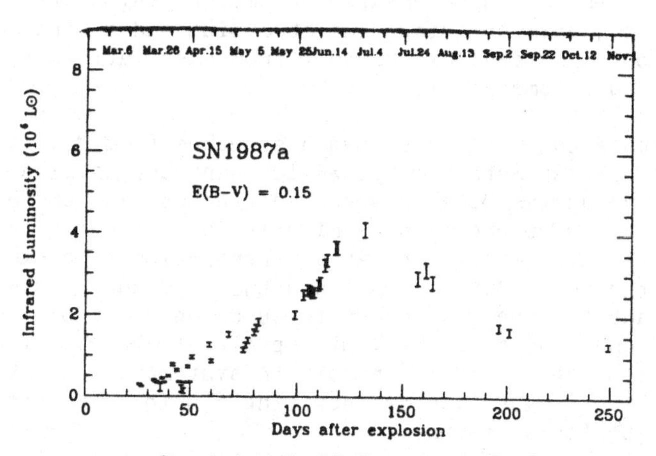

Figure 3. Luminosity of the IR excess as a function of time.

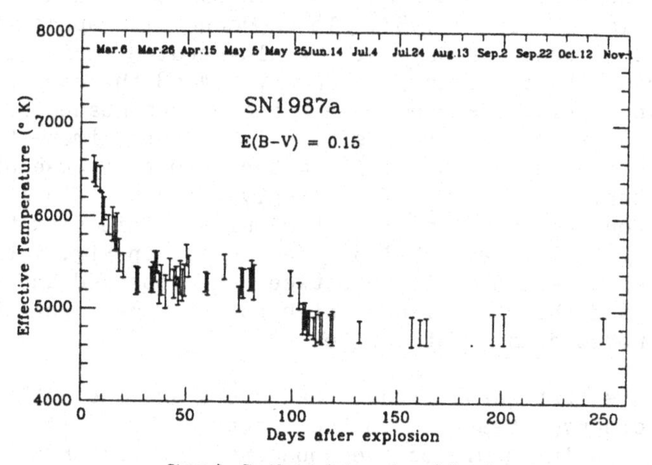

Figure 4. The black body temperature of the main component plotted as a function of time.

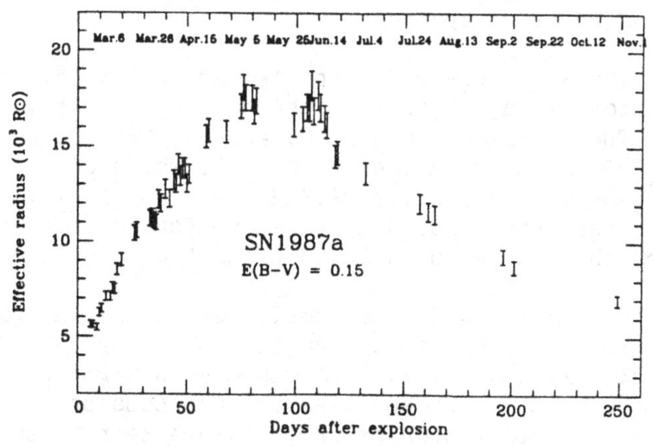

Figure 5. The photometric radius as a function of time.

(1987). All lines from appropriate ionization stages of elements as heavy as nickel, but including also strontium and barium, have been included. Input parameters which come from observations are the colour temperature and bolometric luminosity.

Figure 6 shows a comparison between a spectrum from day 19.78 and the model with 1/4 solar metallicity. While there are still some obvious velocity discrepancies between observed features and those in the model, there are some promising similarities in the pattern of line blocking in various wavelength ranges. Particularly noteworthy are the features in the 4000-4700 Å range, and those in the 5000-5400 Å range. It is important to note that these features result from a contribution of more than one line from one ionic species (indicated in the figure). This being so, it may result in systematic uncertainties in velocities which are derived by ascribing a unique identification and rest wavelength for a feature.

The strong absorption feature near 6100 Å has been identified with the BaII 6141.7 line (Williams 1987). To reproduce its approximate strength in the model for day 19.78 would require increasing the barium abundance by a factor 20. For day 114.68 the comparison of spectra shown in Fig. 7 suggests that the overabundance of barium required to match the observed feature is an approximate factor 5. There is more reason to trust this latter result because at this epoch most of the barium in the model is singly ionized and a large correction factor for unseen stages is not required. This situation does not obtain for day 19.78. Nevertheless one can see in Fig. 7 that there remain discrepancies in position between the observed and computed BaII feature and the other lines which future work ought to resolve using an improved density structure.

Thus it appears that conventional abundances are adequate to explain most of the observed features in the spectrum. There is increasing evidence that barium could be overabundant by a factor 5. Because of line crowding and blending even a qualitative indication of the abundance of strontium has not been possible, but may become clearer with the use of infrared spectra.

A spectrum of the SN at day 227 is presented in Fig. 8. It is clear that the emission lines relative to the continuum have increased quite considerably. The emission peaks in Hα and the NaI D lines are red-shifted by ~ 585 km s^{-1} relative to the LMC, whereas the emission peaks of [OI]6300,63 and [CaII]7291,7323 have no significant redshift, which suggests that the forbidden lines are formed in a region different from that where the allowed transitions are occurring.

The CaII triplet lines now have a complex structure, at least partly because the velocity spread of the region where they are formed has dropped sufficiently that individual components might be resolved. Nevertheless the relative intensities are not consistent with a straightforward interpretation as only 3 lines seen in emission. That OI 8446 may be a significant contributor seems ruled out by an incon-

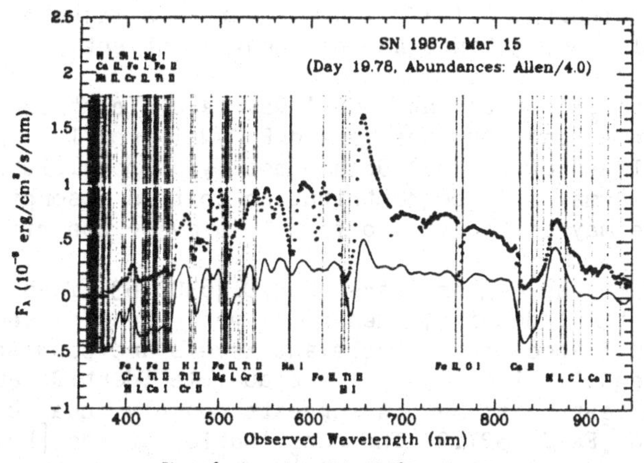

Figure 6. A spectrum at day 19.78 together with a preliminary model spectrum.

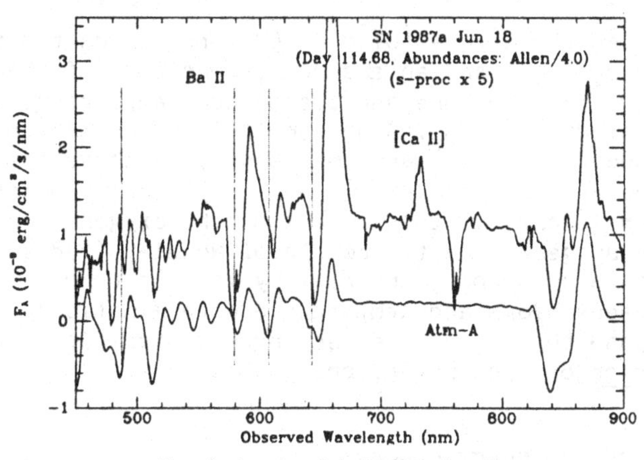

Figure 7. A spectrum at day 114.68 together with a preliminary model spectrum.

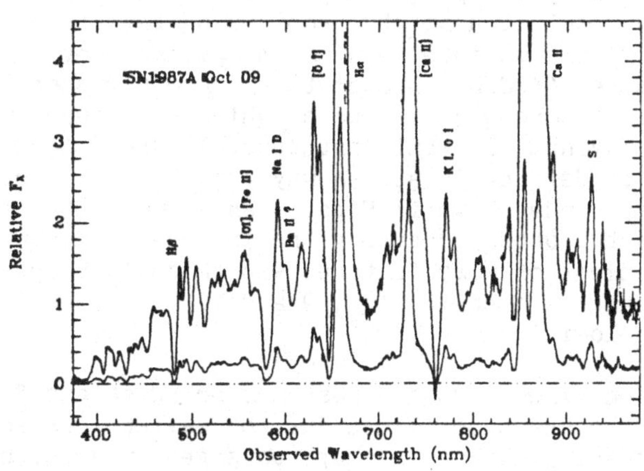

Figure 8. A spectrum at day 227.

sistency in wavelength. [CI]8727 cannot yet be a major contributor because the [CI]9850,23 lines are too weak or absent.

The deep absorption feature at 7600 Å may have a contribution from the strong telluric A band, but the marked P-Cygni profile suggests that the resonance lines of KI 7664,98 are mostly responsible for this feature. OI 7774 may also be contributing to this absorption although the temperature may be too low to make it a major contributor.

If the density in the region where the [OI]6300,63 lines are formed is above their critical density, then one might expect to see a relatively strong [OI]5577 because the relevant levels are populated in thermodynamic equilibrium. In fact a plausible identification of the broad emission feature at 5560 Å suggests roughly equal contributions of [OI]5577 and [FeII]5527.3. The contribution of the [FeII] emission is consistent with identification of [FeII] lines in the IR spectra discussed later.

An observed intensity of the [OI]6300,63 lines corrected for reddening of 7×10^{-9} ergs cm^{-2} s^{-1} and an assumed temperature of 4500°K consistent with the [OI]5577 feature and the colour temperature leads to an observed mass of O° = 0.3 $M_{\odot}$. Since preliminary models show that all oxygen is in the form of O°, and the density is above the critical density i.e. Ne = 5×10^{8} - 10^{9} cm^{-3}, we are probably seeing all of the oxygen in neutral form. If this is all of the oxygen in an envelope of 10 $M_{\odot}$, a value suggested by the models of Woosley (1987), and accepting that there is a factor 3 uncertainty in this number due to uncertainties in observations and temperature determinations, there is no reason to suppose that we are yet seeing any significant enhancement from the <u>interior</u> of the progenitor.

INFRARED SPECTROSCOPY

The early IR spectroscopy has been described by Bouchet et al. (1987). Low resolution spectroscopy using the circular variable filter (CVF) and intermediate resolution spectroscopy (R = 1500) using the IR spectrometer (IRSPEC) showed that the early spectra were characterized by a strong continuum on which were superimposed hydrogen lines having a P-Cygni structure. In June, more emission features began to develop in such a way that by early October the rich emission line spectrum shown in Figs. 9 and 10 was obtained with IRSPEC. CVF spectra obtained during the interval June-October showed that the development of most features was gradual. Proposed identifications of features seen in Figs. 9 and 10 are given in Table 1 together with velocities.

Although hydrogen lines from the Paschen, Brackett and Pfund series still dominate the spectrum, lines due to heavier ions are now apparent. There are at least 9 conspicuous features which so far defy identification. The relative strengths of the 12 cleanest hydrogen lines are consistent with case B recombination strengths within an uncertainty of 30 per cent. Nevertheless there is a total range of

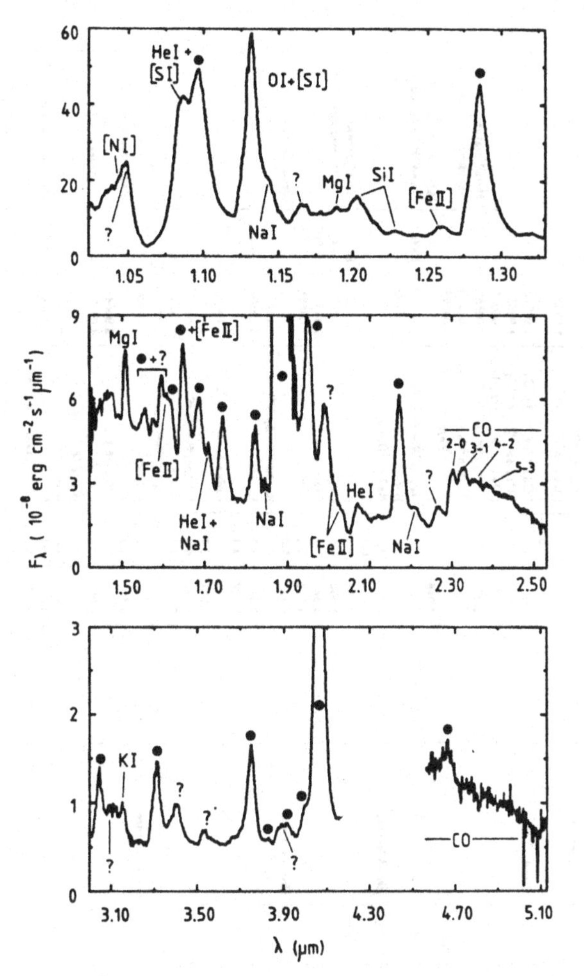

Figure 9. Sections of IR spectra taken during days 223-228 with hydrogen lines marked by dots.

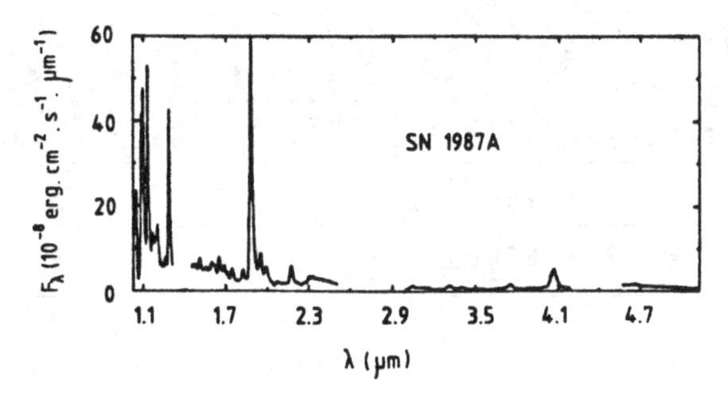

Figure 10. The complete IR spectrum as in the previous figure.

$\lambda_{obs}^{(1)}$	I[2]	Identification[3]	$\lambda_0^{(4)}$	$\Delta v^{(5)}$	Comments
1.047	~20	?			Edge of spectrum
		[N I] $^2D^0 - ^2P^0$	1.040		Large blue shoulder
1.063		?			Absorption feature?
1.086	~30	He I $2s\,^3S - 2p\,^3P^0$	1.08294	800	Blend, P-Cyg. profile?
		[S I] $^3P_2 - ^1D_2$	1.08210		
1.096	~40	H 6–3	1.09379	600	Blend
		([Si I] $^1D_2 - ^1S_0$)	1.09913		
1.132	53	O I $3p\,^3P - 3d\,^3D^0$	1.1287	900	
		[S I] $^3P_1 - ^1D_2$	1.13058		
1.143		Na I $3p\,^2P^0 - 4s\,^2S$	1.139		Shoulder of 1.132
1.162	~3	?			Blend
1.18	~5	K I $4p\,^2P^0 - 3d\,^2D$	1.173		Broad feature
		Mg I $3p\,^1P^0 - 4s\,^1S$	1.182818		
		Ca II $5s\,^2S - 5p\,^2P^0$	1.187		
1.202	~4	Si I $4s\,^3P_1^0 - 4p\,^3D_2$	1.1984		Blend
		Si I $4s\,^3P_0^0 - 4p\,^3D_1$	1.1991		
		Si I $4s\,^3P_2^0 - 4p\,^3D_3$	1.2032		
1.228	0.6	Si I $4s\,^3P_1^0 - 4p\,^3D_1$	1.2104		
		Si I $4s\,^3P_2^0 - 4p\,^3D_2$	1.2227		
1.260	2.4	[Fe II] $a\,^6D_{9/2} - a\,^4D_{7/2}$	1.25666	800	
		(K I $4p\,^2P^0 - 5s\,^2S$)	1.248		
1.285	54	H 5–3	1.28179	700	
1.444	0.5	[Fe I] $a\,^5D_4 - a\,^5F_5$	1.44294		
1.46	~2	H n–4 (Brackett) limit	1.45799		Broad feature
1.488	0.1	Mg I $3d\,^3D - 4f\,^3F^0$	1.487		
1.505	3.2	Mg I $4s\,^3S - 4p\,^3P^0$	1.503	400	
		([Fe I] $a\,^5D_2 - a\,^4F_4$)	1.49820		
1.528	0.1	H 19–4	1.52603		
		K I $3d\,^2D - 4f\,^2F^0$	1.5166		
1.55	1.4	[Fe II] $a\,^4F_{9/2} - a\,^4D_{5/2}$	1.53345		Broad feature
		H 18–4	1.53416		
		[Fe I] $a\,^5D_3 - a\,^5F_5$	1.53510		
		H 17–4	1.54387		
		H 16–4	1.55562		
		CO 3–0 band-head	1.558		
1.573	0.5	H 15–4	1.57004		
		Mg I $4p\,^3P^0 - 4d\,^3D$	1.575		
		CO 4–1 band-head	1.578		
1.60	6.2	H 14–4	1.58803		Broad feature
		Mg I $3d\,^3D - 5p\,^3P^0$	1.5885		
		Si I $4s\,^1P^0 - 4p\,^1P$	1.58884		
		CO 5-3 band-head	1.5990		
		[Fe II] $a\,^4F_{7/2} - a\,^4D_{3/2}$	1.59945		
		[Si I] $^3P_1 - ^1D_2$	1.60679		
		H 13–4	1.61091		
1.644	6.5	H 12–4	1.64070		
		[Fe II] $a\,^4F_{9/2} - a\,^4D_{7/2}$	1.64353		
		[Si I] $^3P_2 - ^1D_2$	1.64453		
1.684	3.5	H 11–4	1.68063	600	
1.705	0.8	He I $3\,^3P^0 - 4\,^3D$	1.70029		
		Na I $3d\,^2D - 5p\,^2P^0$	1.7031		
		Mg I $4s\,^1S - 4p\,^1P^0$	1.71087		
1.740	5.4	H 10–4	1.73619	700	
1.820	5.3	H 9–4	1.81738	500	Bad atm. region
1.851	0.8	Na I $3d\,^2D - 4f\,^2F^0$	1.847		Bad atm. region
1.881	110.	H 4–3	1.87507	900	Bad atm. region
1.948	9.6	H 8–4	1.94453	500	
		(Ca I $4p\,^3P_0^0 - 3d\,^3D_1$)	1.93092		
		(Ca I $4p\,^3P_1^0 - 3d\,^3D_2$)	1.94530		
		(Ca I $4p\,^3P_1^0 - 3d\,^3D_1$)	1.95057		
1.989	6.0	?			
		(Ca I $4p\,^3P_2^0 - 3d\,^3D_3$)	1.97768		
		(Ca I $4p\,^3P_2^0 - 3d\,^3D_2$)	1.98622		
		([Fe I] $a\,^5F_5 - a\,^3F_4$)	1.98045		
2.03		[Fe II] $a\,^4P_{1/2} - a\,^2P_{1/2}$	2.00667		Broad feature on the
		[Fe II] $a\,^2G_{9/2} - a\,^2H_{9/2}$	2.01510		Shoulder of 1.989
		[Fe II] $a\,^4P_{5/2} - a\,^2P_{3/2}$	2.04598		
2.048		?			Absorption feature?
2.068	1	He I $2s\,^1S - 2p\,^1P^0$	2.05810	1500	Blend, P-Cyg profile?
2.172	9.8	H 7–4	2.16550	900	
2.207	0.3	Na I $4s\,^2D - 4p\,^2P^0$	2.207		
2.264	0.6	?			
2.303	~1	CO 2–1	2.294*	1100	Total intensity in
2.330	~1	CO 3–2	2.322*	1000	the 1st overtone
2.36		CO 4–3	2.353*	~1000	I ~ 35*
3.048	2.2	H 10–5	3.03833	900	
3.12	3.6	?			Broad feature
		K I $3d\,^2D - 5p\,^2P^0$	3.149		
3.311	3.5	H 9–5	3.29604	1400	
3.402	2.1	?			
		(Mg I $3d\,^1D - 4p\,^1P^0$)	3.3963		
3.531	0.7	?			
3.659	0.2	K I $5p\,^2P^0 - 6s\,^2S$	3.649		
3.747	4.5	H 8–5	3.73948	600	
		(K I $5p\,^2P^0 - 4d\,^2D$)	3.721		
3.827	0.1	H 16–6	3.81836		
3.89	1.3	?			Broad feature
		(H 15–6)	3.90643		
4.02		H 14–6	4.01971		Shoulder of 4.064
		K I $4f\,^2F^0 - 5g\,^2G$	4.01584		
4.064	25	H 5–4	4.05109	1000	
4.66	~2	H 7–5	4.65244	~500	Noisy spectrum
~4.7	50*	CO fundamental band			

Notes to table:

(1) Observed wavelength in μm. The line center has been defined as the average of the 4 highest points. Center positions of broad features and lines in crowded regions are given with 2 significant digits.

(2) Observed line intensity, in units of $10^{-18}\ erg\,cm^{-2}s^{-1}$.

(3) Identification; transitions which are unlikely to give a substantial contribution to the observed feature are given within brackets.

(4) Rest wavelength of the identified transition; the number of significant digits is given according to the wavelength spread of the multiplet.

(5) Velocity shift in $km\,s^{-1}$ between the observed and the line rest wavelengths. Given only for bright and clearly identified lines. Positive values are for red shifts. Typical uncertainity is 200–300 $km\,s^{-1}$

* CO bands:
The wavelengths of the 1st overtone band peaks around 2.4 μm have been measured in a K-giant star spectrum obtained immediately after the supernova. The integrated intensity of the bands (in units of $10^{-18}\ erg\,cm^{-2}s^{-1}$) includes only the observed wavelength ranges.

Table 1. Infrared Emission Line Intensities measured from spectra obtained during October 5-8.

velocities (all positive as with Hα) of 500-1400 km s^{-1} where the uncertainties in measurement are less than 200 km s^{-1}. These real velocity differences in the hydrogen lines alone pose an interesting problem in modelling the hydrogen line formation mechanism. All of the other lines also occupy a range of positive velocities of 400-1500 km s^{-1}. A redshift bias for the emission component of a strong P Cygni line can arise because of local backscattering due to supersonic turbulence (Lucy 1983). However, the observed effect persists for weak H lines and is also found for the Lyβ-pumped [OI] line discussed below. This suggests a geometrical asymmetry that could perhaps be modelled as a conical region directed away from the observer and in which partial ionization of H and He is maintained by hard radiation leaking out from a central pulsar or its surrounding synchrotron nebula.

One line of interest is OI 1.1287μ, the upper level for which can be populated by cascade following Lyβ fluorescence. Other transitions such as OI 8446 and OI 1302 are expected as a result of this cascade and provide a means of testing this mechanism. It also provides an indicator of the optical depth of Hα in the line forming region.

Two other lines of interest are those of [FeII] 1.533 and 1.644μ, the latter of which has been observed in supernova remnants and may have been observed in SN 1983n, although in the latter case SiI lines provide an equally plausible identification (Oliva 1987). When allowance is made for the contribution of the hydrogen (H 12-4) line to the 1.644μ feature, the relative strengths of these two [FeII] lines are in accord with the theoretical calculations of the relevant transition probabilities (Nussbaumer and Storey 1987). As with oxygen, one can estimate the mass of ionized iron in the line forming region. This gives 0.03 M⊙ of ionized iron which may be the total mass of iron if, as preliminary models suggest, all of the iron is singly ionized. If this amount of iron pertains to the total ejected envelope, the implied mass fraction is 2-3 times the cosmic value, and another factor 2-3 overabundant if the atmosphere of the progenitor star were as underabundant in metals as the HII regions in the LMC. Nevertheless, a better understanding of the region and conditions where these lines are formed is required before far-reaching conclusions are drawn.

The first detection of carbon monoxide CO was made at La Silla with CVF spectra obtained in June. Both the fundamental band at 4.6μ and the first overtone band at 2.3μ were detected. At the earliest times the excess in the M band may have been caused by Pfund β, but at later times the main contributor to the excess must have been the fundamental band of CO. The 2.3μ band seems to have been blended with an unidentified feature at ~ 2.27μ at these early stages. The flux in the 2.3μ band has increased and at day 258 it is evidently decreasing again. The time of maximum flux is difficult to define because of the paucity of the relevant data. The shape of the bands (and the 2.3μ band in particular) is sensitive to temperature (Young 1968) because of the change in the relative population of the vibrational states. A

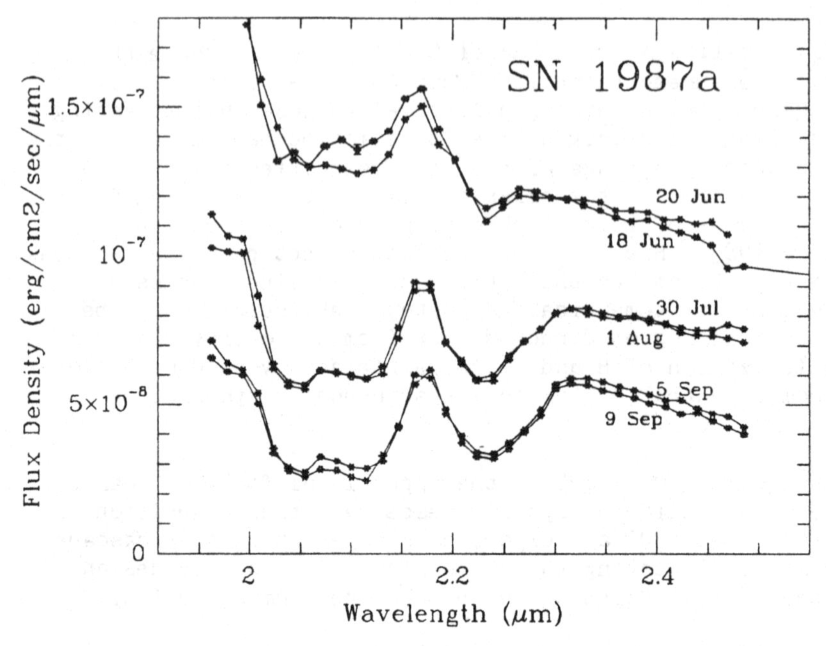

Figure 11. A time sequence of CVF spectra showing the 2.3μ first overtone band of CO.

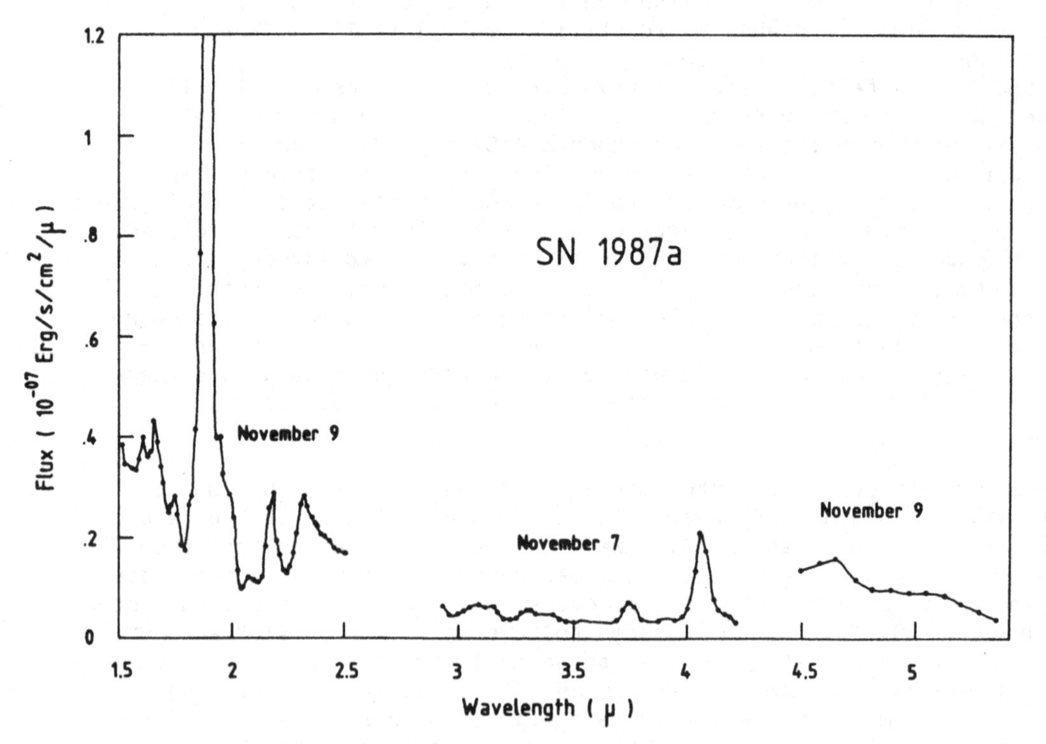

Figure 12. A full CVF spectrum for days 256-258 showing both the fundamental and first overtone bands of CO.

temperature of 2000°K ± 500°K is suggested by the spectra shown in Figure 9. The shape of the band was flatter in earlier spectra and thus a decrease in temperature with time is indicated, provided photon transfer effects within the band are not significant in the context of the differentially expanding envelope. Samples of earlier spectra including the first overtone band at 2.3μ are shown in Figure 11. A complete CVF spectrum for November 7-9 is shown in Figure 12, where the two main bands of CO are still prominent.

The band heads indicated in Figure 9 are resolved and provide an accurate measure of the velocity. These velocities (+1000 km s^{-1}) given in Table 1 are similar to, but slightly higher than, the velocities of the atomic lines seen in the IR spectrum. The presence of CO showing similar velocities to the other species, and apparent at ~ day 100, raises unexpected new questions concerning the physical conditions in the expanding envelope.

INTERSTELLAR LITHIUM

During the early epoch when the SN had a visual magnitude near 3.2, Baade and Magain (1987) were able to obtain high resolution spectra with excellent signal/noise in the region of the LiI lines at 6708 Å. Interstellar lines towards the LMC were not detected and an upper limit of the Li/H ratio $\leq 1.0 \times 10^{-10}$ was found for the abundance in the LMC. This is very similar to the detected amount in old Population II main sequence stars in our Galaxy, a value thought to represent the primordial value produced in the Big Bang. To be able to observe to a sensitivity a factor of 2 better than this might indeed provide a test of whether our current understanding of lithium production is correct.

REFERENCES

Arnett, W.D. 1987, ESO Workshop on SN 1987A, ed. I.J. Danziger, p.373.
Baade, D., Magain, P. 1987, Astron. Astrophys. in press.
Bouchet, P., Stanga, R., Moneti, A., Le Bertre, T., Manfroid, J., Silvestro, G., Slezak, E. 1987, ESO Workshop on SN 1987A, ed. I.J. Danziger, p.159.
Cristiani, S., Bouchet, P., Gouiffes, C., Sauvageot, J.L., Arsenault, R., Francois, P., Barwig, H., Fischerstrom, C., Guenter, T., Haefner, R., Lebertre, T., Loyola, P., Magnusson, P., Manfroid, J., Mekkaden, M.V., Schoembs, R., Slezak, E., Stanga, R. 1987, ESO Workshop on SN 1987A, ed. I.J. Danziger, p.65.
Danziger, I.J., Fosbury, R.A.E., Alloin, D., Cristiani, S, Dachs, J., Gouiffes, C., Jarvis, B., Sahu, K.C. 1987, Astron. Astrophys. **177**, L13.
Fosbury, R.A.E., Danziger, I.J., Lucy, L.B., Gouiffes, C., Cristiani, S. 1987, ESO Workshop on SN 1987A, ed. I.J. Danziger, ESO Workshop on SN 1987A, ed. I.J. Danziger, p.139.

Kirshner, R.P. 1987, ESO Workshop on SN 1987A, ed. I.J. Danziger, p.121.
Lucy, L.B. 1983, Astrophys. J. **274**, 372.
Lucy, L.B. 1987a, Astron. Astrophys. **182**, L31.
Lucy, L.B. 1987b, ESO Workshop on SN 1987A, ed. I.J. Danziger, p.417.
Nussbaumer, H., Storey, P.J. 1987, Astron. Astrophys. in press.
Oliva, E. 1987, Astrophys. J. **321**, L45.
Rufener, F., Maeder, A. 1971, Astron. Astrophys. Suppl. **4**, 43.
Williams, R.E. 1987, Astrophys. J. **320**, L117.
Woosley, S.E. 1987, Astrophys. J., in press.
Young, L.A. 1968, J. Quant. Spectrosc. Radiat. Transfer **8**, 693.

OBSERVATIONS OF SN 1987a AT THE SOUTH AFRICAN ASTRONOMICAL OBSERVATORY (SAAO)

M.W. Feast
South African Astronomical Observatory, P.O. Box 9,
Observatory 7935, South Africa.

A report of work by: L.A. Balona, P. Barrett, B.S. Carter, R.M. Catchpole, A.W.J. Cousins, M.W. Feast, I.S. Glass, D. Kilkenny, J.D. Laing, C.D. Laney, T. Lloyd Evans, F. Marang, J.W. Menzies, G. Roberts, K. Sekiguchi, J. Spencer Jones, P.A. Whitelock (SAAO); A.P. Fairall, H. Winkler (University of Cape Town); G. van Vuuren (University of Natal); W.F. Wargau (University of South Africa); A.S. Monk (Royal Greenwich Observatory); D. Pollacco (University of St Andrews); J. Davies, Q. Parker, R. Wolstencroft (Royal Observatory, Edinburgh); J. Albinson, M.G. Hutchinson, R. Maddison (University of Keele).

SUMMARY

The results ofspectrophotometry, broad band photometry (optical and infrared) and broad band polarimetry are summarized up to to about day 200 after outburst. Problems of consistency between results at different observatories are discussed. The bolometric light curve can be reasonably fitted by the early models of Woosley *et al*. for a blue supergiant progenitor, except for the lack of a plateau phase. This seems to indicate the lack of a thick hydrogen mantle. The decline in m_{bol} is remarkably linear from day 146 onward and yields an e-folding time close to the mean life of Co^{56}. This provides strong evidence that this part of the decline is powered by radioactive heating. An infrared excess has become increasingly prominent since day ~100. It is strongly peaked at ~5 μm and the principal contributor is CO emission. Variable linear polarization indicates departures from spherical symmetry in the ejecta.

INTRODUCTION

Ian Shelton realized towards dawn in Chile on the morning of 1987 February 24 that he had made the momentous discovery of a supernova in the Large Magellanic Cloud. The first ground-based spectroscopic and multicolour photometric observations were made soon after dark that evening at SAAO Sutherland. Since that time our aim has been to make the following observations:

(1) Daily UBVRIJHKL photometry (0.36-3.5 μm)
(2) At least weekly spectra, spectrophotometrically calibrated, at 7A FWHM, covering 3400 to 7600A.

(3) Less regularly;

(a) Spectra at various resolutions
(b) 5, 10, 20 μm photometry
(c) Low resolution 1-4 μm spectra
(d) UBVRI polarimetry (linear and circular)

Some of these observations, and discussions of them, have already been published (Menzies et al. 1987, Catchpole et al. 1987, Menzies 1987, Catchpole 1987). In this paper the results of the observations will be summarized, with particular emphasis on those for the period immediately preceeding the Fairfax meeting.

SPECTROSCOPY

In the earliest spectrum the Balmer lines are quite weak but the emission and absorption components strengthened rapidly (Menzies et al. 1987). Though the SN is thus of type II, the Hα absorption velocities are larger, and the apparent deceleration much more rapid, than in other SNII. The Hα absorption minimum on the first spectrum is at 18500 km s^{-1} (probably a record for an SNII) and the deceleration over the first week was 780 km s^{-1} day^{-1} compared with a mean of 69 km s^{-1} day^{-1} for 6 previously observed SNII (Pskovskii 1978). Hα emission has continued strong. Recent spectra in the Hα region are shown in figure 1 and the complex emission profile, varying with time, can be seen. SAAO spectra are being modelled and discussed in collaboration with Dr L B Lucy and others at ESO (Munich).

A word of caution is in place here. Observers at SAAO and, one has the impression, at most other southern observatories, are aware that the spectrophotometric results especially those obtained at a time of year when all of us are necessarily observing the supernova at large air mass, will require careful scrutiny to make certain that they are of the derived accuracy. It may well be necessary to treat the spectra and simultaneous broad band photometry together in order to achieve satisfactory results.

PHOTOMETRY

Figure 2 shows the SAAO UBVRIJHKL light curves to day 207. Here and in what follows, time is counted from the Kamioka neutrino detection (JD 2446849.82) which is taken as the time of core collapse. These broad bands cover the wavelength range 0.36 μm to 3.5 μm. Since the vast bulk of the radiation is emitted in this wavelength range, the data can be used to derive good bolometric fluxes.

It is important to make a few technical remarks here. Firstly the UBVRI observations are made in the Cousins system. The photometer/telescope combination with which these observations were made is extensively (and almost continuously) used to make observations of a wide variety of stars on this system and indeed has been used to refine and extend the earlier Cousins standards. For normal stars therefore the system is known to reproduce the standard Cousins system with high accuracy. Similarly the JHKL is relative

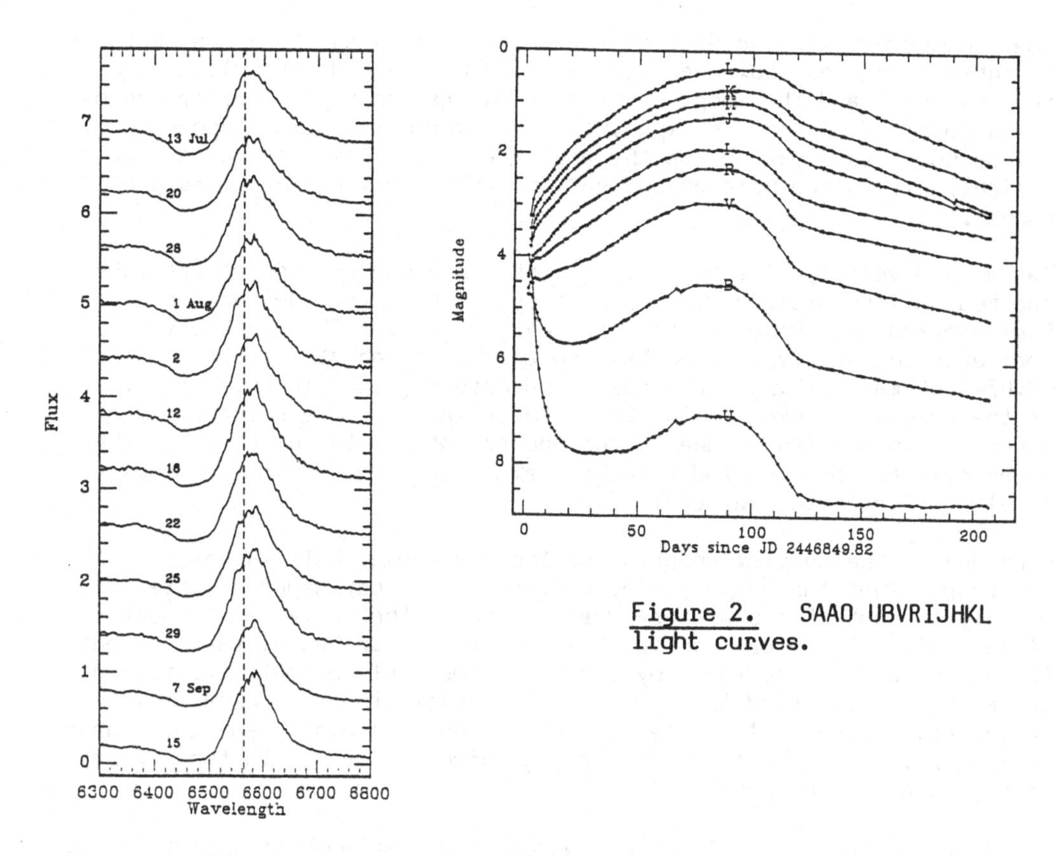

Figure 2. SAAO UBVRIJHKL light curves.

Figure 1. Region of Hα from selected spectra in the period 13 July to 15 September. The rest wavelength of Hα is marked.

to a set of standards set up by Carter (1987) which refine and extend the original system of Glass (1974). Both the original and refined systems have been used at SAAO for a very large number of observations using the same photometer/telescope as for the supernova. However it must be stressed that because of the unusual energy distribution of the supernova, there is no guarantee that other photometers claiming, for instance, to reproduce the Cousins system for normal stars will necessarily accurately reproduce the same results as SAAO for the supernova. This seems to be borne out by the CTIO UBVRI of the supernova (Hamuy et al. 1987) which, despite its being referred throughout to the standards for the Cousins system set up at SAAO, shows systematic differences from the SAAO results. This is especially marked in V-I where the difference amounts to about 0.10 mag. on the average. In these circumstances it is almost certainly best to reduce extensive sets of observations independently of one another and to compare final conclusions rather than attempting to merge observations, unless there is compelling reason to the contrary. So far as the absolute accuracy of derived parameters such as bolometric magnitudes and effective temperatures are concerned, the differences between the different observational system are likely to be small compared with the problems of absolute flux calibrations, methods of integrating the flux over all wavelengths, the adopted distance of the supernova and the adopted interstellar reddening. However trends in the derived quantities, m_{bol}, T_e etc, (which for some purposes may be quite as important as the absolute values) can be traced most reliably within one homogeneous set of data, analysed in a consistent manner.

Based on a variety of lines of argument, SAAO workers and others have adopted an interstellar reddening of $E_{B-V} = 0.20$ for the supernova. Some workers have however preferred $E_{B-V} = 0.15$. This probably indicates that at present we have to accept an uncertainty of ~0.03-0.05 mag in E_{B-V} and hence an uncertainty, due to this cause, in the absolute fluxes of 10-15%. There is a further uncertainty of about 15% in the fluxes due to the uncertainty in the distance of the supernova (taken by us and by most workers to be 50 kpc (true modulus 18.5) e.g. Feast and Walker 1987).

Considering the complex spectrum of the supernova it is rather remarkable that the overall energy distribution corresponds quite closely to that of a blackbody during most of the early development (figure 3). There are considerable deviations in the ultraviolet but these are unimportant for many purposes since only a small fraction of the energy is radiated there. It seems likely that the ultraviolet deficiency is due to line blanketting. There is also a small infrared excess during some of the early stages which may be due to the effects of an extended atmosphere.

The near blackbody distribution suggests that the angular diameters and effective temperatures derived from blackbody fits, are useful quantities. Indeed, Branch (1987) has used these results to show that the early radial velocity results and blackbody angular diameters can be used in the Baade-Wesselink method to obtain a distance modulus for the supernova (18.7±0.2) in good agreement with the LMC modulus adopted above.

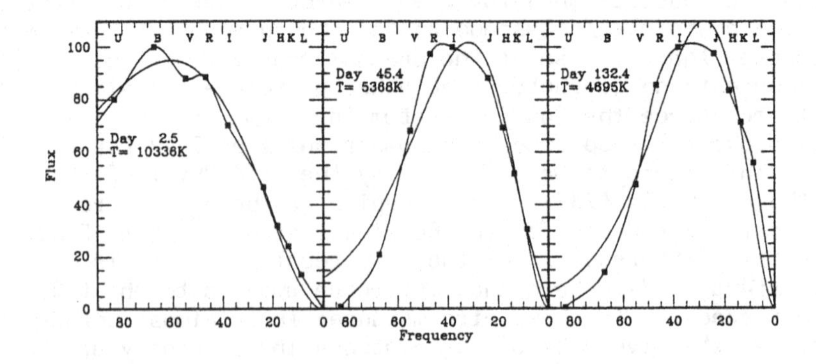

Figure 3. Comparison of blackbody fits to the observations on three dates.

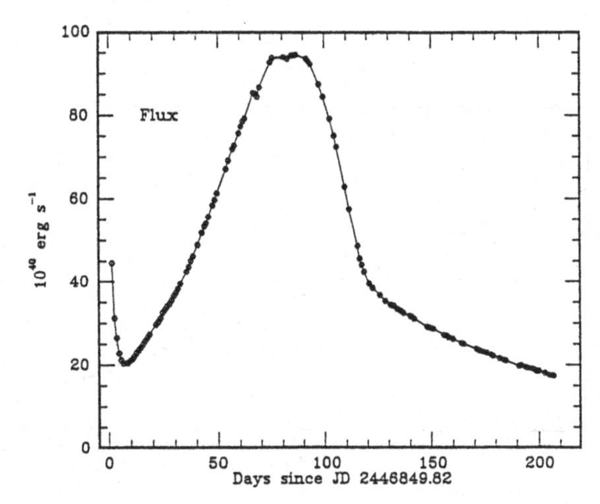

Figure 4. Variation of the bolometric flux with time.

Figure 4 shows the variation of the bolometric flux with time. The fluxes are based on blackbody fits to the SAAO UBVRIJHKL. Perhaps the first thing that strikes one about this curve is its smoothness with no major discontinuities anywhere. It is well known that SN 1987a was visually subluminous compared with other SNeII, reaching only M_V = -14.6 at the time of the initial rise and M_V = -16.2 on about day 75 whereas most SNeII have M_V brighter than -16 at outburst. This underluminosity applies also to the bolometric magnitude (although the bolometric behaviour of other SN is not well known). A comparison with the results for SN 1969l as plotted by Woosley and Weaver (1986 their figure 2) shows SN 1987a to be bolometrically less luminous out to ~140 days but suggested that by day ~200 in the slow (radioactive) decline, the two supernovae were of comparable luminosity at the same times after outburst.

Models based on red supergiant progenitors, such as those illustrated in Woosley and Weaver (1986 figure 2) do not fit the bolometric light curve of SN 1987a. A better fit can be obtained with models based on blue supergiant progenitors (Woosley et al. 1987). The main difference between SN 1987a and these models is that instead of the sharp minimum at about day 7, the models show a rather flat minimum lasting 20-30 days (the "plateau"). In these models the plateau phase can be identified with a period during which the hydrogen mantle is moving out through the photosphere of the supernova. It is obviously attractive from some points of view to suggest that the lack of a plateau implies that the progenitor of SN 1987a lost much of its mass (particularly most of its hydrogen mantle) in an earlier red supergiant phase, probably in the form of a stellar wind. This however may not be the only possible explanation. For instance, mixing of helium into the hydrogen mantle has also been suggested (Woosley 1987). At the time of the minimum there is a change in the rate of change of the angular diameter of the photosphere with time (Catchpole et al. 1987). This suggests a change in structure of the SN atmosphere at this time, perhaps a transition to a more helium rich layer.

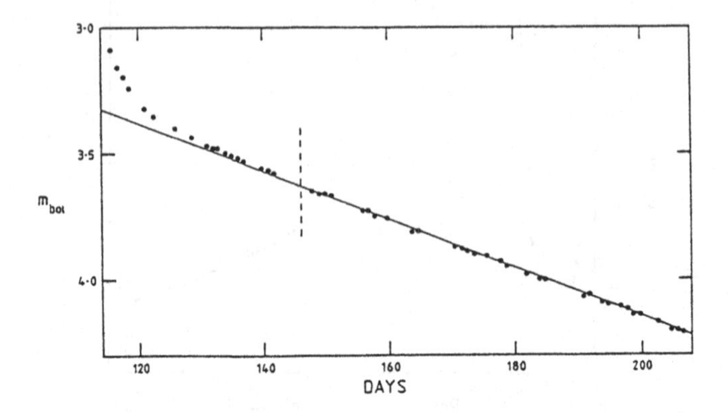

Figure 5. The later part of the light curve (m_{bol}) showing a linear fit to the data from day 146.

The slow decline in the bolometric flux from day ~120 onwards is particularly interesting because it is predicted to be powered by the radioactive decay of Co^{56} (cf Woosley and Weaver 1986 and references there). Figure 5 shows a plot of the bolometric magnitude determined from blackbody fits to SAAO UBVRIJHKL for this later period of the light curve. The straight line is a least squares fit to the data between days 146 and 207. The rapid decline continued to about day 120 and the period 120-140 days is an intermediate phase. After that the points fall very close to a straight line. The root mean square scatter is only 0.005 mag. As table 1 shows there could be a slight decrease in slope with time but on the present data this is not statistically significant. The mean slope corresponds to an e-folding time of 113.6±0.6 days. This is in remarkable agreement with the value for the mean life of Co^{56} (113.6 days, half life 78.76 days) which has frequently been adopted in the astrophysical literature (e.g. Woosley and Weaver 1986). However a value of 111.26 days is now given for this mean life (half life 77.12 days) (cf. Nuclear Data Sheets, May 1987). It is also necessary to investigate how sensitive the derived e-folding time is to the manner in which m_{bol} is determined and to the adopted interstellar reddening. Table 1 shows the results obtained when the flux is calculated using either blackbody or spline fits to the UBVRIJHKL data. Calculations with $E_{B-V} = 0$ are also shown (this is obviously an extreme assumption). Finally an infrared excess was becoming significant during this period (see below). Whether this should be added to the total flux depends on its origin (e.g. flux from an infrared echo, such as some writers have predicted, should not be added to the total flux). Rough estimates show that if one does include the infrared excess m_{bol} would be brightened by ~0.03 mag on day 146 and ~0.06 mag on day 206. The effects of this on the e-folding times of the light curves are shown in Table 1. As the range of values in the table shows the uncertainty in the e-folding time could be as much as 5-10 days. Even so the closely linearly decline in m_{bol} together with an e-folding time so close to the mean life of Co^{56} provides a rather striking verification of the prediction that the slow decline is powered by radioactive heating. Further work may establish whether the tentative indications of a decrease in slope with time are real. Such a decrease is predicted (Arnett 1987) if the contribution to the total energy from a central pulsar, becomes significant at later times. Using the formula of Weaver, Axelrod and Woosley (1980) we can estimate that the mass of Co^{56} and its ancestor Ni^{56}, produced in SN 1987a, was 0.077 $M_{\odot}$.

Table 1

e-folding times (τ) of slow decline

Flux Fit	E_{B-V}	Days	No of Obs	τ (days)	σ (Mag)
Blackbody	0.2	146-170	10	109.7 ± 3.3	.006
Blackbody	0.2	170-190	10	111.9 ± 3.5	.005
Blackbody	0.2	190-207	12	115.5 ± 3.7	.006
Blackbody	0.2	146-207	32	113.6 ± 0.6	.005
Blackbody	0.0	146-207	32	107.1 ± 0.6	.006
Spline	0.2	146-207	32	104.0 ± 0.6	.006
Spline	0.0	146-207	32	99.1 ± 0.6	.008
With infrared excesses added (see text)					
Blackbody	0.2	146-207		119.8	
Spline	0.2	146-207		109.2	

Mean Life of Co^{56} = 111.26 days
σ = Root mean square deviation from a straight line fit to the data

In the early phases of SN 1987a there was only a slight infrared excess over that expected from a blackbody fit to the data. However from about day 100 an infrared excess has been developing which is particularly strong at M (5μ). This is shown in figure 6 which plots SAAO measures of L-M against time. The insert shows that over the period days 100-200, M remained roughly constant. During this same period the SN was decreasing at wavelengths from B to L (figure 2). The excess is clearly seen in the flux distribution on day 203 (figure 7). The infrared energy distribution is very narrow. There is a steep rise from L to M and then a rapid drop to N and Q (10 and 20 μm). This distribution is too narrow to be due primarily to heated dust. Spectra in the 1 to 4 μm region on days 140 and 228 show the presence of emission of the first overtone band of CO. This suggests

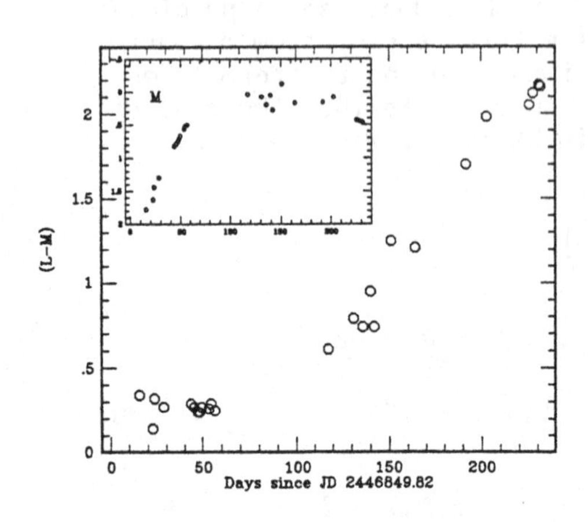

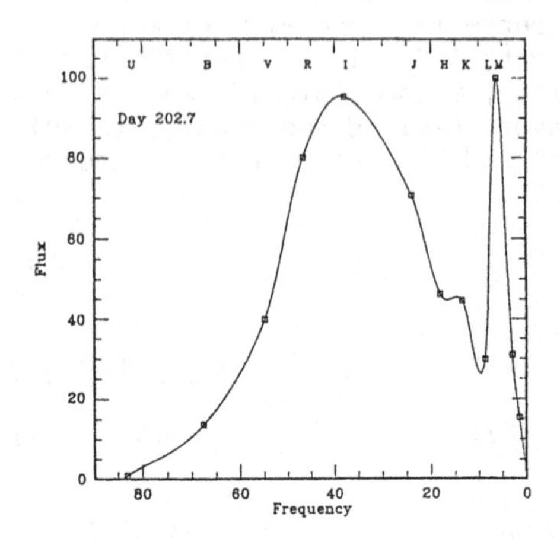

Figure 6. The development of an excess at 5 μm (M) as shown by the (L-M) colour.

Figure 7. Flux distribution ($F\nu$) on Day 202.7 showing M (5 μm) excess. The final two points are N and Q (10 and 20 μm).

that the CO band at 4.8 μm should be very strong and this could be the main contributor to the flux at M. The presence of the 4.8 μm band in emission has in fact been directly observed (Danziger 1987). It is clear that a detailed study of this infrared emission will yield important data on the supernova and its environment.

In conclusion I would like to briefly mention SAAO broad band UBVRI polarimetry. There is no evidence for any circular polarization but linear polarization is present and is variable with time. The interpretation of the data depends rather critically on the value adopted for the interstellar polarization. However it seems clear that at some stages in the development of SN 1987a there were significant asymmetries in the ejecta.

ACKNOWLEDGEMENTS

This paper reports work by the group of astronomers, from a variety of institutions, already named. Several of these have been heavily involved in the reduction and analysis of the data. R.M. Catchpole, J.W. Menzies, P.A. Whitelock, P. Barrett and D. Kilkenny have been particularly involved in this later stage and the conclusions reached in this paper are based primarily on discussions with them.

REFERENCES

Arnett, W.D., 1987 - this volume.
Branch, D., 1987. Preprint.
Carter, B.S., 1987 - in preparation.
Catchpole, R.M., 1987. IAU Colloquium 108 (Tokyo) - in press.
Catchpole R.M., Menzies, J.W., Monk, A.S., Wargau, W.F., Pollacco, D., Carter, B.S., Whitelock, P.A., Marang, F., Laney, C.D., Balona, L.A., Feast, M.W., Lloyd Evans, T.H.H., Sekiguchi, K., Laing, J.D., Kilkenny, R.M., Spencer Jones, J., Roberts, G., Cousins, A.W.J., van Vuuren, G., Winkler, H., 1987. Mon. Not. R. astr. Soc. - in press.
Danziger, J., 1987 - this volume.
Feast, M.W. and Walker, A.R., 1987. Ann. Rev. Astron. Astrophys., 25, 325.
Glass, I.S., 1974. Mon. Not. astr. Soc. Sth. Afr. 33, 53.
Hamuy, M., Suntzeff, N.B., Gonzalez, R. and Martin, G., 1987. NOAO preprint 102.
Menzies, J.W., 1987. ESO Workshop on SN 1987a - in press.
Menzies, J.W., Catchpole, R.M., van Vuuren, G., Winkler, H., Laney, C.D., Whitelock, P.A., Cousins, A.W.J., Carter, B.S., Marang, F., Lloyd Evans, T.H.H., Roberts, G., Kilkenny, D., Spencer Jones, J., Sekiguchi, K., Fairall, A.P., Wolstencroft, R., 1987. Mon. Not. R. astr. Soc., 227, 39p.
Pskovskii, Y.P., 1978. Soviet Astron. J., 22, 201.
Weaver, T.A., Axelrod, T.S. and Woosley, S.E., 1980. Texas Workshop on SNI, Ed. J.C. Wheeler (University of Texas) p. 113.
Woosley, S.E. and Weaver, T.A., 1986. Ann. Rev. Astron. Astrophys., 24, 205.
Woosley, S.E., 1987. Preprint.
Woosley, S.E., Pinto, P.A. and Ensman, L., 1987. Preprint.

OBSERVATIONS OF SN1987A AT THE ANGLO-AUSTRALIAN TELESCOPE

W.J. Couch
Anglo-Australian Observatory, P.O.Box 296, Epping, NSW 2121, Australia.

1 INTRODUCTION

Since outburst, about 15% of all time on the 3.9m Anglo-Australian Telescope (AAT) has been devoted to observing SN1987a. In this report I describe these observations, first discussing the policy and organisation adopted for making them followed by a description of the various types of observation made. In the final section I shall present some of the highlights which have emerged from these observations.

2 POLICY AND ORGANISATION

The AAT is a bi-national facility with the time divided equally between the two large and active British and Australian astronomical communities. Demand for time on the telescope is heavy and very competitive and the telescope is scheduled months in advance. Hence, when SN1987a appeared two problems had to be overcome. Firstly,that of policy : what were the most pertinent observations to make, how frequently should they be made and for how long, and who should have rights to the data? Secondly, that of organisation : how could the observations be fit into a tight and predetermined observing schedule?

Following prompt consultations with the two time assignment committees and a number of supernova experts it was decided that :

(1)An override of one hour per night should apply to all scheduled observers for observing SN1987a whatever instrumentation happened to be on the telescope. Larger overrides (up to half a night) were also to apply from time to time to allow "specialist" observations to be made.

(2)A complete override was to apply should the supernova show a sudden change.

(3)All Director's time (10% of total) and service time was to be dedicated to observing SN1987a. Director's time would also be used at later times to recompense displaced observers.

(4)Unless specialist observations were involved, all data were to be put into a special archive and to be made freely available to the astronomical community.

(5)Our efforts should concentrate on observations which utilised the AAT's large aperture and/or unique instrumentation. The importance of obtaining a consistent set of optical spectra was also recognised.

By taking this approach, it provided for very flexible use of the AAT and ensured that key observations were obtained when required. As mentioned previously, it allowed up to 15% of AAT time to be used for observing SN1987a. Obviously few scheduled observers escaped losing some time as a result of the supernova override, but in general there has been very willing cooperation in allowing the SN1987a observations to proceed. In the last few months overrides have been phased out and taken over by time allotments allocated to specific observing teams by the TAC's for monitoring the supernova and making specialist observations.

3 TYPES OF OBSERVATION

An obvious problem with SN1987a was how we could usefully observe an object of its brightness with a 3.9m telescope. Attenuation of the supernova's light by use of neutral-density filters, defocussing the telescope or stopping down its aperture was always an option but the guiding philosophy has been to concentrate on those types of observations which required instrumentation unique to the AAT or at least rare elsewhere in the southern hemisphere, and those which required the full 3.9m aperture of the AAT. Here now is an outline of the various types of observations which have been made on the AAT :

Optical Spectroscopy : this has involved regular monitoring of the supernova over the range $3000 < \lambda < 10000$Å at a variety of dispersions. The RGO spectrograph + IPCS/CCD combination and the Faint Object Red Spectrograph (FORS) have been the two main work-horses with the former providing spectra at dispersions ranging from 33 to 156A/mm and covering the $3000 < \lambda < 7500$Å region and the latter providing lower dispersion (450Å/mm) spectra over the range $5300 < \lambda < 10900$Å. Our most frequent monitoring has been with FORS for the very simple practical reason that it can be kept mounted on the telescope even when other instruments are in use with the supernova light being fed to it via a fibre-optic image slicer which is installed at an auxillary focus.(A list of all the spectroscopy obtained to date is available fromthe AAO; requests for specific data should be made to the Director.)

Direct Photography : a number of prime-focus plates were taken very soon after SN1987a erupted and provided astrometric confirmation that the progenitor star was Sk -69°202 (White and Malin 1987). They were also used to produce the 'before' and 'after' colour shots shown at the workshop.

Infrared Photometry and Spectroscopy : photometry in the J, H, K, L, L' and M bands has been obtained with the Infrared Photometer-Spectrometer. Some polarimetry in the J, H, K bands was also obtained at very early times. D.A. Allen and W.P.S. Meikle (Imperial College) have been using the Fabry-Perot Infrared Grating Spectrometer to obtain spectra in the atmospheric windows at resolving powers of $\sim$1200 and $\sim$300. Some far infrared (12-20μ) spectra have also been obtained by Aitken *et al* (1987) using the UCL Infrared Spectrometer and show evidence for a long wavelength infrared excess.

Spectropolarimetry : the AAT's RGO spectrograph can be converted to a spectropolarimeter by insertion into the beam of a Pockels cell and calcite prism. It has been used in combination with a GEC CCD detector to provide, for the first time, observations of a supernova with sufficiently high signal-to-noise ratio to show variations in polarization both in the continuum and across the spectral lines. These data will be discussed more fully in section 4.

High-speed Photometry : Manchester (CSIRO Radiophysics) and Peterson (Mt Stromlo and Siding Spring Observatories) have been running observations of this type at regular intervals in an attempt to detect, at the earliest possible time, the pulsar component should it emerge. The AAT is well equipped for such observations from previous studies of all three currently known optical pulsars. In the case of SN1987a, however, only the AAT's data recording systems are being used; a 20cm Celestron telescope is being used to colect the light from the supernova. No periodic signal has yet been detected, but more elaborate data reductions can always be done retrospectively and perhaps help to determine the rate of change of period should a pulsar be discovered in the future.

High to Ultra-high Dispersion Spectroscopy : because of its brightness, SN1987a provides a unique opportunity to probe the interstellar medium (ISM) over the $\sim$50 kpc path-length between us and the Large Magellanic Cloud (LMC) to a level of detail achieved previously only for a few bright stars in the local solar neighbourhood. Two programmes have been undertaken at the AAT : very high signal-to-noise ratio searches for weak lines and very high dispersion studies of strong lines. These will be discussed in the following section.

4 OBSERVATIONAL HIGHLIGHTS

4.1 Optical Spectra

Speakers at this workshop from the other southern observatories have covered comprehensively the nature and time evolution of SN1987a's optical spectrum and so I shall only show here the most recent AAT spectrum of the supernova. In Figure 1 a FORS spectrum obtained on October 9, 1987,is presented. Clearly SN1987a is now in its nebula phase with its spectrum in the $5000 < \lambda < 10900$Å region covered by FORS,being dominated by very strong Hα, CaII and Ca-triplet emission lines. Weaker emission in the OI lines at $\lambda\lambda$6300,6360 is also apparent - it first appeared in July. In order to see the extent of recent evolution in SN1987a's spectrum, I also show in Figure 1 a similar FORS spectrum taken on August 19. It can be seen that there has been a continual strengthening in the CaII and Ca-triplet features while Hα has remained relatively constant.

4.2 Spectropolarimetry

It is thought that radiation transfer in the atmospheres of supernovae is dominated by scattering processes whereby the spectral lines are produced by resonance scattering and the continuum by electron

Figure 1 : FORS spectra of SN1987a taken on 19 August 1987 and 6 October 1987. Prominent emission features are identified. Each spectrum has been normalised according to photometry taken of SN1987a on the same night.

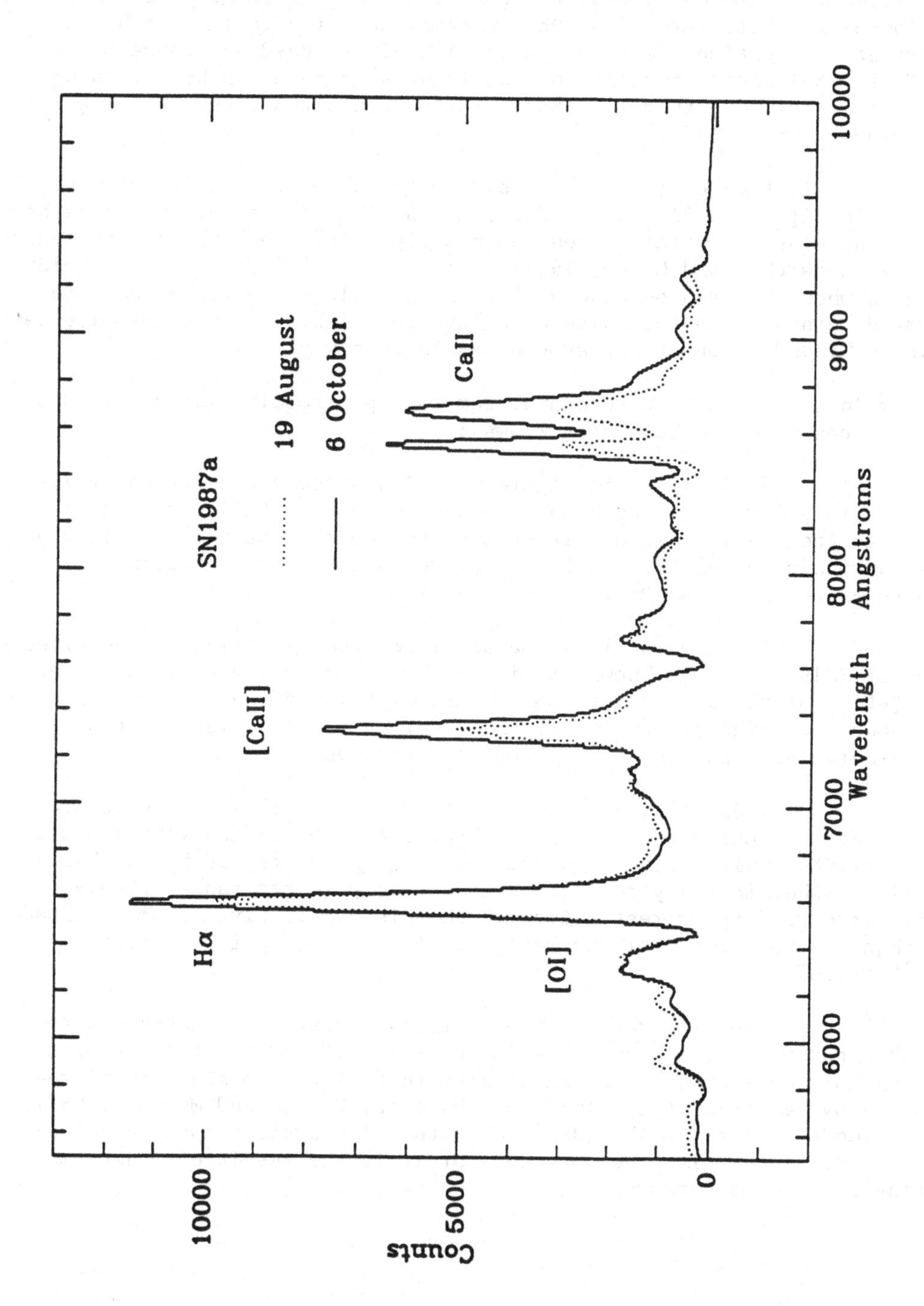

scattering above a thermal continuum-emitting photosphere (Branch 1980). The supernova light entering the line of sight will therefore be polarized but no *nett* polarization will be observed unless the atmosphere departs from spherical symmetry (McCall 1984). Clearly then, spectropolarimetric information had the potential to provide very useful information on the structure of SN1987a's atmosphere (which in turn has important implications for any application of the Baade distance method; McCall 1984) and a programme of obtaining such data has been running on the AAT since outburst. First results have already been published by Cropper *et al* (1987).

A small selection of the spectropolarimetry obtained for SN1987a is shown in Figure 2. Plotted in Fig 2(a) and (b) is the spectrum together with the measured polarization and position angle observed in the region of Hα on March 7 and May 5, 1987; an identical plot for the Ca-triplet region when observed on June 3 1987 is shown in Fig 3(c).As the integrated counts indicate, these data have the highest signal-to-noise ratio ever achieved in observing supernovae in this way.

The main conclusions to be drawn from the spectropolarimetry gathered so far can be summarised as follows :

* As is clear in Figure 2, SN1987a shows significant variations in polarization both in time and in wavelength - particularly through the spectral lines. It must be concluded that the polarization is intrinsic to SN1987a and it therefore must have a scattering atmosphere which is *assymetric*.

* The interstellar (Galaxy + LMC) polarization in the direction of SN1987a is not known precisely. This limits any detailed interpretation of the data since subtraction of this component (which has to be done vectorially) can modify the observed polarization spectrum considerably. Dips can become bumps and vice versa.

*An estimation of the *asphericity* of SN1987a's atmosphere has been made using the models of Shapiro and Sutherland (1982) and McCall(1984) which assume the assymetry can be described by an oblate/prolate spheroid. They indicate axis ratios μ in the range 0.8-0.9. The corresponding overestimation in the distance derived from the Baade method due to incorrectly assuming sperical symmetry is a factor of $\mu^{-1/2} \sim 1.07$.

* When all the spectropolarimetric data are plotted in the Stokes parameter QU plane, they indicate a preferred scattering axis which coincides in position angle with that of the 'mystery spot' reported by Karovska *et al* (1987) and Matcher, Meikle and Morgan (1987). This lends support to the idea that either the mystery spot is physically associated with the supernova or it is the result of a jet emanating from the supernova.

Figure 2 : Spectropolarimetry of SN1987a with *bottom panel* showing integrated counts, *middle panel* showing % polarization and *top panel* showing position angle. (a) Data in the region of Hα for March 7, 1987, (b) Data in the region of Hα for May 5, and (c) Data in the region of the Ca-triplet for July 8.

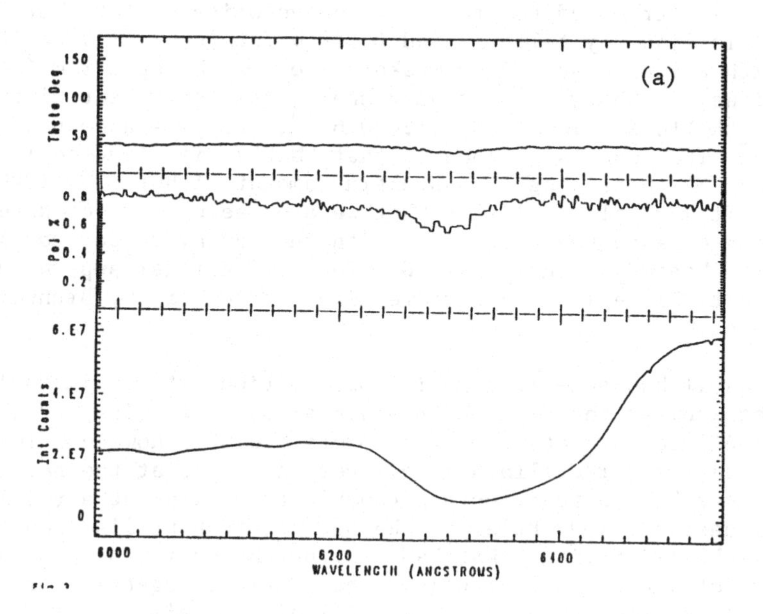

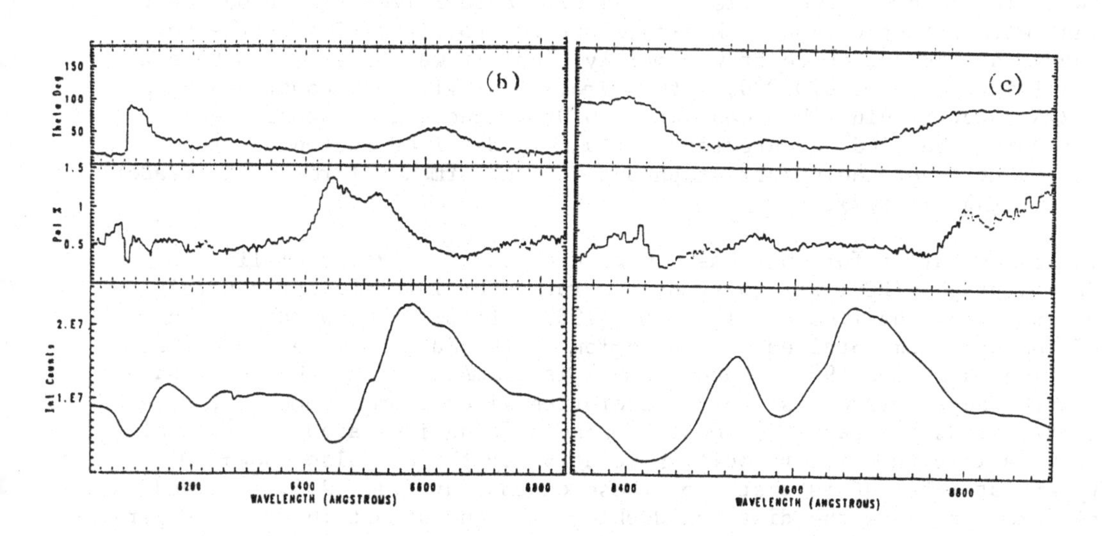

4.3 High and Ultra-high Dispersion Spectroscopy

There have been two particularly exciting results to come from the work done on the ISM via high dispersion spectroscopy of SN1987a. The first, which has resulted from using the AAT's RGO spectrograph at its very highest resolution (0.2Å) in combination with a CCD, involves the first ever detection of the forbidden [Fe X] λ6374.5 line in absorption. This discovery was first reported at the ESO workshop (D'Odorico *et al* 1987); it is to be discussed in more detail by Pettini *et al* (1987). The spectrum crucial to the discovery has been kindly provided by these authors and is shown in Figure 3. Remarkably it has a signal-to-noise ratio of 1170! The broad absorption feature which dominates the spectrum is attributed to [Fe X] at the velocity of the LMC. Contamination of this line by telluric and Diffuse Interstellar Band (DIB) features, which are also seen at a weaker level in the spectrum (see identifications), is thought to be very small; the equivalent width measured for the [Fe X] line is W_λ=16.4±0.6 mÅ, implying a column density of $\sim 10^{17}$ cm^{-2}. The conclusion then is that SN1987a is situated in a region of the LMC containing a substantial amount of hot (T $\sim 10^6$K) coronal gas. It may well be likely that we are seeing part of a very extensive body of such gas associated with the nearby 30 Dor complex which has been heated by the combined effects of earlier supernovae in this supergiant HII region - a picture also put forward by Bruhweiler at this meeting.

The second result has come from making observations at considerably higher dispersion. At the time SN1987a appeared, facilities did not exist at the AAT to work at resolving powers >80,000. However, in the space of ∿6 weeks Peter Gillingham, officer in charge at the AAT, quite ingeniously assembled a spectrograph capable of working at a resolving power of at least 500,000. Briefly, the spectrograph utilised one of the 2 large echelle gratings that had been purchased for a coude spectrograph currently under construction. The grating, together with a 7m focal-length lens (ground by one of the AAT night assistants from a BK7W blank), a borrowed prism and interference filter (for order sorting) were arranged in a double-pass Littrow configuration at the Coude focus. At ∿589nm, where it was mainly used, it was measured to have a resolving power of 570,000, with a V=3.1 star giving 1 count/second per 5.2mA spectral bin. This was with a 0.047arcsec slit, 1 arcsec seeing and using the IPCS as detector. For a more detailed account of the construction of the spectrograph and its performance I refer the reader to Gillingham(1987).

The prime target for what has now been dubbed the 'Peter Gillingham Spectrograph' (PGS) was to study absorption by interstellar neutral sodium along the line of sight to SN1987a. Figure 4 shows a portion of the spectrum obtained in the region of the Na I D_2 line whose rest wavelength is 5889.950Å. Each channel is 3.2mÅ wide and the total wavelength range covered is 1.65Å. Absorption lines from a number of discrete clouds are seen, the most prominent being indicated by vertical tick marks with their velocities relative to the Sun also shown. Of uppermost interest is that 4 of these clouds (at 1.1, 12.5, 18.1 and 64.6 km/sec) show the distinct double component structure due to hyper-

Figure 3 : High resolution spectrum of SN1987a in the region of [Fe X] λ6374.51.

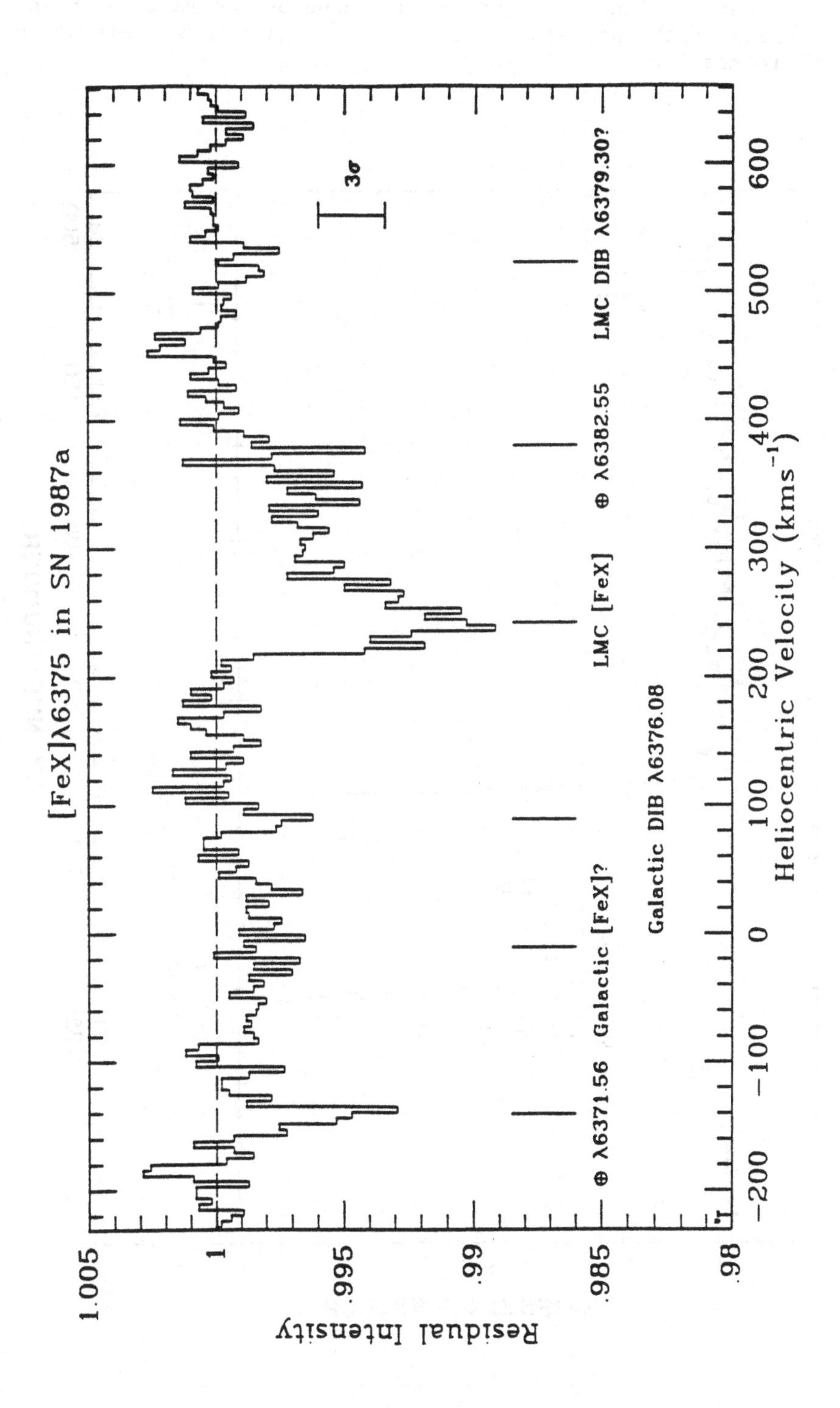

Figure 4 : An ultra-high dispersion spectrum of SN1987a in the region of the Na I D_2 line. Absorption due to clouds along the line of sight is indicated by the vertical bars along with the velocity of each cloud (in km/sec) relative to the Sun.

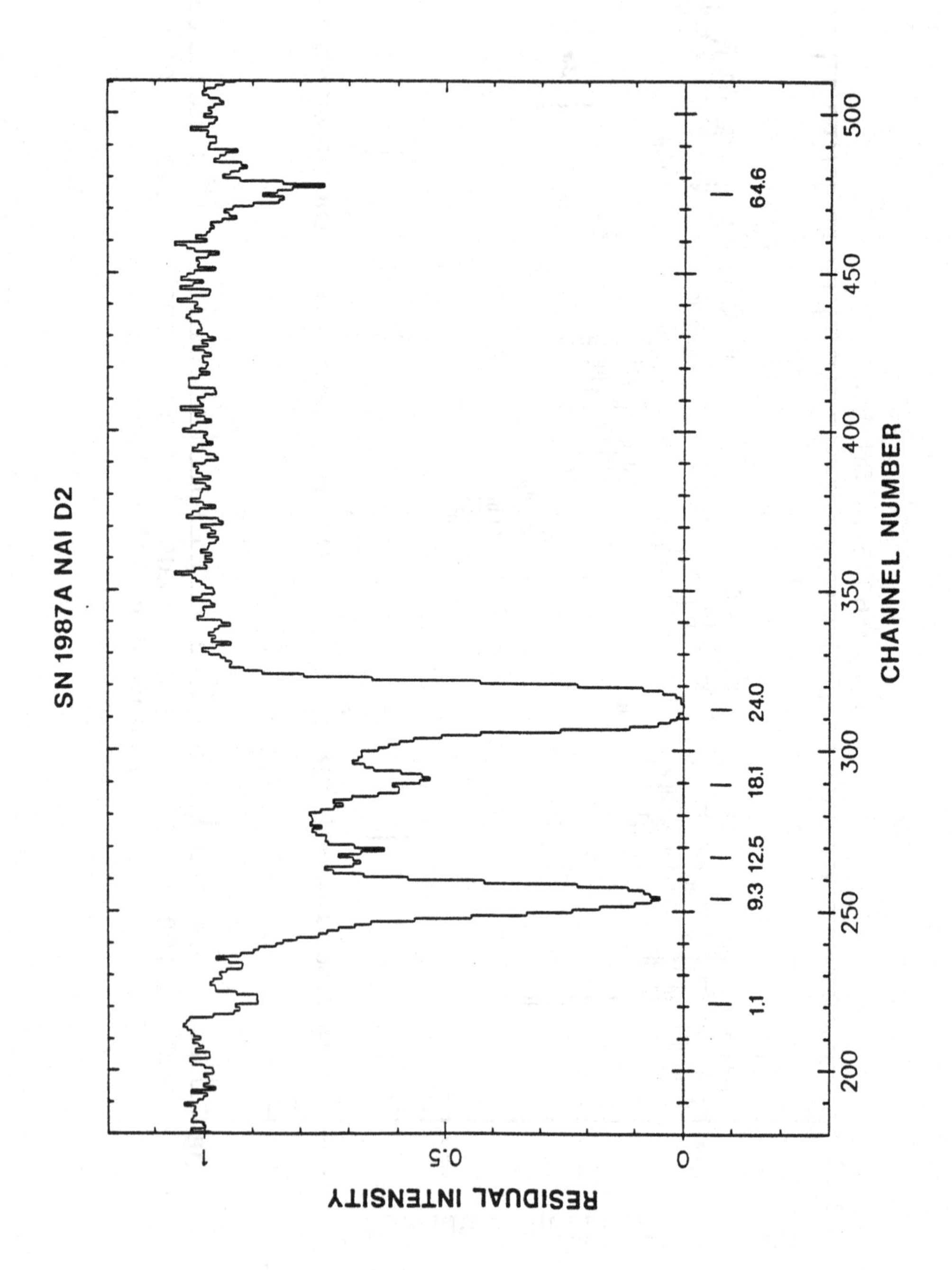

fine splitting of the neutral sodium ground state is clearly resolved. As well as confirming the sensitivity of the PGS (the splitting corresponds to a velocity difference of only 1km/sec) it follows that these clouds are both cold (T < 170K) and non-turbulent (internal velocities of less than 0.18km/sec). For the cloud at 64.6 km/sec, which is thought to be in the galactic halo at a distance of ∿ 5 kpc below the plane of the Galaxy, this is a totally unexpected result in the light of current theories for the production and evolution of the gaseous halo.

Acknowledgements

All the staff of the AAT have contributed greatly to the supernova programme and it is a privilege to report on their behalf. I am particularly grateful to Raylee Stathakis for helping me with the preparation of material presented at the workshop.

References

Aitken, D.K., Smith, C.H. & Roche, P.F. (1987). IAU Circ. 4374.

Branch, D. (1980). In Supernova Spectra, eds. R.E. Meyerott & G.H. Gillespie, p. 39. American Institute of Physics, New York.

Cropper, M., Bailey, J., McCowage, J., Cannon, R.D., Couch, W.J., Walsh, J.R., Straede, J.O.& Freeman, F. (1987), Mon. Not. Royal Astr. Soc. (in press).

D'Odorico, S., Molaro, P., Pettini, M., Stathakis, R. & Vladilo, G. (1987). In Proceedings of the ESO Workshop on SN1987a(in press).

Gillingham, P. (1987). In Present and Future Instrumentation for Large Telescopes (Proceedings of the 9th Santa Cruz Summer Workshop - in press).

Karovska, M., Nisenson, P., Noyes, R. & Papaliolios, C. (1987). IAU Circ. 4382.

Matcher, S.J., Meikle, P.S. &Morgan, B.L. (1987). IAU Circ. 4391.

McCall, M.L. (1984). Mon. Not. Royal Astr. Soc., 210, 829.

Pettini, M., Stathakis, R., D'Odorico, S., Molaro, P. & Vladilo, G. (1987). in preparation.

Shapiro, P.R. & Sutherland, P.G. (1982). Ap.J., 263, 902.

White, G.L. & Malin, D.F. (1987). Nature, 327, 36.

LINEAR POLARIMETRIC STUDY OF SN 1987a

A. Clocchiatti, M. Méndez, O. Benvenuto,
C. Feinstein, H. Marraco, B. García,
and N. Morrell

Observatorio Astronómico, Paseo del Bosque s/n,
1900 La Plata, Argentina

INTRODUCTION

Although extensive amounts of data related to supernovae events existed at the time of the explosion of SN 1987a, little information was available concerning their polarimetric properties (see Shapiro and Sutherland 1982, for a review). Linear polarimetric observations of SN 1987a have been obtained from the very early moments, showing immediately that the polarization was decreasing with time (Benvenuto et al. 1987). Since the polarization produced by the interstellar matter is time independent, it follows that an intrinsic polarization vector (IP) should be responsible for the time dependent characteristic. Also, the wavelength dependence of the observed polarization was far from the well known interstellar relation (Serkowski et al. 1975), supporting the previous conclusion. In addition, as the time scale of the changes of the polarization was in the order of days, it was natural to relate them to the expansion of the external regions of the supernova.

DISCUSSION

The polarization measurements of SN 1987a were carried out with the 0.83 m telescope of La Plata Observatory, and the 2.15 m telescope of CASLEO, San Juan, both in Argentina. In the former we used a rotating half-wave plate photopolarimeter plus BVRI filters, spanning from February 28 up to April 29, while in the latter the Vatican Polarimeter (VATPOL) and the UBVRI filter set was used, from April 2 to April 10.

Because of the small amounts of polarization involved it was necessary to consider contamination by foreground interstellar polarization. This was accomplished by meassuring the polarization of the LMC star Sk $-69^{\circ}203$. When the foreground interstellar polarization vector is subtracted from all measurements, we obtain the results presented in Figure 1.

Fig. 1a: Intrinsic polarization of SN 1987a (corrected for foreground contribution) in the B filter. Solid line, observations from La Plata. Dashed line, observations from San Juan. Numbers at the right of the dots indicate days after February 23. A representative error bar in Q and U is also indicated.

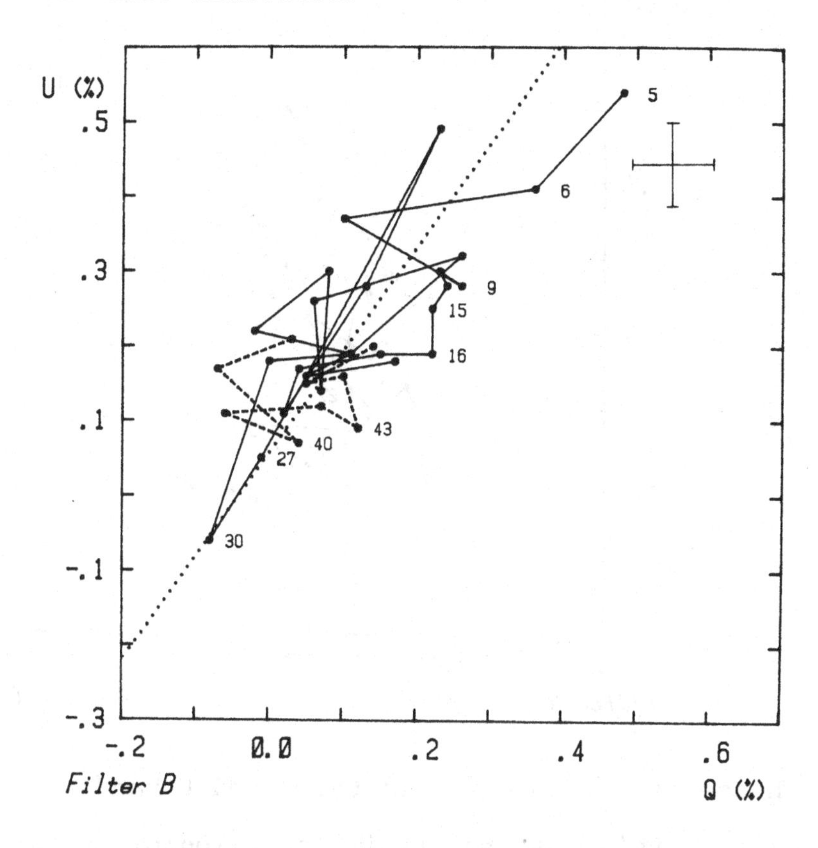

There are some features that should be remarked. First, after removing foreground polarization there still remains an intrinsic component, which is greater than the errors involved, and whose wavelength dependence is clearly not interstellar. Second, the behaviour of the IP is similar in all wavelengths during the first four weeks after explosion. Third, neglecting irregular variations, probably due to some random characteristic of the expansion produced by the explosion, the IP shows a decreasing trend with time, at a position angle of ~ 27 degrees, which is both wavelength and time independent. Fourth, since March 25 approximately a change in the behaviour of the IP in the VRI filters becomes notable. The initial decreasing trend is reversed, and the IP vector begins growing at a different position angle.

Fig. 1b: Same as Fig. 1a but in the R filter. In both figures, the dotted line shows the direction of evolution of the polarization vector in the B filter.

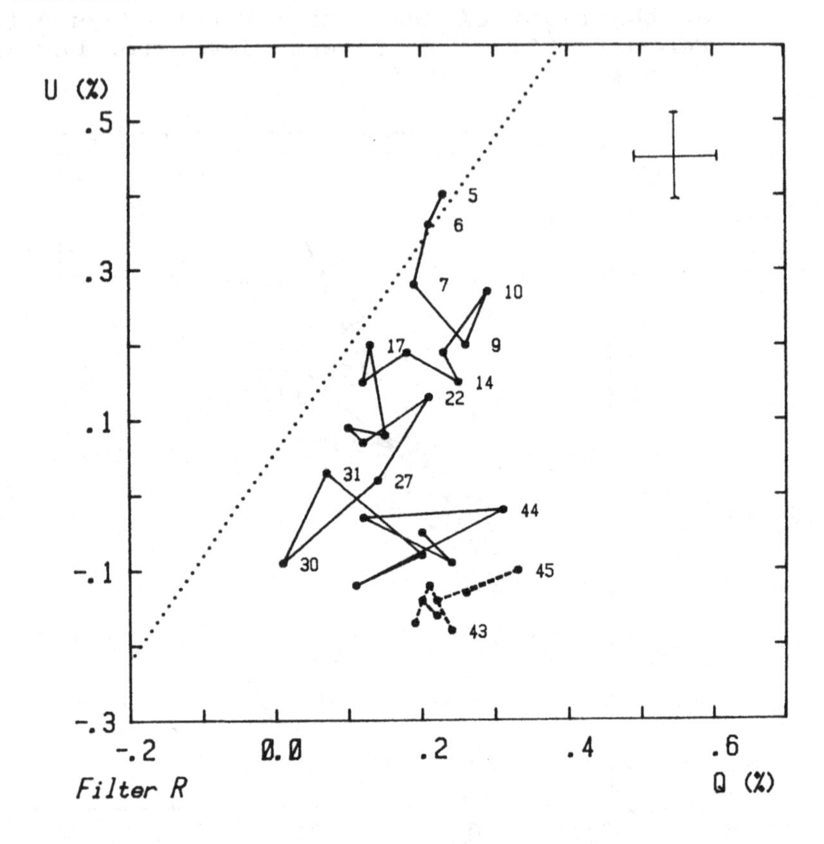

This change is not observed in the U and B bands.

According to this last remark it seems logical to divide the analysis of the IP of SN 1987a in two parts: before and after March 25. The easiest way of explaining this dual behaviour is allowing for the presence of two independent sources of intrinsic polarization. One that dominates the production of polarized light in the first four weeks, and another that becomes important for the longer wavelengths during the second period. Here we deal only with the first component.

According to Shapiro and Sutherland (1982), the wavelength independence of this first component suggest that the opacity would be purely Thompson scattering in a medium where the ratio of the absorption to the total opacity is much less than unity. Obviously, some kind of asymmetry must exist in the scattering medium in order to produce a net polarization.

We assumed that the polarization was produced in an ellipsoidal-optically thin-expanding envelope with an exponential density profile. In this frame, two equations were derived, relating the physical parameters of the shell to the observed polarization, and radial velocity of the mass ejected through the photosphere. These two equations were used in order to study the temporal evolution of the density and typical size of the shell. Notably, disregarding the initial parameters adopted, all the models evolve rapidly to the same solution. For all calculated models the typical size of the shell 30 days after the explosion results 1.65×10^{15} cm.

Although the existence of a non null excentricity is imperative, the observations can be reproduced using very small excentricities. This would support the current assumptions of spherical symmetry in studying global dynamics of supernovae explosions.

After approximately 30 days, the polarization changes both in angle and wavelength dependence. This would suggest that the effect of the absorption on the total opacity becomes important after this moment.

For a more detailed discussion see Méndez et al. (1987).

REFERENCES

Benvenuto, O., Clocchiatti, A., Feinstein, C., García B., Luna H., Méndez, M., Morrell, N., 1987, IAU Circ. 4358.
Méndez, M., Clocchiatti, A., Benvenuto, O., Feinstein, C., Marraco, H., 1987, submitted to Astroph. Journal.
Serkowski, K., Mathewson, D.S., and Ford, V. L. 1975, Astroph. Journal, 196, 261.
Shapiro, P.R., and Sutherland, P.G., 1982, Astroph. Journal, 263, 902.

INFRARED SPECTROSCOPY OF SN 1987a FROM THE NASA KUIPER AIRBORNE OBSERVATORY

H.P. Larson
University of Arizona, Tucson, AZ, 85721, USA

S. Drapatz
Max-Planck-Inst. f. Extraterr. Physik, 8046, Garching, FRG

M.J. Mumma
Goddard Space Flight Center, Greenbelt, MD, 20771, USA

H.A. Weaver
Space Telescope Science Inst., Baltimore, MD, 21218, USA

Abstract. The near-infrared (1.5-3.0 μm) spectrum of SN 1987a was recorded on UT 16.3 April 1987 from the NASA Kuiper Airborne Observatory. The dominant spectral features included multi-component Pα emission at 1.9 μm and a sharp edge in the continuum at 2.7 μm. The interpretation of these features relates to physical conditions in the supernova atmosphere and to the role of dust around supernovae.

INTRODUCTION

The NASA Kuiper Airborne Observatory (KAO) is used primarily for astronomical research in regions of the infrared (IR) and submillimeter spectra that cannot be studied from the ground because of strong molecular absorptions in Earth's atmosphere. When SN 1987a appeared, the KAO had already been scheduled for deployment to Christchurch, New Zealand for cometary astronomy. This fortunate circumstance made it possible for the cometary investigators to use the KAO for studies of SN 1987a also. Our objective was to survey the near-IR spectrum of SN 1987a in wavelength regions between the H, K, and L photometric bands that are obscured by terrestrial H_2O absorption at ground-based telescopes.

OBSERVATIONS

Our airborne spectrum of SN 1987a was acquired with a Fourier transform spectrometer (FTS) (Davis *et al.* 1980) on 16 April 1987 (UT). The total integration time was 2.1 hrs (7:38-9:44 UT). The achieved spectral resolution of 2.0 cm^{-1} is equivalent at 2.2 μm to a resolving power of 2300 and a velocity resolution of 130 km s^{-1}. All features intrinsic to SN 1987a were fully resolved.

Lunar and stellar comparison spectra were also acquired in order to calibrate instrumental and atmospheric transmission factors. Their

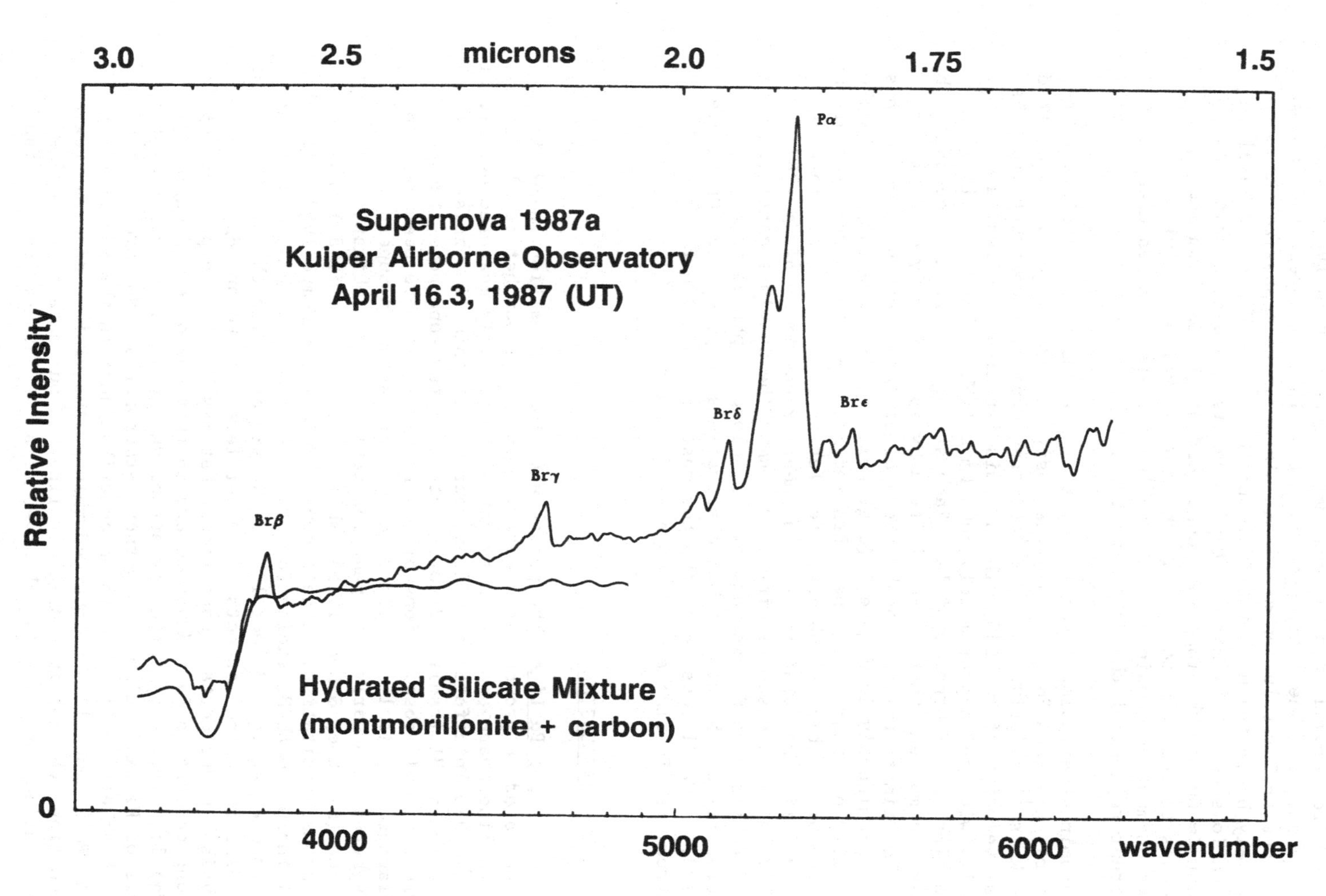

Figure 1. Near-IR spectrum of SN 1987a from the NASA Kuiper Airborne Observatory.

effects were removed from the observed spectrum of SN 1987a by the ratio technique. The ratio spectrum of SN 1987a is presented in Figure 1 at a spectral resolution of 15 cm^{-1} in order to emphasize the relatively broad features intrinsic to SN 1987a. The noise level in this ratio spectrum is not constant; the signal-to-noise ratio is highest (≈400) at the long wavelength end of the spectrum and it decreases to approximately 10 at the short wavelength limit. This variation must be considered when evaluating spectral features in different regions of the ratio spectrum.

The prominent emission lines in Figure 1 are due to HI, including Brβ, Brγ, Brδ, Brε, and Pα. The HI line positions are slightly red-shifted (≈300 km s^{-1}) relative to their rest frequencies, consistent with the radial velocity of the LMC. The HI emission line profiles have a velocity width of approximately 3000 km s^{-1} FWHM; there is little suggestion of a P Cygni profile. Each strong HI line is accompanied by a weaker component that is red-shifted by approximately 4000 km s^{-1} from the primary emission peak. Several of these secondary components appear prominently in Figure 1 (e.g. near Brδ, Pα, and Brε), while the visibility of others (e.g. near Brβ and Brγ) requires inspection of the spectrum at the highest observed resolution because of interference from telluric and solar lines. The near-IR continuum of SN 1987a exhibits a sharp discontinuity at 2.7 μm. The absolute continuum flux at 2.2 μm yields a K magnitude of 1.1, close to photometric measurements at the time of our observations (e.g. $m_k \approx 1.2$; Catchpole *et al.* 1987). We discuss below the two most prominent features in this spectrum: the Pα line and the edge at 2.7 μm.

DISCUSSION

The Pα Line. The presence of this transition in SN 1987a was indicated in ground-based observations (e.g. Bouchet *et al.* 1987a), but its line shape and intensity cannot be reliably measured because of telluric H_2O interference. The properties of this line relate to the excitation of HI itself and to structure in the supernova atmosphere. Previously, only observations of the Balmer lines have been available to diagnose physical conditions and HI excitation in supernova atmospheres. In early work photoionization and recombination were proposed (Kirshner & Kwan 1975), while Branch *et al.* (1981) favored collisional excitation. More recent models (e.g. Hempe 1983; Höflich 1987) include both HI excitation processes, but the results have only been checked against Balmer lines. We are exploring the possibility of including Pα (which connects the upper levels of Hα and Hβ) in the analysis. Under certain conditions that may be model dependent, Hβ photons convert to Pα and Hα photons so that the observed Hα/Pα (or Hβ/Pα) line intensity ratio may establish a useful, new constraint to models of HI excitation in supernova atmospheres. Moreover, confusion (line blanketing, line/continuum contributions) in the spectral region around Pα is much less pronounced than in the visible spectral region. In addition, the multiple Brackett lines in our data characterize HI excitation for $n>4$. The observed relative intensities of these lines are in good agreement with recombination theory for optically thick conditions (Larson *et al.* 1987).

The red-shifted companion to Pα in Figure 1 is the second strongest emission feature in our spectrum. We attribute this line to Pα also because we could not convincingly assign it to another species. Intensity considerations and the lack of corroborating evidence in other spectral regions eliminated the most likely candidates (e.g. HeI, FeII). As noted above, red-shifted components accompany other prominent HI lines in our spectrum. In addition, similar structure is evident in HI line profiles observed in SN 1987a in other spectral regions. Both Hα (16 April; Hanushik & Dachs 1987) and Pβ (26 May; Bouchet *et al.* 1987a) display prominent, partially resolved features at approximately +4000 km s^{-1} displacement from the emission peak. These peaks do not appear to be lines of other species. Thus, at least during April-May, 1987, there were kinematically distinct components in the HI lines in SN 1987a. Moreover, it is interesting to note that similar red-shifted components have been observed in other supernovae. Compare, for example, the Hα line profile in SN 1985p (Chalabaev & Christiani 1987) with that in SN 1987a (e.g. Hanushik & Dachs 1987). Possible interpretations of the multi-component HI line emission lead in many directions; some of them that we are evaluating are listed below.

The interaction of radiation from SN 1987a with neighboring clouds is one plausible scenario. It could produce Doppler-shifted scattered light and, through self-emission, both line and continuum radiation (e.g. IR echoes: Hillebrandt *et al.* 1987; Felton *et al.* 1987) However, quiescent clouds should emit narrow HI lines at their rest frequencies, so self-emission does not appear consistent with the high radial velocity and high velocity dispersion of the red-shifted HI components. Large recessional velocities, perhaps up to several thousand km s^{-1}, may be induced by accumulated radiation pressure on sub-micron grains. Supernova lines scattered from such grains could then exhibit the required Doppler shifts while preserving their velocity dispersion. Because of the light travel time, the scattered intensities would then be characteristic of supernova emission from an earlier epoch when the line-to-continuum ratio was higher. However, intensity dilution from geometrical considerations could severely limit observed flux levels. Another source region currently being evaluated is the extended IR halo detected around SN 1987a (Perrier *et al.* 1987).

Alternately, the red-shifted HI lines could originate from processes or structure within the supernova atmosphere. Radiation transport calculations of HI excitation in supernovae predict only P Cygni profiles for spherical expansion (e.g. Höflich 1987). More complex profiles result from asymmetric expansion (e.g. shell structure; Hutchings 1972) and, in principle, from introducing any radial variation in non-LTE effects. Another possibility worth exploring is Raman scattering which could produce red-shifted lines while preserving their velocity dispersion. Atmospheric structures specific to this supernova may include relativistic (Rees 1987; Piran & Nakamura 1987) or non-relativistic jets. They could supply the observed luminosity and velocity dispersion, although the high radial velocity of relativistic jets would impose restrictions on allowed viewing angles and on the radiative properties of the material in the mantle and behind the jet.

The Feature at 2.7 μm. This feature includes a sharp absorption edge at 2.7 μm and a depressed continuum that apparently extends beyond 3.0 μm. We have considered two interpretations: a photoionization threshold and a molecular absorption band. If the edge at 2.7 μm is an ionization threshold, it cannot be due to HI because the position is wrong. No edges of similar strength are evident elsewhere in spectra of SN 1987a at the time of our observations. There are, however, subtle changes in our continuum, particularly in the region of the Pα line, that may be due to weaker edges. Upon applying simple Rydberg analysis to the 2.7 μm edge and suspected edges elsewhere in the 2 μm spectral region, we require a species with high ionization potential ($\approx$54 eV) and large principal quantum numbers ($\geq$9). One candidate that may satisfy these particular requirements is FeIV, but there is little support for this assignment from other observed spectral properties of SN 1987a.

If the 2.7 μm feature arises from molecular absorption, the absence of rotational structure suggests that the material is solid, i.e. dust. There are good reasons to search for signatures of dust in spectra of SN 1987a: (1) dust should form in the supernova atmosphere; (2) there may be a primordial remnant from the formation of the progenitor star; (3) it may have been ejected from the progenitor star during a mass loss phase; and, (4) it may be interstellar. The presence of dust somewhere near SN 1987a is implied by the IR excess in broadband photometry (Bouchet *et al.* 1987b). In order to proceed with this interpretation we first demonstrate that spectra of cosmochemically acceptable minerals are compatible with the near-IR spectrum of SN 1987a. In Figure 1 we compare the 2.7 μm feature with the reflection spectrum of a mixture of montmorillonite (a terrestrial silicate similar to the hydrated minerals in carbonaceous chondrite meteorites) and carbon black (an opaque mineral common to interstellar grains). This particular mixture was originally prepared to simulate certain asteroid surfaces (e.g. Ceres, Pallas) that were spectrochemically similar to primitive meteorites (Feierberg *et al.* 1981; Lebofsky *et al.* 1981). This band is formed by superimposed modes of structural OH (which is responsible for the sharp edge at 2.7 μm) and bound H_2O (which gives the band its breadth). The fit to the spectrum of SN 1987a is excellent. It implies that low temperature hydrated silicate grains may exist near SN 1987a (note, however, that curve matching is not uniquely diagnostic of specific minerals). It is premature to describe here the origin, location, and amount of silicate dust and its interaction with radiation from SN 1987a. One factor is that observations are still revealing properties of the circumstellar environment of SN 1987a, some of which could be important to our interpretation. Another, more specific, concern involves the lack of independent spectroscopic evidence for silicate grains. In particular, there should be a prominent silicate absorption band at 10 μm, but none has been reported. One possible explanation is that radiative processes fill in this band, but the balance should have changed with time. We therefore prefer to keep open the possibility that the 2.7 μm feature in our spectrum of SN 1987a is unrelated to hydrated silicate grains in spite of the suggestive agreement in Figure 1.

Acknowledgements. We express our gratitude to Mr. L. Haughney and the KAO staff for their support during our southern hemisphere observing program and for their special effort to accommodate our observations of SN 1987a. This research was funded by NASA-Ames Grant NAG2-206.

REFERENCES

Bouchet, P., Stanga, R., Moneti, A., Le Bertre, Th., Manfroid, J., Silvestro, G., Slezak, E., 1987a, to appear in *Proc. ESO Workshop on SN 1987a*, Ed.: I.J. Danziger, ESO, Garching, W. Germany, 1987.

Bouchet, P., Stanga, R., Moneti, A., Le Bertre, Th., Manfroid, J., Silvestro, G., Slezak, E., 1987b, to appear in *Proc. ESO Workshop on SN 1987a*, Ed.: I.J. Danziger, ESO, Garching, W. Germany, 1987.

Branch, D., Falk, S.W., McCall, M.L., Rybski, P., Uomoto, A.K., Wills, B.J., 1981, *Astrophys. J.* 244, 780.

Catchpole, R.M., et al., 1987, *SAAO Preprint No.* 533.

Chalabaev, A.A & Christiani, S., 1987, to appear in *Proc. ESO Workshop on SN 1987a*, Ed.: I.J. Danziger, ESO, Garching, W. Germany, 1987.

Davis, D.S., Larson, H.P., Williams, W., Michel, G., Connes, P., 1980. *Appl. Optics* 19, 4138.

Feierberg, M.A., Lebofsky, L.A., Larson, H.P., 1981, *Geochim. Cosmochim. Acta* 45, 971.

Felton, J.E., Dwek, E., Viegas-Aldrovandi, S.M., 1987, Nature, submitted.

Hanushik, R.W. & Dachs, J., 1987, *Astron. & Astrophys.* **182**, L29.

Hempe, K., 1983, *Mitt. Astr. Ges.* **60**, 107.

Hillebrandt, W., Höflich, P., Schmidt, H.U., Truran, J.W., 1987, *Astron. & Astrophys. Letters*, in press.

Höflich, P., 1987, to appear in *Proc. ESO Workshop on SN 1987a*, Ed.: I.J. Danziger, ESO, Garching, W. Germany, 1987.

Hutchings, J., 1972, *MNRAS* **158**, 177.

Kirshner, R.P. & Kwan, J., 1975, *Astrophys. J.* **197**, 415.

Larson, H.P., Drapatz, S., Mumma, M.J., Weaver, H.A., 1987, to appear in *Proc. ESO Workshop on SN 1987a*, Ed.: I.J. Danziger, ESO, Garching, W. Germany, 1987.

Lebofsky, L.A., Feierberg, M.A., Tokunaga, A.T., Larson, H.P., Johnson, J.R., 1981, *Icarus* **48**, 453.

Perrier, C., Chalabaev, A.A., Mariotti, J.-M., Bouchet, P., to appear in *Proc. ESO Workshop on SN 1987a*, Ed.: I.J. Danziger, ESO, Garching, W. Germany, 1987.

Piran, T. & Nakamura, T., 1987, *Nature*, submitted.

Rees, M.J., 1987, *Nature* **328**, 207.

RADIO OBSERVATIONS OF SN1987A

N. Bartel[1], D. L. Jauncey[2], A. Kemball[3], A. R. Whitney[4], A. E. E. Rogers[4], I. I. Shapiro[1], R. A. Preston[5], B. R. Harvey[2], D. L. Jones[5], G. D. Nicolson[3], R. P. Norris[2], A. Nothnagel[3], R. B. Phillips[4], and J. E. Reynolds[6]

[1] Harvard–Smithsonian Center for Astrophysics, 60 Garden Street, Cambridge, MA 02138, USA
[2] Division of Radiophysics, CSIRO, P.O. Box 76, Epping, NSW 2121, Australia
[3] Hartebeesthoek Radio Astronomy Observatory, P.O. Box 3718, Johannesburg 2000, South Africa
[4] Haystack Observatory, NEROC, Westford, MA 01886, USA
[5] Jet Propulsion Laboratory, California Institute of Technology, 4800 Oak Grove Drive, Pasadena, CA 91109, USA
[6] Mt. Stromlo and Siding Springs Observatories, Private Bag, P.O. Woden, ACT 2606, Australia

The radio burst

Radio emission from SN1987A was detected with Australian telescopes by Turtle *et al.* (1987) two days after the neutrino burst (Aglietta *et al.* 1987; Bionta *et al.* 1987; Hirata *et al.* 1987). But, in contrast to the prominent appearance of the supernova on photographic plates , the supernova was only a relatively weak radio source among neighboring, more luminous radio supernova remnants and HII regions (Figure 1).

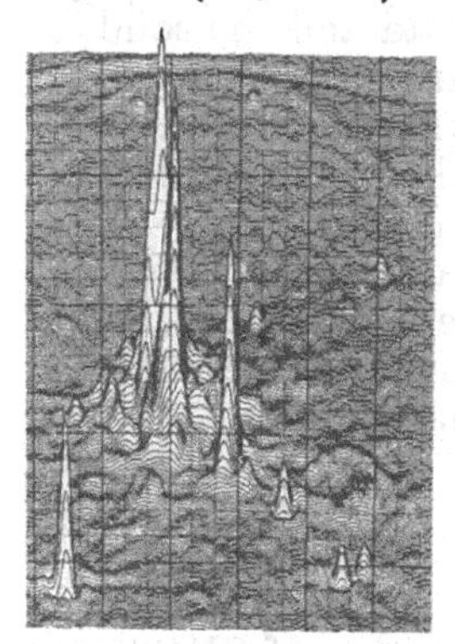

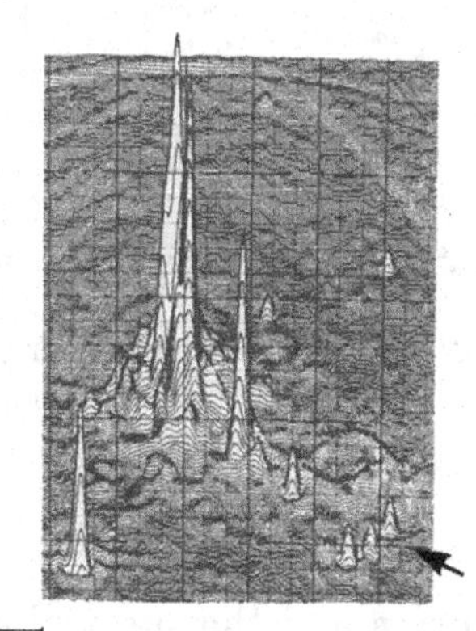

Figure 1. Maps of the radio emission at 0.843 GHz from the environment of SN1987A before (left) and after (right) the explosion. An arrow indicates the location of the supernova. The figure is taken from Turtle *et al.* (1987).

The observed maximum flux density of the burst at radio wavelengths was ~ 140 mJy and occurred at 1.4 GHz at $t = 3\ d$ (by definition, $t = 0$ at the epoch of the Kamiokande–IMB neutrino detection). The maximum radio luminosity between 0.8 and 8 GHz was $\sim 1 \times 10^{33}$ erg s^{-1}, and, for the five days of the burst, the total energy radiated at radio wavelengths was $\sim 1 \times 10^{39}$ erg, both given that the radiation was isotropic. The brightness temperature was greater than 10^7 K. The spectral peak moved from high to low frequencies. After a few days, the flux density at frequencies above 1 GHz became undetectable. Only at 0.843 GHz was the radio emission detected for a further 100 *d*. The supernova's radio "light curves" and spectral evolution are shown in Figures 2 and 3, respectively. These data are consistent with a model in which the radio burst is synchrotron radiation generated by relativistic electrons in an expanding free-free absorbing plasma (see, e.g., Chevalier 1982). The above properties of SN1987A are compared in Figures 2 and 3 with the equivalent radio light curves and spectral evolutions of SN1979C and SN1980K, the only other type II supernovae for which extensive radio data have been obtained. For each of these supernovae and others, Figure 4 displays the spectral luminosity and the time delay between the explosion and the peak of the luminosity.

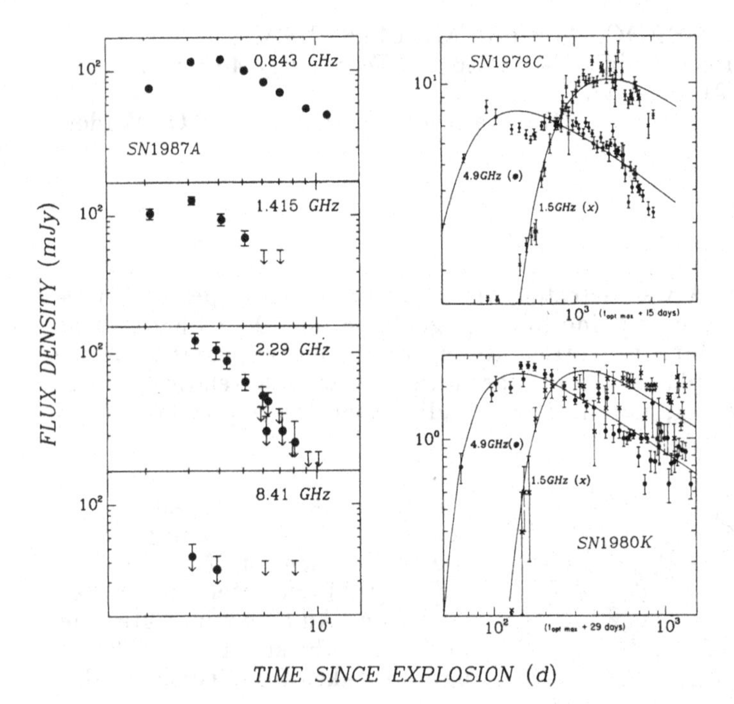

Figure 2. The radio light curves of SN1987A at four frequencies (left) and, for comparison, of SN1979C and SN1980K at two frequencies (right). The data for SN1987A are taken from Turtle *et al.* (1987) but with the flux densities at 2.3 GHz corrected by a factor ~ 0.85 as a result of a recalibration (R. P. Norris 1987, unpublished data). The data and the solid lines from best-fits for the other supernovae are taken from Weiler *et al.* (1986). For more information on the figures here and hereafter, see the original papers.

The shapes of the radio light curves and the spectral evolution of SN1987A are very similar to those of SN1979C and SN1980K; however, the peak of the radio luminosity at, e.g., 1.5 GHz was, respectively, 4 and 2.5 orders of magnitude smaller and the time interval between the explosion and the epoch of the peak of emission, respectively, 2.5 and 2 orders of magnitude shorter. It remains to be seen whether the radio emission from SN1987A is caused by the same physical process responsible for the radio emission from the two other supernovae (e.g. Chevalier 1982; Pacini and Salvati 1981) or whether the weakness of the burst, which would have been undetectable in the other supernovae, implies a different physical mechanism. In the latter case, a second burst may be expected with a flux density reaching 100 to 1000 Jy at, e.g., 5 GHz within a few months.

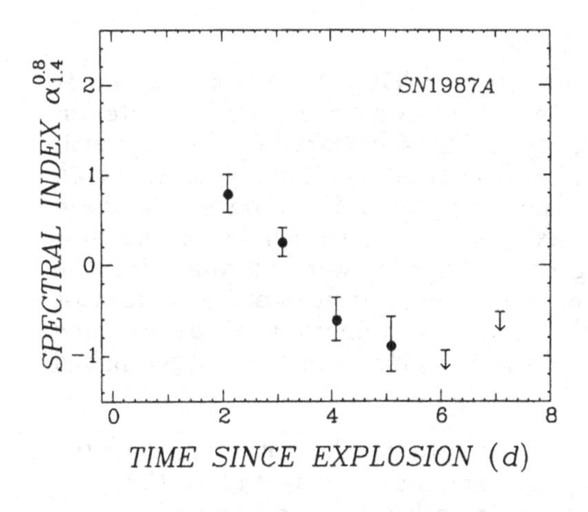

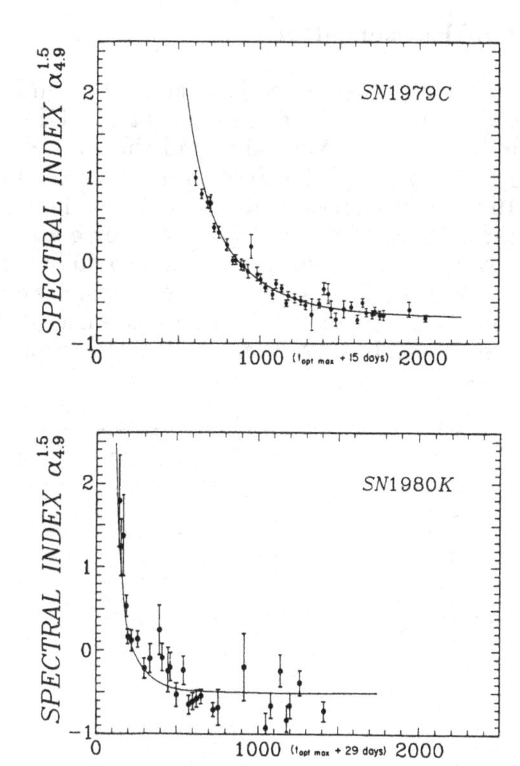

Figure 3. The time dependence of the spectral index, α ($S_\nu \propto \nu^\alpha$), of SN1987A (left) and, for comparison, of SN1979C and SN1980K (right). The data and the solid lines from best-fits for the latter supernovae are again taken from Weiler *et al.* (1986).

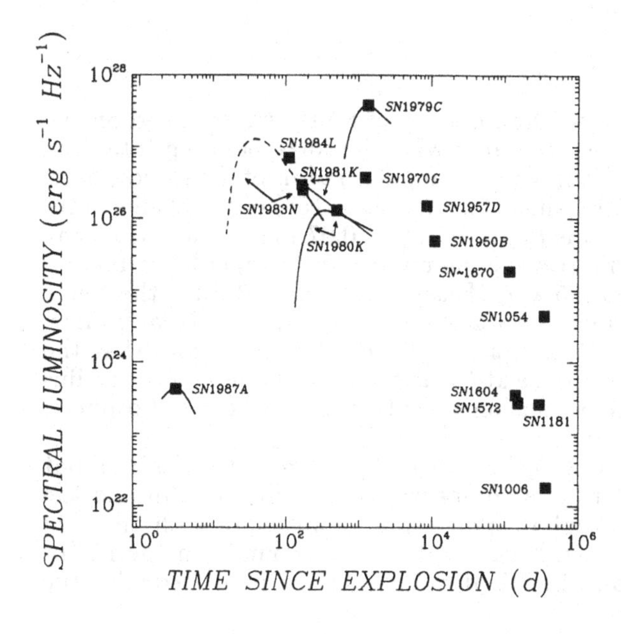

Figure 4. The spectral luminosity at 1.5 GHz and, for SN1987A only, at 1.4 GHz, vs. the age of radio supernovae and young supernova remnants. The squares indicate the largest measured spectral luminosities of the corresponding sources, the solid curves represent all measured spectral luminosities, and the dashed curve predicts spectral luminosities according to Chevalier (1984). The data from the sources other than SN1987A are from Weiler *et al.* (1986) and references therein.

VLBI observations

Following the detection of radio emission from SN1987A, Mark III VLBI observations were conducted at 2.3 GHz with the 34-m diameter DSS42 antenna in Tidbinbilla, Australia, and the 26-m diameter antenna in Hartebeesthoek, South Africa, on days 28 Feb. and 1,2 Mar. 1987 (Jauncey *et al.* 1988; Shapiro *et al.* 1988). No fringes were found from the supernova's radio emission on any of those three days with limits on the correlated flux density of 10 ± 1 mJy on the first day and 11 ± 1 mJy on the two following days. Fringes were obtained for the calibrator sources on all three days. The correlated flux densities of the calibrator sources agree, to within their tolerances, with VLBI measurements made at the same resolution, but five years earlier (Preston *et al.* 1985; G. Nicolson 1987, unpublished data).

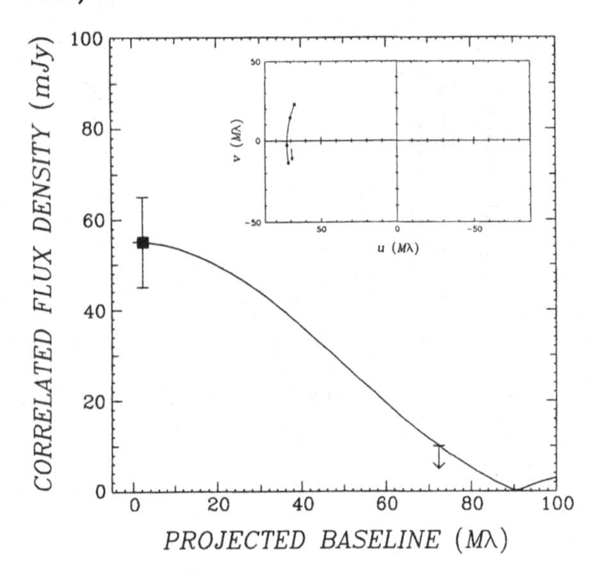

Figure 5. The total flux density and an upper bound on the correlated flux density of SN1987A on ~ 28.5 Feb. 1987, together with the prediction of a model of an optically thin uniform sphere. The prediction is similar to that from a model of an optically thin shell with an outer angular radius of 1.25 mas. The u-v track for the VLBI observations is plotted in the inset.

The upper bound on the correlated flux density of SN1987A obtained on the first day of the VLBI observations was compared with the corresponding total flux densities (Figure 5) to derive a lower bound on the angular radius of the radiosphere. For an optically thin shell model with a shell thickness of ~ 15% of the shell's outer radius, the lower bound on the radius is $\theta_{radio} > 1.25 \pm 0.07$ mas. The use of any other physically plausible model would result in an up to ~ 30% larger lower bound. Given a distance to SN1987A of 50 ± 5 kpc (Feast and Walker 1987), the lower bound on the angular radius corresponds to a lower bound on the linear radius, R_{radio}, of $R_{radio} > 8.3 \times 10^{14}$ cm = 55 AU, at $t = 5.2$ *d*. If one assumes that the radiosphere expanded linearly from zero size at $t = 0$, the lower bound on the radius corresponds to a lower bound on the expansion velocity, v_{radio}, of the radiosphere: $v_{radio} > 19 \times 10^3$ km s^{-1}.

Since the distance to the Large Magellanic Cloud is known to within about 10%, SN1987A allowed for the first time a comparison of the linear radii and the expansion velocities of a supernova's photosphere and radiosphere with those of the supernova's line-forming regions. In Figure 6, the lower bounds on the radius and expansion velocity obtained from the VLBI radio data are compared with the corresponding radii and velocities obtained from optical photometric (Menzies *et al.* 1987) and spectroscopic (Hanuschik and Dachs 1987; Blanco *et al.* 1987) data.

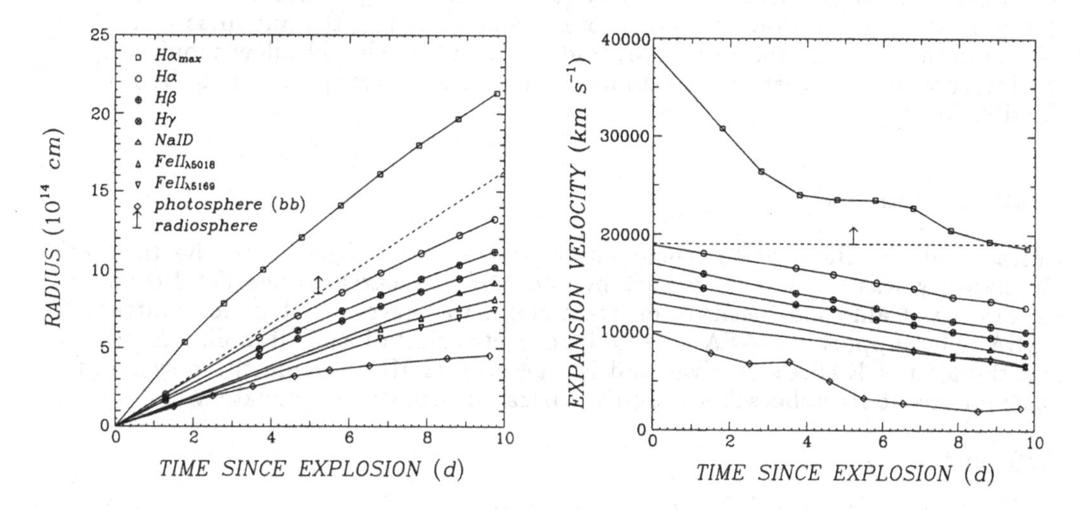

Figure 6. Lower bounds on the radius and the corresponding (assumed uniform) expansion velocity of the radiosphere of SN1987A (dashed lines), compared with radii and expansion velocities of the blackbody photosphere and the line-forming regions. The radii of the line-forming regions were obtained from an integration of the velocities corresponding to the blueshifts of the absorption minima of the indicated lines and the largest observed blueshift, $H\alpha_{max}$, the latter from the blue edge of the $H\alpha$ absorption trough. The expansion velocities of the photosphere were obtained from a differentiation of the photosphere's radii. The uncertainties of the $H\alpha_{max}$ velocities are ~ 1000 km s^{-1} and those of the $H\alpha_{max}$ radii are smaller than the symbols. Other uncertainties were not given in the original papers but are believed to be not larger than those of the $H\alpha_{max}$ velocities and radii, respectively.

At the time of the VLBI observations, the radius and the expansion velocity of the radiosphere were both at least a factor 2.5 larger than those of the blackbody photosphere and, respectively, at least 10% and 25% larger than the radius and velocity inferred from the $H\alpha$-line absorption minimum. The results will not only add to our knowledge of supernovae, but could also be important in limiting the uncertainties accompanying the use of supernovae as distance indicators (see, e.g., Bartel 1985,1986; Bartel *et al.* 1985), if the physical processes responsible for the radio emission from SN1987A are typical for supernovae in general.

Continuing searches for radio emission

Radio monitoring programs continued at several observatories after the apparent demise of the supernova's radio emission. No signal has yet been redetected or detected from SN1987A above a 5-σ bound of ~ 8 mJy at 2.3 and 8.4 GHz (as of July 1987, Norris *et al.* 1987) of 45 mJy at 22 GHz (as of July 1987, Storey *et al.* 1987), and of 500 mJy at 43 GHz (as of July 1987, Norris *et al.* 1987). The most sensitive search for radio emission so far was conducted with the DSS42 and DSS45 antennas at Tidbinbilla at 8.4 GHz, twofold more sensitive than the search at 8.4 GHz mentioned above. Data from each of the antennas were recorded with the Mark III VLBI system on magnetic tapes and sent to Haystack Observatory for processing. No fringes have yet been obtained from SN1987A. The upper bound on the supernova's flux density is ~ 4 mJy (5-σ) as of 22 September 1987. In the case of a substantial increase of the supernova's radio emission, the antennas and VLBI recording facilities in Australia (Reynolds *et al.* 1987) should allow monitoring of the increase of the supernova's radius and perhaps also mapping of its brightness distribution.

Acknowledgment

Research at the Harvard–Smithsonian Center for Astrophysics and the Haystack Observatory was supported in part by the NSF. Research at the Jet Propulsion Laboratory, California Institute of Technology, was carried out under contract to NASA. The support of NASA's Deep Space Network staff at Tidbinbilla, Australia, and the loan of NASA's receiver and NOAA's Mark III data aquisition system for observations at Hartebeesthoek, South Africa, are greatly appreciated.

References

Aglietta, M. *et al.* 1987, *Europhys. Lett.*, **3**, 1315.
Bartel, N. (ed.) 1985, *Supernovae as Distance Indicators*, Lecture Notes in Physics (Springer–Verlag, Berlin), p. 224.
Bartel, N. 1986, in *Highlights of Astronomy*, ed. J. P. Swings (Reidel, Dordrecht),**7**, 655.
Bartel, N., Rogers, A. E. E., Shapiro, I. I., Gorenstein, M. V., Gwinn, C. R., Marcaide, J. M., and Weiler, K. W. 1985, *Nature*, **318**, 25.
Bionta, R. M. *et al.* 1987, *Phys. Rev. Lett.*, **58**, 1494.
Blanco, V. M. *et al.* 1987, *Ap. J.*, **320**, 589.
Chevalier, R. A. 1982, *Ap. J.*, **259**, 302.
Chevalier, R. A. 1984, *Ap. J. (Letters)*, **285**, L63.
Feast, M. W., and Walker, A. R. 1987, *Ann. Rev. Astr. Ap.*, **25**, 345.
Hanuschik, R. W., and Dachs, J. 1987, *Astr. Ap. Lett.*, **182**, L29.
Hirata, K. *et al.* 1987, *Phys. Rev. Lett.*, **58**, 1490.
Jauncey, D.L. *et al.* 1988, to be submitted to *Nature*, .
Menzies, J. W. *et al.* 1987, *M. N. R. A. S.*, in press.
Norris, R. P. *et al.* 1987, IAU Circ. 4432.
Pacini, F. and Salvati, M. 1981, *Ap. J. (Letters)*, **245**, L107.
Preston, R. A., Morabito, D. D., Williams, J. G., Faulkner, J., Jauncey, D. L., and Nicolson, G. D. 1985, *A. J.*, **90**, 1599.
Reynolds, J. E. *et al.* 1987, in *ESO workshop on SN1987A*, ed. I. J. Danziger, (ESO Workshop and Conference Proc., Garching), No.**26**.
Shapiro, I. I. *et al.* 1988, in *IAU Symposium 129, The Impact of VLBI on Astrophysics and Geophysics*, eds. M. J. Reid and J. M. Moran (Reidel, Dordrecht), in press.
Storey, M. C., Conrad, G., Cooke, D. J., Troup, E., Wark, R., and Wright, A. E. 1987, IAU Circ. 4432.
Turtle, A. J. *et al.* 1987, *Nature*, **327**, 38.
Weiler, K. W., Sramek, R. A., Panagia, N., van der Hulst, J. M., and Salvati, M. 1986, *Ap. J.*, **301**, 790.

ULTRAVIOLET OBSERVATIONS OF SN1987A: CLUES TO MASS LOSS

Robert P. Kirshner
Harvard-Smithsonian Center for Astrophysics

Abstract

Ultraviolet observations of SN 1987A provide strong evidence for a circumstellar shell surrounding the explosion. Narrow ultraviolet emission lines result from photoionization of the circumstellar shell by the ultraviolet emission expected when the shock hits the surface of the SK -69 202 star. The gas is nitrogen-rich, has low velocity, and may be located a light-year from the supernova. This gives every sign of having been ejected from the SK -69 202 progenitor during a phase when it was a red supergiant, prior to its life as a blue supergiant. The amount of mass lost, and the mass of hydrogen-rich material on the surface of the star when it exploded are key issues. The UV emission from the circumstellar matter is unlikely to give an accurate mass, but the changes in the hydrogen line profiles during the early evolution may provide a way to estimate the density distribution in the supernova atmosphere, and the mass of hydrogen it contains. A preliminary estimate is that the power-law index of density in the envelope goes as v^{-11} and the mass that lies above a velocity of 6,000 km s^{-1} is between 1 and 6 solar masses.

Ultraviolet Observations

Promptly following the discovery of SN 1987A a target of opportunity program on the International Ultraviolet Explorer was put into action. The IUE was used in both its high and low resolution modes to measure the supernova spectrum from 1200Å to 3200Å. Initial reports can be found in Kirshner et al. (1987), Dupree et al (1987), Panagia et al (1987), Cassatella et al 1987, and Fransson et al (1987).

We have constructed ultraviolet light curves for the supernova from the calibrated IUE spectra obtained through the large (10" x 20") entrance aperture. The observed flux is integrated over rectangular bands. Two features are apparent: the ultraviolet flux was declining before the first optical maximum was reached on 27 February (day 58 of 1987), and the short wavelength bands declined much more rapidly than the long wavelength bands, as might be expected if the photosphere were cooling from a high initial temperature. The decline in the shortest wavelength band is truly precipitous, descending a factor of 1000 in 3 days. The non-zero level reached is the result of other stars in the aperture for the short wavelength, but is due to the supernova itself in the long wavelength bands.

Figure 1. Short wavelength UV light curve. (a) Complete. (b) Recent evolution. Some of the recent increase is due to the circumstellar emission, and some to the increase in the continuum flux from the supernova.

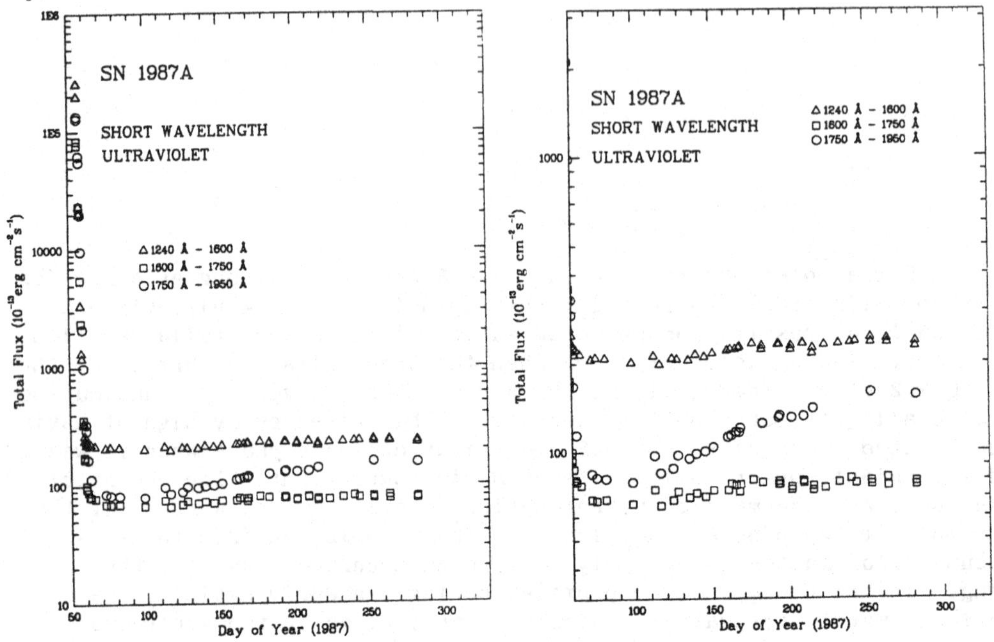

Figure 2. Long wavelength UV light curve. (a.) Complete. (b) Recent evolution. Note the increase in flux from the supernova after the date of the optical maximum (day 140)

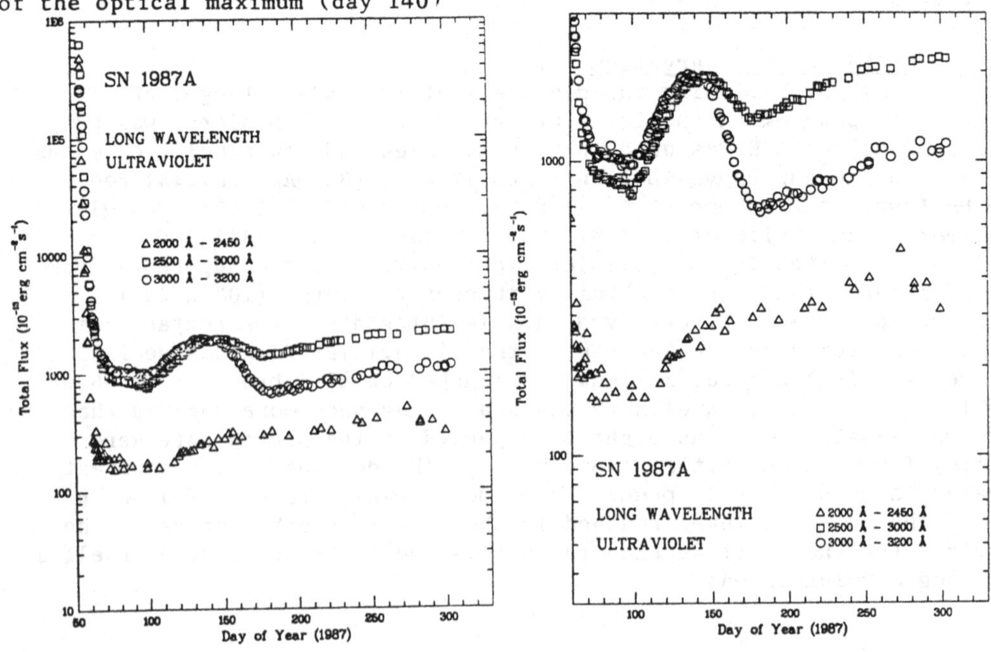

These earliest UV observations can be combined with optical data to derive a bolometric light curve (Blanco et al. 1987). Although the color temperature was only of order 15,000 K at the time of our first data, it must have been far higher on February 23, and it may have been a substantial source of ionizing flux. If we take the moment of the core collapse as Feb 23.316 to agree with the Kamiokande (Koshiba 1987) and IMB (Svoboda 1987) results, then McNaught's observation at 6 mag on Feb 23.443 probably refers to an epoch when the temperature was of order 100 000 K with a photospheric radius of order 10^{13} cm. The ionizing output of the supernova cannot be reliably estimated from the ultraviolet light curves because so much of the ionizing flux that results when the shock hits the surface of the star came before the first ultraviolet observations. A reasonable estimate (Woosley in this volume) might be 10^{47} erg available to ionize and heat the circumstellar gas observed later.

We note that the long wavelength UV reached a maximum at about the same time as the optical in late May (day 140), as shown by the IUE Fine Error Sensor, but that the subsequent behavior has been different. In particular, the UV does not share the exponential decline seen in the optical. This may be due to a modest change in the UV opacity with time, but the UV forms such a small fraction of the flux at these late times that it does not affect the underlying physical mechanism: presumably ^{56}Co decay with a 114 day e-fold time resulting from 0.07 solar mass of ^{56}Ni. The emergence of narrow emission lines in the short wavelength UV is reflected in a very small increase in the flux, because the short wavelength region is dominated by the two background stars discussed below.

Spectroscopy

The detailed spectral evolution includes complex changes over the first few days. In the early spectra, a broad absorption and emission stretching from 2500A to 2800A is likely to be a P-Cygni line of Mg II, with the blue wing of the absorption minimum extending beyond 30,000 km s^{-1} blueshift. Interstellar absorption in both our Galaxy and in the LMC accounts for the sharp feature seen at the rest wavelength of 2800A. In the space of a few days, the spectrum evolves to resemble that seen in the UV spectra of SN I (Panagia 1982, 1984). A reasonable fit to the long wavelength UV spectra for SN I comes from considering the effects of many blended Fe II and Co II lines scattering photospheric light in an expanding atmosphere (Branch and Venkatakrishna 1986) and it is reasonable to expect the same source here for the same effect (Branch 1987, Lucy 1987). It seems likely that the ordinary photospheric abundance of iron peak elements is capable of producing the observed lines.

In contrast, the successful models for the ultraviolet spectra of SN II consider the emission from a substantial circumstellar layer (Fransson et al 1984), such as might be expected around a red supergiant with a slow dense wind. The absence of these features makes it unlikely that SK -69 202 had such a wind immediately before the explosion, although it does not rule out a wind at an earlier stage of evolution. The picture in which SN 1987A has a low density circumstellar envelope is consistent with the weak, brief radio emission (Turtle et al 1987) and the absence of early X-ray flux (Makino 1987) as described by Chevalier and Fransson (1987).

Survivors

The persistence of UV flux in the short wavelength range is the result of stars which are present in the aperture when the satellite is pointed at the supernova. With accurate position measurements and image synthesis of the SK -69 202 field in hand (Walborn et al 1987, West et al 1987), it is possible to conduct a careful deconvolution of the IUE data and to establish that the two stars that are still present are stars 2 and 3. As described by Sonneborn in this volume, the 12 mag B3 I supergiant is not the source of the UV light. It has disappeared and it may be identified as the progenitor of SN 1987a (Sonneborn, Altner, and Kirshner 1987, Gilmozzi et al 1987).

Circumstellar Emission

Starting in 1987 May, narrow emission lines began to appear in the short wavelength IUE spectra (Wamsteker et al 1987, Kirshner et al 1987). The lines are narrow (<1500 km s^{-1}) and near zero velocity. It is hard to see how they could emerge from the fast moving debris, especially as the opacity of that material is very high. A more likely picture is that they arise from circumstellar material surrounding the supernova. This idea is strengthened by the actual line identifications. Strong lines of N III, N IV, and N V are present, and the corresponding carbon and oxygen lines are weak. This suggests that the material is nitrogen-rich, as might be expected for the envelope of a star which has undergone extensive CNO hydrogen burning. While the details depend on a proper understanding of the origins of the emission, the ratio C/N implied by the observations must be down from normal values by a factor of 30 or more. A plausible picture would be that the B3 I star which exploded spent some time as a red supergiant, and that the material which we see in the short wavelength UV results from mass loss during that interlude. The investigations which link mass loss for massive stars to their surface compositions are especially important here (Maeder 1987). The high nitrogen abundance we see requires very substantial mass loss, essentially peeling the star down to the hydrogen-burning zone where nitrogen is enhanced by the CNO cycle. It is interesting to note that a very similar UV spectrum is seen in the nitrogen-rich material ejected from Eta Carina (Davidson et al 1986), which is another star that is presumably on the path to becoming a supernova.

If we are seeing the results of a wind, the material would be at a distance that corresponds to the red giant wind velocity multiplied by the time the star spent as a blue supergiant. A reasonable estimate would be 10 km s^{-1} for 30000 years, or a distance of 1 light year. The ionizing flux from the supernova outburst would ionize and heat some of this circumstellar gas. The emission that we see would come either from the excitation of these ions, or possibly from recombination. If the

Figure 3. Averaged short wavelength spectrum, formed by subtracting the March 87 spectrum, which contains only flux from the two background stars, from the observations. Note the strong emission lines of nitrogen in three ionization stages: N III, N IV, and N V. In a gas of normal composition, C IV 1550 is the strongest line in this part of the spectrum.

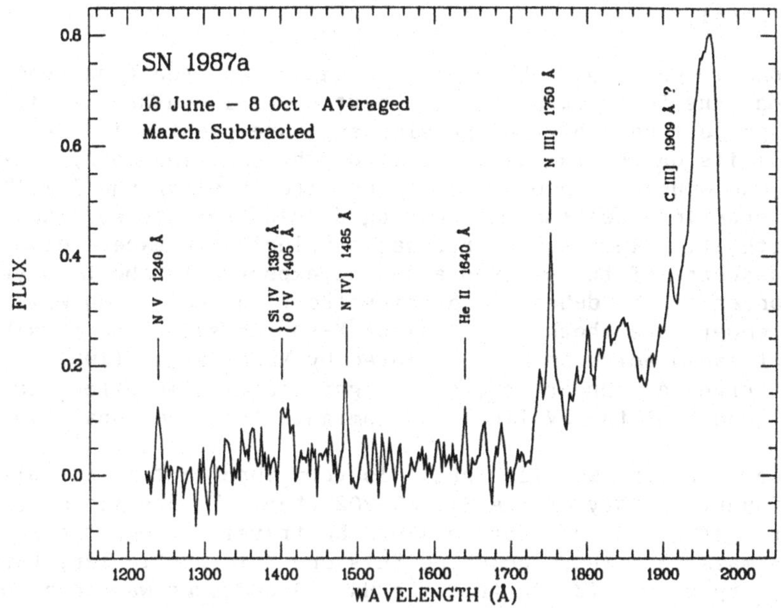

Figure 4. Observed line profiles at the blue wing of H alpha from CTIO data, fit by a simple model. Note the shift to the red of the absorption, the result of the outer, faster moving layers becoming transparent.

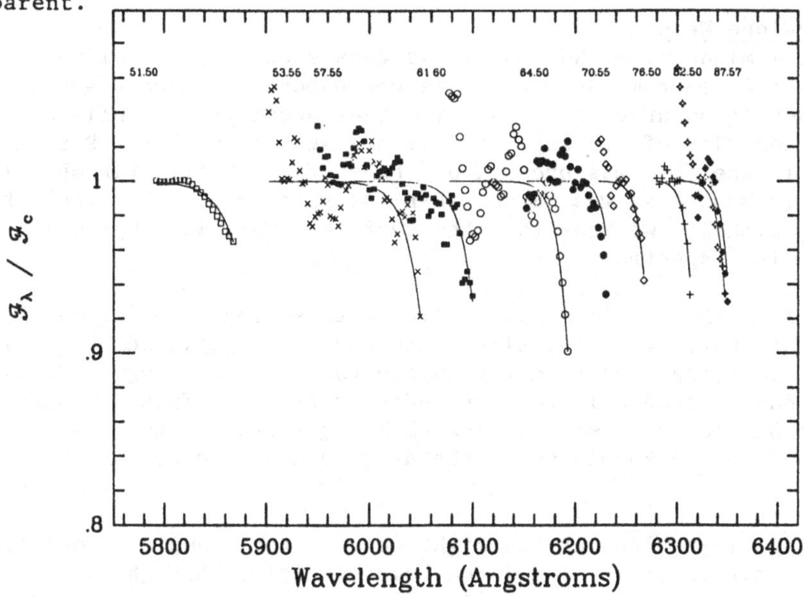

recombination time is long, then the flux we see should grow with time as the light travel effects allow us to see more of the circumstellar shell. The available radius should grow as $t^{1/2}$; its area, and the flux should grow as t. In fact, the increasing flux of N III] 1750Å matches that rate, although the data are still quite sketchy. The flux should continue to increase until the age is comparable to the light travel time across the shell.

For a reasonable guess of 1 LY scale, we can expect the flux from this shell to continue to increase for the next several months. If it grows strong enough to make a high-dispersion exposure practical, very stringent limits on the radial velocity of the emitting material will be measured. One amusing consequence of the circumstellar shell will be a violent interaction when the fast moving debris from the supernova collides with it. Since the debris has a velocity in excess of 0.1c, this recrudescence of the supernova can be expected in about 10 years, or a little longer if the debris is decelerated. The collision will result in a high-temperature shock with copious X-ray emission and possibly a new burst of radio emission as calculated by Masai et al (1987). If these new sources do appear, the same light travel time effect that shapes the growth of the UV lines may dominate their temporal evolution.

The presence of a circumstellar shell is very important in establishing the evolutionary history of the SK -69 202 star. Theory alone does not provide a robust guide to understanding the travel through the H-R diagram for massive stars, including this one. In particular, this shell seems like very good evidence that the SK -69 202 star was once losing mass, presumably in a slow dense wind like that seen around red supergiants.

Envelope Mass

The luminosity of SK -69 202 is consistent with a helium core mass of about 6 solar masses, and this corresponds to a main sequence mass of order 15 solar masses. An important question that will affect the future behavior of the supernova is how much mass of the 9 solar masses not in the core was present on the surface of the star when it exploded. If mass loss was large, then the mass on the star could have been small, although we know that the surface layers were hydrogen-rich from the optical spectra.

One way to find the envelope mass would be to estimate the circumstellar mass, and subtract. This is unlikely to lead to a satisfactory estimate because the ionizing flux from the supernova is not so large that we can be sure it has ionized all the surrounding material. This approach is also vulnerable to large errors from clumping effects, and in any event requires an accurate knowledge of the density which we are unlikely to possess.

Another avenue is to look at the light curves, for example as modelled by Woosley (in this volume). The large bump extending through May implies the energy deposited by radioactivity or other sources was trapped and

Figure 5. Density distribution derived from the simple model used to fit the lines in Figure 4. The slope has density falling as v-11, and the integrated mass above 6000 km s^{-1} is approximately 1.8 solar masses.

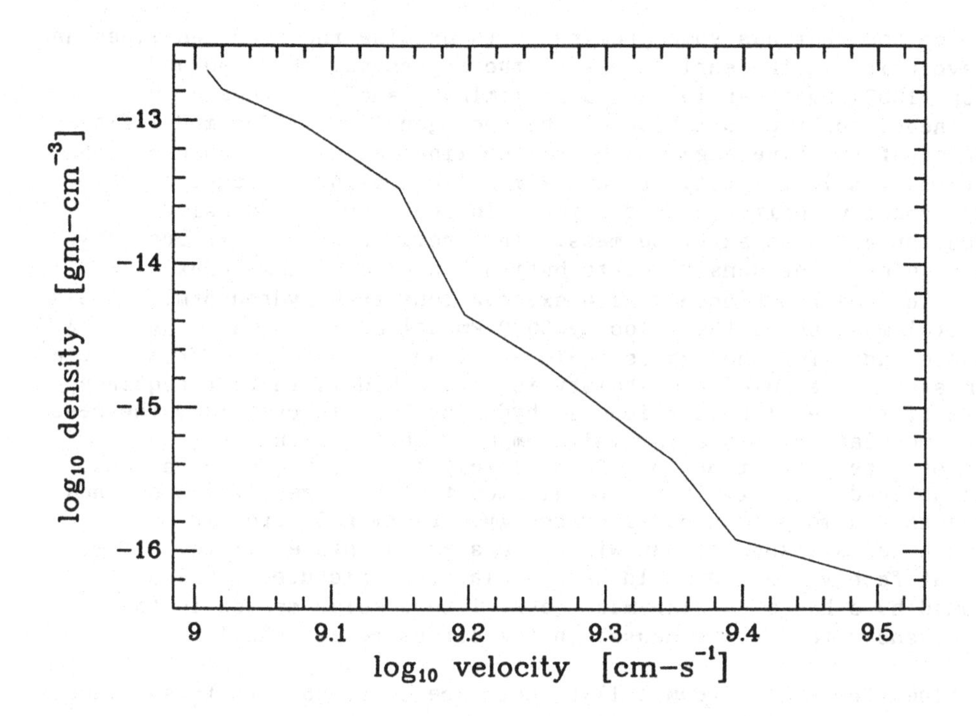

had to diffuse out. But the details of the energy sources (which could include a neutron star or perhaps a substantial recombination energy) and of the opacity (which must depend on the composition and ionization of the stellar mantle) are not necessarily easy to model, so we may not have an accurate way to find the hydrogen mass from this approach.

A third method that has some promise is to examine the hydrogen lines and their evolution in the early weeks of the supernova. Eastman and Kirshner (1987) have carried out a preliminary analysis, based on a simple theory for the formation of the hydrogen lines. The model fits the shapes of the blue edge of the Balmer lines at various epochs. The rapid recession in velocity of the Balmer lines allows a model of the density versus velocity to be derived. Integrating that density distribution gives an envelope mass. The preliminary results are shown in the figures. The density distribution has a power law index of v^{-11}, which is in reasonable accord with expectations from hydrodynamic models, and a total mass above the velocity 6000 km s^{-1} of 1.8 solar masses. A reasonable range for the errors would admit an envelope mass from 1 to 6 solar masses. The chief uncertainty in the method is that it requires an estimate of the level populations in hydrogen. It is certainly the case that the population in n=2 is small compared to the ground state population. The population in n=2 is closely tied to the observations, but the desired quantity is the total amount of hydrogen, which depends on doing this atmospheric model correctly. Eastman is creating a detailed model atmosphere that will help sharpen this estimate. If we take it at face value, we would have a plausible picture in which the star would have had a 6 solar mass core, 1 to 6 solar masses in the envelope, and 8 to 3 solar masses in the circumstellar shell.

If the clues from the circumstellar shell are combined with these lines of reasoning from the supernova's behavior, it is hard to escape the conclusion that mass loss played an important role in the evolution of SK -69 202, and helped to determine its properties as SN 1987A.

I am very grateful for the outstanding cooperation of the IUE Observatory staff and for the tolerance of all the IUE observers whose work has been dislocated by the observations of SN 1987A. The special contributions of George Sonneborn, George Nassiopoulos, Eric Schlegel, and Ronald Eastman were essential to the work reported here. This work on supernovae is supported by NASA grants NAG5-841 and NAG5-645 and by NSF grant AST85-16537 to the Harvard College Observatory.

References

Blanco, V.M. et al. Ap.J. **320**, 589.
Branch, D. 1987, Ap.J. (Lett.) **320**, L121.
Branch, D. and Venkatakrishna, K.L. 1986, Ap.J. (Lett.) **306**, L21
Cassatella,A. et al 1987 Astr. Ap. **177**, L29.
Chevalier, R.A. and Fransson, C. 1987 Nature **328**,44.
Davidson, K., et al. 1986 Ap.J. **305**, 867.
Dupree, A.K., Kirshner, R.P., Nassiopoulos, G.E., Raymond, J.C., and Sonneborn, G. 1987, Ap.J. **320**, 597.
Eastman, R. and Kirshner, R.P. 1988 B.A.A.S. Vancouver late papers.
Fransson, C. et al. 1984, Astr. Astrophys. **132**, 1.
Fransson, C. et al. 1987, Astr. Astrophys. **177**, L33.
Gilmozzi, R. et al. 1987, Nature **328**, 318.
Kirshner, R.P., Sonneborn, G., Crenshaw, D.M., and Nassiopoulos, G.E. 1987, Ap.J. **320**, 602.
Kirshner, R.P. et al. 1987 IAU Circular 4435.
Koshiba, M. 1987, IAU Circular 4338.
Lucy, L. 1987, Astr. Astrophys. **182**, L31.
Maeder, A. 1987, Astron. Astrphys. **173**, 287.
Makino, F. 1987 IAU Circular 4336.
Masai, N. et al in IAU Colloquium 108 (K. Nomoto, ed.) Tokyo.
Panagia, N. 1982, 3rd European IUE Conference, ESA SP-176, p.31.
Panagia, N. 1984, 4th European IUE Conference, ESA SP-218, p.15.
Panagia, N. et al. 1987, Astr. Astrophys. **177**, L25.
Sonneborn, G., Altner, B.A., and Kirshner, R.P. 1987, Ap.J. (Lett.) **323**, L35.
Svoboda, R. 1987, IAU Circular 4340.
Turtle, A.J. et al. 1987, Nature **327**, 38.
Walborn, N.R., Lasker, B.M., Laidler, V.M., and Chu, Y-H 1987, Ap.J. (Lett.) **321**, L41.
Wamsteker, W. et al. 1987, IAU Circular 4410.
West, R.M., Lauberts, A., Jorgensen, H.E., and Schuster, H-E, 1987, Astr.Astrophys. **177**, L1.

ON THE ENERGETICS OF SN 1987A

N. Panagia[1]
Space Telescope Science Institute
and
University of Catania

Abstract. The photometric observations of SN 1987A are discussed in the context of the supernova energetics. Adding the information from spectroscopy and neutrino observations, it is concluded that SN 1987A is a rather "normal" Type II explosion with an compact initial configuration, just as expected for the progenitor Sk -69°202. The advanced FES light curve displays a steady linear decay with an e-folding time of 136 days. It is argued that this may be the evidence of energy supplied by both ^{56}Co radioactive decay and a young pulsar.

Introduction: the European IUE Observations

Since my talk will substantially lean on them, let me first summarize the results of the IUE observations made from the other side of the Atlantic. The observations, carried out as part of the ESA Target-of-Opportunity Program for observing bright supernovae, were started in the early morning of the 25th of February, on the first European shift available after the supernova discovery had been announced by an IAU telegram (Wamsteker *et al.* 1987a, Panagia *et al.* 1987, Cassatella *et al.* 1987). The main results obtained so far are:

1. Initially the UV flux was rather high indicating high photospheric temperatures (T > 14000 K).

2. The UV flux dropped by orders of magnitude in a few days (Panagia *et al.* 1987) implying a small initial radius, say, of the order of few tens solar radii.

3. The star Sk -69°202 has been ascertained to be the SN progenitor: this was first suggested in Panagia *et al.* (1987) and subsequently confirmed by the more detailed analyses presented by Gilmozzi *et al.* (1987) and Sonneborn *et al.* (1987).

4. Emission lines of highly ionized species (N V 1240Å, Si IV 1397Å, O IV] 1405Å, N IV] 1485Å, He II 1640Å, N III] 1750Å and C III] 1909Å) were first

[1] Affiliated with the Astrophysics Division, Space Sciences Department of ESA.

detected on late May - early June (Wamsteker *et al.* 1987b) and found to increase in intensity at later epochs (Kirshner *et al.* 1987). Their analysis (Cassatella 1987, Fransson *et al.* 1987) shows that they originate from a circumstellar shell, which is several light-months in size and whose nitrogen abundance relative to both carbon and oxygen is enhanced by a factor of 50 or higher as compared with solar values. Because of its characteristics, this shell is likely to be the remnant of the wind ejected by the star when it was a red supergiant.

5. The strong UV flux of the early epochs has offered a unique opportunity to obtain high dispersion spectra to study the interstellar medium in our Galaxy and in the LMC toward SN 1987A. A large number of components for both highly ionized and low ionized species have been detected (de Boer *et al.* 1987, Grewing *et al.* 1987, Blades *et al.* 1987a, b), indicating a rather complex structure for the intervening matter.

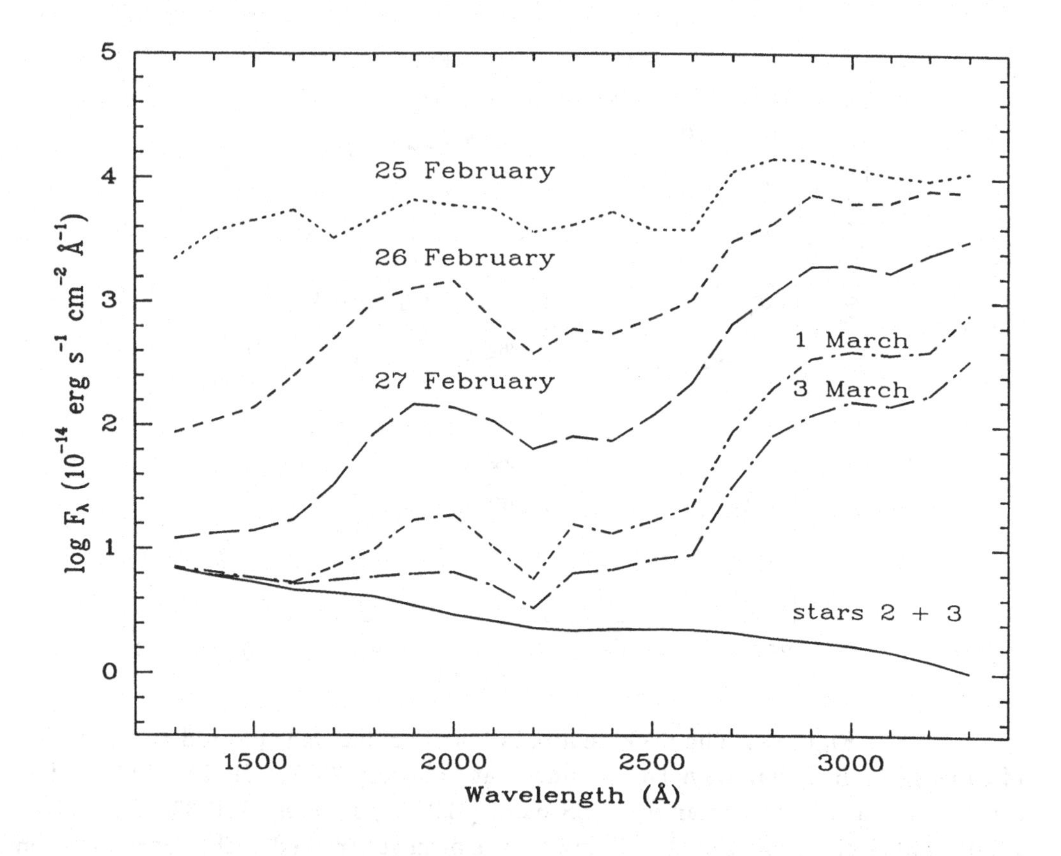

Figure 1 . The IUE spectra, rebinned into 100Å bands, are shown for the first observing week, February 25 through March 3. For comparison, the total spectrum of stars 2+3 is also shown as a solid line.

UV and Optical Photometry with IUE

Figure 1 displays the IUE spectra obtained in the first observing week. In order to better appreciate the overall evolution, the spectra were rebinned into 100Å bands. It is apparent that the flux decays very quickly with time, the more so at the shortest wavelengths (see also Fig. 2). Moreover, it is clear that the two "neighbor" stars (star 2 and 3; Walborn *et al.* 1987a, b) soon account for most of the emission at short wavelengths (see also Panagia *et al.* 1987, Gilmozzi *et al.* 1987). In fact, at all epochs later than March 1 their radiation dominates the continuum shortward of 1700Å and their contribution is comparable to the intrinsic SN emission at wavelengths shorter than 2600Å (see also Panagia 1987a). To study the energetics of the UV emission of SN 1987A we have binned the low resolution spectra into three "photometric" bands, i.e. 1250-1750Å, 1750-2750Å and 2750-3350Å.

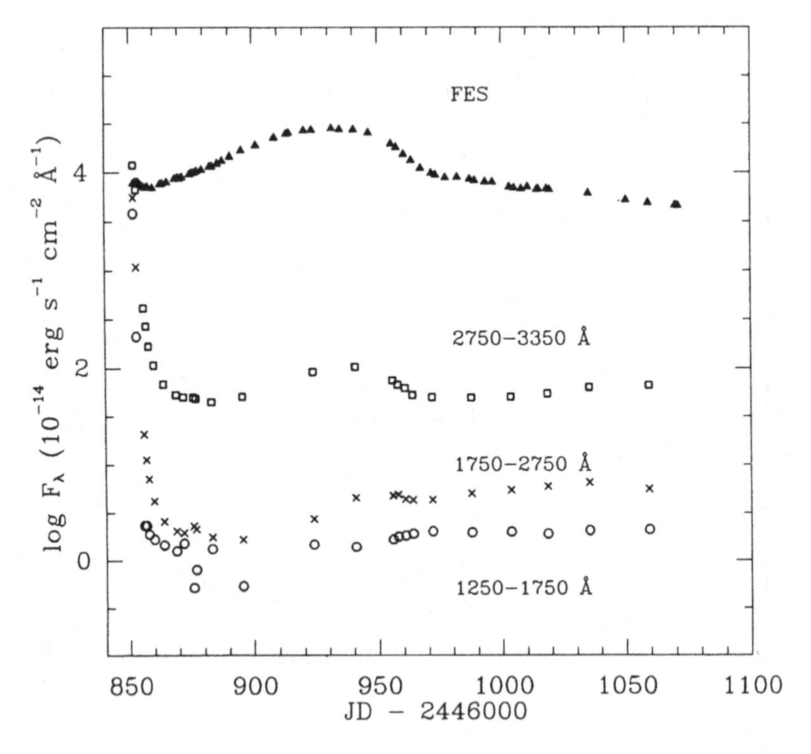

Figure 2. The average flux in four IUE bands is plotted as a function of time, given in Julian days (we remind that February 23.0 is JD 2446849.5). The symbols are as follows: triangles FES band (5200Å), squares 2750-3350Å, x 1750-2750Å and circles 1250-1750Å. All fluxes have been subtracted of the contamination by stars 2 and 3 (see text). The apparent fluctuations, which are most conspicuous at short wavelengths in the period (JD-2446000) 870-885, are not real and are the result of the subtraction procedure.

The interval 1150-1250Å was not considered because it is strongly contaminated by geocoronal Lyman-alpha. In all bands the contamination due to star 2 and 3 was corrected for according to the flux estimates of Gilmozzi *et al.* (1987).

The light curves in the three UV bands together with that of the FES band are displayed in Figure 2. For each of them we plot the average flux as a function of time. The flux in the FES band (effective wavelength 5200Å) was estimated from the FES counts adopting the standard transformation into magnitudes and assuming the flux for zero magnitude to be 4.65 10^{-9} erg cm^{-2} s^{-1} A^{-1}.

Looking at Figure 2 it is apparent that the flux evolves quite differently in the different bands. At all IUE wavelengths the flux decreases since the earliest epoch (February 25.15; Wamsteker *et al.* 1987a) at a rate that is higher at shorter wavelengths. In particular, the initial decay rates (25 February - 1 March) are 0.766 dex/day, 0.610 dex/day and 0.371 dex/day in the bands 1250-1750Å, 1750-2750Å and 2750-3350 Å, respectively. This type of an evolution is the result of a temperature decrease in the photospheric layers of the supernova combined with an increase of the line blanketing with time. Although this may be a general feature of all supernovae, SN 1987A displays a particularly fast evolution. For comparison SN 1979C, a type II-L which was discovered around maximum and promptly observed with IUE, had initial decays of 0.072, 0.064, 0.032 and 0.030 dex/day at 1300, 1800, 2400 and 3000Å, respectively (Panagia *et al.* 1980): these rates are about ten times slower than SN 1987A's. This indicates that the temperature evolution of SN 1987A has been unusually fast and, therefore, had an initial radius unusually smaller than average Type II supernovae. During the rest of March the SN emission is still decreasing at all UV wavelengths but at substantially slower rates (i.e. 0.021, 0.077 and 0.065 dex/day in the three bands, respectively). In April the flux starts rising again: at the longest wavelengths it reaches a maximum around May 25, followed by a decline of 0.34 until about June 20 when it starts rising again at a steady rate of 0.0014 dex/day. At short wavelengths the evolution is qualitatively the same but with smaller amplitude and with some time delay relative to longer wavelengths.

In the optical the evolution is rather different. After a very quick rise as reported in the IAU circular No. 4316 (no detection on Feb 22, V = 6.1 on Feb 23.44 and V = 4.5 on Feb 24.33), our FES data show that (Fig. 2; see also Fig. 4) the flux first reached a local maximum around the 27th of February with m(FES) = 4.4 (or, correspondingly, M(FES) = -14.7), declined until March 4 by about 0.15 magnitudes and then started rising again at an average rate of 0.021 mag/day. Around May 20 the FES flux reached a maximum: the apparent magnitude was m(FES) = 3.0, which after allowance for reddening implies an absolute magnitude of M(FES) = -16.1. Since then, the FES flux declined, initially at an average rate of 0.03 mag/day, until about the end of June when, rather abruptly, the decay became much shallower with a steady rate of 0.0080 mag/day (see also Fig. 4) which corresponds to an e-folding time of 136 days.

The Energetics

To realize how much the various spectral bands contribute to the total luminosity as a function of time it is convenient to consider the behaviour of the quantity λF_λ which is a good tracer of the energy content in each band. Figure 3 presents the λF_λ curves for the UV, now lumped into two bands (SWP 1250-1950Å and LWP 1950-3350Å) and the FES, as obtained from our measurements, combined with the data of the I, H and L bands as found in the literature (Bouchet *et al.* 1987, Menzies *et al.* 1987, Catchpole *et al.* 1987, Hamuy *et al.* 1987). The fluxes were corrected for reddening adopting E(B-V) = 0.20. This is the value appropriate for Sk -69°202 (Panagia *et al.* 1987) and, therefore, for the SN as well. The reddening

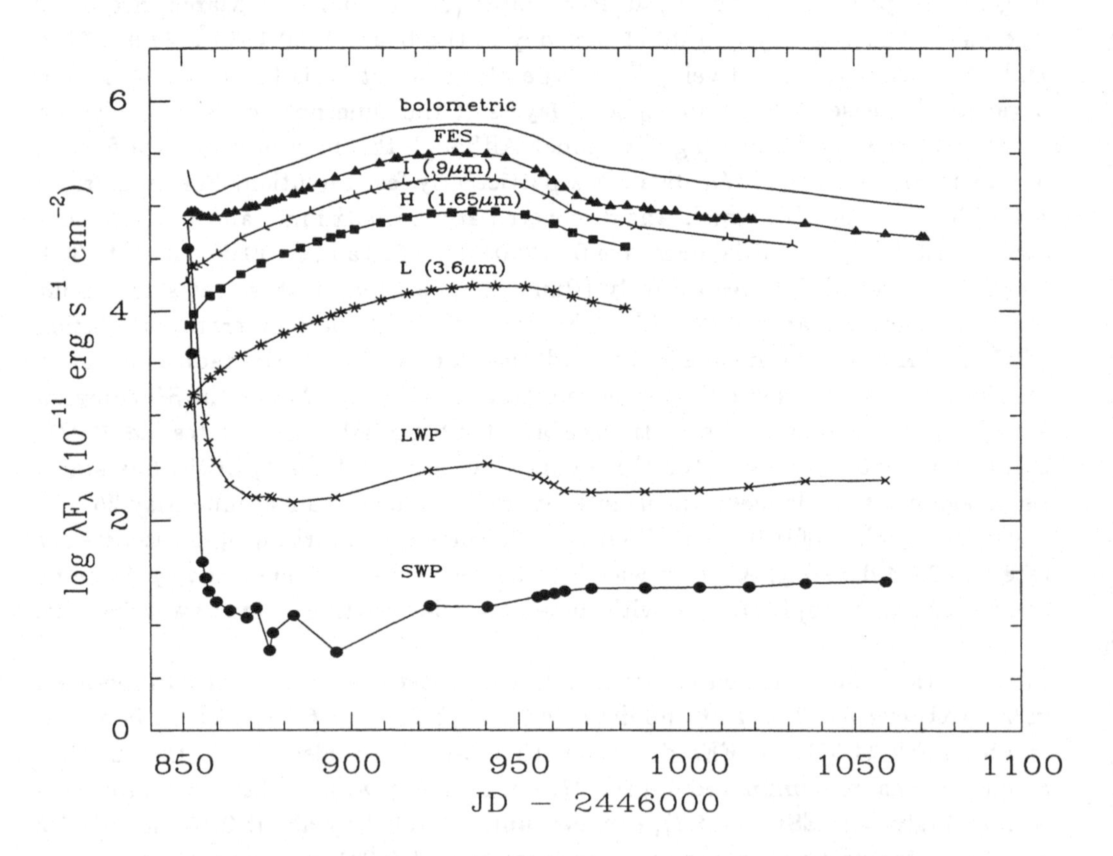

Figure 3. The quantity λF_λ for several bands from UV to IR is plotted as a function of time. The fluxes have been subtracted of star 2 and 3 contamination and dereddened adopting E(B-V) = 0.2. The apparent fluctuations in the SWP band during the period (JD-2446000) 870-885 are not real and are the result of the subtraction procedure.

correction has been computed attributing E(B-V) = 0.05 to our own galaxy, with an extinction law as compiled by Savage and Mathis (1979), and E(B-V) = 0.15 to the LMC for which we adopted the extinction law determined by Fitzpatrick (1986) for the 30 Dor region.

It is clear that the UV emissiom, which is important at the earliest epochs (about 50% of the total emission on February 25.15), becomes rapidly a minor fraction of the whole luminosity of the supernova. In fact, around the 1st of March the optical emission already dominates the total SN emission. On the other hand, it is clear that at epochs earlier than our first IUE observation the UV flux is likely to have been much higher than that in the optical, as one can easily realize by naively extrapolating back in time with the rates determined for the early epochs. As for the longer wavelengths only the very near IR (i.e. I band; $\lambda = 0.9\ \mu m$) emits a comparable amount of energy as in the optical, whereas in the IR proper, the emitted energy is of the order of 20% essentially at all epochs later than early March.

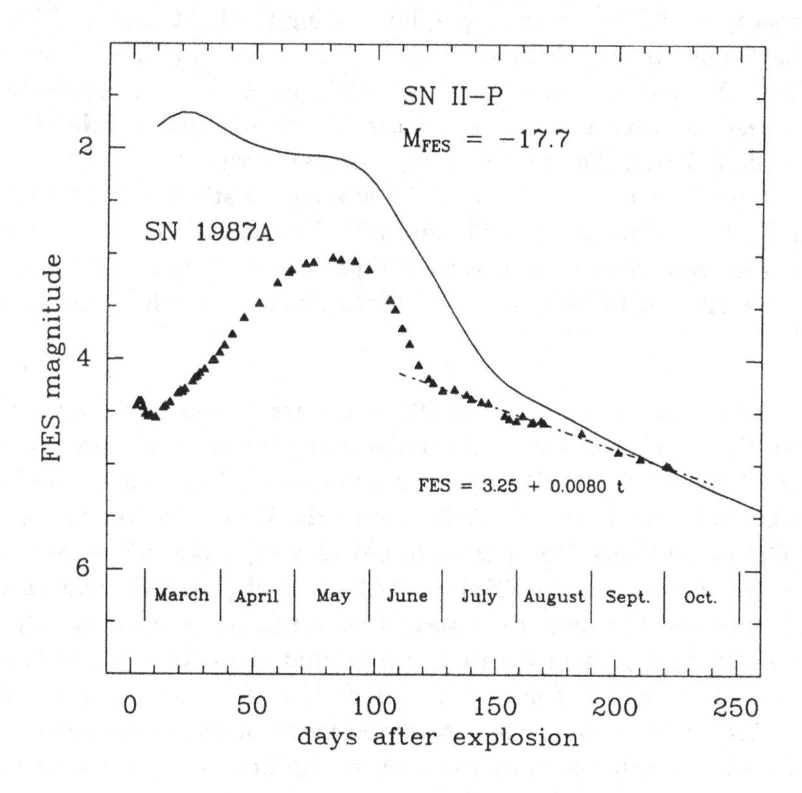

Figure 4. The FES light curve of SN 1987A is compared with the average light curve a type II-P supernova (adapted from Doggett and Branch 1985) whose FES magnitude at maximum is -17.7 and which is assumed to be at the LMC distance and to be reddened by E(B-V) = 0.2.

Comparing the light curve of this SN with those of classical type I or II SNe (e.g. Doggett and Branch 1985) it is clear that SN 1987A does not follow any of them: the initial rise is much steeper than "normal" and the time evolution is also quite different, because at high frequencies the decay is much faster than that of any SN studied with IUE (Panagia, 1987b) and in the optical the SN has had two maxima, the first one at a very early epoch (only 4 days after explosion!) and the main one as late as three months after explosion. However, even if the light curve does not resemble that of any classical type, the presence of hydrogen in the envelope and the conspicuous release of neutrinos speaks in favor of the collapse of a relatively massive star. And the massive star origin is confirmed by the identification of the progenitor with Sk -69°202 (Gilmozzi *et al.* 1987) whose initial mass was about 20 $M_{\odot}$ (e.g. Maeder 1987, Maeder and Meynet 1987). Therefore, it is most reasonable to consider it as an unusual kind of Type II supernova.

Integrating over the entire spectrum and over time from February 25 to October 12 we obtain a total radiative energy of about 8 10^{48} erg. This value is comparable to that of *normal* type II SNe: for example, integrating the light curves of SN 1980K (Barbon, Ciatti and Rosino 1982) we find that it has radiated an energy of $\sim$2.5 10^{49} erg during the first four months of its evolution. Also, the neutrino flux is in very good agreement with the theoretical predictions for type II SNe (Bahcall *et al.* 1987). And similarly, the kinetic energy associated with this event is perfectly *normal* for a type II event. In fact, since the envelope is still optically thick at the time of this Conference we can roughly estimate the mass of the ejecta to be more than two solar masses. Adopting an average expansion velocity of 10000 km s^{-1} the kinetic energy turns out to be 2 10^{51} erg, at least, which is of the "right" order for a type II SN.

Actually, even the radiative flux of SN 1987A at late epochs is *normal.* Figure 4 compares the FES light curve with the average one for type II-P SNe (this average curve was derived by suitably interpolating between the average B and V light curves from Doggett and Branch's [1985] compilation). We see that at the present time (1987 October 12) SN 1987A is as bright as a type II-P SN whose absolute magnitude at maximum was M(FES) = -17.7. Clearly, at early epochs the SN 1987A flux is considerably lower than *normal* , say a factor of ten, but only for the first couple of months. It is interesting to note that the main peak of SN 1987A occurs at the plateau epoch of normal type II-P SNe. So, in naive terms, it looks as if initially SN 1987A lacks some substantial input of energy and concludes all of its first phase of the light curve evolution in the first week or ten days, when it reaches its first maximum, but at later epochs it tends to more canonical conditions. As profusely discussed in the theoretical section of this Conference (e.g Arnett; Nomoto; and Woosley; in addition to the classical paper of Chevalier and Klein, 1979) most of these features can easily be accounted for by admitting that the initial radius was unusually small, of the order of few 10^{12} cm. This is just the radius of a B3 supergiant such as Sk -69°202 was classified. Therefore, we conclude that SN 1987A is essentially a *normal* SN explosion which happened to occur at a *peculiar* time of the evolution of its progenitor. A possible clue as for why this star was a B supergiant when it exploded is provided by the strong overabundance of nitrogen relative to both carbon and oxygen as deduced from the UV emission lines observed since early June (Wamsteker *et al.* 1987b, Cassatella 1987, Fransson *et al.* 1987). As summarized by Walborn (1987) evidence of CNO-cycled material at the surface of a massive star indicates that that star has gone through the red supergiant phase. And indeed a strong overabundance was measured in the ejecta of SN 1979C (Fransson *et al.* 1984) whose direct progenitor was a red supergiant. Then, it appears that Sk -69°202 was a blue supergiant that had returned back from the red supergiant region, possibly because of efficient mass loss.

The late evolution of the FES light curve deserves some special consideration. As mentioned at the end of Section 2, since the end of June the FES flux has been observed to decline at a steady rate with an e-folding time of 136 days. An exponential decay is strongly suggestive of a very regular, self-clocked physical mechanism such as the ^{56}Co radioactive decay. However, the FES decay is 20% slower than the ^{56}Co radioactive decay, and, for that matter, is also 20% slower than the decline of the bolometric light curve as computed on the basis on photometric data (e.g. Feast, this Conference). This is a puzzling result because the FES bandwidth is so broad (e.g. Holm and Crabb 1979) that it should include a substantial fraction of the total SN flux at all epochs. As a consequence, the FES magnitude should evolve quite closely like the bolometric one. In addition, the observed colors seem not to have varied appreciably over the last four months (e.g. Hamuy *et al.* 1987) and, therefore, one can hardly ascribe such a discrepancy to a systematic change of the SN continuum shape which could affect the effective flux measured in the FES band. On the other hand, the determination of the bolometric luminosity by integrating a smooth continuum derived by interpolating over the average fluxes provided by photometric measurements may be affected by systematic errors in the presence of strong emission lines, as observed in SN 1987A (e.g. Danziger, this Conference). In fact, with such a procedure one tends to overestimate/underestimate the actual flux if most of the lines falls within/outside the photometric bands. In view of this, we are inclined to believe that the FES light curve provides a better estimate of the actual decay of the bolometric luminosity. This fact, in turn, implies that the ^{56}Co decay alone cannot account for the entire energy output of SN 1987A as observed during the summer and a contribution by another source of energy is required. Since the decay time is only 20% slower than the ^{56}Co decay, the mass of Ni produced in the explosion should remain around a value of 0.07 $M_\odot$ as suggested by Woosley and Arnett (this Conference). On the other hand, the luminosity of the next most plausible energy source, a central rotating neutron star, would not be constrained to be less than 5 10^{39} erg s^{-1} but could be appreciably higher, say up to few times 10^{40} erg s^{-1} or so. This would represent the first *detection* of a newborn pulsar after a stellar core collapse. A more definite answer to this fascinating problem will be provided by both a direct integration of spectrophotometric fluxes and future optical and infrared measurements which eventually shall display the true asymptotic trend of the light curve or, even, reveal the presence of the central collapsed stellar remnant.

REFERENCES

Bahcall, J. N., Piran, T., Press, W. H. and Spergel, D. N., 1987, *Nature*, **327**, 682.

Barbon, R., Ciatti, F., Rosino, L., 1982, *Astron. Astrophys.* **116**, 35.

Blades, J. C., Wheatley, J. M., Panagia, N., Grewing, M., Pettini, M., Wamsteker, W., 1987a, this Conference.

Blades, J. C., Grewing, M., Panagia, N., Pettini, M., Wamsteker, W., and Wheatley, J. M., 1987b, in preparation.

Bouchet, P., Stanga, R., Le Bertre, T., Epchtein, N., Hamann, W. R. and Lorenzetti, D., 1987, *Astron. Astrophys.*, **177**, L9.

Cassatella, A., 1987, ESO Workshop "SN 1987A", ed. J.I. Danziger, in press.

Cassatella, A., Fransson, C., van Santvoort, J., Gry, C., Talavera, A., Wamsteker, W. and Panagia, N., 1987, *Astron. Astrophys.*, **177**, L29.

Catchpole, R. M. *et al.*, 1987, preprint.

Chevalier, R. A., Klein, R. I., 1979, *Astrophys. J.* **234**, 597.

de Boer, K., Grewing, M., Richtler, T., Wamsteker, W., Gry, C., and Panagia, N., 1987, *Astron. Astrophys.* **177**, L37.

Doggett, J. B. and Branch, D., 1985, *Astron. J.* **90**, 2303.

Fransson, C., Benvenuti, P., Gordon, C., Hempe, K., Palumbo, G. G. C., Panagia, N., Reimers, D. and Wamsteker, W., 1984, *Astron. Astrophys.* **132**, 1.

Fransson, C. *et al.*, 1987, in preparation.

Fitzpatrick, E. L., 1986, *Astron. J.*, **92**, 1068.

Gilmozzi, R., Cassatella, A., Clavel, J., Fransson, C., Gonzalez, R., Gry, C., Panagia, N., Talavera, A. and Wamsteker, W., 1987, *Nature*, **328**, 318.

Grewing, M., Blades, J. C., Panagia, N., Pettini, M., and Wamsteker, W., 1987, ESO Workshop "SN 1987A", J.I. Danziger, in press.

Hamuy, M., Suntzeff, N.B., Gonzalez, R. and Martin, G., 1987, preprint.

Holm, A. V., and Crabb, W. G., 1979, *NASA IUE Newsletter* No. 7, 40.

Kirshner, R. P. , Sonneborn, G., Cassatella, A., Gilmozzi, R., Wamsteker, W. and Panagia, N., 1987, *I.A.U. Circ.* No. 4435.

Maeder, A., 1987, ESO Workshop "SN 1987A", ed. J. I. Danziger, in press.

Maeder, A. and Meynet, G., 1987, *Astron. Astrophys.*, **182**, 243.

Menzies, J. W. *et al.*, 1987, *M.N.R.A.S.*, **227**, 39P.

Panagia, N., 1987a, ESO Workshop "SN 1987A", J. I. Danziger, in press.

Panagia, N., 1987b, in "High Energy Phenomena around Collapsed Stars", ed. F. Pacini (Dordrecht: Reidel), p. 33-49.

Panagia, N. *et al.*, 1980, *M.N.R.A.S.*, **192**, 861.
Panagia, N., Gilmozzi, R., Clavel, J., Barylak, M., Gonzalez Riestra, R., Lloyd, C., Sanz Fernandez de Cordoba, L. and Wamsteker, W., 1987, *Astron. Astrophys.*, **177**, L25.
Savage, B. D. and Mathis, J. S., 1979, *Ann. Rev. Astron. Astrophys.*, **17**, 84.
Sonneborn, G., Altner, B., and Kirshner, R. P., 1987, *Astrophys. J.*, in press.
Walborn, N. R., 1987, I.A.U. Colloquium No. 108 "Atmospheric Diagnostics of Stellar Evolution", in press.
Walborn, N. R., Lasker, B. M., McLean, B., and Laidler, V. G., 1987a, *I.A.U. Circ.* No. 4321.
Walborn, N. R., Lasker, B. M., Laidler, V. G., Chu, Y. -H., 1987b, *Astrophys. J. (Letters)*, **321**, L41.
Wamsteker, W., Panagia, N., Barylak, M., Cassatella, A., Clavel, J., Gilmozzi, R., Gry, C., Lloyd, M., van Santvoort, J., and Talavera, A., 1987a, *Astron. Astrophys.*, **177**, L21.
Wamsteker, W., Gilmozzi, R., Cassatella, A. and Panagia, N., 1987b, *I.A.U. Circ.* No. 4410.

ON THE NATURE AND APPARENT UNIQUENESS OF SN 1987A

Alexei V. Filippenko
Department of Astronomy
University of California
Berkeley, CA 94720 USA

Abstract. I show that objects such as SN 1987A may be relatively common, but have previously been missed for various reasons during photographic surveys of the sky. I also conclude that the metallicity of the LMC is higher than generally assumed in theoretical models of SN 1987A, and is unlikely to be responsible for most of the observed peculiarities of the SN. Finally, a spectrum of the Type II SN 1986I is illustrated. It is remarkably similar to that of SN 1987A at a comparable time ($\sim$ 9 months) past outburst.

1 INTRODUCTION

Many properties of SN 1987A in the Large Magellanic Cloud (LMC) seem to make it a unique object among Type II supernovae (SNe II). It was subluminous at maximum brightness, had an unprecedented light curve during its first four months, evolved very rapidly in color and spectral properties at early times, and exhibited spectral anomalies such as high expansion velocities and strong absorption lines of barium and other *s*-process elements. Moreover, its progenitor was a blue supergiant star (B3 I), rather than a red one.

Some theoretical studies of SN 1987A suggest that the low metallicity of the LMC was largely, or entirely, responsible for its peculiar characteristics. Indeed, it may even be said that Shklovskiĭ (1984) "predicted" the existence of SNe like 1987A. He argued that the low heavy-element abundance of the LMC would prevent massive stars from generating extended atmospheres through strong stellar winds; consequently, they would produce faint SNe II.

It is important to establish whether objects such as SN 1987A are intrinsically rare or quite common. If Shklovskiĭ (1984) is correct, numerous subluminous SNe II may occur in irregular galaxies and very late-type spirals. Why, then, have they not been detected in *nearby* galaxies ($d \lesssim 20$ Mpc) during systematic searches conducted over the past 30 – 50 years? Here I examine this question, as well as the hypothesis that the properties of SN 1987A were determined largely by its low metallicity. Also, a few comparisons are made between the spectral characteristics of SN 1987A and other SNe II. All distance moduli of galaxies are taken from the compilation by Rowan-Robinson (1985).

2 HOW SUBLUMINOUS WAS SN 1987A?

The apparent visual brightness of SN 1987A reached its maximum ($V = 2.97$) on 20 May 1987 UT (Hamuy *et al.* 1987), roughly 90 days after core collapse (23 Feb. 1987 UT). Since the distance modulus of the LMC is 18.7

mag, the corresponding absolute luminosity was $M_V = -15.7$. Correcting for extinction ($A_V \approx 0.6$ mag), I find that $M_V^0 = -16.3$ at maximum.

An average of many SNe II discussed by Barbon *et al.* (1979), Branch & Bettis (1978), and others shows that $M_V^0 \approx -17.0 \pm 1$ at maximum if $H_0 = 100$ km s^{-1} Mpc^{-1}. In this case SN 1987A cannot be considered too subluminous in the visual band. This conclusion is weaker if the long distance scale ($H_0 = 50$ km s^{-1} Mpc^{-1}) is adopted, since typical SNe II then have $M_V^0 \approx -18.5$ at maximum. The observed dispersion is so large, however, that SN 1987A falls at the tail end of the distribution, rather than in a distinct class by itself.

At blue wavelengths, on the other hand, the story is different. Although the stellar surface was hot and therefore very blue immediately after the main shock wave broke through, it cooled dramatically during the following two weeks. Only after three weeks did the shapes of the B and V light curves closely resemble each other (Hamuy *et al.* 1987), but by this time SN 1987A was much fainter in B than in V. When it reached secondary maximum (hereafter referred to simply as "maximum"), $B = 4.58$ whereas $V = 2.97$, and $M_B^0 = -14.9$ ($A_B = 0.8$ mag). This is one-twelfth the average luminosity ($M_B^0 = -17.6$) of other SNe II observed at maximum (Barbon *et al.* 1979; $H_0 = 100$ km s^{-1} Mpc^{-1}). The discrepancy becomes a factor of four (1.5 mag) larger if the long distance scale is used.

3 WHERE ARE OTHER SUPERNOVAE LIKE 1987A?

Given that SN 1987A was fairly luminous in V, it is natural to wonder why more examples of SNe II with peculiar light curves and rapidly evolving spectra have not been found. Was SN 1987A unique in these respects, or have previous searches somehow missed similar SNe II?

A key point to realize here is that a majority of searches for SNe have been done with emulsions sensitive to *blue* ("photographic"), rather than to visual, wavelengths (Zwicky 1964; Trimble 1983; Barbon *et al.* 1984). Since the blue/UV outburst of SN 1987A was so ephemeral, the possibility of discovering many other SNe of this subclass shortly after core collapse is remote. Instead, they are most likely to be noticed several months later, near the apparent (yet secondary) maximum. At this time, SN 1987A had $M_{pg}^0 \approx -15.2$, from the relation ($m_{pg} \approx B - 0.3$) quoted by Branch & Bettis (1978).

If it were located in the Virgo cluster ($m - M = 31.3$), SN 1987A would have been no brighter than $m_{pg} = 16.1$ at maximum, neglecting extinction. A more realistic estimate that includes extinction is $m_{pg} \approx 17$. Most such SNe would probably not have been noticed, for example, by Zwicky during his extensive search with the Palomar 18-inch Schmidt camera (1936–1940; Zwicky 1958, 1965), which had a limiting magnitude of $m_{pg} \approx 16.5 - 17$. Nor would they have been found in other photographic surveys conducted with small telescopes. However, they should easily have appeared on plates obtained with the 48-inch Schmidt telescope and examined by Zwicky and colleagues (1950–1975); the limiting magnitudes of the deepest blue and red plates were ~ 21 and ~ 20, respectively (Abell 1958). Can we, therefore, conclude that SNe like 1987A are uncommon?

This is definitely not the case. Several factors can account for the failure to find these SNe *even if* they were recorded on the 48-inch Schmidt plates of the Virgo cluster and other galaxies at comparable distances ($d \lesssim 20$ Mpc). For example, SNe as faint as 1987A should often be hidden by the overexposed portions of their host galaxies, especially if they occur in, or near, bright H II regions such as 30 Doradus. Moreover, the sheer number of stars at faint magnitudes means that the plates must be searched extremely carefully in order to notice "new" stars having $m_{pg} = 17 - 18$, and to distinguish them from multitudes of tiny plate flaws. Schmitz & Gaskell (1988) clearly demonstrate the abysmal discovery rate of faint SNe. Also, photographs through *red* filters, on which objects like SN 1987A would have looked brighter, were much less frequently procured after completion of the Palomar Sky Survey.

An equally important consideration is the failure to *recognize* subluminous and peculiar SNe on the Schmidt (and other) photographs. Many of the SNe catalogued by Zwicky were actually detected on archival Palomar Sky Survey plates, years after the objects had faded away (Zwicky 1974). Some of these could well have been similar to SN 1987A, but we will never know. Also, few good light curves and spectra were generally obtained in various surveys, even of those SNe noticed shortly after the discovery plates or films were developed. This is clearly evident from the latest catalogue of Barbon *et al.* (1984), which shows that only 32% of all known SNe have been classified. Finally, as pointed out by Schmitz & Gaskell (1988), the light curves of *some* faint SNe II (e.g., SN 1923A; Lampland 1936) did indeed resemble that of SN 1987A, if one assumes that these objects were first seen near the time of their *secondary* maximum. This is likely, since the initial blue outburst of SN 1987A, at least, was exceedingly short lived.

Perhaps it can be argued that if SNe like 1987A were common, they should have been discovered in the Local Group and in other relatively nearby galaxies. After all, the expected magnitude at (secondary) maximum in M31 ($m - M = 24.2$) is $m_{pg} = 9 - 10$, depending on the extinction. The corresponding range in the M81 group ($m - M = 27.3$) is $12 - 13$, while that in the M101 group ($m - M = 29.2$) is $14 - 15$. It is quite possible, though, that no such SNe have occurred in the Local Group during the past 30 – 50 years, since the number of galaxies is so small. Moreover, those few that may have exploded in other nearby groups could easily have been missed, especially if they were superposed on bright H II regions. A good example is the Type Ib SN 1985F, which was actually recorded near maximum brightness ($B = 12.1$) on plates taken in the Soviet Union (Tsvetkov 1986), but was not recognized until after Filippenko & Sargent (1985) discovered it 8 months later during a spectroscopic survey of galactic nuclei. In any case, we certainly do not expect highly subluminous SNe to be *very* common, because the observed rate of normal SNe II and SNe Ib already accounts for the demise of most stars more massive than $\sim 8M_{\odot}$ (e.g., Trimble 1982).

During the next decade, we have a wonderful opportunity to quantify the production rate of SNe like 1987A. Red-sensitive CCDs are now routinely used to monitor many galaxies (e.g., in the Berkeley Automated Supernova Search), and sharp-eyed observers such as the Rev. Robert Evans are contributing valuable visual observations of the nearest galaxies. It should also be relatively straightforward to periodically obtain CCD images of irregular and late-type spiral galaxies in the Virgo Cluster, hundreds of which have been catalogued and illustrated (Sandage & Binggeli 1984; Binggeli *et al.* 1985). I hope to soon start a program of this kind with the 1-m Nickel reflector at Lick Observatory.

4 THE ROLE OF METALLICITY

It has been suggested that the low metallicity of the LMC can account for many of the strange properties of SN 1987A (e.g., Arnett 1987; Hillebrandt *et al.* 1987; McCray *et al.* 1987; see also Shklovskiĭ 1984). The metallicity quoted in these theoretical studies is usually ~ 0.25 solar. I believe, however, that this must be questioned when dealing with SN 1987A. The observed metallicity in the LMC exhibits a wide range ($\sim 0.2 - 1$), depending on the types of objects examined. In particular, relatively young LMC stars such as Cepheid variables are hardly at all metal deficient (Smith 1980; Harris 1983). H II regions, open clusters, and planetary nebulae also do not have large deficiencies (Pagel *et al.* 1978; Becker *et al.* 1984), and B-type supergiants shouldn't be too different.

Thus, the metallicity of the LMC is *not* remarkably low in comparison with that of typical H II regions in spiral galaxies, and it is much higher than in the most metal-poor blue compact galaxies (Kunth & Sargent 1983). A great number of normal SNe II have occurred in H II regions; it would be incredible if *all* of these had markedly higher metallicity than young stars in the LMC. I am currently involved in a spectroscopic study of the H II regions in which SNe II have occurred. Several sites of normal SNe II may show lower abundances than the LMC, and the average metallicities are not unusually high. Definitive conclusions, of course, must await detailed analysis of a larger data set.

My suspicion is that the metallicity of the LMC had at most a peripheral effect on the nature of SN 1987A. It is also unclear which measure of metallicity is most significant. Should we concentrate on true metals (Fe, etc.), N, N/O, or some other quantity? Indeed, at this workshop Woosley has stated that high He abundance, rather than the metals themselves, may be of critical importance. Other potentially significant factors include mass loss (e.g., Chiosi & Maeder 1986; Woosley *et al.* 1987) and the treatment of convection (Schwarzschild versus Ledoux criteria; e.g., Chiosi & Summa 1970; Truran 1988).

5 SPECTRA OF TYPE II SUPERNOVAE

It appears that the spectra of SNe II exhibit a range of properties considerably greater than that of SNe Ia. Space does not permit a full discussion of different objects here, although numerous spectra will soon be published (Filippenko 1988). In many respects, however, the spectra resemble one another. Absorption lines of Ba, Sc, and Sr identified by Williams (1987) in SN 1987A, for example, are also present to a lesser extent in other SNe II (e.g., SN 1986I). Chalabaev & Cristiani (1987) mention that a second emission component could be seen on the red side of the Hα profile in SN 1985P; this may be similar to the small "bumps" displaced by $\sim \pm 4000$ km s^{-1} from Hα and other H lines in SN 1987A (Phillips 1988). Until recently, spectra of SNe II having high resolution and large S/N ratios have not routinely been obtained, but now it is possible to begin making detailed comparisons.

Of the few individual SNe II that have been closely studied, none has shown short-term spectral variations as dramatic as those of SN 1987A. This could simply be a consequence of having preferentially observed bright SNe II, if evolutionary rate turns out to be inversely proportional to luminosity. It seems, though, that the rapid evolution of SN 1987A was largely confined to early phases. The recent spectra (Sept., Oct. 1987) shown by Phillips and Danziger at this workshop are very similar to that of SN 1986I about nine months past core collapse (Fig. 1), and to that of SN 1985P at a comparable phase (Chalabaev & Cristiani 1987). Besides Hα, strong emission from intermediate-mass elements is present. Many of these emission lines are also visible, with different relative strengths, in the spectra of SNe Ib long past maximum (Filippenko & Sargent 1986), but the typical expansion velocities in SNe Ib are much larger than in SNe II.

It will be very interesting to follow the spectral development of SN 1987A between one and five years past outburst, since this is essentially unexplored territory. Also, measurements of the Hα flux should be continued as long as possible to see if radioactive decay of ^{56}Co is the dominant source of energy in the ejecta at very late times (Uomoto & Kirshner 1986). Meanwhile, high-quality spectra of other SNe II must frequently be obtained and compared with those of SN 1987A. Only in this way will we gain a better understanding of the general properties of SNe II, and of SN 1987A as well.

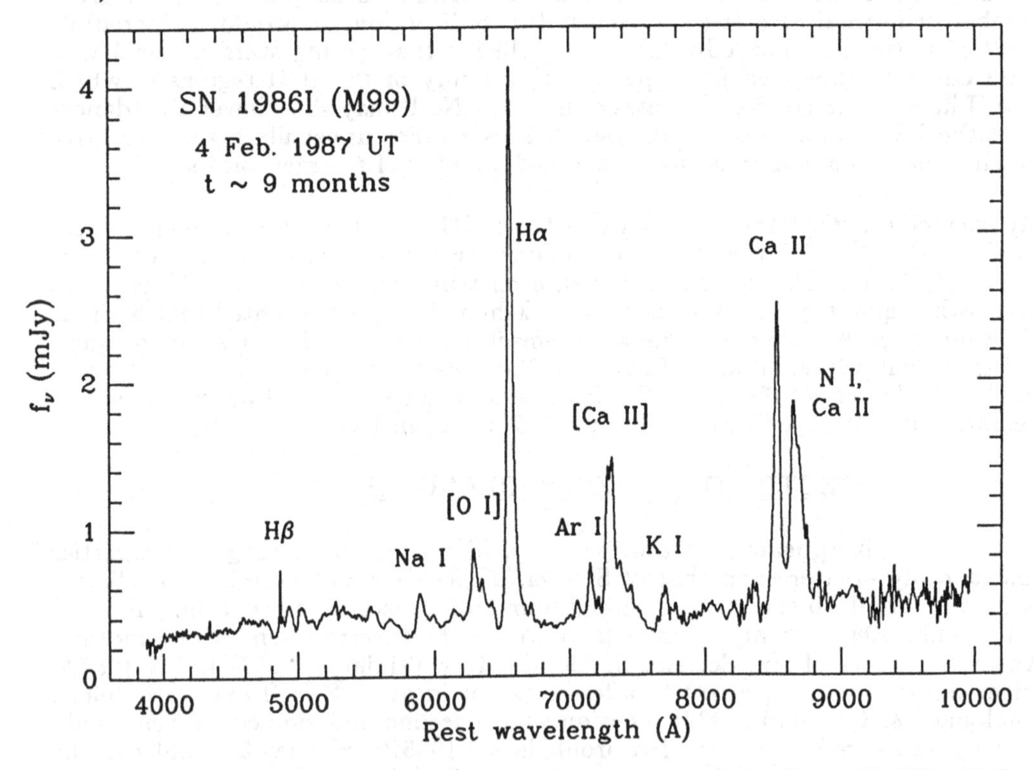

Figure 1: Late-time (~ 9 months) optical spectrum of the Type II SN 1986I. This "supernebular" phase is characterized by extremely strong Hα and Ca II emission superposed on a weak continuum. The unresolved emission spikes at Hα, Hβ, and Hγ are produced by H II regions.

ACKNOWLEDGMENTS

C. M. Gaskell, Ph. Podsiadlowski, A. C. Porter, M. F. Schmitz, and J. C. Shields are thanked for interesting conversations. I obtained the spectrum shown in Figure 1 as a Guest Observer (with W. L. W. Sargent) at Palomar Observatory, which is owned and operated by the California Institute of Technology. This research is partially supported by CalSpace grant CS–27–87.

REFERENCES

Abell, G. O (1958). *Ap. J. Suppl.*, **3**, 211.
Arnett, W. D. (1987). *Ap. J.*, **319**, 136.
Barbon, R., Cappellaro, E., Ciatti, F. & Kowal, C. T. (1984). *Astr. Ap. Suppl.*, **58**, 735.
Barbon, R., Ciatti, F. & Rosino, L. (1979). *Astr. Ap.*, **72**, 287.
Becker, S. A., Mathews, G. J. & Brunish, W. M. (1984). In *IAU Symp. 105, Observational Tests of the Stellar Evolution Theory*, ed. A. Maeder and A. Renzini, p. 83. Dordrecht: Reidel.
Binggeli, B., Sandage, A. & Tammann, G. A. (1985). *Astr. J.*, **90**, 1681.
Branch, D. & Bettis, C. (1978). *Astr. J.*, **83**, 224.
Chalabaev, A. A. & Cristiani, S. (1987). In *ESO Workshop on SN 1987A*, ed. I. J. Danziger, in press.
Chiosi, C. & Maeder, A. (1986). *Ann. Rev. Astr. Ap.*, **24**, 329.
Chiosi, C. & Summa, C. (1970). *Astrophys. Space Sci.*, **8**, 478.
Filippenko, A. V. (1988). In preparation.
Filippenko, A. V. & Sargent, W. L. W. (1985). *Nature*, **316**, 407.
Filippenko, A. V. & Sargent, W. L. W. (1986). *Astr. J.*, **91**, 691.
Hamuy, M., Suntzeff, N. B., Gonzalez, R. & Martin, G. (1987), preprint.
Harris, H. C. (1983). *Astr. J.*, **88**, 507.
Hillebrandt, W., Höflich, P., Truran, J. W. & Weiss, A. (1987). *Nature*, **327**, 597.
Kunth, D. & Sargent, W. L. W. (1983). *Ap. J.*, **273**, 81.
Lampland, C. O. (1936). *Pub. A.S.P.*, **48**, 320.
McCray, R., Shull, J. M. & Sutherland, P. (1987). *Ap. J. Lett.*, **317**, L73.
Pagel, B. E. J., Edmunds, M. G., Fosbury, R. A. E. & Webster, B. L. (1978). *M.N.R.A.S.*, **184**, 569.
Phillips, M. M. (1988). These proceedings.
Rowan-Robinson, M. (1985). *The Cosmological Distance Ladder.* New York: Freeman.
Sandage, A. & Binggeli, B. (1984). *Astr. J.*, **89**, 919.
Schmitz, M. F. & Gaskell, C. M. (1988). These proceedings.
Shklovskiĭ, I. S. (1984). *Pis'ma Astr. Zh.*, **10**, 723. (*Sov. Astr. Lett.*, **10**, 302.)
Smith, H. A. (1980). *Astr. J.*, **85**, 848.
Trimble, V. (1982). *Rev. Mod. Phys.*, **54**, 1183.
Trimble, V. (1983). *Rev. Mod. Phys.*, **55**, 511.
Tsvetkov, D. Yu. (1986). *Pis'ma Astr. Zh.*, **12**, 784. (*Sov. Astr. Lett.*, **12**, 328.)
Truran, J. W. (1988). These proceedings.
Uomoto, A. & Kirshner, R. P. (1986). *Ap. J.*, **308**, 685.
Williams, R. E. (1987). *Ap. J. Lett.*, **320**, L117.
Woosley, S. E., Pinto, P. A., Martin, P. G. & Weaver, T. A. (1987). *Ap. J.*, **318**, 664.
Zwicky, F. (1958). In *Handbuch der Physik*, **51**, ed. S. Flügge, p. 766. Berlin: Springer-Verlag.
Zwicky, F. (1964). *Annales d'Astrophysique*, **27**, 300.
Zwicky, F. (1965). In *Stars and Stellar Systems*, Vol. 8, ed. L. H. Aller and D. B. McLaughlin, p. 367. Chicago: University of Chicago Press.
Zwicky, F. (1974). In *Supernovae and Supernova Remnants*, ed. C. B. Cosmovici, p. 1. Dordrecht: Reidel.

A COMPARISON OF THE SN 1987a LIGHT CURVE WITH OTHER TYPE II SUPERNOVAE, AND THE DETECTABILITY OF SIMILAR SUPERNOVAE.

Mark F. Schmitz and C. Martin Gaskell
Astronomy Department, Dennison Building, University of Michigan, Ann Arbor, MI 48109-1090.

Abstract. We have compared the B and photographic light curves of SN 1987a with a large number of type II light curves. We have also examined the number of reported SNe discoveries as a function of maximum photographic magnitude. We find that the detection probability falls off by about a factor of ten per magnitude below magnitude 14. We suggest that if SN 1987a were to have been in a galaxy outside the local group it would be most likely to have been detected on the broad secondary peak rather than at the initial outburst. We have searched the available light curves and found three SNe which might be similar to SN 1987a but detected during the secondary peak. These three SNe all turn out to be of low luminosity. We discuss the luminosity function for type II SNe and argue that low luminosity SNe like 1987a are actually very common. SN 1987a itself would probably not have been detected at the distance at which most SNe detected in the last 50 years were found.

INTRODUCTION

The photometric behaviour of SN 1987a was different from expectations at the time of discovery. It failed to reach 1st magnitude, the blue and UV light faded extremely rapidly but the visual light curve continued to rise and peaked about 80 days after the neutrino burst. In this paper we will address the question "How different is SN 1987a compared with other type II SNe?".

DETECTABILITY OF SNE WITH LIGHT CURVES LIKE 1987a

We have examined the numbers of SNe detected as functions of apparent B magnitude It is well known (e.g. Zwicky 1965) that the numbers do not rise as fast as the expected factor of 4 per magnitude. We plotted the numbers for both type I and type II SNe and found that the curves have the same shape. Normalising to a 100% detection probability for SNe with B $<$ 12, we obtained the following approximate detection probabilities as a function of B for SNe studied over the past 60 years:

TABLE 1

11.5	12.5	13.5	14.5	15.5	16.5
100%	63%	19%	5%	0.5%	0.05%

SN 1987a had a maximum B magnitude of 4.5 and the LMC has a distance modulus (m - M) of 18.5. If we ignore the probable extinction of A(B) = 0.8 (since presumably other type II SNe are reddened as well!) the absolute B magnitude is -14.0. The average detected type II SN is in a galaxy with cz = 1000 km s^{-1} (about the distance of Virgo) and taking H_o = 50 gives (m - M) = 31.5 and B = 17.5 at that distance. With the detection (and follow-up) techniques used for the past 60 years or so SN 1987a would therefore not have been detected at that distance (unless H_o is very large). Filippenko (1988) gives additional discussion of this question of detectability.

Notice (in Table 1) that the detection probability is falling off very rapidly at these magnitudes (almost a factor of 10 per magnitude). A type II SN like 1987a in a galaxy outside the local group is therefore most likely to be detected when B is at maximum. Since the initial flash a few hours after core collapse decays very rapidly, detection is most likely to occur during the secondary peak 80 days later. We assert therefore that in searching for light curves like 1987a, one should not align them at t = 0 in 1987a, but should fit the secondary peak instead. A search of about 30 available type II SN light curves with more than a few points (see Schmitz and Gaskell 1988 for the light curves) produced the three shown in Fig. 1 as those most closely

Fig. 1. Photographic light curve for SN 1987a and three other similar type II SNe.

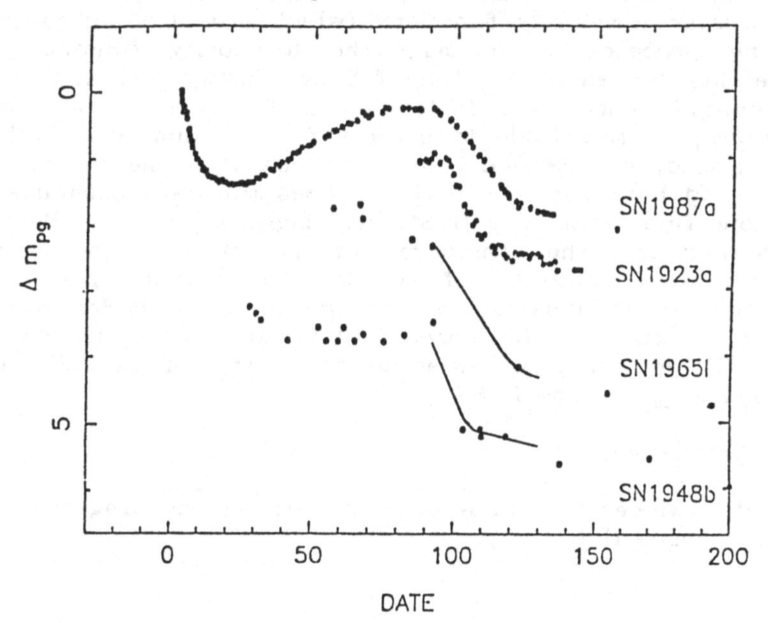

resembling 1987a. The 1987a photographic light curve was generated from the data of Menzies et al (1987), Catchpole et al (1987) and Blanco et al (1987) using the colour transformation of Arp (1961). The data for 1923a are from Lampland (1936). 1948b data are from Mayall & Sill (1949) and 1965l data are from Ciatti & Barbon (1971).

Our selection was done purely on the basis of the shapes of the light curves, but these three SNe turn out to be three of the faintest type II SNe known. For 1923a, 1948b and 1965l the approximate absolute photographic magnitudes are respectively -12.5, -12.9, and -14.7. Note that two of these are fainter than 1987a. Since these Sne all occured in nearby galaxies the absolute magnitudes are independent of the Hubble constant. We would also draw attention to Panagia (1988) who has independently pointed out that if one fits the plateau of type II SNe then the difference in absolute magnitude between SN 1987a and what has previously been regarded as a "normal" type II is greatly reduced.

THE LUMINOSITY FUNCTION OF TYPE II SUPERNOVAE

Luminosity functions for SNe have been given (e.g. by Tammann 1982) but so far they have been calculated under the assumption that there is a preferred absolute magnitude with a relatively small scatter about the mean (as indeed seems to be the case for type Ia SNe). Tammann (1982) calculated his type II luminosity function simply by giving equal weight to each observed SN. Barbon, Ciatti & Rosino (1979) similarly quoted a mean and standard deviation (not including, it should be noted, their least luminous type II SN). Branch and Bettis (1978) added a small Malmquist correction. We believe that a better way of calculating the luminosity function is analagous to that used to find quasar luminosity functions (which are also not narrow Gaussians). One proceeds to estimate the luminosity function by calculating weights for each SN II as follows: firstly one applies a (usually substantial) correction factor as a function of observed maximum photographic magnitude to correct for the number of other identical SNe missed, and secondly one normalises to the volume in which the SN could have been seen. If one does this (see Gaskell and Schmitz 1988) one finds that type II SNe as bright as type Ia SNe are extremely rare and that the luminosity function rises to the lowest luminosities (absolute magnitudes of -14 or -13) at which point the statistics become poor because of the small number of such SNe. There is no evidence of any low luminosity cutoff to the SN luminosity function. Type II SNe with luminosities as low as that of SN 1987a are the most common kind of type II SNe.

Acknowledgement

C.M.G. wishes to acknowledge the strong encouragement of Jim Lattimer to pursue this study.

REFERENCES

Barbon, R., Ciatti, F. & Rosino, L. (1979), Ast. Ap., 72, 287.
Blanco, V. M. et al. (1987) preprint.
Branch, D. & Bettis, C. (1978) A.J., 83, 224.
Catchole, R. M. et al. (1987) preprint.
Ciatti, F. and Barbon, R. (1971) Mem. Soc. Ast. Ital., 42, 145.
Filippenko, A. V. (1988) this conference.
Gaskell, C. M. & Schmitz, M. F. (1988) in preparation.
Lampland, C. O. (1936), P.A.S.P., 48, 320.
Mayall, N. U. and Sill, R. C. (1949) P.A.S.P., 61, 97.
Menzies, J. W. et al. (1987), M.N.R.A.S., 227, 39P.
Panagia, N. (1988) this conference.
Tammann, G. A. (1982) in Supernovae: A Survey of Current Research, ed. M. J. Rees & R. J. Stoneham, p371. Dordrecht: Reidel.
Schmitz, M. F. and Gaskell, C. M. (1988) in preparation.
Zwicky, F. (1965) in Stellar Structure, ed. L. H. Aller & D. B. McLaughlin, Chicago: University of Chicago Press.

P-CYGNI FEATURES AND PHOTOSPHERIC VELOCITIES

Lars Bildsten
Center for Radiophysics and Space Research, Cornell University
J. C. L. Wang
Dept. of Astronomy and Astrophysics and the Enrico Fermi Institute
University of Chicago

ABSTRACT

The photospheric velocities implied from the P-Cygni features of $H\alpha$, $H\beta$, and $H\gamma$ in SN 1987A *differ*. This implies that the "photosphere" relevant to P-Cygni formation may be frequency dependent. To obtain a P-Cygni feature from above a fixed surface, the photons leaving there must not undergo other frequency dependent absorption processes (bound-free or free-free) or be smeared in frequency due to multiple Thomson scattering. We delineate different regimes when the frequency dependence of this surface should be manifest, and those when it would not be. We introduce a natural way to define the surfaces relevant to P-Cygni line formation.

I. INTRODUCTION

The optical spectrum of SN 1987a showed strong P-Cygni absorption features of Balmer Hα, Hβ, and Hγ (Danziger *et al.*, 1987, Menzies *et al.* 1987, and Blanco *et al.* 1987). The photospheric velocities implied from the P-Cygni features of Danziger *et al.* (1987) gave $v_{H\alpha} > v_{H\beta} > v_{H\gamma}$ at all times. These velocities were determined from the average velocity between the emission peak and the absorption trough. This implies a frequency dependent surface, above which line formation takes place. We give a prescription for defining the surface relevant to P-Cygni line formation.

II. DEFINITION OF P-CYGNI LINE FORMATION SURFACES

P-Cygni line formation arises from resonant bound-bound scattering of line photons in a thin shell above a photosphere. The simple calculation of the problem in the Sobolev approximation depends on the photon travelling freely from the emitting surface and being resonantly scattered in this thin shell. If frequency dependent continuum absorption processes are important between this emitting surface and the observer, then it will be necessary to define different surfaces for different lines. In addition, in a differentially expanding envelope, a photon suffering multiple Thomson scatters will be Doppler shifted. Hence, we must choose a surface above which both of these processes are unimportant.

A photon of frequency ν_0 emitted by a shell moving at a velocity v relative to an observer will be measured by this observer at a blueshifted frequency $\nu = \nu_o(1 + v/c)$. If the Thomson optical depth between this shell at v and another at $v + \Delta v$ is τ_{es}, then a simple one-dimensional model gives an accumulated redshift of $-\frac{2\Delta v}{c}\tau_{es}$. This arises from the Doppler shifting in Thomson scattering during the diffusion across a differentially expanding envelope. Hence the photon's frequency in the observer's frame would be $\nu = \nu_o(1 + \frac{v}{c} - \frac{2\Delta v}{c}\tau_{es})$. Therefore, for the frequency smearing to be small, $\tau_{es} \ll \frac{v}{\Delta v}$. If $\frac{v}{\Delta v} \lesssim \tau_{es} \lesssim \frac{c}{\Delta v}$, photons are able to escape from the shells by diffusion but are strongly redshifted by multiple Thomson scatters. If $\tau_{es} \gtrsim \frac{c}{\Delta v}$, the diffusion time for photons across the layer of thickness Δv is longer than the expansion time, so photons are tied to the matter and are redshifted with the expansion. For an envelope with a density profile $\rho \propto r^{-N}$, a characteristic value for Δv is that thickness which gives rise to a factor of two change in the column density. Using this criterion, $\frac{v}{\Delta v} \sim N - 1$.

A second constraint on the surface comes from not allowing an absorption due to other continuum processes (bound-free or free-free). Free-free absorption is only important at early times, when the electron density is high. If the optical depth to the bound-free transition from a surface is greater than 1, then this absorption process would disallow P-Cygni line formation. Since bound-free absorption is frequency dependent, the surface at which $\tau_{bf} = 1$ will vary with frequency. Thus, at the relevant surface, $\tau_{bf} \leq 1$.

The conditions for the surface are then: 1) that a photon will not have its inherent blueshift smeared out and 2) no other continuum absorption processes will occur. For intermediate regimes, where $\tau_{es} \sim 1$ and $\tau_{bf} \sim 1$ but $\tau_{es} > \tau_{bf}$, we must use the effective optical depth $\tau_{eff} = (\tau_{es}\tau_{bf})^{1/2}$ since the photon's random walk path length to escape from the surface is longer than the free streaming length. To determine the relevant surface, start at the outer edge and integrate inwards, accumulating the two optical depths, τ_{es} and τ_{bf}. If τ_{bf} reaches 1 prior to τ_{es}, then the appropriate surface is at $\tau_{bf} = 1$ and these surfaces are frequency dependent (case 1). If τ_{es} grows much faster than τ_{bf}, then the surface should be placed at the spot where negligible smearing would occur, e.g., at $\tau_{es} \approx \sqrt{\frac{v}{\Delta v}} \approx \sqrt{N-1}$. These surfaces are frequency independent (case 3). In the intermediate case, $\tau_{es} > \tau_{bf}$ and τ_{es} is only slightly greater than one at the $\tau_{bf} \approx 1$ surface. The elongation of the path length due to the random walk must then be accounted for. Thus, the relevant surface lies at $\tau_{eff} \sim 1$ and a slight frequency dependence may exist (case 2).
The three cases are denoted by the ratio τ_{bf}/τ_{es}.

Case 1 :
$$\frac{\tau_{bf}}{\tau_{es}} \geq 1,$$

$\tau_{bf} = 1$ determines the surface, frequency dependence manifest.

Case 2 :
$$\left(\frac{\Delta v}{v}\right)^2 \leq \frac{\tau_{bf}}{\tau_{es}} < 1,$$

$\tau_{eff} = 1$ determines the surface, mild frequency dependence possible.

Case 3 : $$\frac{\tau_{bf}}{\tau_{es}} < \left(\frac{\Delta v}{v}\right)^2$$

τ_{es} determines the surface, no frequency dependence expected.

The blueshift of the absorption maximum in the P-Cygni feature is determined by both the bound-bound opacity and the position of the photosphere (Branch 1980). Although our prescription defines an appropriate photosphere above which a line can form, the velocity of this photosphere is *not* necessarily the velocity of the absorption maximum. Furthermore, while observationally the velocities of the absorption maxima within the Balmer series differ, this *does not* imply that the photosphere is frequency dependent. It is possible to have a frequency independent photosphere and have the velocity differences be caused by non-resonant scattering or differences in the bound-bound optical depths. One should therefore be very careful in relating calculated "photospheric" velocities to the velocities of the observed P-Cygni absorption maxima.

III. CONCLUSIONS

We have outlined the optical depth regimes and provided a prescription for defining the surfaces relevant to P-Cygni line formation. Without a detailed atmospheric model, it is not possible *a priori* to equate the velocities of these surfaces with the velocities of the observed P-Cygni absorption maxima.

We would like to thank Profs. D. Chernoff, E. Salpeter, and I. Wasserman for much advice and encouragement. L. B. acknowledges NSF grant AST 84-15162, while J.C.L.W. acknowledges grant NAGW-830 and a Robert McCormick Fellowship.

REFERENCES

Blanco, V. M. *et al.* 1987, *Ap. J.*, **320**, 589.

Branch, D. 1980 in *Supernovae Spectra*, ed. R. Meyerott & G. H. Gillespie (New York: AIP) p. 39.

Danziger, I. J. *et al.* 1987, *Astr & Astrop.*, **177**, L13.

Menzies, J. W. *et al.* 1987, *Mon. Not. R. Astr. Soc.*, **227**, 39p.

THE NEUTRINO BURST FROM SN 1987a DETECTED IN THE MONT BLANC LSD EXPERIMENT

M.Aglietta[a] G.Badino[a] G.Bologna[a] C.Castagnoli[a] A.Castellina[a] V.L.Dadykin[b] W.Fulgione[a] P.Galeotti[a] F.F.Kalchukov[b] V.B.Kortchaguin[b] P.V.Kortchaguin[b] A.S.Malguin[b] V.G.Ryassny[b] O.G.Ryazhkaya[b] O.Saavedra[a] V.P.Talochkin[b] G.Trinchero[a] S.Vernetto[a] G.T.Zatsepin[b] V.F.Yakushev[b]

a) Istituto di Cosmogeofisica del CNR, Torino, Italy, and Istituto di Fisica Generale dell'Università di Torino, Italy

b) Institute of Nuclear Research, Academy of Sciences of USSR, Moscow, USSR

(Presented by P.Galeotti)

ABSTRACT.

In this paper we discuss the event, (5 interactions recorded during 7 seconds) detected in the Mont Blanc Underground Neutrino Observatory (UNO) on February 23, 1987, during the occurrence of supernova SN 1987A. The pulse amplitudes, the background imitation probability, and the energetics connected with the event are reported. It is also shown that some interactions recorded at the same time in other underground experiments, with a lower detection efficiency, are consistent with the Mont Blanc event.

1. INTRODUCTION

An Underground Neutrino Observatory (UNO) has been built[(1)] by our two Insitutes with the main aim to search for bursts of low energy neutrinos from stellar collapses. The UNO is a Liquid Scintillation Detector (LSD), running[(2)] since October 1984 in the Mont Blanc Laboratory, at a depth of 5200 hg/cm^2 of standard rock underground. The very large coverage of rock, and an additional shield, provide very good background reduction, and allow us to operate the UNO at a very low energy threshold.

An event, considered as a candidate of a neutrino burst, was detected[(3)] on real time during its occurence, and identified before optical observations from the computer print-out on February 23.12

($2^h 52^m 36^s$ UT), 1987. On February 24.23 Shelton in Las Campanas Observatory (Chile) reported observation of an optical supernova (SN 1987A) in the Large Magellanic Cloud, ~ 50 kpc faraway. Optical data indicate that no star brighter than magnitude 12 was present at February 23.08, and that the supernova had a magnitude 6.1 at February 23.44. Preliminary results of our detection have been published[4-6] previously; here we report the energy values of the 5 pulses in the burst, and discuss the background imitation probability of the event. We compare also our data with the results obtained in other underground experiments, and discuss the connected neutrino outflow from the stellar collapse which originated supernova SN 1987A. It is shown that the Mont Blanc event, besides being self consistent, it is also additionally supported by all the other experimental evidence available.

2. THE MONT BLANC UNDERGROUND NEUTRINO OBSERVATORY

The Mont Blanc neutrino telescope, located at the depth of 5,200 hg/cm^2 of standard rock underground in a cavity along the road tunnel linking Italy and France, is a Liquid Scintillation Detector (LSD) consisting of 72 counters (1.5 m^3 each) in 3 layers, arranged in a parallelepiped shape (see fig.1) with 6x7 m^2 area and 4.5 m height.

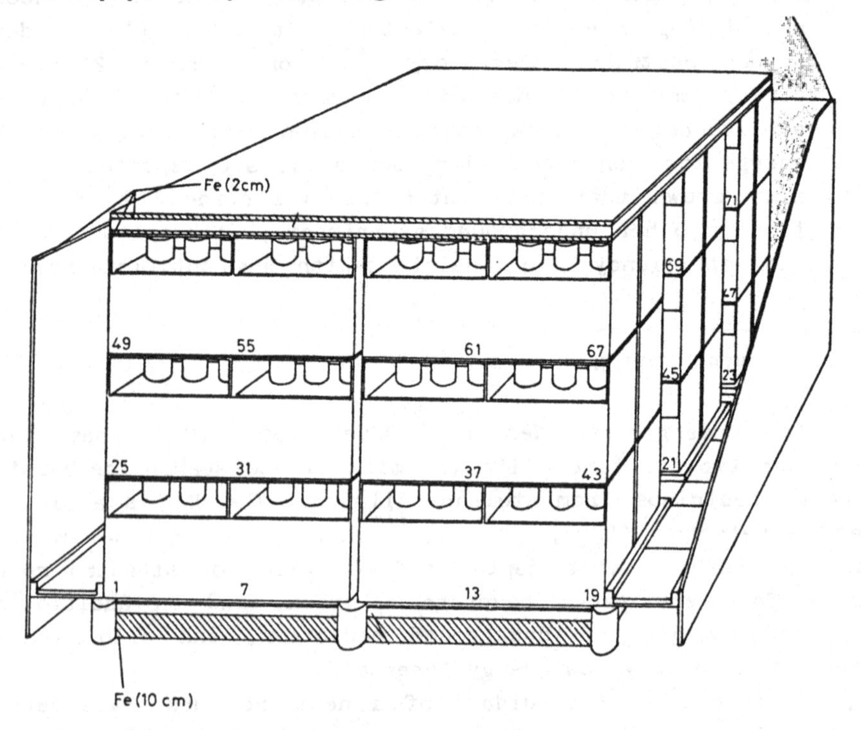

Fig. 1. – The 90 ton liquid scintillation detector in the Mont Blanc Laboratory

The total active mass is 90 tons of the liquid scintillator C_nH_{2n+2} (with $\bar{n}$ = 10), containing 8.4 10^{30} free protons. The low energy local radioactivity background from the surrounding rock has been reduced by shielding each scintillation counter and the whole detector with more than 200 tons of Fe slabs.

In underground experiments, the main source of background is due to cosmic ray muons and their interactions in the rock surrounding the detector, which may induce contained pulses from secondary neutrons or gamma rays. This source of background is very low at the large depth underground of the Mont Blanc Laboratory, where about 3.5 muons per hour are recorded on the average in the whole LSD detector.

The liquid scintillator is watched from the top of each counter by 3 photomultipliers, in a 3-fold coincidence within 150 ns. Our calibrations[1], both from muons and with a ^{252}Cf source, show that a 1 MeV energy loss yields on the average 15 photoelectrons in 1 scintillation counter.

The low background of the LSD experiment and the high energy resolution of the scintillation counters allow us to operate the UNO at very low energy thresholds: fig. 2 shows the LSD detection efficiency as a

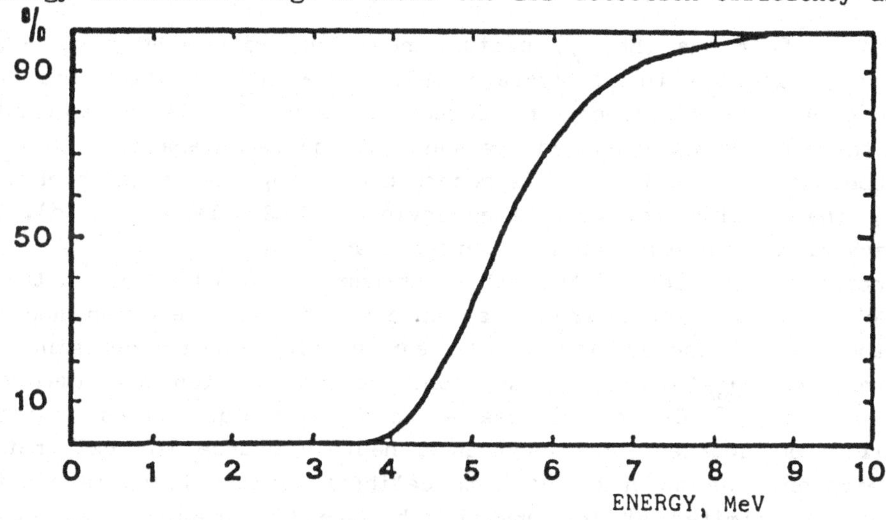

Fig. 2 - The LSD detection efficiency

function of the visible energy of the detected pulses. The electronic system consists of 2 levels of discriminators for each scintillation counter. A high-level (HEL) discriminator for pulses above the energy threshold ~ 6-7 MeV for the 56 surface counters, and ~ 5 MeV for the 16 internal ones, with a total trigger rate of 0.012 Hz. A low-level (LEL) discriminator for pulses above the energy threshold 0.8 MeV is active only during a 500 μs wide gate, opened for all the 72 counters by the main high-level trigger. Two ADCs per counter measure the

energy deposition in the scintillator in 2 overlapping energy ranges. A TDC gives the relative time of the interactions with a resolution of 100 ns. Three memory buffers, 16 words deep, for the 2 ADCs and the TDC of each scintillation counter, allow us to record all pulses without dead time. On-line software prints any burst of pulses satisfying our operational definition of a neutrino burst, namely a burst of pulses above a given multiplicity in a given time.
This recording system allows us to detect both products of $\bar{\nu}_e$ interactions with the free protons of the scintillator (namely, positrons and gammas in a delayed coincidence within 500 μs), through the capture reaction:

$$\bar{\nu}_e + p \rightarrow n + e^+$$
$$\llcorner\, n + p \rightarrow d + \gamma$$

which gives the main signal in detecting neutrinos from collapsing stars. In addition, also electrons from elastic scattering of neutrinos of other species with the electrons of the scintillator can be detected in the LSD. For positron detection, the pulse amplitude is given by:

$$E_e = T_e + m_e c^2 \simeq E_\nu - 1.3 \text{ MeV}$$

The gammas from the (np,dγ) capture reaction, with energy E_γ = 2.2 MeV, are emitted with an average delay of ~ 200 μs after the main interaction; the efficiency to detect a gamma, in the same counter where the neutron was produced, is about 40% on the average.
The absolute time in the LSD is recorded by using the signal broadcasted by the Italian Standard Time Service (IEN Galileo Ferraris). The accuracy of the absolute time is better than 2 msec.
The energy calibration of the LSD experiment is made by using both the 2.2 MeV peak of gammas from the (np,d γ) capture reaction and the distribution of energy losses of near vertical muons crossing the detector, peaked at ~ 150 MeV. Neutrons from the spontaneous fission of a low activity ^{252}Cf source, placed in different positions inside a scintillation counter, are used as a neutron source for calibrating the detector. The accuracy of both calibration techniques is about $\pm$ 10%, and is limited at low energies by the ADCs resolution, and at high energies by the low muon statistics (on the average, 3.5 muons per hour cross the LSD experiment). The photomultiplier amplification and their stability are checked up about once per month, by using a ^{60}Co source placed onto the scintillation counter.

3. THE NEUTRINO BURST DETECTED IN THE LSD EXPERIMENT ON FEB. 23.12

Since January 1, 1986, the LSD experiment has been running

with an average efficiency of 90% (and almost 99% since October 1986). Recently, the detector shielding has been partly increased for test purposes, with paraphin and lead in order to further decrease the low energy background from the surrounding rock. Trigger pulses are analysed in order to have a long term statistics and to search for bursts. The experimental time distributions of these pulses, grouped in bursts above a given multiplicity, are plotted in fig. 3 as a function of their duration and with a bin width of 10 sec. The

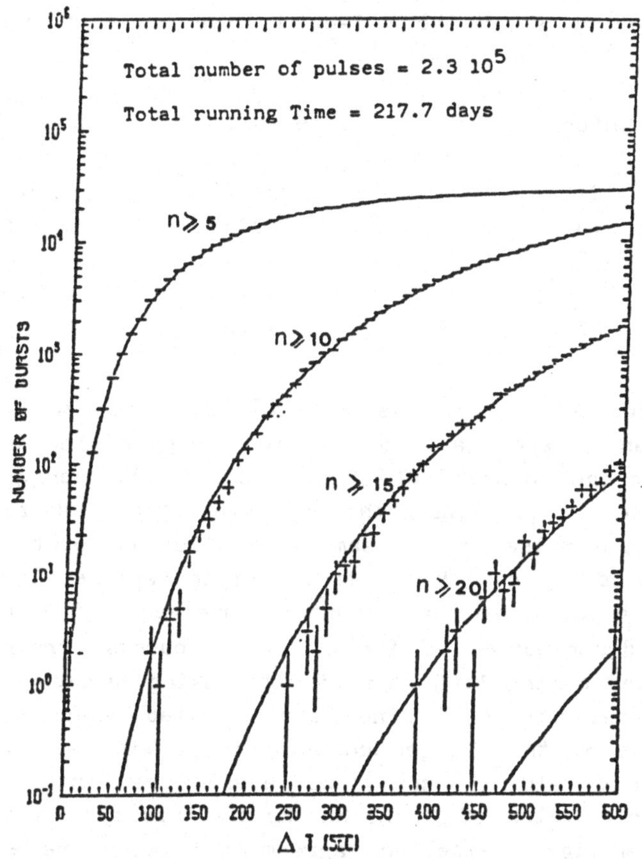

Fig.3.- Experimental distribution of bursts of pulses with multiplicity above 5,10,15 and 20 as a function of their duration (sec), with a binning of 10 sec, measured during 217.7 days.

distributions of fig. 3 refer to a data-taking period of 217.7 days (from September 28, 1986, to May 23, 1987). The smooth behaviour as well as the agreement with the predicted Poisson distributions (represented by the continuous curves) show that the trigger counting rate is stable and the detector is properly working during this time.

For monitoring purposes the detector counting rate is also checked on-line every 100 triggers. By analysing these data, and from the tape analysis several days before and after the event discussed here, we have been able to verify that the apparatus was running properly throughout the entire period.

On February 23.12, 1987, ($2^h52^m36^s$ UT), an event, consisting of a burst of 5 pulses and printed on the computer output in real time at the occurence, was recorded in 5 different counters (3 of them internal) during 7 seconds. Table I gives the event number,

Table I - Characteristics of the pulses in the burst detected on February 23rd, 1987

Event no.	Counter no.	Time (UT)	E_{vis} (MeV)
994	31	$2^h52^m36^s.79$	6.2
995	14	40.65	5.8
996	25	41.01	7.8
997	35	42.70	7.0
998	33	43.80	6.8

the counter number, the absolute universal time (with an accuracy better than 2 msec), and the visible energy of the detected pulses (with an average accuracy of about 15%). A low energy pulse, with energy E = 1 MeV, accompanying the 3rd interaction, was recorded 278 μs after the main pulse in the same scintillation counter. From the measured efficiency to detect γ's from neutron capture, we expect on the average 2 such pulses in the 5 counters involved in the burst.

Fig. 4 shows the distribution of the number of bursts versus their multiplicity recorded during 1.96 days of data-taking encompassing the event discussed here. The full lines are computed according to a Poisson distribution of the trigger counting rate, with a binning of 10 seconds, and mean value given by our average raw trigger rate, which has the stable value of 0.012 Hz during this run as for the previous ones. From fig. 4 excellent agreement between the expected and measured distributions is found, except for the point at t = 7 sec, which has been added just to show the event considered here. The imitation rate of this event, evaluated in the on-line analysis from the raw trigger rate, is 0.7 per year, or $6.5\ 10^{-4}$ in the time interval corresponding to the uncertainty of the instant of collapse, i.e. the 8.2 hours interval from February 23.08 to 23.44, as suggested by the optical observations.

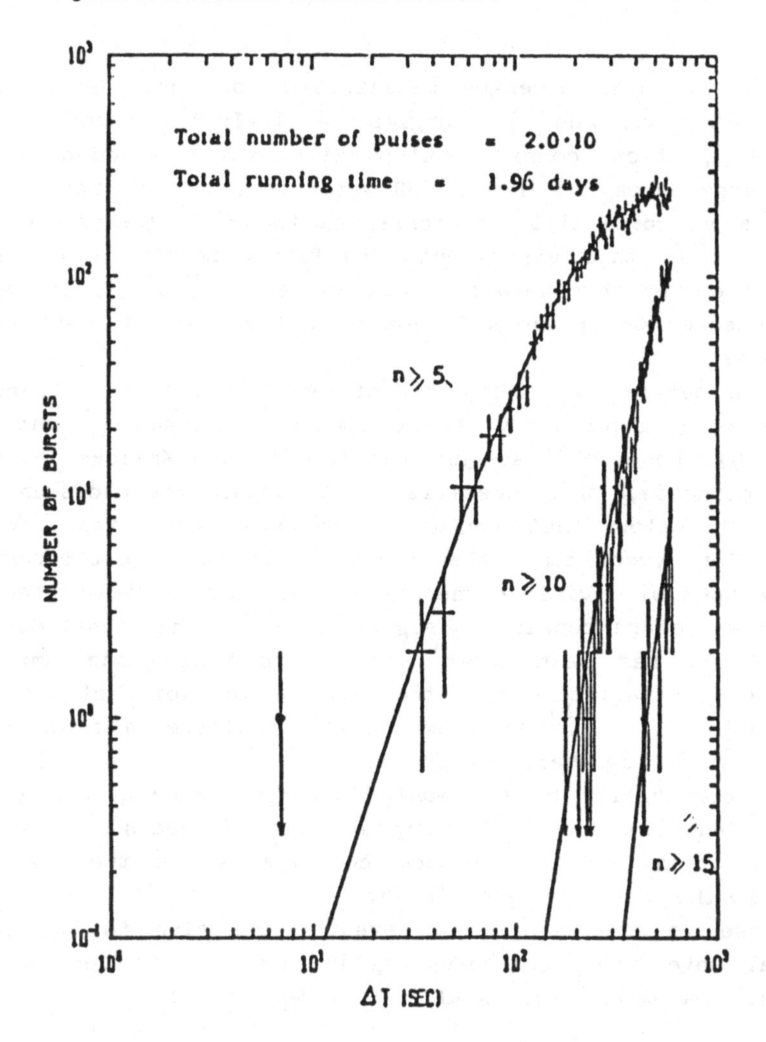

Fig. 4.- The same as for fig. 3, but for the run (of duration 1.96 days) including the event of Feb.23,1987, shown by a large dot.

4. COMPARISON WITH OTHER EXPERIMENTS

Neutrino detection from supernova SN 1987 A was reported also by the Kamiokande II[7], IMB[8], and Baksan[9] detectors. The main signal of neutrinos from collapsing stars is due to positrons produced through the ($\bar{\nu}_e$,p) capture reactions; in water Cerenkov detectors the visible energy of the recorded pulses is given by the e^+ kinetic energy only, while in scintillation detectors also the 1 MeV from e^+e^- annihilation contributes to the visible energy. Hence, the signal recorded in the Mont Blanc experiment in the range (5.8-7.8) MeV of measured energies corresponds to visible energies in the range

(4.8-6.8) MeV in water Cerenkov detectors if one considers only e^+ annihilation at rest, and at energies ~ (10-15)% smaller, i.e. of order (4-6) MeV, if one considers also the e^+ annihilation on flight. Since the energy threshold in the IMB water Cerenkov detector is ~ 20 MeV, there is no possibility to correlate the Mont Blanc event with pulses recorded in this experiment. The Baksan Scintillation Telescope, with an energy threshold of about 10 MeV, and a trigger rate of 0.033 Hz, detected one pulse with energy 10.8 MeV at February 23, UT $2^h 52^m 34^s \pm 2$ sec.

The comparison between the LSD and Kamiokande II data is hampered by the high degree of uncertainty in the Kamiokande linkage to universal time ($\pm$ 1 min). However, assuming that the IMB and Kamiokande events at UT $7^h 35^m$ coincide, it is possible[10] to obtain the necessary time correlation. It follows that during a 10 seconds time interval closed to the Mont Blanc event time, the Kamiokande II experiment recorded 4 pulses, 2 of them with visible energies ~ 12 MeV and ~ 8 MeV respectively, energies which are considered significative in the signal detected at $7^h 35^m$ UT. It has been shown[6] that the Mont Blanc-Kamiokande events are not contradictory from the experimental point of view, and the two signals fit[10,11] the same stellar collapse emitting a high luminosity burst of low energy neutrinos.

The combined probability of the Mont Blanc-Kamiokande events gives a random coincidence rate of one every at least 77 years[12], or to a rate of less than $1.2\ 10^{-5}$ random coincidences in the 8.2 hours uncertainty in the time of the collapse.

Finally, close to the Mont Blanc event time, also the 2 running gravitational wave antennas, operating in Rome and Maryland at room temperatures, detected[13] a signal not due to seismic noise.

5. DISCUSSION

The energy spectrum of the neutrinos emitted from a collapsing stellar core can be approximated[14] by a distribution similar to a Fermi Dirac one, namely:

$$\Phi\ (\bar{\nu}_e/\text{sec MeV}) \propto \frac{\varepsilon^2\ e^{-\alpha\varepsilon^2}}{1+e^{\varepsilon}} \qquad (\varepsilon = E_\nu/KT)$$

where E_ν is the $\bar{\nu}_e$ energy (in MeV), and T the temperature of the neutrinosphere. The correction factor $\exp(-\alpha\varepsilon^2)$ takes into account neutrino absorption in the stellar envelope above the neutrinosphere.

The central temperature KT of a type II presupernova is supposed to be less than 1 MeV, and the temperature of the neutrinosphere, after neutrino trapping, is of the order of a few MeV. Depending on the values of the absorption parameter, the energy spectrum of the neutrinos emitted from the collapsing stellar core can be different from the spectrum produced during the collapse. Because of the energy dependence of the neutrino cross section, high energy neutrinos may be more absorbed than the low energy ones, and bigger the mass of the star envelope stronger is the shift of the spectrum towards low energies. The neutrino interactions recorded in the LSD experiment agree with a temperature of the neutrinosphere of ~ 2 MeV and absorption parameter $\alpha \sim 0.1$.

The total energy involved in the burst can be estimated assuming that the 5 pulses recorded in the LSD are due to $\bar{\nu}_e$ capture processes. This assumption seems natural as the $\bar{\nu}_e$ capture cross section is about 2 orders of magnitude higher than the scattering cross section of ν_e with the electrons of the scintillator. Adopting a distance of 52 kpc for SN 1987A, and using the cross section:

$$\sigma(E_\nu) = 9.45\ 10^{-44}(E_\nu - \Delta M)\left[(E_\nu - \Delta M)^2 - m_e c^2\right]^{1/2} \quad cm^2$$

where ΔM is the neutron to proton mass difference, and m_e is the electron rest mass, the total energy recorded in the LSD experiment is $6 \cdot 10^{53}$ erg. The total energy outflow in $\bar{\nu}_e$, integrated over the energy spectrum (3), is:

$$1.2\ 10^{54} \leq E_\nu \leq 6.3\ 10^{54} \quad erg \qquad \text{(at 90\% c.l.)}$$

The limited statistics of the number of pulses in the burst may explain why this value is slightly higher than the predicted theoretical expectation.

The expected number of $\bar{\nu}_e$ interactions in the other underground experiments can be estimated by using their experimental detection efficiency and the energy spectrum (3). Table II gives the expected and recorded number of pulses in the experiments which reported neutrino detection from SN 1987 A. In spite of the larger number of target protons in comparison with the LSD, the number of recorded interactions is rather small because of the lower detection efficiency of these experiments at low energies. Hence, from the experimental point of view, there is no contradiction between the event detected in the Mont Blanc Neutrino Observatory and in the other detectors at the same time.

Table II - Expected and detected $\tilde{\nu}_e$-interactions in other experiments.

Experiment	Expected	Detected
BAKSAN	1.1 $\pm$ 5	1
KAMIOKANDE	$\leqslant$ 10 $\pm$ 8	2
IMB	0	0

6. CONCLUSIONS

The neutrino burst recorded on real time in the Mont Blanc Underground Neutrino Observatory, during the occurence of supernova SN 1987 A, is self consistent and indicates that the collapse had a duration of several seconds, during which a high luminosity burst of neutrinos was emitted by a low temperature neutrinosphere. Even if at large distance, this stellar collapse produced a significant signal in the LSD experiment because of its very low energy threshold and high efficiency to detect low energy pulses. A second burst, detected 4.7 hours after the Mont Blanc one, is not contradictory from the experimental point of view, and indicate that the stellar collapse in the Large Magellanic Cloud developed on at least two stages.
This conclusion is strongly supported by the luminosity and spectral evolution of the supernova at the very early stages, which agrees[15] with the start time of the collapse as given by the Mont Blanc event time. Neutrinos detected at 7^h35^m UT may have been emitted in a delayed pulse from the neutron star already existing. Also the presence of a very bright companion star[16,17], caused by this supernova explosion, is hard to be explained within the framework of the standard theoretical models, based on spherical simmetric collapse, without angular momentum and magnetic field. The two bangs scenario[10,18,19] seems to fit better the experimental evidences available; this scenario can be a simple, natural explanation of the two neutrino pulses, which, because of the different energy spectra, have induced different information in detectors of different type.

REFERENCES

1. G.Badino et al., Nuovo Cim., 7C,573,1984
2. M.Aglietta et al., Nuovo Cim., 9C,185,1986
3. M.Aglietta et al., IAU Circ.no.4323 (Feb.28, 1987)
4. M.Aglietta et al., Europhys.Lett.3,1315,1987
5. V.L.Dadykin et al., JETP Lett., (in russian), 45,464,1987
6. M.Aglietta et al., Europhys.Lett., 3,1321,1987
7. K. Hirata et al., Phys.Rev.Lett., 58,1490,1987
8. R.M.Bionta et al., Phys.Rev.Lett., 58, 1494, 1987

9. L.N.Alexeyeva et al., JETP Lett. (in russian), 45, 461, 1987
10. A. De Rujula, Phys. Lett. B, 193, 514, 1987
11. D.N.Schramm, Comm. on Nuclear and Particle Physics, 1987, in press
12. A. De Rujula, CERN TH-4839, 1987
13. G.Pizzella, Proc. 4th G.Mason Workshop, 1987, in press
14. D.K.Nadjozhin, I.V.Ostroschenko, Sov. Astron., 24, 47, 1980
15. E.J.Wampler et al., Astr. and Astrophys. Lett., 1987, 182, L51
16. C.Papaliolios et al., Proc. 4th G.Mason Workshop,1987,in press
17. W.P.S.Meikle et al., Nature, 1987, in press
18. W.Hillebrandt et al., Astron. Astrophys. Lett., 1987, 180, L20
19. L.Stella and A.Treves, Astron. Astrophys. Lett., 1987 in press

TOWARD OBSERVATIONAL NEUTRINO ASTROPHYSICS

M. Koshiba
CERN/(University of Tokyo)

First of all, the credits to my collaborators. The collaborators of this experiment are; K. Hirata, T. Kajita, M. Koshiba, M. Nakahata, Y. Ohyama, N. Sato, A Suzuki, M. Takita, and Y. Totsuka, (Dept. of Physics and ICEPP, Univ. of Tokyo): T. Kifune and T. Suda, (Inst. Cosmic Ray Research, Univ. of Tokyo): K. Takahashi and T. Tanimori, (Nat. Lab. High Energy Physics (KEK): K. Miyano and M. Yamada, (Dept. of Physics, Univ. of Niigata): E.W. Beier, L.R. Feldsher, S.B. Kim, A.K. Mann, F.M. Newcomer, R. Van Berg, and W. Zhang, (Dept. of Physics, University of Pennsylvania): B.G. Cortez, (AT&T Bell Laboratories, Holmdel).

Some of you may say: "Why toward? It's already here to work on."

It is true that (1) the first observation of the neutrino burst from the supernova SN1987a by KAMIOKANDE-II which was immediately confirmed by IMB and (2) the first real-time, directional, and spectral observation of solar ^{8}B neutrinos also by KAMIOKANDE-II could perhaps be considered as signaling the birth of the observational ν astrophysics. The field, however, is still in its infancy and is crying out for tender loving care.

Namely; while the construction of astronomy requires the time and the direction of the signal and that of astrophysics in addition the spectral information, the observations of (1) could not give the directional information and the results of both (1) and (2) are still suffering from the meager statistics.

How to remedy this situation to let this new born science of observational neutrino astrophysics grow healthy? This is what I should like to address in this talk.

We have three families of elementary particles. The first family consists of 15 particles; u-quark and d-quark each in 3 colors and in 2 helicites, left-handed and right-handed, electron in 2 helicities, and left-handed electron-neutrino. If the neutrino had a non zero mass one has to add also right-handed neutrino to make the total number of particles in the family equal to 16. The second and the third families consist of (c-quark, s-quark, muon, and mu-neutrino) and (t-quark still to be discovered, b-quark, tau, and tau-neutrino), respectively. In weak interactions the left-handed two quarks in a family form a doublet, (u,d), (c,s), and (t,b), and so do the left-handed lepton and the corresponding neutrino, (e,e-neutrino), (mu, mu-neutrino), and (tau,tau-neutrino). The so-called Standard Theory based on SU(2), doublet behavior, for weak interaction and U(1) for electro magnetic interaction received a dramatic confirmation by the discovery of Z^0 and W^{+-} and the unification of weak- and electro-magnetic-interactions was accomplished in the form of a local gauge field theory of SU(2)xU(1). The strong interaction is known to be described by another local gauge field theory based on SU(3), triplet behavior of tri-colored quarks, called Quantumchrodynamics. It is then a natural consequence to consider a new local gauge theory based on a larger symmetry containing SU(3)xSU(2)xU(1) describing strong-, weak-, and electromagnetic-interactions, Grand Unified Theories; GUTS. Such theories in general predict the baryon-number violating process of proton decay and the existence of magnetic monopole. Some of the GUTS predict also another baryon-number-violating process of neutron-antineutron oscillation and the existence of heavy neutrino with the possibility of the lepton-number-violating neutrino oscillation. These baryon-number-violating processes and those of lepton-number-violation play an essential role in explaining the matter-antimatter disparity of the Universe. The extrapolations to high energy of the three interaction strengths, strong, electro-magnetic, and weak, seem to converge to a same value at around the energy scale of 10^{14}GeV. This development of theory is in line with the prevailing trend of particle theory which reduces the dynamics to the geometry of space-time; Einstein's general relativity may be considered as precurser. The Superstring theory may be the ultimatum of this line of approach, though it does not give any specific results yet which can be tested by experiment.

The two experiments, IMB in USA and KAMIOKANDE in Japan, were both conceived at the end of 1978 to search for proton decay. The main motivation came from the grand unified theories published in mid 1970's, one of which based on SU(5) predicted the life time of proton in the range of some 10^{29} years. Among the several nucleon decay experiments so far performed, these two are unique in the sense that they both use the Imaging Water Cherenkov method. Both the experiments have been running for several years -- except for some months spent in upgrading -- and IMB, with its larger mass of 7000 tonnes, contributed most towards killing the original (SU(5) theory of placing the lower limit in some 10^{32} years for the partial lifetime of the decay (proton) to positron plus pi-zero), whilst KAMIOKANDE of 3000 tonnes, but with a better resolution, contributed most towards pushing up the lower limit to 10^{31} years range for the partial lifetime of the decay (proton to anti-neutrino plus K^{+}), among others, thereby making the Super Symmetric mass scale larger than originally thought.

The experiment KAMIOKANDE as shown schematically in Fig. 1 is a detector of 3000 tonnes water surrounded by a set of photomultipliers. Any electrically charged particle of velocity in excess of 3/4c produces Cherenkov light in water and the observation of this light pattern gives the vertex, the direction of motion, and the energy of the particle. If we can collect a sufficient number of photons we can identify the particle in most cases and this reduces the background due to the atmospheric neutrinos. The aim of the experiment was not only to search for the possible existence of proton decays but also to determine the branching ratios into all the possible decay modes, if the protons decay at all, because these branching ratios will tell us the type of the underlying symmetry group of the grant unification. In order to facilitate this within the rather limited funding available a little more than a year was spent for developing really large photomultipliers of 50cm diameter. This was done in collaboration with a Japanese firm, HAMAMATSU PHOTONICS, and enabled us to cover 20% of the entire inner surface by photocathode. The aim of the experiment was still wider. Our Universe is full of mysteries and its observation has been limited, except the pioneering work of R. Davis, to those with electromagnetic signals, visible light, radio-waves, X-rays, infrared radiation and gamma-rays. These electromagnetic signals interact rather strongly with matter and the information they convey to us is limited to that on the thin surface of stellar body. On the other hand, the neutrinos will be able, due to the fact that they have only weak interaction with matter, to tell us about the deep interior of the stellar bodies, if we can build and operate a large sensitive detector capable of detecting them real-time and with directionality; the neutrino astrophysics. In fact, besides the solar neutrino problem, there have been predictions[1]), among others, on the neutrino burst emission at the time of neutron star formation and on the emission of high energy neutrinos from active stellar sources where cosmic ray accerration is expected. Furthermore, the extremely high-temperature and high-density state right after the Big Bang could have produced any particles theoretically imaginable irrespective of their masses; e.g., Monopoles, Super-Symmetric particles and/or anything. These heavy particles would have been decelerated by the expansion of the Universe and eventually get trapped gravitationally by our neighbor star, the Sun, where they catalyse nucleon decays, Rubakov-effect of Monopole, or annihilate each other, the lightest, and hence non-decaying, of S-S particles. These processes are expected to produce the neutrinos deep inside the Sun of energies higher than the ordinary neutrinos from the fusion process of solar energy generation; around 35MeV from mu-plus decays in the case of Monopopole Rubakov-effect and still higher energies depending on the mass in the case of S-S particle annihilations. The search for these particles is important not only for particle physics but also for cosmology where the problem of Dark Matter is accute. The elementary particle picture of the Universe; this was what we aimed at when designing this experiment.

The detector was installed 1000m underground in the newly excavated cave of KAMIOKA MINE, about 300Km west of Tokyo; 36.42°N and 137.31°E. The data-taking began on 4 July 1983 and it was immediately recognized that the energy spectrum of mu-e decay electrons can be observed down to about 12MeV where the low energy background sets in. This implies that there is a possibility of making the real-time, directional, and spectral observation of solar ^8B neutrinos of energies up to 14.06 MeV by means of their elastic scattering off electrons in the water, if we can reduce the background sufficiently to lower the detection threshold down to several MeV. One remembers that the pioneering experimental work of R. Davis measuring the number of Ar atoms produced from Cl in his underground detector by the bombardment of the highest energy component, from ^8B decays, of solar neutrinos revealed that the observed neutrino flux is only 1/3 of the Standard Solar Model prediction[2]. It is disconcerting to encounter such a discrepancy in the only stellar source which we can study "up-close" and a number of explanations were proposed such as the lowering of temperature and/or of helium content of the core and resonant flavor flipping of finite mass neutrinos in traversing matter; Mikeyev-Smirnov-Wolfenstein effect. Definite conclusion has not been reached yet. This experiment was until recently the only neutrino-astrophysical observation.

This plan of aiming at observing the recoil electrons due to solar ^8B neutrinos was announced at ICOBAN '84, Park City, Utah, January 1984, and A.K. Mann of University of Pennsylvania immediately showed interest. We decided to form a collaboration; KAMIO-KANDE-II. The American team was to provide new electronics of ADC+TDC with multi-hit capability while the original KAMIOKANDE team was to install additional anti-counter layer completely surrounding the main detector and to improve the purity of the water. The KAMIOKANDE-II began data-taking in January of 1986. The detector performed as expected and the trigger rate for low energy event dropped drastically as compared with the early 1985 run. At the beginning of 1987 the trigger rate as 7.5 MeV threshold was 0.60Hz, of which 0.36Hz was due to cosmic ray muons. Therefore KAMIOKANDE-II was laready taking data of recoil electrons produced by solar neutrinos at the expected rate of 0.3 event per day and hence ready to observe neutrino burst from stellar gravitational collapse, if it occured within our Galaxy or in its vicinity, because such neutrinos are expected to be of higher energy and bunched in time which make its detection much easier than that of solar neutrinos.

When we learned of the optical discovery[3] of the supernova explosion in the Large Magellanic Cloud, SN1987a, the latest data tapes at KAMIOKA site were immediately sent by express to ICEPP where all the analyses are made.

Fig. 2-a shows a part of the raw data of low energy events as come out of our laser printer and this was how the supernova neutrino signal was first detected. Nhit is the number of hit photomultipliers; hit threshold at 1/3 of one photoelectron. The events of Nhit fewer than 20 are classified as background in this experiment; 7.5 MeV detection threshold. The time-zero of the burst was given to be 07:35:35UT of 23 February 1987 with a very generous error estimation of plus/minus 1 minute. In Fig. 2-b the IMB data[4] which immediately confirmed our observation, open circles, were combined with our data, filled circles, assuming that the IMB burst correspond to the second subcluster, of the highest energies, of KAMIOKANDE-II burst.[5] The better calibrated clock of IMB, plus/minus 50ms, then indicates that the time-zero of our burst was 07:35:4 OUT instead of 07:35:35UT. IMB had 5000 tonnes of water to observe 5 events during the time interval of 1 to 4 seconds, whilst KAMIOKANDE-II of 2140 tonnes observed 3 events in the same period. In Fig. 2-c is shown the data at quiecent time, which contains the time interval of the supernova neutrino signal claimed by Mont Blanc experiment, .124UT of 23 February 1987.[6] The 12MeV event between 4 and 5 min in Fig.2-c occured at 02:52:40UT of 23 February 1987.

These supernova neutrino data provided remarkable verifications of the basic concepts of the neutrino astrophysics.[7] Namely, the time structure, the flux, and the average energy of the observed neutrinos confirmed decisively the predictions of theoretical neutrino astrophysics. In particular, the production of anti neu-e in the hot plasma by the neutral current processes and the release, in the form of neutrino burst, of the gravitational potential energy of about 1.4 solar mass iron core, Chandrasekhar mass limit, have been verified. The temperature as estimated from the observed average event energy is in good agreement with the theoretical expectation. The time elapsed between the neutrino burst and the optical luminosity growth is also in line with the theoretical estimation for the blue giant progenitor star. The data have not only supplied the astrophysical results; they have elicited more than scores of theoretical papers also in particle physics: on the mass, the electric charge, the life time, the magnetic moment, the number of species, the oscillation of neutrino, etc. It should be noted that the KAMIOKANDE-II data of Fig.2-a are essentially the raw data as printed out by a laser printer and hence that, if the computer had been at the site, it could have issued an advanced warning of the supernova occurrence even before the optical sighting. This is just about what the data tells us.

Let me now describe what is involved in the solar neutrino observation because they will clarify the points one should be careful in working in this field. The most difficult part of this solar neutrino observation is the struggle with backgrounds. There are three major sources of background. The first is the low energy radiation, gammas and neutrons, coming from the surrounding rocks. Even with the completely encasing anti-counter layer of 1.5m thickness water, we had to restrict the fiducial volume to be the innermost 680m^3 where the volume distribution of low energy events was statistically checked to be uniform. The second source of background is the radioactive trace elements, like U and Ra, in the detector water. This was dealt with by installing special ion-exchange columns to absorb these elements.

Rn in air also causes considerable trouble and we had to make the entire inner circulating water system airtight. In the meantime the inner water was kept very transparent by circulating through a series of fine filters and UV-irradiater. The light attenuation length reached 50m. The third background source is the nuclear interaction of muons with the O-nuclei in the water, resulting sometimes in a variety of long lived, msecs to seconds, radioactive nuclear fragments. This type of background are deleted in the off-line analysis by noting their temporal as well as spatial correlations with the preceding muon event. Even after these background eliminations the overall signal-noise ratio is still about 1 to 10 and only after making use of the directional correlation with respect to the Sun we could extract the solar neutrino signal; our present signal-noise ratio around the sun-earth direction is 2 to 1 for the signal expected from Standard Solar Model, but it is still improving at a considerable rate.

The result[8)] shown in Fig. 3 is the integral energy spectrum, after background subtraction, of recoil electrons produced in water by neutrinos coming from the direction of the Sun. It is based on the 127.8 live days of data-taking this year and is still preliminary in nature. The result in Fig. 3 already shows that the bare Standard Solar Model is negated at 90%CL, confirming the result of R. Davis. It should be noted that this confirmation has been done with the directional, real-time, and spectral observation. It is then either that the resonant flavor flipping in matter of finite mass neutrinos (M-S-W effect) is actually taking place, or that the lowering of the temperature and/or of He abundance in the solar core is responsible. The latter is now being extensively studied by helio-seismology. Also in Fig. 3 are shown schematically the predictions of M-S-W effect. In view of the fact that the signal/noise ratio of the observation is still improving, one can reasonably expect to reach a definite decision, say within a years time, between the two solutions, adiabatic or non-adiabatic, of M-S-W effect.

These two observations, supernova neutrino burst and solar ^{8}B neutrinos, could perhaps be considered as signaling the birth of observational neutrino astrophysics.

The KAMIOKANDE and KAMIOKANDE-II have produced not only the clean results on proton decay[9)] but also the most stringent lower limit, 1.2×10^8 sec at 90% C.L., for the lifetime of neutron-antineutron oscillation[10)]. They have given also the best experimental upper limit for the Monopole flux[11)] by using the Rubakov effect in the Sun and the result is shown in Fig. 4 which contains also the upper limits obtained by other methods as functions of Monopole velocity.

The searches for Dark Matter candidates and for Super Symmetric particles were done along a similar approach[12)]. Namely we look for the high energy neutrinos from the direction of the Sun. The preliminary results as of April 1987 of KAMIOKANDE from the analysis of the contained events, 1.5GeV to 15GeV, from the direction of the Sun are; at 90% C.L., the 4.5 to 25GeV mass range is excluded for Dirac neutrinos, the 15 to 27GeV mass range is excluded for Majorana neutrinos, and the 3 to 15GeV mass range is excluded for scaler neutrinos.

The observations by the both KAMIOKANDE's on the non-accelerator neutrinos are schematically summarized in Fig. 5; the supernova neutrinos, the solar ^{8}B neutrinos, an upper limit for the flux of low energy neutrinos from the past supernovae, the neutrinos produced in the atmosphere by cosmic rays, and the upper limits for the high energy muons produced by the neutrinos from point sources in the sky.

The advantages of the Imaging Water Cherenkov method in low energy neutrino astrophysics have been, I believe, amply demonstrated by the performances of IMB and KAMIOKANDE-II and they lie in its capability of real-time, directional, and spectral observation of neu-e, also neutrinos of other flavors and their anti's for that matter though with smaller cross-sections, by means of its scattering off electrons in the water. The anti neu-e is readily detected with a larger cross-section, real-time and spectral, by means of the inverse beta decay of protons in the water, as exemplified in the SN1987a neutrino burst observation of Fig. 2-a. Furthermore, the Imaging Water Cherenkov method is at present the only way, technically and economically, to realize the next generation detectors of much larger masses, say tens of thousands of tonnes or still larger, which are keenly desired if we were to obtain quantitative understanding of solar interior and/or supernova explosion. The effort of still improving the characteristics of 50 cm diameter photomultiplier resulted, as of March this year, in the time-zitter at 1 photoelectron level of 4.67nsec FWHM, a factor of 2 improvement, and in the 80% collection efficiency of photoelectrons as compared with the previous 43%. The length of the tube is shortened by about 20cm.

The Super-KAMIOKANDE[13)] plan was first announced at the ICOBAN'84 meeting. It is schematically shown in Fig. 6. It is a scale-up of KAMIOKANDE-II by a linear factor of 2.5, as well as doubling of the photomultiplier density on the surface. This detector with a total of 32,000 tonnes of inner water, will not only serve as the thermomenter for registering the variation of the solar core temperature, with better than 1% accuracy over a week, but also give about 4,000 events for a supernova explosion occuring in our Galactic center, which is optically dark. Those 4000 events can certainly provide us with the detailed information on the dynamics of the gravitational collapse of a star (the Neu-e events from the initial neutronization giving also the directional information); the change of neutrino energy spectrum with time, and so forth. For a better directionality, we should have a network with good timing accuracy of such detectors, not only of imaging water-Cherenkov but also of liquid scintillator and of liquid Ar, etc. A 10 microsec timing accuracy of a world-wide network will give a minute-of-arc directional accuracy. Such detectors will also explore proton decay in the 10^{34} years range.

Another very interesting field of neutrino astrophysics is that of high-energy point sources.[14] There have been claims to have observed point sources-Cyg-X-3 and Vela Pulser, etc.-with high-energy gamma rays of energy 10^{12}eV to 10^{15}eV or with neutrino-produced muons, but the situation is far from conclusive. KAMIANDE-II searched also for the upward-going muons from the direction of SN1987a and found none so far as of 28 July 1987.

The physical process utilized in high-energy neutrino astronomy is the charge-current interaction with nucleons, in which the produced leptons--muons and electrons--give the directionality. Here the importance lies in the sensitive area and in the angular resolution. We need at least 10^4m^2 sensitive area and better than 1° angular resolution. The change of threshold from 6MeV in SuperKAMIOKANDE to 1GeV in this high energy neutrino astronomy overcompensates in cost the surface ratio. Namely, instead of 0.71m spacing of photomultipliers in the former one can employ 0.71m times (square root of 1000/6)=9m in the latter, if we neglect the light attenuation in water.

LENA, Lake Experiment on Neutrino Astronomy,[15] was proposed along these lines. It is a two-dimentional expansion of SuperKAMIOKANDE by a linear factor of 4; namely: 150m diameter times 30m depth of water surrounded by an array of 50cm diameter photomultipliers on a 5m lattice, as is schematically shown in Fig. 7. There is an outer layer, on the top and sides of the 5x5x5m^3 water Cherenkov modules, each containing one 50cm diameter photomultiplier. Thanks to the directionality of the Cherenkov light, we do not have to install this gigantic detector deep underground, and a part of a natural lake or a surface pit can be utilized to observe clearly the upward-going muons produced by neutrinos. About 4,000 such muons, plus some electrons, can be expected yearly for a 10^4m^2 sensitive area and the expected angular resolution of the detector is about 0.8^o. Since this detector can differentiate between muon events and electron events, it is possible to study, in the form of appearance experiment, also the M-S-W effect in the Earth by using the cosmic ray neutrinos. The large inner volume (300,000m^3) of this detector will give some 100 times more sensitivity as compared with existing experiments in the search for relic heavy Super Symmetric particles annihilating in the Sun.

It should also be noted that this detector can observe also high energy gamma ray shower clearly separated from the large background of cosmic ray hadron showers. This is because the top layer of 5x5x5m^3 modules act as the energy flow detector and the lower main detector acts as the muon detector covering 100% of the shower detection area. In the conventional air-shower type experiment the muon detectors cover only about 10^{-4} of the shower area and this causes a factor of 100 increase in the observed muon number fluctuation, thereby making the gamma/hadron separation rather difficult. The angular resolution of gamma ray observation with LENA is estimated to be around 1°.

Let me now describe a dream of mine for the future, as schematically shown in Fig. 8. It now seems clear that a world network of neutrino-astronomical observatories with good timing accuracy is to be installed. Besides the Super-KAMIOKANDE and LENA which are now being seriously considered in Japan, the ones in the United States (Super-AMERICANDE), in Europe (Super-EUROPEANDE), and in Australia (Super-AUSTRALIANDE) would not only serve as the real-time thermometer for the center of our Sun, with 1% accuracy daily, but could also give the advanced warning of supernova explosion, with a minute-of-arc directional information, to optical, radio, and/or space stations throughout the world. In high-energy neutrino astronomy, the network of LENA-ANDES, LENA-URALSKAYA, LENA-ABORIGINALE, LENA-TIBET, LENA-AFRICA, and so on, will cover the entire sky continuously, and the important thing here is that this network can observe the high-energy gamma-rays simultaneously with the high-energy neutrinos from an active stellar object at the time of its radio outburst.

The cost, about 40 MUS dollars for each Super-NDE (Neutrino Detection Experiment) and about 5MUS dollars for each LENA, is of course very substantial but, in view of its impact on the neighboring scientific disciplines, contributions of 1% each from Particle Physics-, Astronomy-, Space-, and Nuclear Physics- budgets may hopefully be expected. The project is ideally suited for a peaceful international collaboration.

In conclusion, I should like to express my sincere gratitude for the far-sighted support of the Kamioka project by the Ministry of Education, Culture and Science of Japan, and my thanks to the collaborators of KAMIOKANDE-II. This talk was prepared while the author was staying at CERN as a guest professor and I should like to acknowledge the very warm hospitality I was received with there.

NOTES

1) S. Chandrasekhar, 1939, An Introduction to the Study of Stellar Structure, reprinted by Dover, NY 1957.
Ya. Zeldovich and O. Guseinov, Sov. Phys. JETP Lett. 1 (1965) 109.
S.A. Colgate and R.H. White, Astrophys. J. 143 (1966) 626.
S.E. Woosley et al, Astrophys. J., 302 (1986) 19.
A. Burrows and J.M. Lattimer, Astrophys. J., 307 (1986) 178.,
D.N. Schramm and W.D. Arnett, Astrophys. J., 198 (1975) 629.
W.D. Arnett, Can. J. Phys. 45 (1967) 1621.
For a review of theoretical neutrino astrophysics see:
D.Z. Freedman, D.N. Schramm, and D.L. Tubbs, Ann. Rev. Nucl. Sci. 27 (1977) 167.
S.E. Woosley and T.A. Weaver, Ann. Rev. Astron. Astrophys. 24 (1986) 205.
For a review of earlier experimental neutrino astrophysics see:
K. Lande, Ann. Rev. Nucl. Part. Sci., 29 (1979) 395.

2) J. N. Bahcall et al, Astrophysical Journal L79 (1985) 292.

3) I. Shelton. International Astronomical Union Circular No. 4316.

4) R.M. Bionta et al; Phys. Rev. Lett. 58 (1987) 1494.

5) K. Hirata et al; Phys. Rev. Lett., 58 (1987) 1490.
The suggestion of associating the IMB burst with the second sub-cluster of the KAMIOKANDE-II burst was made by H. Suzuki and K. Sato, Univ. of Tokyo preprint, UTAP-53/87, 12 May.

6) M. Aglietta et al, Internat. Astron. Union Circular No. 4323.

7) See for instance the paper by D. Schramm to appear in the proceedings of 1987 International Symposium on LEPTON and PHOTON INTERACTIONS at High Energies, Hamburg, July 27-31, 1987.

8) K. Hirata et al., 1987 Univ. of Tokyo preprint, UT-ICEPP-87-04, or Univ. of Pennsylvania preprint, UPR-0144E

9) The latest result of KAMIOKANDE-II on proton decay was reported by T. Kajita at Rencontre de Moriond, March 1987.

10) M. Takita et al, Phys. Rev., D34 (1986) 902.

11) T. Kajita et al, Journ. Phys. Soc. Japan, 54 (1985) 4065.

12) See for instance;
J.S. Hagelin et al, Phys. Lett., 180B (1986) 375.
T.K. Gaisser et al, Phys. Rev. D34 (1986) 2206.

13) M. Koshiba, ICOBAN '84, Park City, Utah, January 1984, and Proc. XXII Int. Conf. on High-Energy Physics, Leipzig, 1 (1984) 244.
Y. Totsuka; Proc. Seventh Workshop on Grand Unification/ICOBAN '86, Toyama, Japan, April 1986; Proc. by World Scientific of Singapore (1987) 118.

14) M.L. Marshak et al, Phys. Rev. Lett., 54 (1985) 2099 and 55 (1985) 1965.
G. Battistoni et al, Phys. Lett., 155B (1985) 465.
Y. Oyama et al, Phys. Rev. Lett., 56 (1986) 991.
Ch. Berger et al, Phys. Lett., 174B (1986) 118.

15) M. Koshiba; Workshop on Future Projects, Inst. of Cosmic Ray Res., Univ. of Tokyo, March 1986.
M. Koshiba, S. Orito, and K. Kawagoe; Workshop on Non-Accelerator Physics, KEK, September 1986.

Figure captions

Fig. 1 The detector of KAMIOKANDE-II. The numbers give the dimensions in millimeter.

Fig. 2-a. The supernova signal, as it was first detected, of the KAMIOKANDE-II experiment. It is a part of the laser printer output of the low energy raw data. Nhit is the number of hit photomultipliers.

Fig. 2-b. A combined display of IMB- and KAMIOKANDE-II data.

Fig. 3. The integral energy spectrum of the recoil electrons produced by the neturinos from the direction of the Sun. It is still preliminary in nature.

Fig. 4. The Monopole flux limit obtained by KAMIOKANDE. The method of search here is to look for the neutrinos of energy around 35MeV from the direction of the Sun; Rubakov-effect in the Sun. Other flux limits obtained by various methods are also shown for comparison.

Fig. 5. The results obtained so far by KAMIOKANDE and KAMIOKANDE-II on the neutrinos of non-accelerator origin are schematically shown.

Fig. 6. The Super-KAMIOKANDE detector in vertical cross-section. The dimensions are given in millimeters.

Fig. 7. A schematic cross-sectional view of the LENA detector. The small squares surrounding the main detector are the $5x5x5m^3$ modules.

Fig. 8. The dream of the author; a world network of neutrino astrophysical observatories.

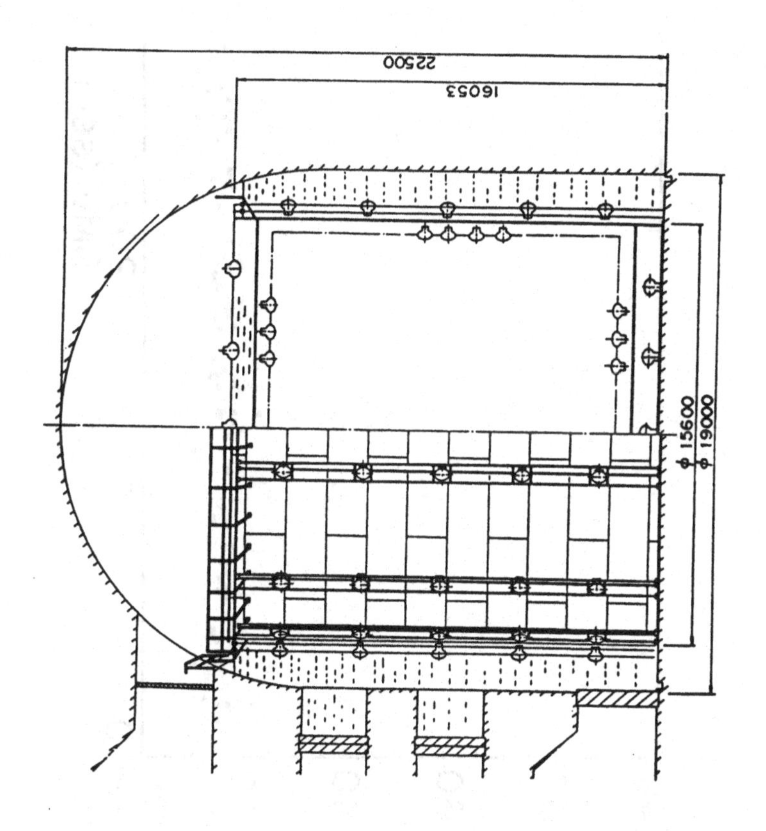
22500
16053
ϕ 15600
ϕ 19000

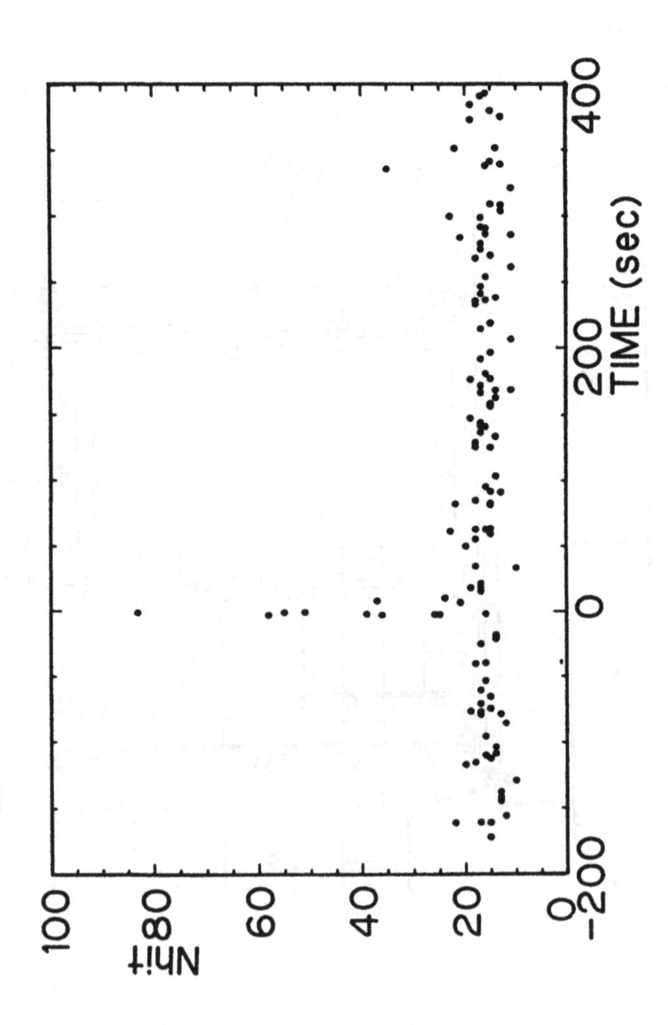
Nhit
100
80
60
40
20
0
−200
0
200
400
TIME (sec)

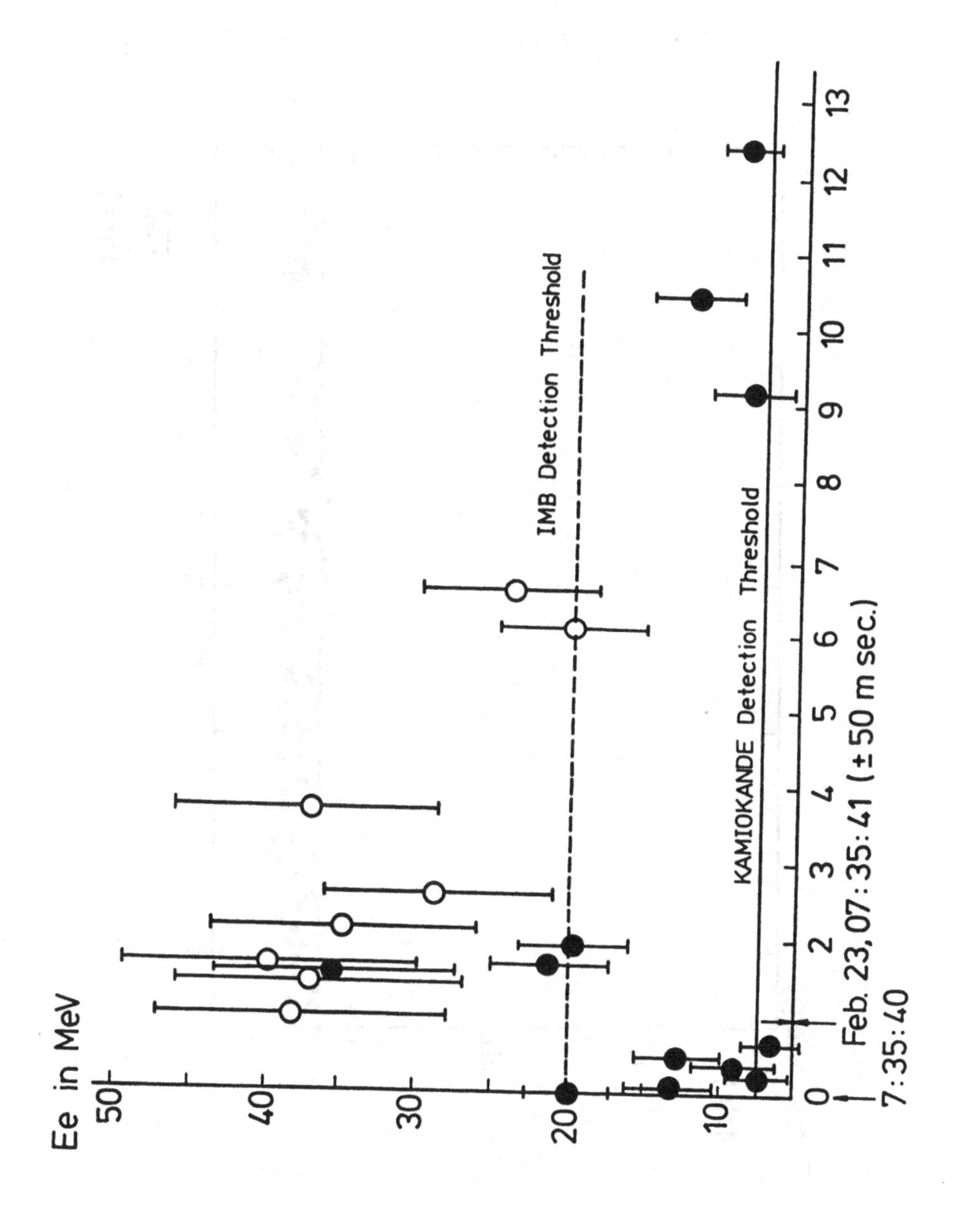
Ee in MeV
50
40
30
20
10
IMB Detection Threshold
KAMIOKANDE Detection Threshold
0
1
2
3
4
5
6
7
8
9
10
11
12
13
Feb. 23, 07:35:41 (±50 m sec.)
7:35:40

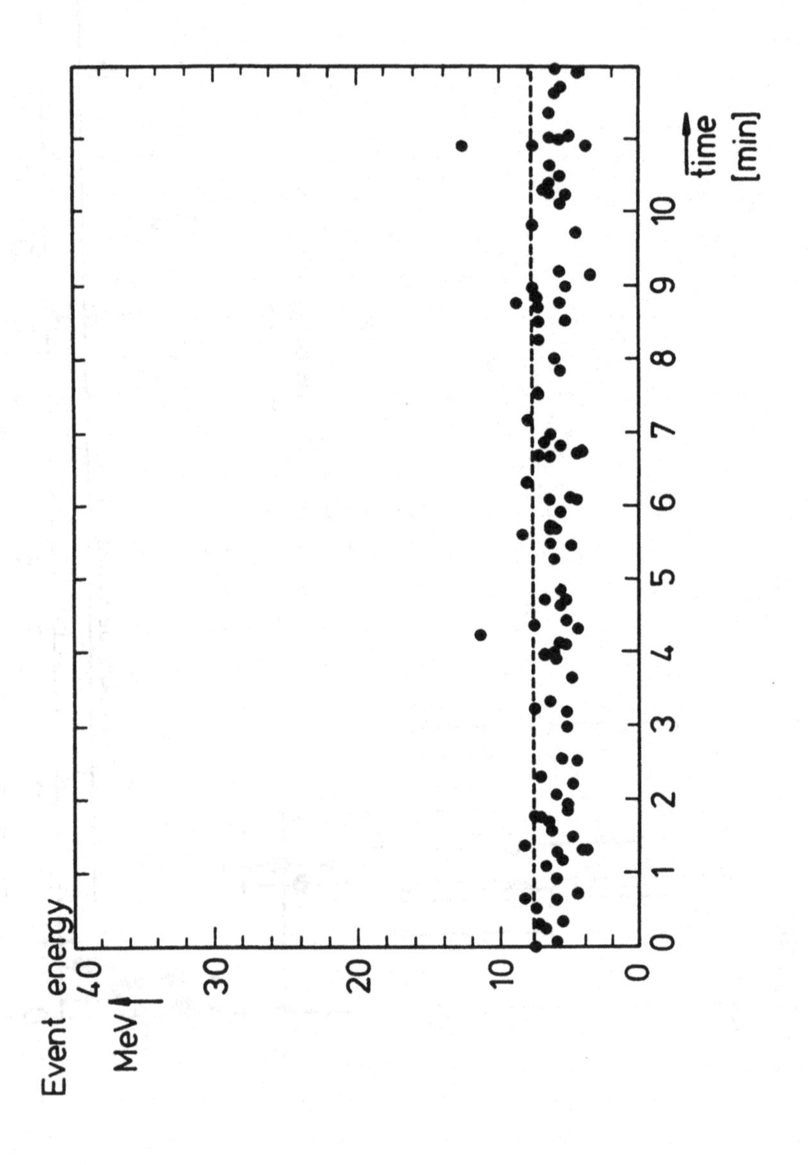
Event energy
MeV
40
30
20
10
0
0
1
2
3
4
5
6
7
8
9
10
time
[min]

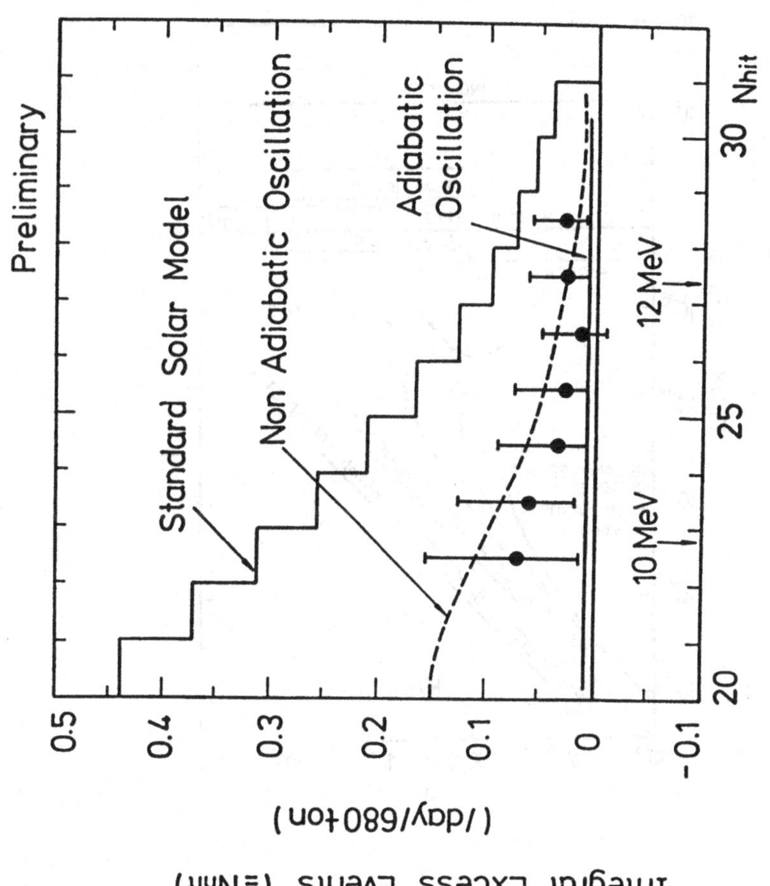
Preliminary
Standard Solar Model
Non Adiabatic Oscillation
Adiabatic Oscillation
10 MeV
12 MeV
Integral Excess Events (≥ Nhit)
(/day/680 ton)
0.5
0.4
0.3
0.2
0.1
0
-0.1
20
25
30
Nhit

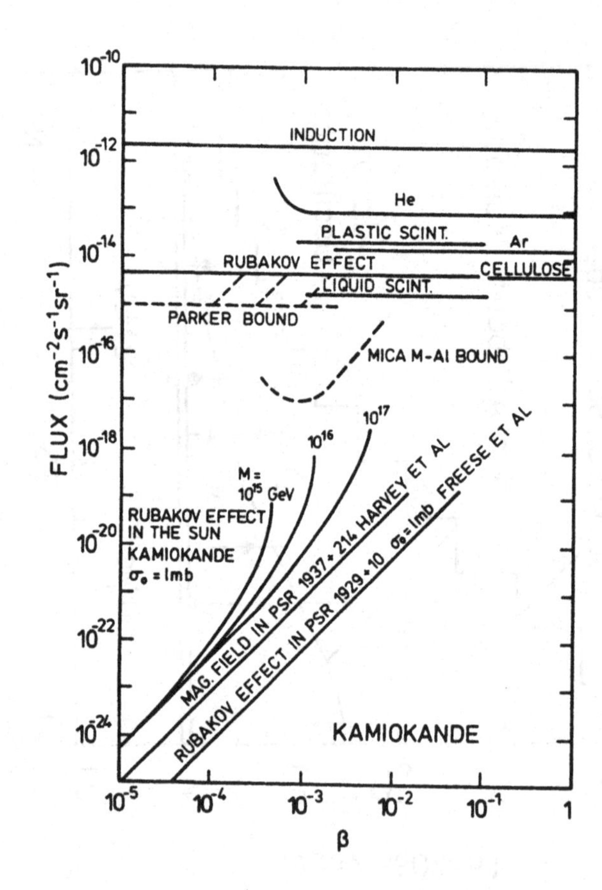
INDUCTION
He
PLASTIC SCINT.
Ar
RUBAKOV EFFECT
CELLULOSE
LIQUID SCINT.
PARKER BOUND
MICA M-Al BOUND
M = 10^15 GeV
10^16
10^17
RUBAKOV EFFECT IN THE SUN KAMIOKANDE σ₀ = 1mb
MAG. FIELD IN PSR 1937+214 HARVEY ET AL
RUBAKOV EFFECT IN PSR 1929+10 σ₀ = 1mb FREESE ET AL
KAMIOKANDE
FLUX (cm⁻²s⁻¹sr⁻¹)
β

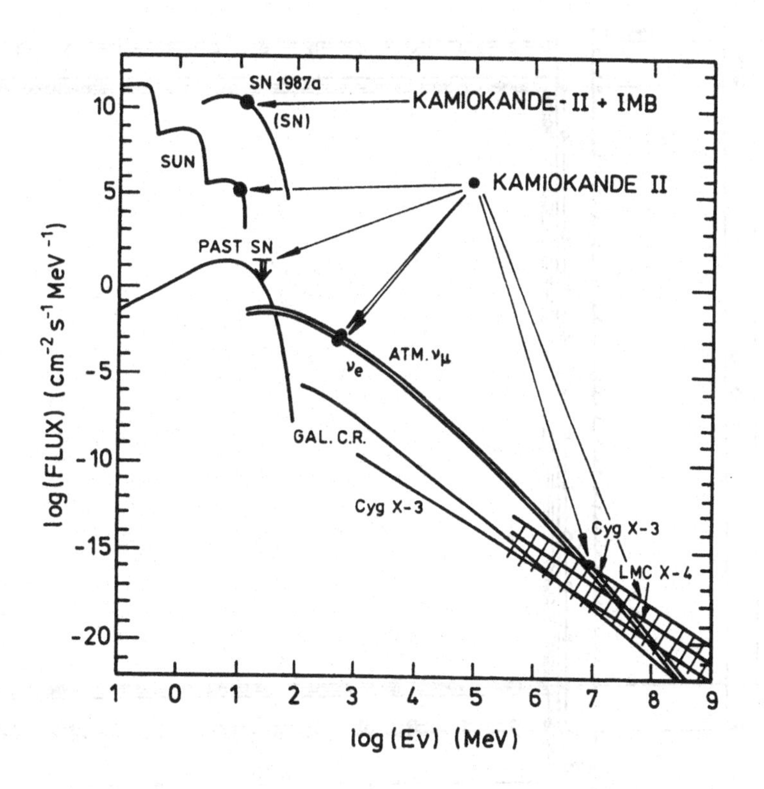
SN 1987a
(SN)
KAMIOKANDE - II + IMB
SUN
KAMIOKANDE II
PAST SN
ATM. ν_μ
ν_e
GAL. C.R.
Cyg X-3
Cyg X-3
LMC X-4
log (FLUX) (cm^{-2} s^{-1} MeV^{-1})
log (Ev) (MeV)
10
5
0
-5
-10
-15
-20
-1
0
1
2
3
4
5
6
7
8
9

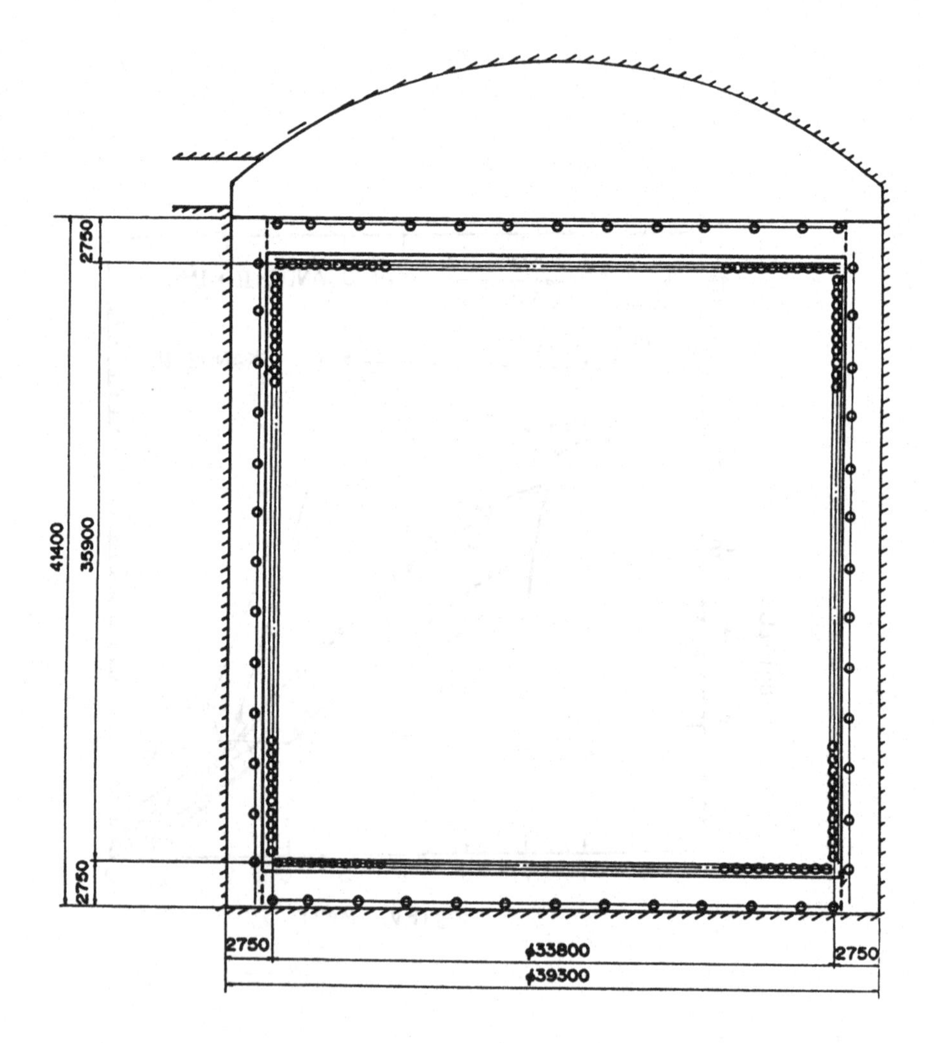
2750
35900
41400
2750
2750
ϕ33800
2750
ϕ39300

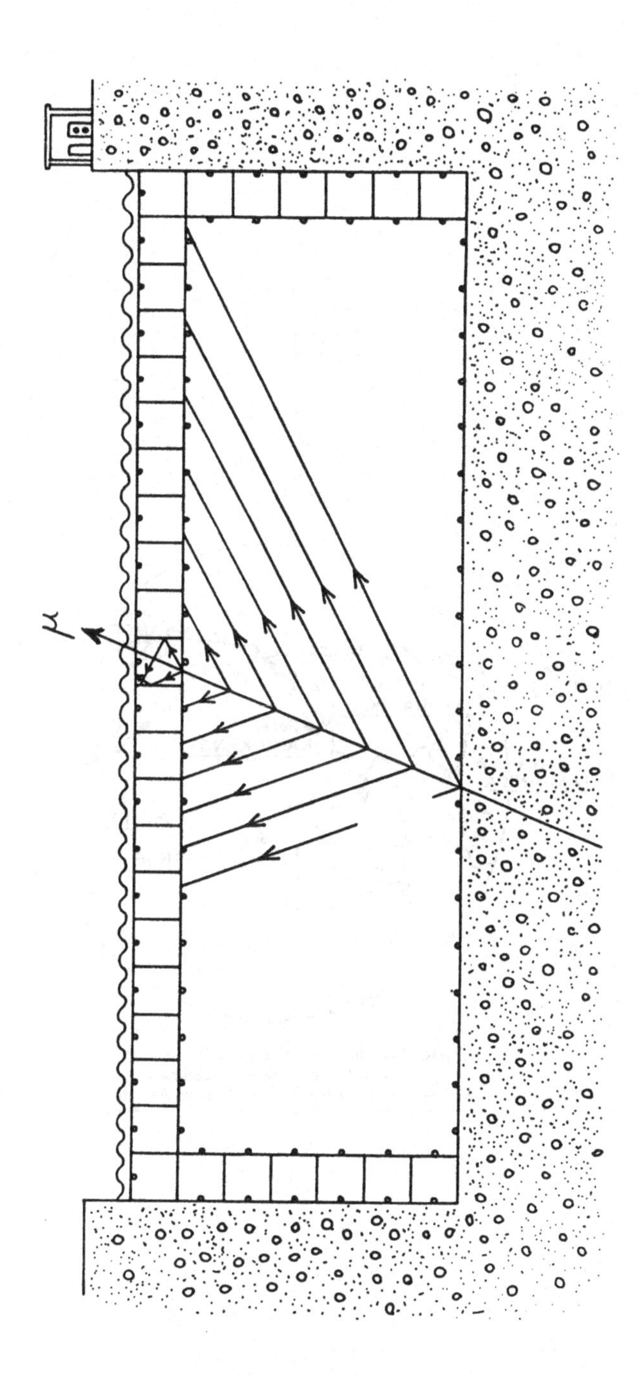
μ

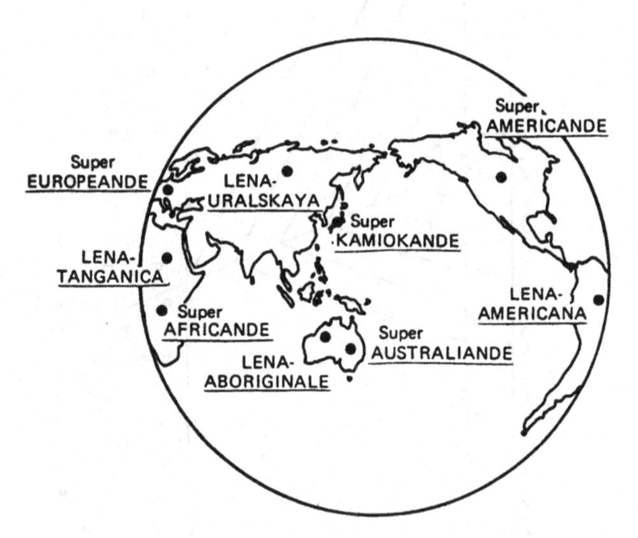

...NDE: Neutrino Detection Experiment
(Alias. Nucleon Decay Experiment)
LENA: Lake Experiment on Neutrino Astrophysics

The Discovery of Neutrinos from SN1987a with the IMB Detector

James Matthews
Physics Department, University of Michigan
Ann Arbor, MI, 48109

FOR THE IMB COLLABORATION

C.B. Bratton,[4] D. Casper,[2,11] A. Ciocio,[11] R. Claus,[11]
M. Crouch,[8] S.T. Dye,[5] S. Errede,[9] W. Gajewski,[1] M. Goldhaber,[3]
T.J. Haines,[13] T.W. Jones,[6] D. Kielczewska,[1,7] W.R. Kropp,[1]
J.G. Learned,[5] J.M. LoSecco,[10] J. Matthews,[2] R. Miller,[1] M. Mudan,[6]
L.R. Price,[1] F. Reines,[1] J. Schultz,[1] S. Seidel,[2,11] E. Shumard,[12]
D. Sinclair,[2] H.W. Sobel,[1] J.L. Stone,[11] L. Sulak,[11] R. Svoboda,[1]
G. Thornton,[2] J.C. van der Velde,[2]

[1] The University of California, Irvine, Irvine, California 92717
[2] The University of Michigan, Ann Arbor, Michigan 48109
[3] Brookhaven National Laboratory, Upton, New York 11973
[4] Cleveland State University, Cleveland, Ohio 44115
[5] The University of Hawaii, Honolulu, Hawaii 96822
[6] University College, London WC1E6BT, United Kingdom
[7] Warsaw University, Warsaw, Poland
[8] Case Western Reserve University, Cleveland, Ohio 44106
[9] The University of Illinois, Urbana, Illinois 61801
[10] The University of Notre Dame, Notre Dame, Indiana 46556
[11] Boston University, Boston, Massachusetts 02215
[12] AT&T Bell Laboratories, Summit, New Jersey 07901
[13] University of Maryland, College Park, Maryland 20742

ABSTRACT. A burst of eight neutrino events preceeding the optical detection of the supernova in the Large Magellanic Cloud has been observed by the IMB detector. The events span an interval of six seconds and have visible energies in the range 20-40 MeV. Revised energies and reconstructed angles are presented and comparisons to standard supernova theory made.

INTRODUCTION

According to conventional supernova theory, the recently observed supernova explosion SN1987a in the Large Magellanic Cloud (LMC) should have released approximately 3 x 10^{53}ergs of gravitational binding energy in a burst of 10^{58}neutrinos in a time interval of a few seconds (e.g., Colgate & White, 1960; Wilson, et al., 1986; Woosley, et al., 1986). We have observed a burst of neutrino events contained within the IMB water-Cerenkov detector and occurring close to the estimated time of the supernova collapse. The observed signal consists of eight neutrinos with energies in the range of 20-40 MeV, spaced over an interval of 6 seconds. The background rate of contained events from cosmic ray produced neutrinos interacting in the detector is ~2/day in the range 20-2000 MeV. I discuss here the analysis which obtained the events, the determination of their energies and kinematics, and some inferred properties of the neutrino beam.

THE SIGNAL FROM SN1987A

The IMB detector (Bionta, et al.,1983; Haines, 1986), which was designed to search for proton decay, is located in the Morton-Thiokol salt mine near Fairport, Ohio (41.7°N, 81.3°W) at a depth of 1570 m of water-equivalent (mwe). It consists of a rectangular tank (22.5 x 17 x 18 m^3)filled with purified water. The six sides are instrumented with 2048 eight-inch photomultiplier tubes (PMTs) arranged on an approximate 1 m grid. The water serves both as a target for incoming neutrinos and a Cerenkov radiator for the charged products of such interactions. The timing, pulse-height, and geometry of PMT hits are used to reconstruct the vertex, direction, and energy of a charged particle tracks. We consider the active volume for this search to be the entire detector, or about 6800 metric tons of H_2O.

The signal of a supernova is expected to be a burst of low-energy neutrinos occurring over a period of seconds. After the optical discovery of SN1987a (Kunkel & Madore, 1987), triggers with less than 100 PMT hits ($<\sim$ 75 MeV) in a 60 hour period surrounding the time of the optical sighting were tabulated and examined for such bursts (Bionta, et al., 1987). All triggers were subsequently analyzed to determine if they were due to charged particles entering the detector, particles originating within the apparatus (a "contained event"), or spurious triggers of instrumental origin.

Eight contained events in a 6 second span starting at 7:35:41 (UTC) were identified as electron neutrino or antineutrino interactions in the detector volume. We estimate the efficiency for finding the neutrino-burst events, once the detector has triggered, is essentially 100% based on a double scanning technique. None of the events can be identified as a muon track since no decay electron followed. The time of this burst coincides with a similar observation in the Kamioka Nucleon Decay Detector (Hirata, et al., 1987).

Any trigger reconstructing as a contained event is interpreted as a neutrino event: there is no background above our threshold of 20 MeV. No other low energy ($<\sim$75 MeV) neutrino events were found in the 60 hour window. The normal rate for contained events over our entire energy range (20-2000 MeV) is <2 day^{-1}; low energy events occur only about once every three days. The burst structure of these events is remarkable: even two neutrino events (over our entire energy range) have never been recorded in such close time proximity since we began operating in 1982.

The visible energies and directions of these events are determined by correcting the observed PMT pulse heights for noise and geometric effects and comparing with detailed simulations of Cerenkov light production in the detector; these Monte Carlo programs have been cross-checked with samples of cosmic ray muons and muon-decay electrons. We have refined these analyses since our original publication (Bionta, et al., 1987) and give revised characteristics of the observed events in Table 1 below.

Table 1

Event	(a) Time (UT) 23 Feb. 1987	(b) Energy (MeV)	(c) Polar Angle (deg)
1	7:35:41.374	38 ± 7	80 ± 10
2	7:35:41.786	37 ± 7	44 ± 15
3	7:35:42.024	28 ± 6	56 ± 20
4	7:35:42.515	39 ± 7	65 ± 20
5	7:35:42.936	36 ± 9	33 ± 15
6	7:35:44.058	36 ± 6	52 ± 10
7	7:35:46.384	19 ± 5	42 ± 20
8	7:35:46.956	22 ± 5	104 ± 20

(a) Absolute UT is accurate to ± 50 ms. Relative times are accurate to the nearest millisecond.

(b) Visible energy; Additional systematic error in energy scale estimated to be $\pm 10\%$.

(c) Angle with respect to direction away from SN1987A. Angle errors include multiple scattering and event reconstruction.

If we assume these events are caused by electron anti-neutrinos interacting on free protons, then the visible energy is about 1.3 MeV less than the neutrino energy.

ANGULAR CHARACTERISTICS

The angular distribution of the directions of detected electrons and positrons is determined by the composition of the neutrino beam (ν or $\bar{\nu}$). The IMB detector is sensitive to both scattering- and absorption-type nuetrino interactions. Scattering events (e.g., $\nu_e e^- \rightarrow \nu_e e^-$) should exhibit electron directions which are strongly inclined in the forward direction. Absorption events on free protons ($\bar{\nu}_e p \rightarrow ne^+$) are essentially isotropic, having only a slight energy dependent forward asymmetry at the energies considered here.

In evaluating the angular distribution of the eight IMB events we needed to account for one quarter of the detector being systematically inoperative: For a seven hour period of time containing the observed burst, one-quarter of the PMTs, representing regions of the detector's south and top walls, were not operational due to the failure of one of four power supplies. The local position of the supernova, at the time of the burst was 42° below the horizon and 28° west of south.

A detailed analysis of the detector's response has been made and we find there is no significant bias in detection efficiency with respect to the polar angle θ measured against an axis along the direction from the LMC. By computer simulation, samples of low energy electrons are generated isotropically in the detector and the triggering criteria is applied with the appropriate PMTs removed. The efficiency for triggering is essentially independent of $\cos(\theta)$ as shown in Figure 1 for samples of 20 and 30 MeV electrons. One reason is that a low energy electron travels only a few centimeters in water and scatters quite a bit. The inoperative patches of the detector have only a small effect since their subtended solid angle is very small.

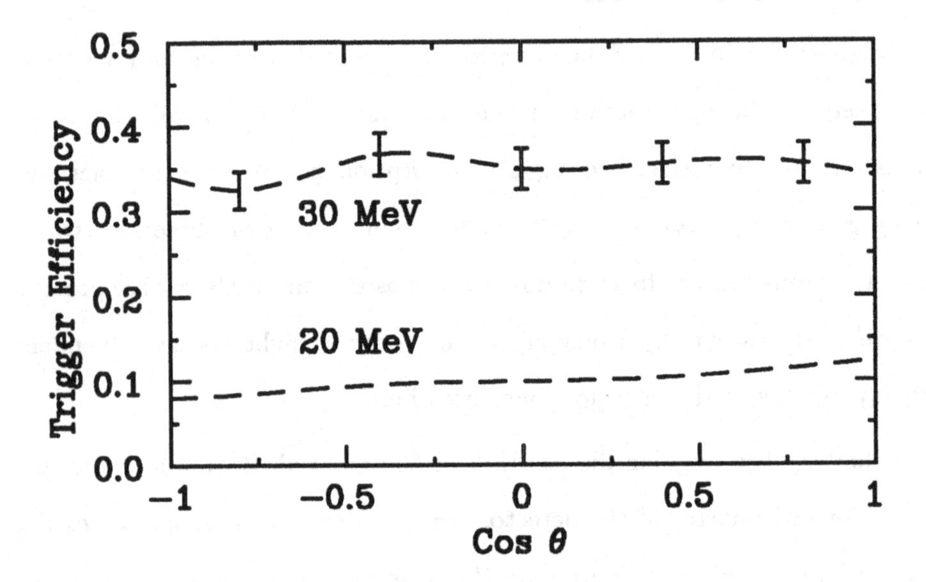

Figure 1. Triggering efficiency vs. polar angle

Although the polar angle variable $\cos(\theta)$ is unbiased, there is an asymmetry in detection efficiency with respect to the azimuthal angle about an axis from the LMC. This causes the net lowering of the detection efficiency from the fully operational detector. For calculations involving $\cos(\theta)$ we assume an isotropic response.

The overall trigger efficiency as a function of visible energy has been determined this way, averaging over angles. The results are shown in Figure 2.

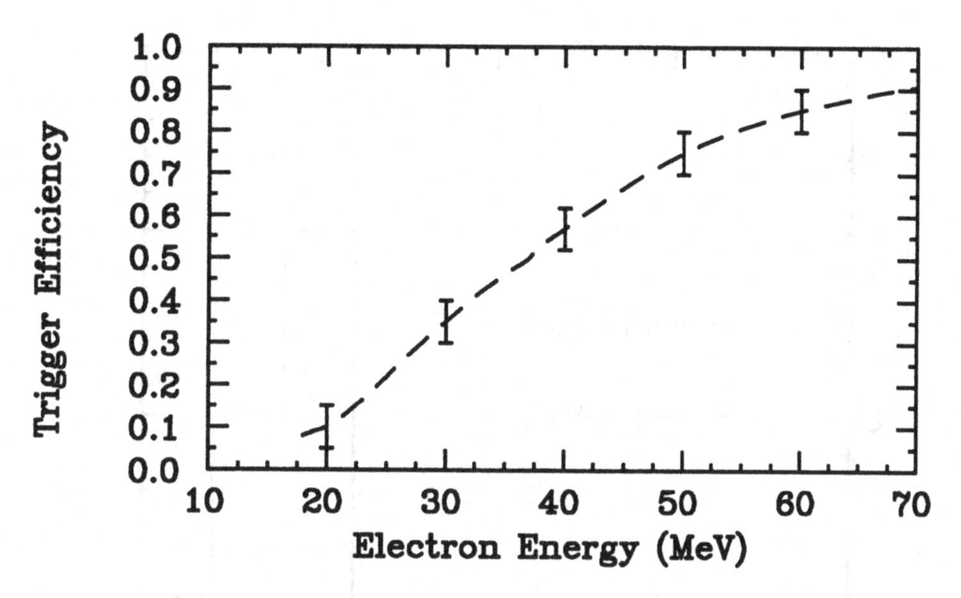

Figure 2. Triggering efficiency vs. electron energy

These calculations have been been checked by simulating Cerenkov tracks in the detector using an artificial light source.

The angular distribution is histogrammed in Figure 3 and exhibits the 8 IMB events. There appears to be forward peaking. However, as noted above, the angular distribution expected from $\bar{\nu}_e p \rightarrow ne^+$ is only approximately isotropic. The distribution expected from 30 and 50 MeV $\bar{\nu}_e$'s interactions is shown.

The likelihood that the events came from a parent population of isotropic events can be quantified. A Monte-Carlo simulation using the Kolmogorov statistic gives a 5% probability that a fluctuation in the forward direction as large as this could result for eight events from a parent sample which is isotropically distributed in the variable $cos\theta$. The probability is slightly higher for a population assumed to to have a small forward assymetry. It is thus not unreasonable to identify the eight events as $\bar{\nu}_e p \rightarrow ne^+$.

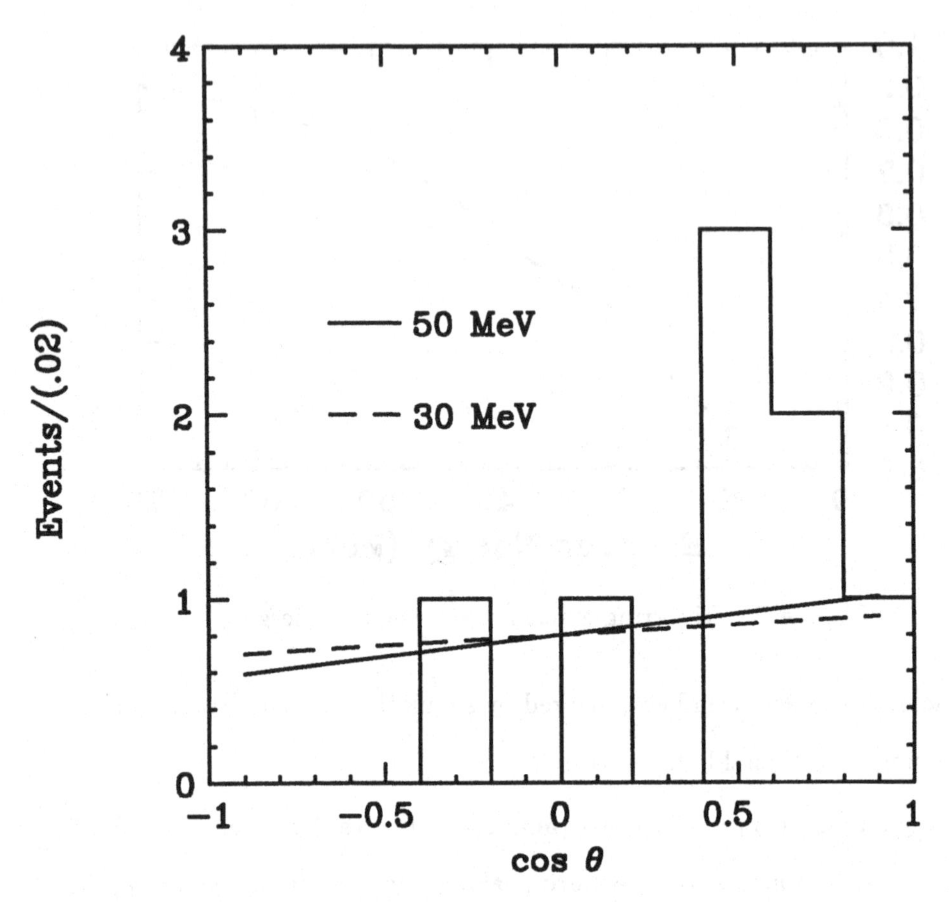

Figure 3. Polar angle distribution of events. Also shown are distributions expected for $\bar{\nu}_e p \to n e^+$ at two energies.

CHARACTERISTICS OF THE BEAM

Estimates of the parameters of the neutrino beam can best be obtained by using maximum likelihood techniques to calculate temperature and composition (ν_ℓ and $\bar{\nu}_e$) within the "standard model" of supernovae. This approach makes use of all kinematical and experimental information in the data. The following treatment and notation has been described elsewhere (Matthews, et al., 1987) and is similar to that of other authors (Bahcall, 1987; Burrows, 1987).

The neutrinos are assumed to come from a single thermal distribution. I consider only the two most probable reactions: $\bar{\nu}_e p \to n e^+$ on free protons and the scattering reaction $\nu_\ell e^- \to \nu_\ell e^- (\nu_\ell = \nu_e, \bar{\nu}_e, \nu_\mu, \bar{\nu}_\mu)$. Cross sections used take into account angular and energy distributions in detail.

We calculate the probability that each event came from a Fermi-Dirac energy distribution of temperature T and a mixed population of ν and $\bar{\nu}$ characterized by the ratio

$$R = \frac{\sum_\ell \sigma_\ell \Phi_\ell}{[\sum_\ell \sigma_\ell] \times \Phi_{\bar{\nu}_e}}$$

This is the ratio of average flux of neutrinos to the flux of $\bar{\nu}_e$, weighted by electron-scattering cross section appropriate to each neutrino flavor ℓ. For equal fluxes of all neutrino flavors, $R = 1$; for a ν_e dominated beam, $R \to\sim \frac{1}{2}\Phi_{\nu_e}/\Phi_{\bar{\nu}_e}$

The joint probability for the set of all events is computed and maximized by varying values of T and R. A flux is calculated from the normalization using these most likely values. The results of this calculation are:

Temperature	$4.2^{+1.0}_{-0.8}$ MeV
Ratio R	< 40 (90% CL)
$\bar{\nu}_e$ Flux	$(0.79 \pm .28) \times 10^{10} cm^{-2}$
Total $\bar{\nu}_e$ Energy	$(4.8 \pm 1.7) \times 10^{52}$ergs

SUMMARY

The IMB detector has recorded a burst of neutrinos originating in the stellar collapse seen in the LMC. The interactions are consistent with $\bar{\nu}_e p \to n e^+$ on free protons. If interpreted within conventional models of supernova explosions, the data are in good agreement with predictions of flux and energy expected from such events.

REFERENCES

J.N. Bahcall, T. Piran, W.H. Press, and D.N. Spergel, *Neutrino Temperatures and Fluxes from the LMC Supernova* (submitted to *Nature*, 7 April 1987)

R.M. Bionta, et al., *Phys. Rev. Lett.* **58**,1494 (1987)

R.M. Bionta, et al., *Phys. Rev. Lett.* **51**, 27 (1983).

A. Burrows and J.M. Lattimer,*Neutrinos from SN1987a*, Steward Observatory Preprint No. 725, University of Arizona (1987)

S.A. Colgate and R.H. White, *Ap. J.* **143**, 626 (1966).

T.J. Haines, et al., *Phys. Rev. Lett.* **57**, 1986 (1986).

K. Hirata, et al., *Phys. Rev. Lett.* **58**,1490 (1987)

W. Kunkel and B. Madore, IAU Circular 4316, Feb. 24, 1987.

J. Matthews, et al., *Proc. U. of Minnesota Workshop on SN1987a*, in press (1987)

J.R. Wilson, et al., *Ann. N.Y. Acad. Sci.* **470**, 267 (1986).

S.E. Woosley, J.R. Wilson, R. Mayle, *Ap.J.*,**302**,19 (1986)

PEERING INTO THE ABYSS: THE NEUTRINOS FROM SN1987A

Adam Burrows
Departments of Physics and Astronomy, University of Arizona,
Tucson, AZ 85721, USA

Abstract. The multi-continent detections of the neutrinos from the core of SN1987a open a new chapter in high energy astrophysics. For the first time, we have penetrated the supernova ejecta and glimpsed at the violent convulsions that attend stellar collapse and the birth of a neutron star. The neutrino emissions are the only good diagnostic of implosion physics, and the new data allow us to test "supernova" theories in a unique way. I will compare the theory developed over the last 20 years with these observations to extract what information there is in these epochal detections.

INTRODUCTION

Though the LMC supernova (SN1987a) is even now continuing to reveal fascinating and unexpected phenomena and will be scrutinized for many years in the optical, UV, infrared, X-ray, and, perhaps, γ-ray, the neutrino pulse, lasting only ~10 seconds, is already history. The epochal neutrino data culled by the IMB and Kamiokande II collaborations (Bionta et al. 1987; Hirata et al. 1987) provide us with our first test of the basic theories of stellar collapse and neutron star formation. Much has already been said about these events, some useful and some not, but in these months since February 23.316 (UT) 1987, a consensus has emerged concerning their general interpretation. In this contribution, I will first summarize the pre-SN1987a theory of collapse, etc. and then will discuss what we have determined about the fiery death of the cores of massive stars and the birth of neutron stars. Some of these conclusions can be found in a still burgeoning literature, but I have been quite selective in this synopsis. Since we have only 19 authenticated events, we cannot settle many of the outstanding issues in supernova theory. That must await a nearby, galactic Type II supernova. However, these data are allowing us to verify the basic tenets of stellar collapse theory developed over the last 20 years. The cynic will retort that it is quite natural for those closest to the theory to see in the detections what they had expected. Exotics such as a neutrino mass, neutrino oscillations, hypothetical particles, untoward source behavior, and the like are not impossible. It is simply that these sparse data do not allow us to probe these issues with confidence. A 5 kpc, rather than a 50 kpc, supernova would be most welcome in this regard.

DISCUSSION

The essentials of neutron star structure determine the basics of the neutrino signature of its formation. The canonical neutron star has a radius of 10 kilometers and a gravitational mass of ~1.4 $M_{\odot}$. The corresponding matter densities range between 10^{14} and 10^{15} gm/cm^3, in the general vicinity of the density of the nucleus. As the name suggests, neutron stars are composed predominantly (~95%) of neutrons, with a small admixture of electrons and protons to ensure beta equilibrium ("$n \rightleftarrows p + e^-$"). There are now probably no less than 10^8 neutron stars in our galaxy, of which $\sim 10^5$ are radio pulsars and ~20 are x-ray pulsars. Simple arguments suggest that $\sim 10^3$ of these collapsed objects are within 10^2 parsecs of the earth.

When the core of a massive star, composed of either iron-peak elements or O-Ne-Mg, reaches the effective Chandrasekhar mass ($M_C \sim 1.2$-$1.8\ M_\odot$) (Woosley & Weaver 1986) through a combination of photodisintegration (Hoyle & Fowler 1964) and/or electron capture (Nomoto 1984), it collapses dynamically and pulls away from the rest of the star. The flow separates into an inner subsonic core and an outer supersonic mantle (Brown et al. 1982). Compression further raises the electron Fermi energy above the electron capture thresholds and copious electron neutrino (ν_e) emission ensues. Concomitantly, however, the opacity of the core to neutrinos increases and beyond a density of $\sim 10^{12}$ gm/cm^3, these neutrinos are trapped (Mazurek 1974; Sato 1975) in the flow. Net neutronization ceases before nuclear densities are reached. A chemical equilibrium between electrons and electron neutrinos is established at a fixed lepton number (electron-type) of between 0.35 and 0.4, far above the canonical 0.05 of a cold neutron star. A Fermi sea of degenerate electron neutrinos with a chemical potential near 200 MeV develops.

Upon reaching nuclear densities, the subsonic inner core stiffens, rebounds, and drives a shock wave into the outer mantle. Though it may have taken a full second to reach nuclear densities, the subsequent dynamics is on millisecond time scales. Either the shock overcomes the debilitating effects of nuclear dissociation and electron neutrino radiation after the breakout of the neutrinosphere to become a prompt Type II supernova (Colgate & Johnson 1960) or it stalls, only to be revived later after a short pause of between 0.1 and 1.0 seconds (Wilson et al. 1986; Bethe and Wilson 1985). In either case, shortly after bounce, the residue is in hydrostatic equilibrium. It is hot, lepton-rich, and only marginally bound. Electron neutrino losses during the collapse amount to only $\sim 10^{51}$ ergs and those accompanying breakout to less than 10^{52} ergs (Burrows & Mazurek 1983). Most of the energy and leptons that must be radiated to form a neutron star, whose binding energy is 2-4×10^{53} ergs, has yet to be released.

Neutrinos are not only copiously produced at the high temperatures and densities achieved subsequent to stellar collapse, but can escape on a timescale that is significantly shorter than that of the photons that are the traditional work horses of stellar evolution. The standard model predicts that neutrinos of all species (ν_e, $\bar{\nu}_e$, ν_μ, $\bar{\nu}_\mu$, ν_τ, $\bar{\nu}_\tau$), not just ν_e's, carry away the neutron star's binding energy in roughly equal amounts in, not milliseconds, but seconds (Burrows & Lattimer 1986), as these neutrinos diffuse out of the hot, opaque quasi-hydrostatic protoneutron star. This time scale is set by the opacity of dense matter. The major sources of opacity are charged-current absorption (involving both ν_e and $\bar{\nu}_e$), electron scattering, and neutral-current scattering off of free neutrons and protons (involving all neutrino species). The initial integrated depth of the star to ν_e's is $\sim 10^4$, whereas that for the mu and tau neutrinos is $\sim 10^3$.

The initial emission is dominated by the cooling and neutronization of the shocked outer core. This emission may be enhanced by convective transport (Burrows 1987a). The early phase lasts no more than half a second and blends into the long-term phase of diffusion from the inner core. As the neutrinos escape, they downscatter in energy. The average energy of the ν_e's and $\bar{\nu}_e$'s should be 10-20 MeV and that of the ν_μ's, $\bar{\nu}_\mu$'s, ν_τ's, and $\bar{\nu}_\tau$'s (hereafter "ν_μ's") should be 15-25 MeV (Bowers and Wilson 1982; Mayle et al. 1987; Woosley et al. 1986). It is not expected that 100 MeV neutrinos will be seen in great numbers, that ν_e's will dominate, or that the signal will last only milliseconds. It is expected that the $\bar{\nu}_e$'s, by their large absorption cross section on protons, will dominate the signal in water Cherenkov detectors. Though the ratios of the total energy emitted are approximately ν_e:$\bar{\nu}_e$:ν_μ::1.2:1:4, the ratios of the signals are expected to be ν_e:$\bar{\nu}_e$:ν_μ::2:40:1. Since photons cannot escape from the core during this phase, neutrinos are the only diagnostic of the event.

The pre-SN1987a list of predictions for the neutrino signal are given in Table 1. Note that the total emitted neutrino energy is expected to dwarf by two orders of magnitude the energy of the supernova explosion itself. From the energetic point of view, the supernova of a sideshow to the main event: neutron star birth. There is not expected to be much signal from the dynamical phase of collapse, rebound, and shock formation. The core energy is trapped on collapse timescales and must be released on neutrino diffusion timescales after the dense core has formed.

Table 1. Sketch of Standard Neutrino Signature.

Phase	Duration	Energy
- Collapse:	10 - 10^2 milliseconds	$\sim 10^{51}$ergs
- Prompt (breakout) ν_e burst	1 - 3 milliseconds	$\sim$1 - 3 x 10^{51}ergs
- Cooling and Neutronization	seconds	$\sim$2 x 10^{53}ergs

total energy: $\nu_e : \bar{\nu}_e$: "ν_μ" *:: 1.2 : 1 : 4; all species

BUT

total signal (in H_2O): $\nu_e : \bar{\nu}_e$: "ν_μ" *:: 2 : 40 : 1

40 ← $\bar{\nu}_e + p \rightarrow n + e^+$

$\langle \epsilon_{\nu_e}, \epsilon_{\bar{\nu}_e} \rangle \sim$ 10 - 20 MeV; not 100 MeV or 0.1 MeV

2 x 10^{53}ergs » 10^{51}ergs » 10^{49}ergs

(ν) (SN K.E.) (light)

"Neutrino star" embedded in core of massive star

(*"ν_μ" ≡ $\nu_\mu, \bar{\nu}_\mu, \nu_\tau, \bar{\nu}_\tau$)

The neutrino data from both the Kamiokande (KII) and IMB collaboration (Hirata et al. 1987; Bionta et al. 1987) are shown in Table 2 and depicted in Figure 1. Not in the table is the important fact that the neutrinos were observed before the light. Collapse is supposed to initiate the sequence of events that leads to the disassembly of the star and the optical pyrotechnics.

Fig. 1. The evolution of the Kamiokande II and IMB electron energies versus time.

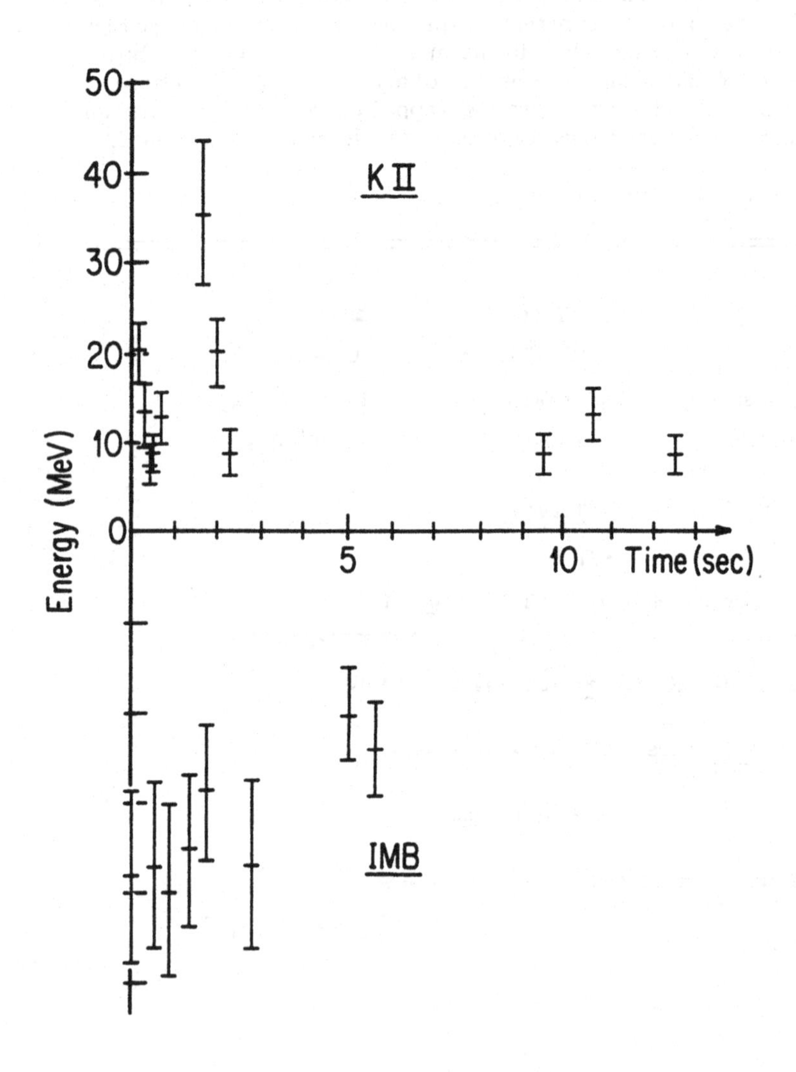

The large angles with respect to the LMC in Table 2 indicate that, indeed, most of the events were $\bar{\nu}_e$ events, since they lead to a roughly isotropic distribution of detectable positrons. Electron neutrino scattering events are much more forward-peaked. The first one or two events may be ν_e events, but the statistics are such that his can never be proven. It need not be the case. Pre-SN1987a calculations would have predicted a total of no more than one ν_e event in either detector (Burrows & Mazurek 1983, and references therein). Perhaps the most important fact that can be extracted from Table 2 is that the signal lasted for seconds, not milliseconds. Some (too many to reference) have suggested that this long duration indicates that the neutrino must have a mass. A nonzero neutrino mass and a spectrum of energies will result in a spread of arrival times if the intrinsic burst duration at the source is short. However, as I have stated, the long signal duration is a predicted and straightforward consequence of the large opacity of the residue to neutrinos. The signal in KII at 9-12 seconds is easy to fit without a mass. Furthermore, much has been made of the seven-second gap in the KII data. Pulses at the source, neutrino masses, and neutrino oscillations, and combinations thereof, have all been evoked. To my mind, the best explanation is that the gap is simply a statistical fluctation. Not only do the IMB data

Table 2. Data from the Kamiokande II and IMB detectors.

Event	Time (sec)	Electron energy (MeV)	Angle with respect to LMC (degrees)
Kamiokande II			
1	0.000*	20.0±2.9	18±18
2	0.107	13.5±3.2	15±27
3	0.303	7.5±2.0	108±32
4	0.324	9.2±2.7	70±30
5	0.507	12.8±2.9	135±23
6**	0.686	6.3±1.7	68±77
7	1.541	35.4±8.0	32±16
8	1.728	21.0±4.2	30±18
9	1.915	19.8±3.2	38±22
10	9.219	8.6±2.7	122±30
11	10.433	13.0±2.6	49±26
12	12.439	8.9±1.9	91±39
IMB			
1	0.00*	38 (±25%)	74 (±15)
2	0.42	37 "	52 "
3	0.65	40 "	56 "
4	1.15	35 "	63 "
5	1.57	29 "	40 "
6	2.69	37 "	52 "
7	5.59	20 "	39 "
8	5.59	24 "	102 "

*Time for initial event set equal to zero.

**Excluded by Kamiokande II collaboration.

fill the gap, but small-number statistics suggests that with a total of 11 events, gaps are expected (Lattimer 1987). An a priori probability calculation with a reasonable underlying distribution function reveals that the probability of some sort of bunching and gaps of seconds is quiet large. Many workers have been fooled by the low probabilities of a posteriori calculations. Since the stellar evolution calculations predict steady signal out to tens of seconds, Occam's razor suggests that, on occassion, one can be both conservative and correct.

Figure 2 depicts the anti-electron neutrino luminosity versus time for the first five seconds after bounce, as calculated by Bruenn (1987), Mayle and Wilson (1987) and Burrows (1987b). The broad dark swathe represents my attempt to guess the range of inferrable luminosities from an analysis of both the IMB and KII neutrino data and is not meant to be definitive. However, as Figure 2 suggests, the source luminosity seems to vary from $\sim 4\times 10^{52}$ ergs/s, initially, to $\sim 10^{51}$ ergs/s after five seconds. This behavior is roughly consistent with what was expected, as the model luminosities indicate. The code I used to calculate the neutrino signals was developed in 1985-1986 (Burrows & Lattimer 1986) to address precisely the problem of the formation of neutron stars. It is a fully implicit Henyey-like stellar evolution code with general relativity, general relativistic transport of all six neutrino species in the diffusion limit, and a detailed nuclear equation of state. The spectra and luminosities are, and always have been, appropriately redshifted. A virtue of this code is that a

Figure 2. The anti-electron lumnosity ($L_{\bar{\nu}_e}$) in ergs/s versus time for the first five seconds for a sample of model calculations and as inferred (in shade) from the data. See the text for discussion.

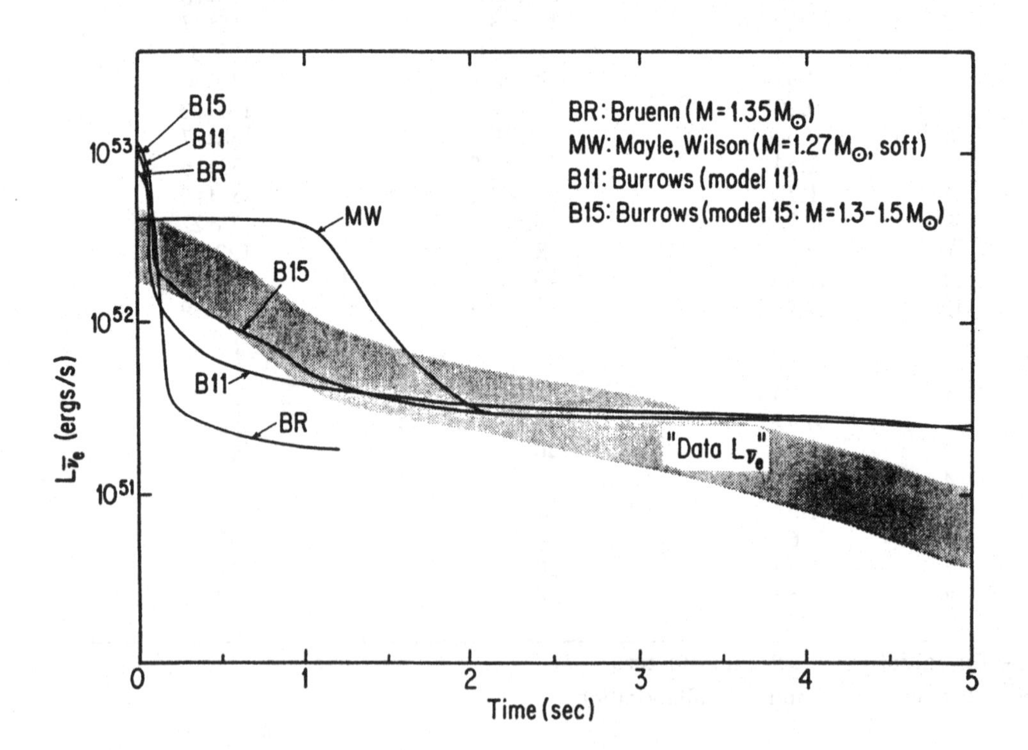

whole range of initial models, accretion regimes, equations of state, etc. can be tested quite quickly. Execution of a complete run takes only 1 hour of VAX 8650 time. This is ~100 times faster than other codes that are being applied to the same problem. A drawback of this approach is that the full hydrodynamic problem is not being solved, nor is the supernova mechanism being addressed. Fortunately, or unfortunately, the supernova mechanism itself does not affect the neutrino burst in ways that 19 events can test.

The models shown (B11 and B15) are merely representative and many behaviors can be accommodated. Note that variations or ambiguities in the initial core mass, the neutron star equation of state, the amount and rate of accretion after bounce, the degree of convection, or the neutrino cross sections at supranuclear densities imply that there is significant latitude in fitting the KII and IMB data and determining the parameters of the young neutron star in the LMC.

Figure 3 shows the evolution of the average anti-electron neutrino energy during the first 12 seconds according to various groups. The difference between B11 and B17 is in the more realistic anti-neutrino cross sections that were incorporated in the code after February 23. The difference is not large, but since the IMB detector has a high threshold ($\epsilon_{e^+} \sim 20$ MeV), a change in the effective temperature of the $\bar{\nu}_e$'s

Figure 3. The average anti-electron energy ($\epsilon_{\bar{\nu}_e}$) in MeV versus time for a sample of model calculations. The dark region denotes the region that best fits the data. See the text for an explanation.

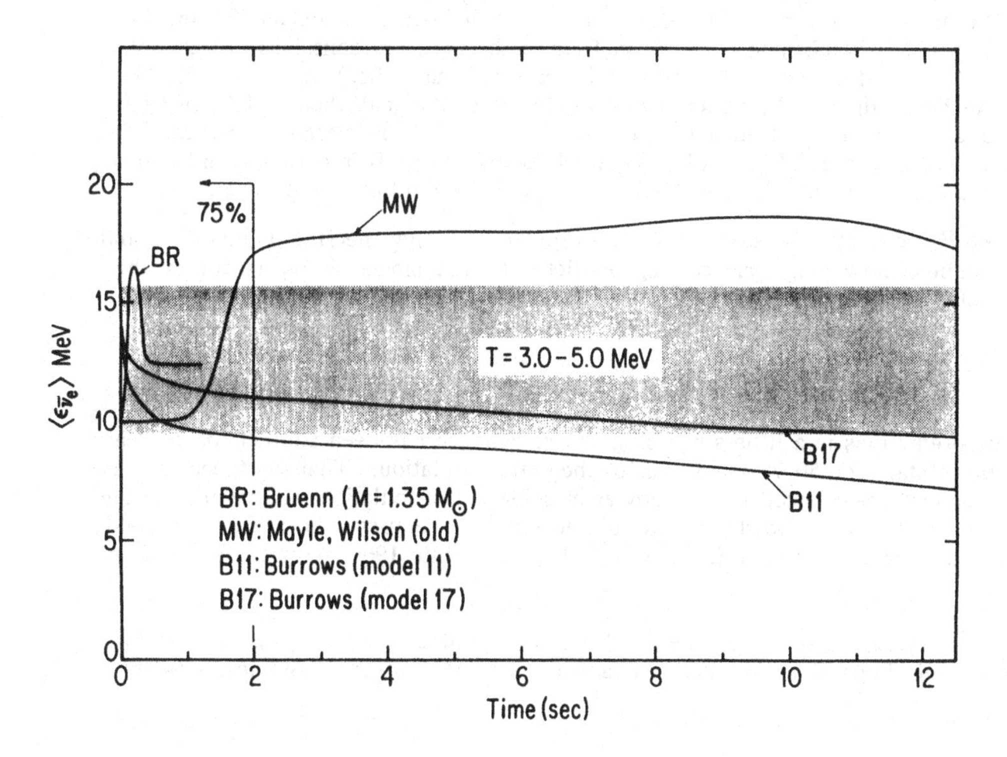

of 10% can have a non-trivial effect on the neutrino signal it registers. The swathe in Figure 3 depicts the range in the redshifted temperature (3-5 MeV) that the data imply (Burrows & Lattimer 1987). Models BR and B17 seem to fit quite well, but no attempt was made to fit the models to the data in this figure. These models are simply representative of the characteristic behavior of the calculated signals.

I performed a whole series of model calculations with different initial core masses, accretion rates, equations of state, etc. in an effort to determine what models can be eliminated by the data. The predicted integrated event number versus time for the first 18 seconds for a subset of this series is shown in Figure 4. Superposed are the Kamioka and IMB data. The detection efficiencies, thresholds, and fiducial masses for each detector were folded into the calculation and a distance of 50 kpc was used. No normalization was imposed. The dashed curves marked I#, denote signals in the IMB detector and the solid lines marked K# denote signals in the Kamiokande II detector. Models 27, 29, and 30 incorporated accretion at the level indicated in the figure legend. The total $\bar{\nu}_e$ energy ($E_{\bar{\nu}_e}$) radiated in each model during the first 20 seconds is shown to the left of the model numbers in the legend. One "foe" is 10^{51} ergs in the jargon of the field. Stiff means that the cold neutron state equation of state is similar to Bethe-Johnson I, and soft means that it is similar to Bethe-Johnson V (see Figure 5).

Figure 4 indicates that models 27 (Black hole, soft), 28 ($M = 1.2\ M_\odot$), 29 ($\sim 0.4\ M_\odot$ of accretion, stiff), and 34 (Black hole, stiff) do not fit the two data sets well. If the neutron star recently formed in the LMC were massive ($\gtrsim 1.7\ M_\odot$), the accretion component to the luminosity would have been too large, too early to explain what we saw. We arrive at a similar conclusion, if a black hole formed (models 27 and 34). Note that when a black hole forms (* in figure), the neutrino emissions are abruptly (<1 millisecond) truncated. If the residue had a small mass ($\lesssim 1.2\ M_\odot$, model 28) there would not be enough binding energy to provide the IMB signal, though KII can be fit. The above considerations lead me to conclude that the baryon mass of the residue is between 1.2 $M_\odot$ and 1.7 $M_\odot$, with 1.3-1.6 $M_\odot$ being the preferred range, and that a black hole did not form. $E_{\bar{\nu}_e}$ is $\sim 3\times 10^{52}$ ergs and the binding energy radiated was $2.0\text{-}2.5\times 10^{53}$ ergs. It is gratifying that the rapid rise during the first 1-2 seconds and the subsequent slow long-term cooling predicted by the models is borne out by the data. Note, however, that 19 events do not a certainty make.

CONCLUSION

That the characteristics of these neutrino events are consistent with previous predictions is a little surprising. Core collapse has been studied for 30 years and neutron stars for 50 years in blissful theoretical isolation. That the theorists were "on the money" speaks well for the power of scientific speculation and calculation and is a testament to the stalworth pundits of the last half century who nurtured a field that only now is being tested at its core (Colgate & White 1966; Arnett 1967; Wilson 1971).

What have we learned? Though we did not have enough events to illuminate the Type II supernova mechanism itself, the inferred luminosity ($\sim 10^{52}$

Figure 4. The total integrated number of events in IMB and Kamioka versus time for the first 18 seconds after bounce for various recently calculated models. Superposed is the actual data (solid: Kamioka; dashed: IMB). Threshold and efficiency corrections have been incorporated and a distance of 50 kpc to SN1987a was assumed. A more detailed discussion of the figure and its legend are given in the text.

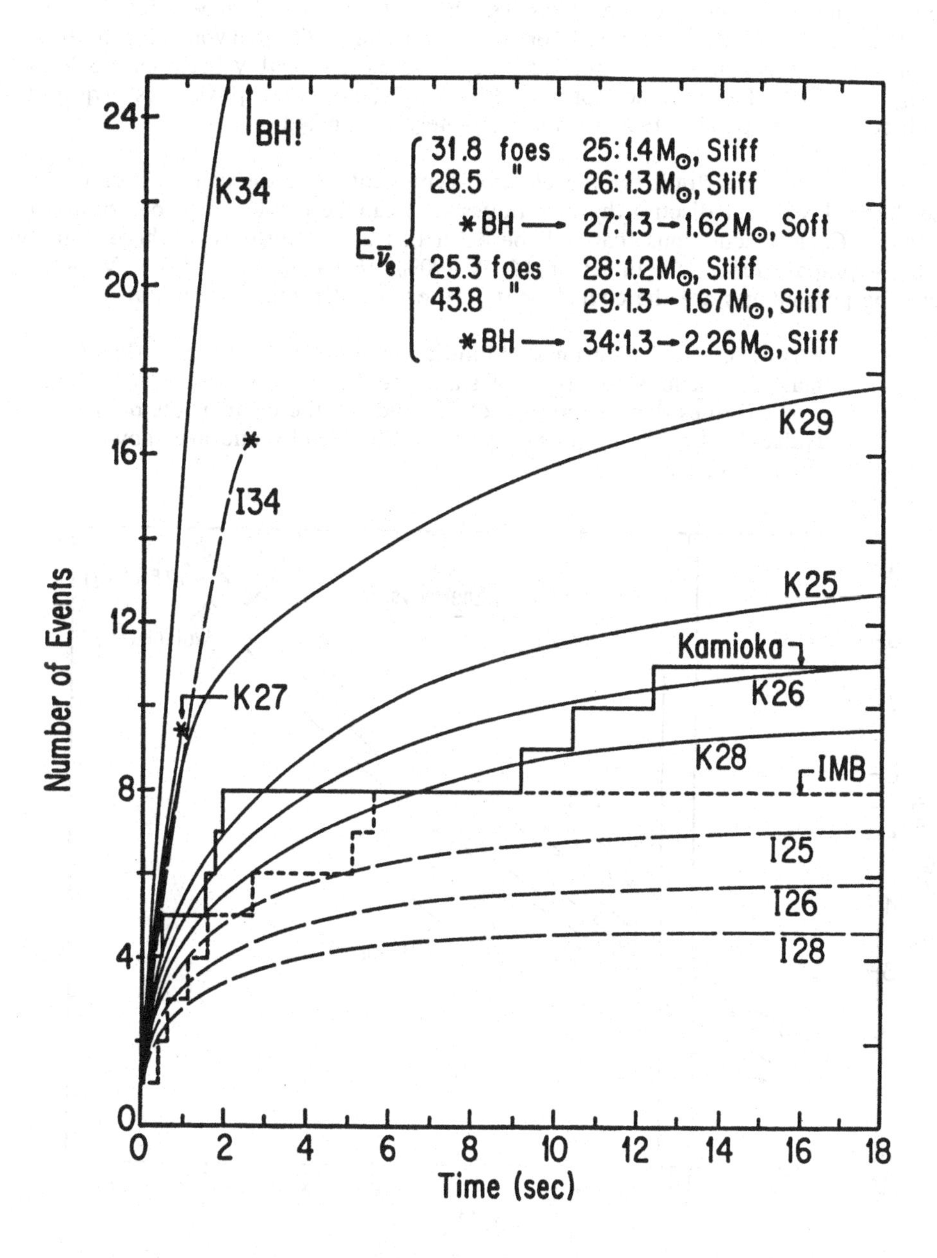

ergs/s in $\bar{\nu}_e$'s) is adequate to power the long-term neutrino mechanism if the prompt mechanism aborts (Bethe & Wilson 1985; Burrows & Lattimer 1985). Most or all of the events were $\bar{\nu}_e$'s, as had been predicted. The burst duration was a few seconds, not milliseconds. The average effective $\bar{\nu}_e$ temperature was 3-5 MeV. $E_{\bar{\nu}_e}$ was $\sim 3\times10^{52}$ ergs, and E_{total} was $\sim 2\text{-}3\times10^{53}$ ergs. Anti-neutrino luminosities between 4×10^{52} ergs/s and 10^{51} ergs/s seem to be indicated. The above temperature and luminosities imply that the radius of the object was less than 50 km, probably between 10 and 20 km. This is just what is expected for a neutron star. The baryon mass of this neutron star is between 1.2 $M_\odot$ and 1.7 $M_\odot$, but is most probably between 1.3 $M_\odot$ and 1.6 $M_\odot$. A black hole did not form. Interestingly, no exotic physics is required, though some exotic possibilities are not completely eliminated.

An upper limit for the electron-type neutrino mass can be derived from these data. I will not discuss the techniques that can be employed to determine it. My Monte Carlo calculations that will be described and reported elsewhere (Burrows 1987b, in preparation) yield limits of 14 eV (90%), 16 eV (95%), and 19 eV (99%), where the percentages are the confidences. To arrive at these numbers, all the

Figure 5. Cold neutron star binding energies versus gravitational mass for various equations of state (see Arnett & Bowers 1977). The vertical lines denote the masses derived for the components of the Hulse-Taylor binary pulsar system (PSR1913+16) and are shown only for reference.

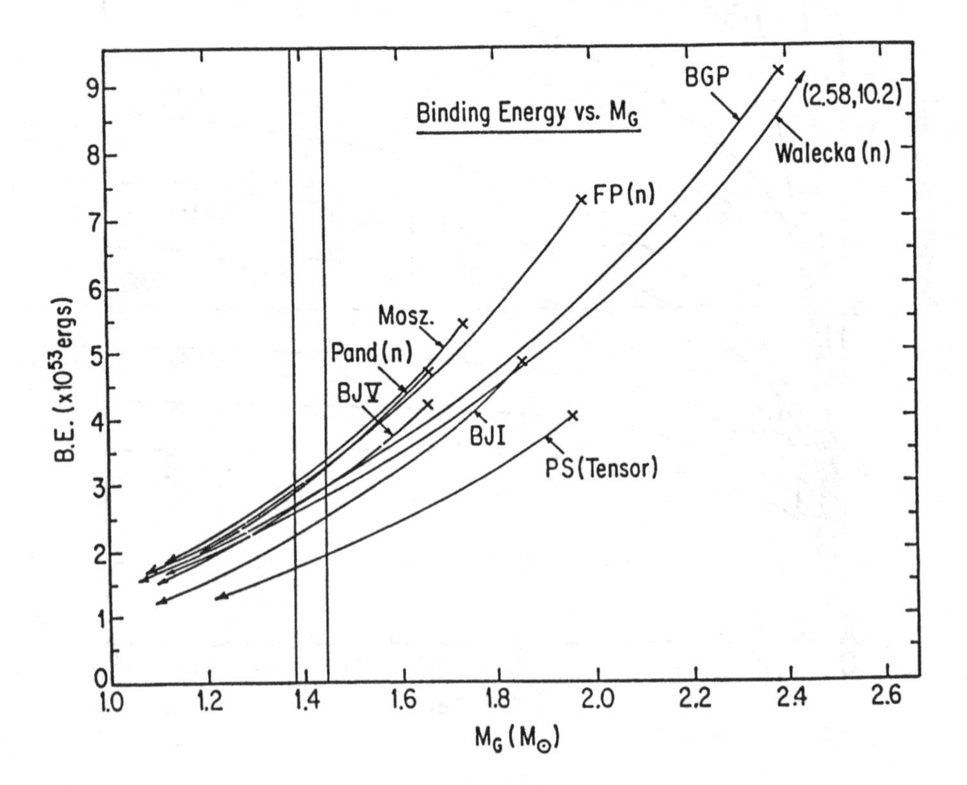

Kamioka data were employed, conservative assumptions were made thoughout the calculation in the spirit of a mass limit, and a reasonable source model was employed. Curiously, the 95% number (~ 16 eV) is exactly what is derived by Spergel and Bahcall (1987, as reported by Bahcall at this meeting) by a completely different and independent technique. If these calculations stand up, the electron neutrino mass is very probably lower than the lower limit quoted by Boris et al. (1985).

ACKNOWLEDGMENTS

I would like to acknowledge Dave Arnett, Jim Lattimer, Ron Mayle, Jim Wilson, and Stan Woosley for useful and interesting conversations and the Alfred P. Sloan Foundation for financial support.

REFERENCES

Arnett, W. D. (1967). Can. J. Phys., 45, 1621.
Arnett, W. D. & Bowers, R. L. (1977). Ap. J. Suppl., 33, 415.
Bethe, H. & Wilson, J. R. (1985). Ap. J., 295, 14.
Bionta, R. M. et al. (1987). Phys. Rev. Lett., 58, 1494 (IMB Collaboration).
Brown, L. E. et al. (1982). Nucl. Phys., A375, 481.
Burrows, A. & Lattimer, J. M. (1986). Ap. J., 307, 178.
Burrows, A. (1987a). Ap. J. (Letters), 318 L57.
Burrows, A. (1987b). In Proceedings of the UCLA Workshop on Extra Solar Neutrino Astronomy, Sept. 30-Oct. 2, UCLA, ed. D. Cline.
Burrows, A. (1987c). In preparation.
Burrows, A. & Lattimer, J. M. (1985). Ap. J. (Letters), 299, L19.
Burrows, A. & Lattimer, J. M. (1987). Ap. J. (Letters), 318, L63.
Burrows, A. & Mazurek, T. J. (1983). Nature, 301, 315.
Boris, S. D. et al. (1985). Sov. Phys.-JETP Letters, 42, 130.
Bowers, R. L. & Wilson, J. R. (1982). Ap. J. Suppl, 50, 115.
Bruenn, S. W. (1987). Phys. Rev. Lett., 59, 938.
Colgate, S. A. & Johnson, H. J. (1960). Phys. Rev. Lett., 51, 235.
Colgate, S. A. & White, R. H. (1966). Ap. J., 143, 626.
Hirata, K. et al. (1987). Phys. Rev. Lett., 58, 1490 (Kamiokande II collaboration).
Hoyle, F. & Fowler, W. (1964). Ap. J. Suppl., 9, 201.
Lattimer, J. M. (1987). In Proceedings of the Minnesota conference on SN1987a, eds. T. Walsh & K. Olive, Minneapolis, Minn., June 4-6.
Mayle, R. et al. (1987). Ap. J., 318, 288.
Mayle, R. & Wilson, J. R. (1987). Preprint.
Mazurek, T. J. (1974). Nature, 252, 287.
Sato, K. (1975). Prog. Theor. Phys., 54, 1325.
Nomato, K. (1984). Ap. J., 277, 791.
Wilson, J. R. (1971). Ap. J., 163, 290.
Wilson, J. R. et al. (1986). In Proceedings of the 12th Texas Symposium on Relativistic Astrophysics, ed. M. Livio & G. Shaviv, Ann. N.Y. Acad. Sci., 470, 267.
Woosley, S. E. (1986). Ap. J., 302, 19.
Woosley, S. E. & Weaver, T. A. (1986). In Radiation Hydrodynamics in Stars and Compact Objects, eds. D. Mihalas and K.-H. A. Winkler, p. 91. Berlin: Springer-Verlag.

PHENOMENOLOGICAL ANALYSIS OF NEUTRINO EMISSION FROM SN 1987A

J. N. Bahcall and D. N. Spergel
Institute for Advanced Study, Princeton, NJ 08540

W. H. Press
Harvard University, Cambridge, MA 02138

Presented by John Bahcall

Abstract
Monte-Carlo and maximum likelihood analyses of the neutrino emission from SN 1987a are used to establish the allowed range of the $\bar{\nu}_e$ emission temperature, rise time, cooling time, flux, and total energy. The acceptable range of the temperature is: $T_0 = 4.2^{+1.2}_{-.8}$ MeV, of the cooling time: $\tau = 4.5^{+1.7}_{-2.0}$ sec, and of the total emitted energy: $6.1^{+3.5}_{-3.6} \times 10^{52} N_{\rm all}$ erg, where $N_{\rm all}$ is the ratio of energy emitted in all neutrino forms to that emitted only in the form of $\bar{\nu}_e$. The combined data from Kamiokande and IMB are sufficient to reject masses of ν_e that exceed 16 eV at the 5% significance level. For fun and instruction, we "interpret" *post facto* some simulated data drawn from a distribution that describes well the actual data.

I. INTRODUCTION

We need some fun at this conference. Therefore, let us omit the details of the statistical analyses that we have done, so as to save some time for discussing simulated data: what Kamiokande II might have seen had we been unlucky (or lucky, you will have to make that judgment). First, we describe the simplest, single temperature model (§II); then, present the results when neutrino cooling is taken into account (§III). Next we describe the analysis of the neutrino mass limit (§IV). The Fun and Games are discussed in §V. The summary is given in §VI.

Fortunately, the statistical methods that we introduced (Bahcall, Piran, Press, and Spergel 1987, hereafter Paper I) in this connection have been adopted by a number of authors; several of the subsequent talks will describe results obtained by these techniques. Subsequent discussions will give the important technical details as well as many relevant references. In order not to duplicate references and equations, we will only emphasize here a few general aspects of the analysis.

It is important to analyze simultaneously the Kamiokande and the IMB results. If one ignores, for example, the Kamiokande data one obtains too high a neutrino emission temperature and if one ignores the IMB data one infers too much ν−e scattering. The combined data are fit satisfactorily although not perfectly by a very simple phenomenological model (which can't be rejected at the 5% significance level, see below). One must include both the measured energies and angles in the likelihood function; both quantities carry significant information about the incident neutrinos. The measurement uncertainties are crucial (both for energy and angle). For example, including the measurement errors in energy increase (for our best-fit single temperature model) the expected number of events in the IMB detector that are seen to be above 35 MeV by 50%, and double the total number of events above 40 MeV. Of course, one must include explicitly the energy sensitivity of each detector and use accurate values for the neutrino and anti-neutrino interaction cross sections. Our values for all of these quantities are specified in Paper I.

We have performed estimation-of-parameters *and* goodness-of-fit tests on the data. Maximum likelihood parameters were estimated using a single likelihood function that incorporates for both experiments the expected energy and angular distributions, as well as the quoted detector efficiencies and the measurement errors. The joint likelihood function also takes into account the absolute number of events in each detector. Use of a joint likelihood function gives parameters that are most consistent with the observations in both detectors. To test whether this consistency is credible, without reference to likelihood, we find the range of parameters allowed by the Kolmogorov-Smirnov (KS) measures of the goodness-of-fit to the observed energies and angles of the individual events. Finally, we have verified also that the maximum likelihood and KS tests are consistent by doing Monte Carlo simulations with the maximum likelihood parameter values. The simulations do indeed produce synthetic data sets that are strongly concentrated in the KS-allowed region.

The Kamiokande II and the IMB experiments and data have been well described by Koshiba and Matthews in the preceding talks (see also Hirata et al. 1987 and Bionata et al. 1987). There is no need for us to repeat what all of you now should already know about this historic detection. We will give you a "visual examination" at the end of the talk which will test, among other things, your familiarity with the experimental data.

II. THE SIMPLEST MODEL

We report first a *single temperature* analysis of all the observed 8 events detected by the IMB collaboration and the first 8 events seen by the Kamiokande detector. The subsequent 3 Kamiokande events, with somewhat lower average energy, may reflect a cooling of the neutrino photosphere with time. We include all 19 of the events in the "cooling solution" that will be discussed in the next section ; the two analyses give essentially the same results for the total energy emitted and the peak temperature.

In principle, we would like to consider solutions in which each type of neutrino (i = electron, muon, tau, and their anti-neutrinos) is described by a thermal spectrum, with an associated temperature, T_i, and a flux, F_i , at Earth. Unfortunately, the data are not sufficient to permit this detailed an analysis; the number of parameters would be almost comparable to the number of events. Instead, we take an orthogonal tack and adopt a "minimum" model, in which all neutrinos and anti-neutrinos have the same temperature (with no high-energy cutoff). We also collapse the problem to only two incident fluxes. We separate electron neutrinos, ν_e 's, from all other neutrinos and anti-neutrinos (ν_μ , ν_τ , and their anti-particles). At the temperatures considered here, these other neutrinos and anti-neutrinos can only scatter off the electrons in the water targets. We lump everything except $\bar{\nu}_e$ into one flux of "average" scatterers. We expect (see Paper I), but cannot prove, that $F_{\nu_e} \sim 0.5 \times F_{scatt.}$.

The best constant value for the temperature is

$$T = 4.1 \text{ MeV}. \tag{1}$$

The corresponding value of $F_{\bar{\nu}_e} = 0.5 \times 10^{10}\text{cm}^{-2}$.

The single temperature model represents a satisfactory fit to the observations. For $T = 4.1$ MeV and $F_{scatt.}/F_{\bar{\nu}_e} = 10$, the probability of obtaining a worse KS measure than was found for the observed energy distribution is 33% for the Kamiokande data and 65% for the IMB data.

Monte Carlo simulations show that (5% significance) between 3 and 7 of the first 8 events in Kamiokande were due to absorption of electron anti-neutrinos. The KS test can only reject the possibility of 0 scattering events at 2% significance, which is small but not completely negligible. There were between 6 and 8 absorption events in IMB (95% confidence level).

The analysis shows that there is a well-defined range of anti-neutrino fluxes,

$$F_{\bar{\nu}_e} = (0.15 \text{ to } 0.7) \times 10^{10}\text{cm}^{-2}, \tag{1b}$$

(95% confidence level) that is consistent with both the observed event rates. If we require that the count rates in IMB and Kamiokande are compatible *and* assume that the detector efficiencies are accurately known then the neutrino temperature must exceed 3.7 MeV.

For the temperature and flux ranges inferred here, we would not expect a statistically significant signal in either the Mount Blanc or the Baksan scintillator detectors (see, however, the discussion by P. Galeotti at this conference). Assuming a detection efficiency of 100%, we estimate $0.8 \times (T/4.1 \text{ MeV})^2 \left(F_{\bar{\nu}_e}/0.5 \times 10^{10}\text{cm}^{-2}\right)$ absorption events per 100 tons of scintillator.

The flux of $\nu_{scatter.}$ is poorly determined; this flux is constrained by a small number of events in the forward peak of either detector. The $\nu_{scatter.}$ flux is harder to determine than the $\bar{\nu}_e$ flux because the scattering cross section is much smaller than the absorption cross section. The strongest experimental limit on $F_{scatt.}$ comes from the detection of 1 to 5 (forward-peaked) scattering events in Kamiokande. We find:

$$F_{scatt} = (0.1 \text{ to } 5) \times 10^{10}\text{cm}^{-2}. \tag{1c}$$

We do not expect a detectable number of events in the ^{37}Cl experiment of Davis. The calculated number of events varies from 0.02 to 2. *According to our estimate, the background from solar neutrinos would have swamped Davis's supernova signal!*

III. COOLING NEUTRON STAR

A simplified analytic model of a cooling hot neutron star, motivated by detailed computer calculations, describes well *all* the neutrino events detected including their time dependence. The observations do not require explanations that invoke exotic physics or complicated astrophysics. As we shall see , the parameters in even this simple model are not severely constrained.

There is one unexpected feature of the LMC supernova neutrino data: the 7.3 seconds gap between the first 8 and the last 3 events in the Kamiokande II data. The Kamiokande II detector observed 8 events in the first 1.9 seconds, followed by a quiet period of 7.3 seconds, and then three events were detected within 3.2 seconds. The IMB detector observed 6 events in the first 2.7 seconds, followed by a quiet period of 2.4 seconds, and then two events were detected within 0.6 seconds. There are two IMB events in the Kamiokande "time gap."

Figure 1 of Spergel, Piran, Loeb, Goodman, and Bahcall (1987) shows that the average energy of an event declines with time and demonstrates the agreement of the data with a cooling blackbody model. A cooling hot neutron star model fits well all the observed data and provides an estimate of the radius of the hot neutron star. We focus on this simple model to show that it is not necessary to invent new physics to explain the data: The observations can be fit by a simple model motivated by detailed calculations performed (by many investigators) before the occurrence of SN 1987a.

We combine the IMB and Kamiokande data sets assuming that the first neutrino observed by IMB arrived at the same time as the first neutrino detected by Kamiokande; the offset time is not known precisely. Given the observed rates, the expected time lag is $\approx 1/4$ second. Our conclusions do not change if we include a time lag of this order. For simplicity in presentation, we also neglect neutrino scattering events; the inclusion of scattering would not alter the estimates of neutrino temperature nor change any of the conclusions summarized here.

The temporal structure of the combined data can be fitted by an exponentially decaying flux $F \propto \exp[-(\ln 2)t/t_{1/2}]$ with $t_{1/2} = 2^{+0.9}_{-0.8}$ seconds (see Fig. 2 of Spergel et al. 1987). (All numbers in this section are quoted with 95% confidence limits.) Monte-Carlo simulations of data drawn from this function show that a worse fit for the Kolmogorov-Smirnov (KS) measure would be obtained in 10% of the cases. The exponential decay is not unique and other functional forms can provide even better fits to the data.

When the core of a massive star can no longer support itself, it collapses rapidly on a dynamical timescale of milliseconds. When the density in the core reaches nuclear densities, nuclear pressure stops the collapse and in this core bounce, the gravitational binding energy is converted into thermal energy. The hot neutron star is so dense that it can cool only by neutrino emission. The densities achieved during core collapse are so large that the core is no longer transparent to neutrinos and the characteristic anti-neutrino emission time is not the collapse time of milliseconds, but rather the neutron diffusion time, seconds. Detailed calculations of neutrino transport of the cooling hot neutron star suggest (e.g., Burrows and Lattimer 1985; Wilson, Mayle, Woolsey, and Weaver 1986) that the bulk of the energy is emitted in all flavors of neutrinos over several seconds.

We fit the observed data pairs (E_i, t_i) to a phenomenological model where the neutrino source is a black body with an exponentially decaying temperature: $T = T_0 \exp(-t/4\tau)$. (The energy density at the surface is proportional to T^4, thus τ is the cooling timescale for the hot neutron star). The likelihood function is maximized at $T_0 = 4.2$ MeV, $\tau = 4.6$ secs, and a total fluence $F = 1.3 \times 10^{10}$ $\bar{\nu}_e$ cm^{-2}. Using the multidimensional KS test and Monte Carlo simulations to determine the significance of this solution, we find that the observed KS measure is better than 55% of the cases obtained from the synthetic data. At 95%

confidence level we obtain :

$$T_0 = 4.2^{+1.2}_{-.8} \text{ MeV} \;;\; \tau = 4.5^{+1.7}_{-2.0}\text{s} \;;\; F_0T_0^2 = 4.0^{+2.4}_{-2.0} \times 10^{10}\text{cm}^{-2}\text{MeV}^2\text{s}^{-1}. \quad (2)$$

Because of the strong correlation of the estimated flux with temperature, we have quoted $F_0T_0^2$. The Kamiokande data alone yields a peak temperature of $2.9^{+1.3}_{-0.5}$ and the IMB data alone yields a peak temperature of $4.9^{+4.2}_{-1.9}$.

Both the cross-section, σ, and the energy dependent detector efficiencies influence the event rate in the detectors. Thus, the event rate $\propto T^{\eta}$ with $\eta \approx 5-8$ in the temperature range considered. The characteristic decay time of the event rate is $4\tau/\eta$ and we expect that half of the events will arrive in the first 2.0 seconds.

Since the event rate drops much faster than the temperature, the simplest model (constant temperature, see previous section) gives a reasonable fit to the data.

The luminosity of a black-body of radius R can be equated to the product of the detected flux, F and the average energy of a neutrino, $3.15T$,

$$L_{\bar{\nu}_e} = 4\pi D^2 F(3.15T), \quad (3)$$

where D is the distance to the LMC. This yields an estimate of the radius of the cooling neutron star,

$$R = 27^{+17}_{-15} D_{50} \text{ km}, \quad (4)$$

where $D_{50} \equiv D/50$ kpc. This neutrino-sphere radius will decrease as the neutron star cools.

The total energy emitted by the hot neutron star,

$$\int L_\nu dt = N_\nu F_0(3.15T_0)(4\pi D^2)\tau = 6.1^{+3.5}_{-3.6} \times 10^{52} N_{\text{all}} D_{50}^2 \text{erg}, \quad (5)$$

where N_{all} is the ratio of energy emitted in all neutrino species to the energy emitted in electron anti-neutrinos. The quantity N_{all} is very uncertain due to the small number of detected scattering events. The need to invoke theory in order to estimate N_{all} exacerbates the uncertainties in the total luminosity. Several authors have suggested that since N_{all} is expected to exceed 6, the observations are consistent only with very hard equations of state or with the formation of a black hole, These authors have not included the uncertainties in the energy due to small number statistics. Equation (5) shows that even if we assumed $N_{all} = 8$, the total energy emitted would still be consistent with the binding energy of a $1.4M_\odot$ neutron star and a wide range of equations of state.

The success of this simplified model implies that it will be difficult to use the observed neutrino flux to confirm more detailed models. The supernova has confirmed the general picture of core collapse; however, it has not provided sufficient data to discriminate between equations of state or to validate specific detailed models. There is no need to evoke new particle physics or complicated astrophysical scenarios to explain the observed data. When a supernova is observed in our own galaxy, the detectors should record many hundreds of events and neutrino spectroscopy may then reveal surprises about stellar collapses and weak interaction physics.

IV. THE MASS OF ν_e

We will not review the already extensive literature on supernova neutrino mass limits (or determinations). To review the published and preprint literature would require by itself an entire invited talk. Instead, we will describe briefly the results of an extensive set of Monte Carlo simulations by Spergel and Bahcall (1987) which answer the following question. Suppose nuclear physicists measure a finite value for m_{ν_e}. How large does the claimed laboratory value of m_{ν_e} have to be in order that it is in conflict (5% significance) with the supernova observations?

The answer to this question is not identical to the answers to related questions that have been discussed in the literature.

Our basic procedure is to calculate confidence limits by performing Monte Carlo simulations that take account of the complexity of the actual experimental measurements. We find for each assumed value of m_{ν_e} the best fit of the observed data to a model of the time-dependence of the neutrino emission. We take account of the possible satanic tendency of nature by allowing the supernova temperature to first rise and then fall as a function of time, permitting low energy neutrinos to be emitted first. We use the best-fit parameters of the model for each value of m_{ν_e} to generate Monte Carlo simulations of what might have been observed, assuming both the best-fit model and the assumed mass m_{ν_e} are correct. The confidence level for this value of the mass is then determined by how often the observed data fit the model better than the simulated data.

Our basic assumption is: the neutrino emission temperature can be fit by a smoothly varying function of time. The functional forms we use in our calculations represent well the published values for the neutrino emissivity calculated from detailed models of supernova explosions. Of course, the functions we use are more general than the results of any specific theoretical calculation and therefore allow a potentially wider range of acceptable neutrino masses.

In practice, the mass limit does not depend sensitively on the assumed form of the temperature vs. time curve; we find a variation of about ± 1 eV over the range of models we consider. The strongest limits are obtained when the Kamiokande and IMB data sets are combined. Because of the uncertainty in the absolute time of arrival of the events in the Kamiokande data set, we determine for each Monte Carlo (in which of course the absolute times are known explicitly) the difference in arrival times, Δ, between the first IMB and the first Kamiokande event. This distribution of offset times is used in obtaining the mass limit.

We investigated six models (or parameterizations) of the neutrino emission. Most of the models have six free parameters: a peak temperature, T_{Max}, with a characteristic rise time, t_{rise}, and a characteristic fall time, t_{fall}. We studied models with both linear and exponential temperature rises and decays. In addition, the models included a neutron star radius, R, and a cooling time for the radius, τ_{radius}, as well as an offset time, Δ, between the first observed events in the two detectors. The parameters in each model were determined by maximizing the likelihood function describing the real observations, using a two-dimension generalization (energy and arrival time) of the K-S test.

The range of models and parameters included in these simulations is larger than in any of the other discussions we have seen. This causes our derived mass limit to be more conservative.

If the electron neutrino is massive, then the times at which neutrinos are detected will be a function of energy, $t_{detect} = t_0 + 0.26(m_\nu/E_\nu)^2$ s, where t_0 is the corresponding detection time of a massless neutrino. Because of the energy-dependence of the arrival times and the uncertainty in a given event's energy, the expected event rate is a complicated function of energy and time. The expected rate of events at a given time and energy will depend on the parameters in each phenomenological model. The model parameters were determined separately for each realization in each model by maximizing the joint likelihood function, which is a product over all events in both the Kamiokande and IMB detectors.

We determine the limits on m_{ν_e} by finding the best fit to the observed data for an assumed neutrino mass. If the KS measure of the best fit was worse than that obtained in 95% of the Monte Carlo simulations, the assumed neutrino mass was rejected at the "5% significance level".

Figure 1 of Spergel and Bahcall (1987) shows the fraction of simulated data that has a worse fit then the observed data averaged over offset times. This fraction drops well below 5% at 16 eV in all of the models we have investigated. *Hence, a mass m_{ν_e} in excess of 16 eV can be rejected at the 5% significance level.*

How much do the special assumptions made by other authors affect the final derived limits? We answered this question by repeating our Monte Carlo simulations making the same assumptions as other authors.

*) *Only Kamiokande Data*: yields a mass limit of 21 eV (with a maximum neutrino temperature of 2.5 MeV). The parameters determined by this solution are in conflict with the IMB data. [Burrows (preceding talk) avoided obtaining such a high mass limit by assuming a temperature of 4 MeV , which fits the combined data, although he analyzed only Kamiokande data.]

*) *Omitting without justification the lowest point in the experimental data*: the mass limit increases to 20 eV.

*) *Adding the highest energy background event, "event 6," (6.5* MeV*)*: the mass limit decreases to 12 eV.

*) *Simultaneous First Events in Kamiokande and IMB*: 12 eV mass limit for $\Delta = 0$ secs.

*) *Rise Time* = 0 secs : 12 eV [This assumption was made by Abbott, De Rujula, and Walker (1987).]

*) *Distinct Sub-pulse of Kamiokande*: 6 eV if assume, as in Burrows (1987), that the initial events in the first 0.5 secs were physically distinct.

*) *Neutronization Pulse*: 4 eV if one assumes one of the first two Kamiokande events was a ν_e (not a $\bar{\nu}_e$) from a physically distinct process that proceeded the main neutrino cooling (Dar and Dado 1987).

Several authors have deduced specific values of the neutrino mass by focusing on a special pattern in the data. This naturally leads us to the *fun* part of this talk.

V. FUN AND GAMES

Now we can enjoy ourselves! Let our imaginations take over. We can play the game of: "What if...?"

Figure 1a shows the distribution of events in an energy-time plane (MeV-seconds) that was found for 100 simulated Kamiokande detections of SN 1987a. †

The Monte Carlo simulations were created using a best-fit to the observed distribution with a temperature of the form $T(t) = 4\text{MeV}\exp(-t/4 \times 4.5\text{ sec})$. The actual temperature and time histories of the simulations take account of the sensitivity of the detector as a function of energy, the measurement errors, and the energy dependence of the cross section. The simulated luminosity has an exponential time dependence of 4.5 sec, but the simulated event rate has a mean life of about 2.9 sec (as in the observed data) because the detection efficiency involves additional powers of the temperature (through the interaction cross section and the detector sensitivity). In order to make the simulations as "bland" as possible, we have set the rise time equal to zero and required that the total number of events be exactly eleven as in the actual data.

The remaining eleven figures (Fig.1b-1d, Fig.2a-2d, and Fig. 3a-3d) represent "interesting" simulations of this featureless energy-time dependence. Each of these simulations occurred within the first batch of 100 tried. What would we have said about each of these simulations had they been the actual realization of the bland distribution we are sampling?

Fig. 1b shows clearly a "neutronization" pulse, 5 events within the first 1 second. The remaining 6 events are stretched out over 10 seconds. Had this realization been observed we would have been sure that the standard picture of neutrino processes in supernovae was wrong. The neutronization pulse shows more than an order of magnitude too many initial ν_e 's, implying that the total binding energy exceeds that of a neutron star and contradicting the MSW solution of the solar neutrino problem.

Fig. 1c shows a neutrino temperature that is clearly increasing with time. We theorists would have had to turn intellectual cartwheels to explain that one.

Fig. 1d is a beauty: it shows two separate temperatures, one with typical electron energies of 20 MeV and one with typical electron energies of 10 MeV. This clearly manifests the MSW effect: the high energy events represent MSW converted mu and tau anti-neutrinos while the lower energy events are $\bar{\nu}_e$'s that are unconverted.

Fig. 2a shows two separate groups of events, the first group corresponds to a zero mass neutrino and the second group to a different flavor neutrino with mass 15 eV (conveniently just below all existing laboratory measurements).

Fig. 2b shows a 7 sec gap between the initial and final group of events. The late arriving events must represent massive particles (perhaps ν_μ or ν_τ with a mass of

---

† We also created 1000 additional simulations, which we have not examined. The first batch of 100 was so rich in possible scenarios that we did not consider it necessary to look at the next 1000. Anyone with access to a Bitnet account and wishing to receive a machine readable copy of these realistic simulations can do so by contacting Schuver@IASSNS.

Fig. 1 . Fig. 1a shows the distribution of events in the energy-time plane for 100 simulated Kamiokande detections of SN 1987a. The "bland" Monte Carlo simulations are described in §V. Fig. 1b exhibits a neutronization pulse; Fig. 1c shows a neutrino temperature that increases with time; and Fig. 1d indicates that the MSW effect is occuring.

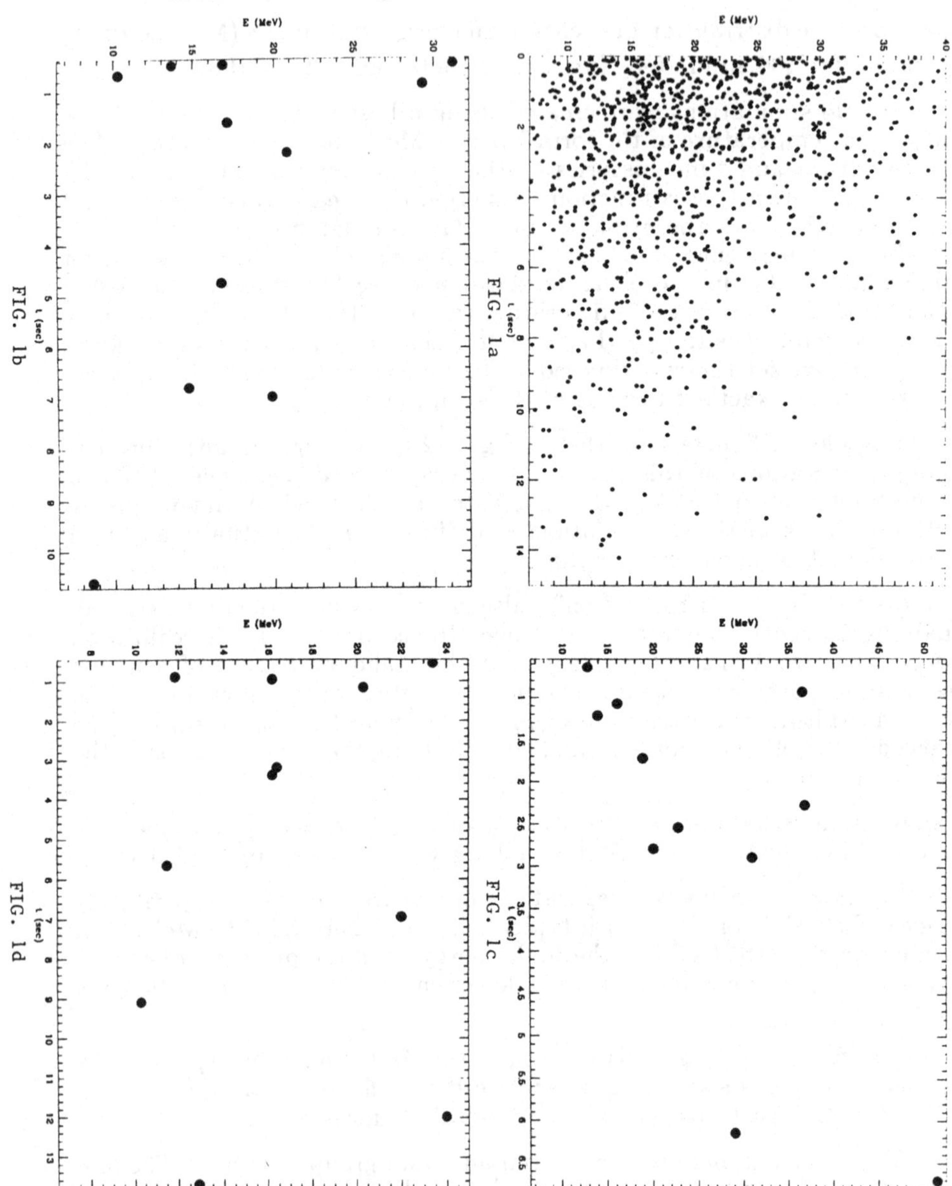

Fig. 2. Fig. 2a shows two separate events, perhaps corresponding to neutrinos of different masses. Fig. 2b shows shows a 7 sec gap. Figs. 2c and 2d show both a neutronization pulse and an approximately 7 sec gap.

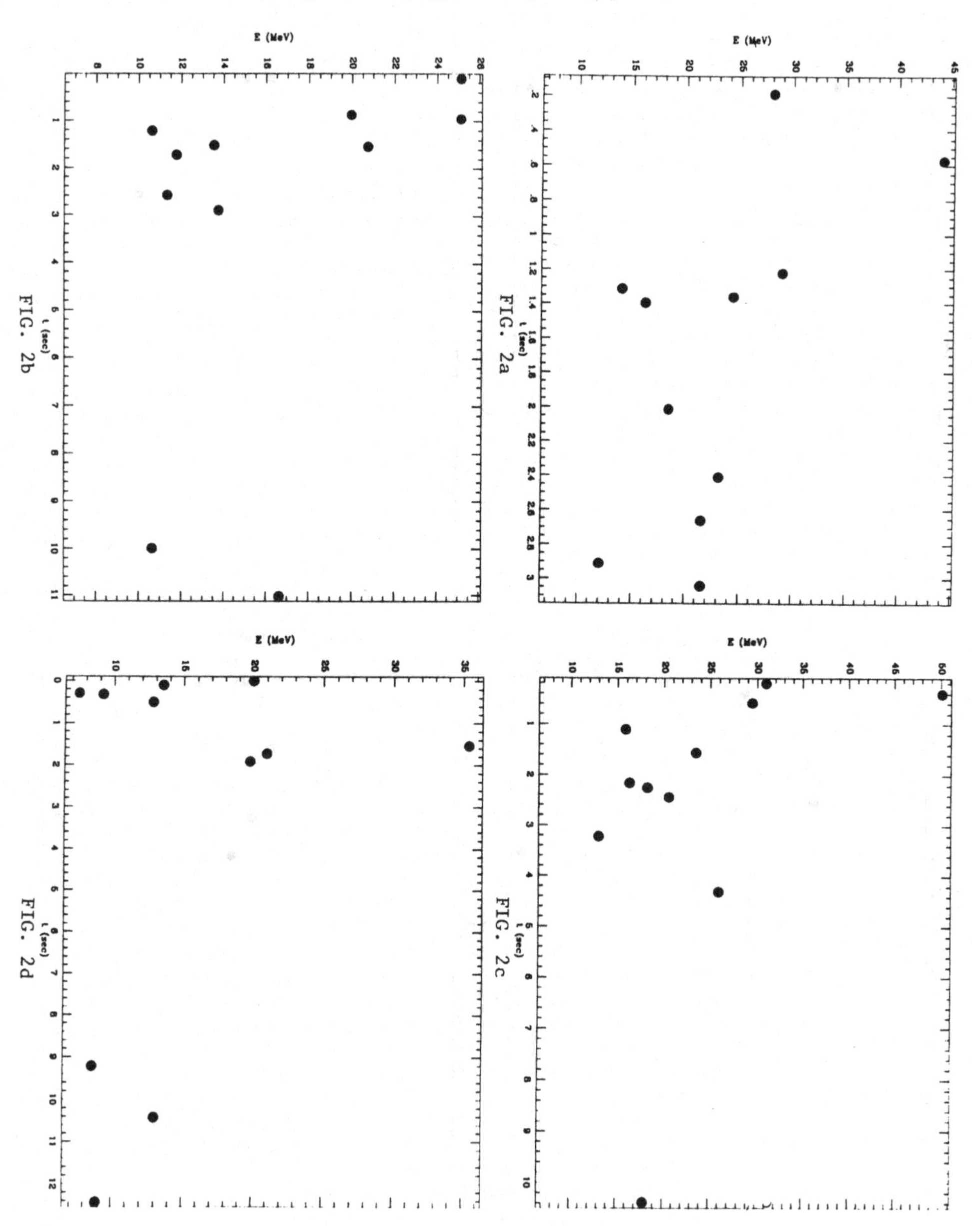

Fig. 3. Fig. 3a shows three separate bursts. Fig. 3b shows a neutronization pulse followed by a 1.5 sec gap. Figs. 3c and 3c show both a neutronization pulse and a gap.

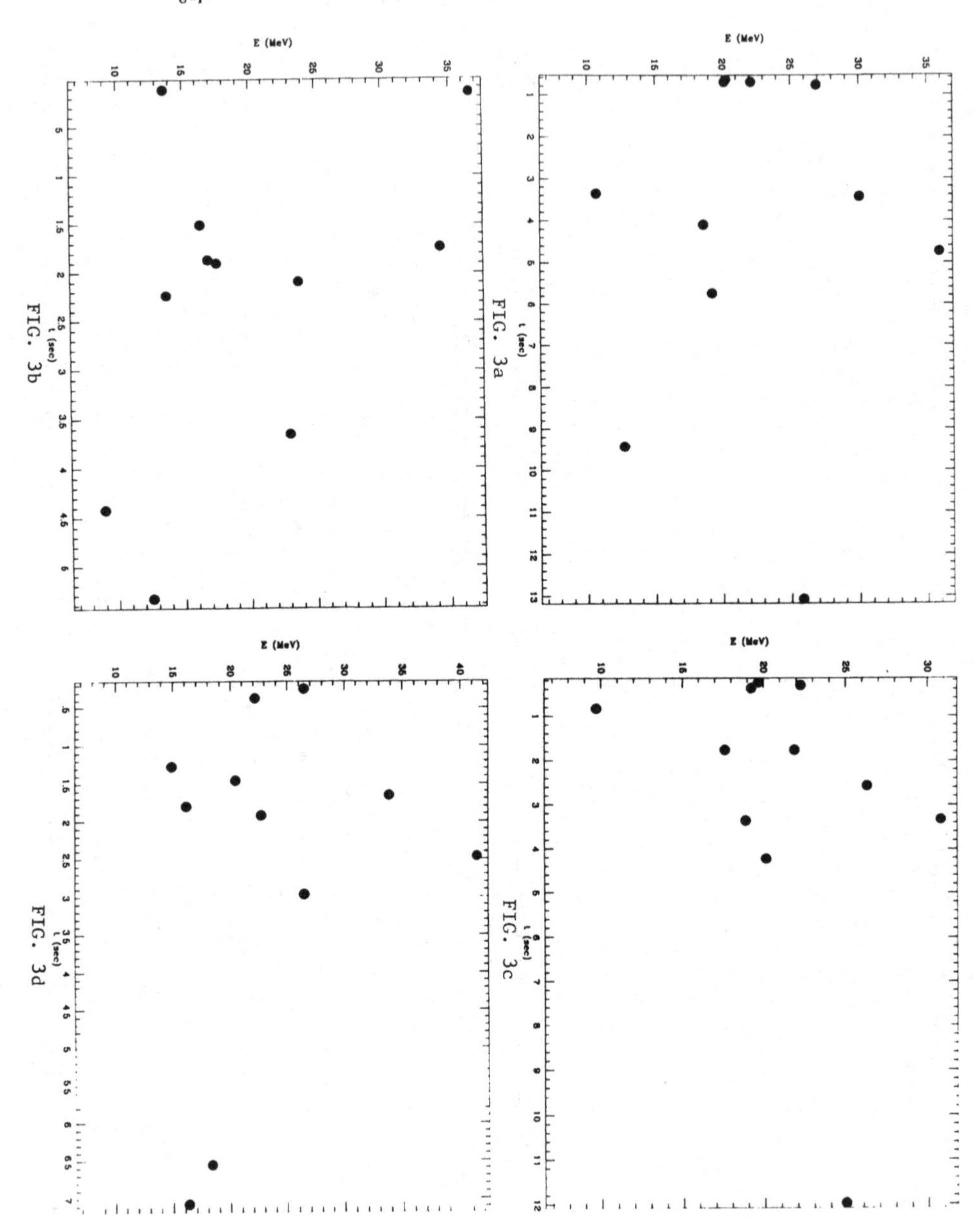

25 eV) that close the universe and solve the missing mass problem. Figs. 2c and 2d show both an initial neutronization pulse and an approximately 7 sec gap.

Fig. 3a shows events that clearly come in three bursts: a neutronization pulse of 4 events in less than 1 sec, followed by two pulses of decreasing intensity. Fig. 3b shows an unusual phenomenon: a neutronization pulse followed by an immediate 1.5 sec gap. Fig. 3d shows a neutronization pulse followed by a 3.5 sec gap. In Fig. 3c, a single high energy event arrives after an 8 second gap. Supernova models predict that only a few low energy events should arrive after the first few seconds. The 8 second delay is due to the 40 eV τ neutrino rest mass—a tremendous success for cosmology and the solution to the missing mass problem!

These simulations confirm Mark Twain's observation: "There is something fascinating about science. One gets such wholesale returns of conjecture out of such a trifling investment of fact."

Incidentally, I forgot to mention that Fig. 2d is the actual data.

VI. HAMLET PARAPHRASED

The physics of the explosion was complicated and requires detailed models for a correct description. However, the success of the simplified models described here suggests that the data from the LMC supernova are too sparse to discriminate among more sophisticated models or to justify inventing exotic new physics. In order to test in more detail our understanding of stellar collapse, we must await detection of a galactic core collapse [$\sim$ 50 times as many events expected at a rate of order 1 collapse event every 8 years, see Bahcall and Piran 1983] or await the availability of much larger detectors to observe stellar collapses in other galaxies.

Our basic conclusion may be stated by slightly rephrasing Hamlet's famous comment:

"There are more things about SN1987A in heaven and earth, Horatio, than are dreamt of in the *Astrophysical Journal*, but they are hard to prove at the 5% significance level."

ACKNOWLEDGMENTS

This research is supported partially by NSF grant PHY-8620266 and by NJ High Technology Grant 86-240090-2 at IAS, and NSF grant PHY-8604396 at Harvard. J. Bahcall acknowledges the grant of a Regent's Fellowship from the Smithsonian Institution which allowed part of this work to be performed at the Harvard-Smithsonian Center for Astrophysics, Cambridge, Massachusetts.

REFERENCES

Abbott, L., de Rujula A., and Walker, T. (1987). Submitted to *Phys. Letters*.

Bahcall, J. N., Piran, T., Press, W. H., and Spergel, D. N. (1987). *Nature* **327**, 682. (Paper I)

Bionata, R. M. et al. (1987). *Phys. Rev. Lett.* **58**, 1494.

Burrows, A. (1987). Anjona Theoretical Astrophysics preprint 87-12 (to be published).

Burrows, A., and Lattimer, J. M. (1985). *Ap. J.* **307**, 178.

Dar, A., and Dado, S. (1987). Preprint Technion (submitted to *Phys. Rev. Letters.*

Hirata, K. et al. (1987). *Phys. Rev. Lett.* **58**, 1490.

Spergel, D. N., and Bahcall, J. N. (1987). Submitted to *Phys. Letters.*

Spergel, D. N., Piran, T., Loeb, A., Goodman, J., and Bahcall, J. N. (1987). *Science* **237**, 1471.

Wilson, J. R., Mayle, R., Woolsey, S. E., Weaver, T. E. (1986). *Ann. N. Y. Acad. Sci.* **470**, 267.

Mass Determination of Neutrinos

Hong-Yee Chiu
Goddard Space Flight Center
Greenbelt, MD 20771, USA

Abstract. We have developed a time-energy correlation method (called the correlation mass method) to determine the signature of a nonzero neutrino mass in a small sample of neutrinos detected from a distant source. We apply this method to the Kamiokande II (KII) and Irvine-Michigan-Brookhaven (IMB) observations of neutrino bursts attributed to Supernova 1987a in the Large Magellanic Cloud (LMC). We obtain a neutrino rest mass of 3.6 eV. A further analysis, using Monte Carlo simulations that allow the energy of each event in the KII data set to vary with a Gaussian distribution that has the experimentally quoted mean and deviation, yields a further estimate for the neutrino rest mass of $2.8^{+2.}_{-1.4}$ *eV*. The same analysis also suggests that the KII data describe an initial neutrino pulse of less than 0.3 *sec* full width, followed by an emission tail lasting at least ten seconds.

1. **Time Correlation.** Consider a group of particles of equal, nonzero mass that are emitted at a source with different energies and times of origin. Let these particles propagate along a certain direction over a distance $L >>$ the dimension of the source. Consider the time dispersion between two particles after propagation. Let the times of origin be t_1, t_2, the times of detection be t'_1, t'_2, their energies be E_1, E_2 (assume, for the time being, $t'_1 > t'_2$ and $E_2 > E_1$), and the true mass be m_ν, then:

$$t'_1 - t'_2 = t_1 - t_2 + (L / v(E_1, m_\nu) - L / v(E_2, m_\nu)) \quad (1)$$

where v is the velocity, given by the well known equation:

$$E = m_\nu / \sqrt{1 - \beta^2}, \quad \beta = v/c \quad (2)$$

and c is the velocity of light. In the relativistic limit, Eq.(1) can be written as:

$$\Delta t'_{12} = \Delta t_{12} + (L / 2c)(m_\nu/E_1)^2(1 - (E_1/E_2)^2) \qquad (3)$$

where Δt_{12} is the difference $(t_1 - t_2)$, etc. First consider the case that $\Delta t_{12} = 0$ (that is, the two particles $1, 2$ originated at the same instant), then Eqs. (1) and (2) [or Eq.(3)] will yield a unique mass from the observed energies E_1, E_2 and the times of detection, t'_1, t'_2. If the two particles did not originate at the same instant, then an uncertainty will be introduced in the value of the mass derived from Eq.(3). The uncertainty is governed by the magnitude Δt_{12}. The dependence of the uncertainty in m_ν on Δt_{12} may be estimated from Eq.(3). Generally speaking, for a meaningful mass determination from *a pair* of particles, Δt_{12} must be smaller than (or at most comparable to) the energy dependent term in Eq.(3):

$$\Delta t_{12} \leq (L / 2c)(m_\nu/E_1)^2(1 - (E_1/E_2)^2) \qquad (4)$$

Neglecting Δt_{12} and applying Eq.(3) to *all* pairs of particles in the group at the point of detection, for those pairs of particles which satisfy condition (4) (independent of the actual time of origin), the mass obtained, $m_\nu^{(c)}$ (which is not the true mass on account of the uncertainty in Δt_{12}, and which will be referred to as the *correlation mass*) will concentrate around the true mass m_ν. However, for pairs which do not satisfy Eq.(4), $m_\nu^{(c)}$ may be of any value, including being imaginary [if the causality condition (5) is not satisfied]. In other words, if the true mass m_ν is not zero, in the correlation mass space there may exist a high concentration of pairs near the true mass. If the true mass m_ν is zero, then there will be no correlation in the $m_\nu^{(c)}$ space. Conversely speaking, if a mass correlation is observed, and if chance coincidence can be excluded, then we may conclude (a) there is a true time correlation at the source, and (b) the true mass m_ν is not zero. The following discussions will center around the correlation mass, and the superscript *(c)* will henceforth be dropped.

The assumption invoked just before Eq.(1) that

$t'_1 > t'_2$ and $E_2 > E_1$,

may be generalized into the so-called causality condition:

$$(E_2 - E_1)(t_1 - t_2) > 0. \quad (5)$$

In simple words, Eq.(5) requires that for a pair to be considered correlatable, it is necessary that the higher energy particle of the pair arrives *before* the lower energy one. Thus Eq. (5) is a selection rule to pick out those pairs which can *possibly* show a correlation. The inclusion of pairs violating Eq.(5) (yielding imaginary correlation masses) does not generate any new information. On the other hand, the inclusion of noncausal pairs can be used to test the validity of data, as we will see.

Another necessary condition to observe correlation is that observationally neutrinos (or other particles) must be detected with adequate time resolution; obviously if the detection of a neutrino is followed by a dead time Δt_d so that the detection of a subsequent neutrino is possible only after the elapse of a time Δt_d, correlation can only be observed if $m_\nu > m_\nu^{(d)}$ where $m_\nu^{(d)}$ is given by:

$$\Delta t_d = (L / 2c)(m_\nu^{(d)}/E_{min})^2 (1 - (E_{min}/E_{max})^2) \quad (6)$$

where (E_{min}, E_{max}) are the the minimum and the maximum energies of the data set. Thus, the observability of a correlation is dependent on the time resolution and the energy range of the detection system.

The advantage of correlation mass analysis is in the improvement of data structure. If *n* events have been observed, the total number of independent pairs that may possibly be subjected to mass correlation analysis is $n(n-1)/2 >> n$, [the actual number of pairs that can be studied is subject to the causality condition (5)]. Thus, this method is most advantageous for the case where n is small. Since the correlation mass method is a statistical method, its conclusions must be subjected to statistical tests, as will be discussed later.

2. **Simulations**. In the absence of analytical proofs, we carried out a number of computer simulations to study the applicability of the correlation mass method. The energies and times of arrival of 50 neutrinos are first chosen at random. These random values are scaled to an energy range and a time window of emission corresponding to the data structures of the KII (Hirata *et al*, 1987) and the IMB (Bionta *et al*, 1987) sets. These random values are then time shifted according to Eq. (3), using an input value for the true mass m_ν (in following examples, the input value is 4 *eV*). The distribution functions in the correlation mass space are then obtained for all eligible pairs [i.e., those satisfying the causality condition (5)]. If our reasonings in §*1* are correct, then for the time shifted data a high concentration of masses around the input mass m_ν should be observed, while the distribution function for the original data set should exhibit random behavior, i.e., it should follow a Gaussian or Poisson distribution.

We studied two cases, using the same set of random data. These two cases were tailored to the energy ranges of the KII and the IMB data sets. For reasons that will become clear in §*3*, a value of 4 eV is chosen for the input mass m_ν. For the distance L we use that to the Large Magellanic Cloud (165,000 light years). In the KII group the energy range is from 6.3 to 35 MeV and in the IMB group the energy range is from 20 to 40 MeV. Changing the lower limit in the KII correlation studies from 6.3 to 7.5 MeV results in virtually no change in conclusion. A time window of 0.35 seconds is used for both cases. Several random data sets were studied, yielding virtually the same results. Discussions of other time windows will follow shortly.

Figures 1 (a) and (b) show the distribution functions for one of our studies. In both figures, curves marked with x represent the distribution functions for the original random "data". A discernable peak appears in for the distribution function for the KII energy range in the case of the time-shifted data set, but no clear peak is observed for the time-shifted data set corresponding to the IMB energy range. In both energy ranges the distribution functions for the original simulation data exhibit the character of a random distribution.

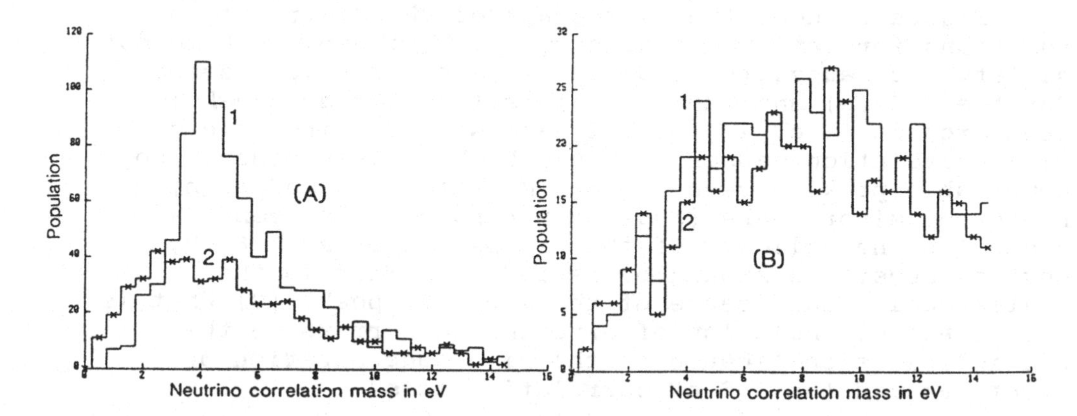

Figure 1. Distributions of neutrino pairs in the correlation mass space in a simulation using 50 neutrinos with the KII energy range (a) and with the IMB energy range (b), with a time window of 0.35 seconds. Unmarked curves refer to time-shifted data and curves marked with (x) refer to the original random simulation data.

Further numerical experiments show that for the KII energy range time correlation is observable even with a time window as long as 1 second, while for the IMB energy range time correlation is observable only for a time window < 0.08 second. Note that the 'dead time' in the IMB experiment is 0.1 second while that in the KII experiment is only 50 nanoseconds.

It should also be noted that in these simulations the time window is not necessarily the duration of the observed events; it gives an idea of the observability of a correlation in relation to the time density of events; in the KII data two events occuring as far as 1 second apart can still be correlated (if the energy difference is the same as the maximum energy range) while in the IMB case only events less than 0.08 second apart will show a correlation. Thus, the long experimental dead time in the IMB data set can allow the observation of a correlation for a true mass greater than, say, 5 *eV* at best.

3. **Data Analysis.** We now apply correlation mass analysis to both KII and IMB data sets for the neutrino events attributed to SN1987a. The Mt. Blanc data (Aglietta *et al*, 1987) set is not considered because the time of detection of this experiment is very different

from the KII and IMB experiments, which are synchronous to within timing accuracy (less than 1 minute).

Figure 2 shows the histograms of the distribution functions for real and imaginary [i.e., those which do not satisfy the causality condition (5)] correlation masses for the KII (a) and IMB (b) data sets. A sharp peak is seen around 3.6 eV in the KII data with 6 pairs of events, with correlation masses within 0.5 eV of each other. No comparable peak is seen in the IMB data. In addition, a number of minor peaks (less than 4 pairs) are seen. With regard to the validity of the six-pair peak at 3.6 eV, several questions should be raised: (1) What is the statistical significance of the six-pair peak? (2) What is the effect of inclusion of experimental errors on the statistical significance of the high concentration of events around the 3.6 eV correlation mass?

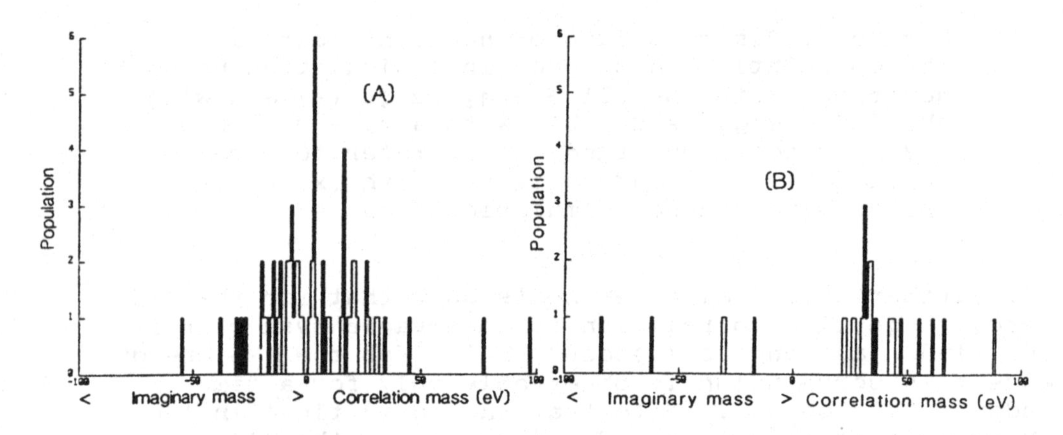

Figure 2. Distribution of neutrino pairs in the correlation mass space for the KII data (a) and the IMB data (b). The negative sides of the abscissa represents imaginary correlation masses.

3.1a. *Statistical significance of the 3.6 eV peak.* It has been shown to us that changing cell size in the histogram can drastically modify the appearance of the histogram (Elitzur 1987). To avoid this subjective difficulty, a proper question is: Is the correlation within 0.5 eV of each other among 6 pairs statistically significant? To answer this question, the following statistical tests were carried out: we looked for the probability for a random set of 12 events within the KII energy range that will give rise to at least one case of 3, 4, 5, or 6 pairs whose correlation masses are within 0.5 eV of each other. The answer is as follows: for 3 pairs, 99 %, for 4 pairs, 40 %, 5 pairs, 10 %, 6 pairs,

3 %. We thus conclude that, statistically, the most significant correlation is found in the 6 pair peak for the KII data set, giving rise to a mass signature at 3.6 eV. The remaining peaks in the KII data and in the IMB data are more likely statistical coincidences. The distribution function in the imaginary correlation mass space is consistent with no correlation at all, as should be the case. This point will be further explored below.

3.1b. *Other statistical considerations.* In *§3.1a* the generated random data set is assumed to be evenly distributed over the entire energy range. The observed energy of neutrino events appears to decline with time. Can a physical process mimic the character of a mass correlation derived from Eq.(3)? To study this possibility, we chose an energy time relationship that can mimic Eq.(3), as follows:

$$E \propto t^{-1/2} . \quad (7)$$

Since the mass correlation found in *§3.1a* refers to the first 6 events within the first 1 second of time, a statistical study that is expected to give the best probability for coincidence is made, as follows: 6 values of time t are randomly chosen, the temperature of a Maxwellian distribution is then chosen according to Eq.(7), from which the energy of the neutrino is randomly generated from the Maxwellian distribution function. The probability for 6 correlation pairs with masses within 0.5 eV of each other, as obtained from 10,000 studies, is 0.016.

3.2. *Inclusion of Experimental Errors.* Since the average error in the energies is around 20 %, the correlation obtained in *§3.1* can be made to disappear if the energies of the neutrinos are selectively chosen, within the allowable experimental errors. To evaluate objectively the effect of experimental errors on the observed correlation, a large number of data sets (over 10,000) were artificially generated from the KII data set as follows: A Gaussian distribution of probable values of energy is assumed around the nominal values, with the half width at half maximum determined by the experimental errors. Data sets are then generated randomly using these Gaussian distribution functions. Correlation mass analysis is then applied to each data set generated. The cumulative distribution function, shown in Figure 3 (a), then represents the probability distribution function of the population in the correlation mass space. A sharp maximum is observed near 2.8 *eV*, with half maximum points at 4.8 *eV* and 1.4 *eV* respectively. This sharp maximum is largely contributed by neutrinos 1, 2, 3, 4, and 6. Neutrino 6 has an energy of 6.3 *MeV*, which is in the noise regime. Removing Neutrino 6 from this analysis results in Figure 3(b), which also shows a maximum at about 2.8 eV. although somewhat reduced in magnitude. Thus the presence of Neutrino 6 does not critically affect a correlation around 2.8 *eV*. On the basis of the presence of a strong correlation of Neutrino 6 with others, it is reasonable that Neutrino 6 is a true supernova event.

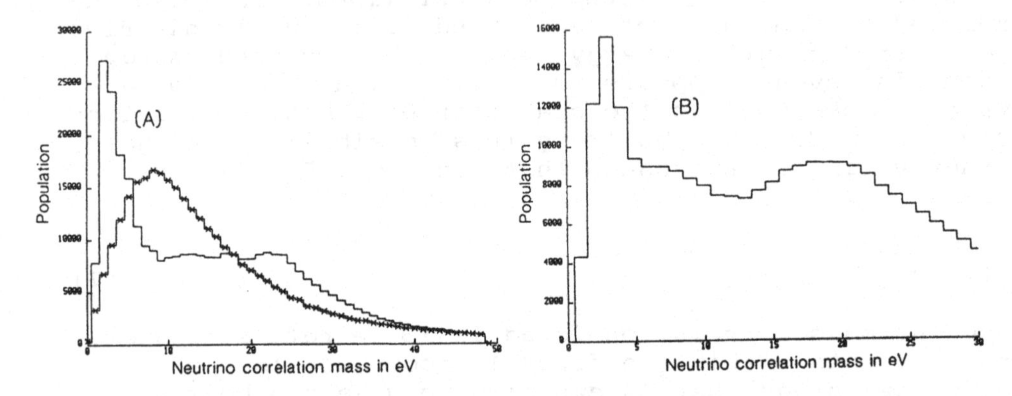

Figure 3. Cumulative distribution of cases using KII energy data but allowing the energy to vary according to a Gaussian distribution of errors. Neutrino 6 is included in (a) but not in (b). In curves marked with (x) the time of arrival is randomized, and the random character is seen for both cases.

In addition, all other correlations with 3, 4 and 5 pairs in the original histogram disappeared entirely in the cumulative distribution. This is expected from the statistical analysis *§3.1*. The only mass signature in Figure 1(a) that survived inclusion of experimental errors is the one at 3.6 *eV*. The cumulative distribution yields a mass signature around 2.8 *eV*, with half width at half maximum points at 1.4 and 4.8 *eV* respectively.

4. Conclusions. Based on a correlation mass study of the KII neutrino detections, we conclude:

(a) A single mass signature, at about 3 eV, has been found. The probability that this mass signature is statistically significant is 97 %. This mass signature is found only in the KII data set, consistent with our analysis that a mass signature below 5 *eV* should not be found in the IMB data. Our analysis also showed that the probability that this mass signature is < 0.5 eV, is less than 1 %.

(b) The probability that this mass signature is produced by random coincidence is 0.03.

(c) The rest mass energy of the electron neutrino is either around 3 *eV* (with a probability of 97%), or 0 (with a a probability of 3 %).

(d) Neutrinos 1, 2, 3, 4, and 6 show a strong correlation indicating that they were all emitted within a short time interval around 0.1 second. Two time scales thus exist, a fast initial time scale (~ 0.1 *sec*), possibly caused by neutronization, and a slower process possibly due to thermal cooling.

Details of this work are published elsewhere (Chiu 1987, Chiu, Chan and Kondo 1987a, 1987b, 1987c).

References

Aglietta M. *et al* (1987), Europhysics Ltrs, *3*, **12**, 1315.
Bionta, R. M. *et al*, (1987). Phys. Rev. Ltrs. *58*, 1494.
Chiu, H. Y., (1987). "Supernova, Stellar evolution and Neutrinos", in C. C. Lin Symposium, Ed. C. Yuan, *et al* (World Scientific, Singapore).
Chiu, H. Y., Chan, Kwing L., and Kondo, Y., (1987a). BAAS, *19*, **2**, 740.
Chiu, H. Y., Chan, Kwing L., and Kondo, Y., (1987b). "Analysis of Neutrinos from Supernova 1987a", to appear in Proceedings of IAU Symposium 108 (held in Tokyo, Japan, Sept. 1 - 4), Ed. K. Nomoto (Springer-Verlag, Berlin, New York).
Chiu, H. Y., Chan, Kwing L., and Kondo, Y. (1987c). "Correlation Mass Method for Analysis of Neutrinos from Supernova 1987a", to appear in *Astrop. J.*.
Elitzur, M. (1987). Private communication.
Hirata, K. *et al* (1987), Phys. Rev. Ltrs., *58*, 2023.

In addition, many references to this subject exist within this volume.

NEUTRINO TRANSPORT IN A TYPE II SUPERNOVA

Donald C. Ellison
Physics Dept., North Carolina State Univ., Raleigh, NC 27695
Peter M. Giovanoni
Astronomy Program, Univ. of Maryland, College Park, MD 20742
Stephen W. Bruenn
Physics Dept., Florida Atlantic Univ., Boca Raton, FL 33431

1. INTRODUCTION. We use a Monte Carlo simulation to calculate neutrino production and transport in the vicinity of the core bounce shock during a type II supernova explosion. The sources of neutrino opacity include scattering off electrons, nucleons, and nuclei, the absorption of electron neutrinos, and neutrino pair annihilation. The neutrino production is from electron capture and pair processes behind the shock. Energy gains and losses produced by first-order Fermi shock acceleration and inelastic collisions with nuclei and electrons are included. Using existing spherically symmetric supernova models to obtain the core conditions, we calculate the emitted neutrino energy spectra paying particular attention to the transport in the semi-transparent regions where the neutrino opacity ~ 1.

The core bounce shock is followed from bounce until it passes the neutrinosphere. The density outside the shock is taken to be independent of time and given by $\rho \propto r^{-3}$, where r is the radius (e.g., Bethe et al. 1980; Bethe and Wilson 1985). Behind the shock the density is determined by mass conservation. We have approximated the simulation results of Burrows and Lattimer (1985) for the infall velocity, v_{in}, with $(v_{in}/c) \approx (2.78 \times 10^5/r)^{0.62}$, where r is given in cm and c is the speed of light. In a like manner, an analytic expression was obtained for the postshock velocity, v_{post}, i.e., $(v_{post}/c) \approx 0.1[1 + (10/3)log_{10}(r/r_s)]$, where r_s is the radius of the shock. The pre-shock temperature is approximated by $T_1 \approx T_c(\rho/\rho_c)^{1/3}$ (Shapiro and Teukolsky 1983), and the post-shock temperature is calculated using the conservation of energy assuming that the shock completely dissociates the heavy nuclei. Unshocked material consists mainly of heavy iron group nuclei, while dissociated shocked material will be mainly free protons and neutrons. For simplicity of calculation we have taken the lepton number per baryon, Y_L, to be constant with $Y_L = 0.39$. The number of electrons per baryon, Y_e, and the number of neutrinos per baryon, Y_ν, are likewise taken to be constant everywhere with $Y_\nu = 0.04$.

We are concerned only with those neutrinos produced by the shock heated material after core bounce and before the shock passes the neutrinosphere. The majority of the ν_e's are produced by electron capture behind the shock with a considerably smaller contribution resulting from thermal electron-positron annihilation. Muon and tau neutrinos are only produced by annihilation and hence are much less copiously produced. The production rates for pair annihilation are taken from the numerical work of Bruenn (1985).

2. MONTE CARLO SIMULATION. The scattering and absorption processes that each neutrino undergoes are represented by an interaction rate; $R_1, ..., R_n$. A particular rate, R_i, is determined by, $R_i = cn_i\sigma_i$, where n_i is the number density of scatterers or absorbers and σ_i is the cross section for the ith

process. The total interaction rate is, therefore, $R = \sum_{i=1}^{n} R_i$, and the neutrino mean free path is $\lambda_\nu = ct$, where $t = 1/R$ is the average time between interactions. We consider the following interactions: (a) coherent scattering off nuclei, (b) scattering off nucleons, (c) absorption by nuclei, (d) absorption by nucleons, (e) scattering off electrons, and (f) neutrino pair annihilation. Interactions (a) and (b) proceed only by neutral currents and the cross-sections are the same for all neutrino types. We have used cross-sections from Tubbs and Schramm (1975) for interactions (a) through (d). The cross-section and energy exchange for electron scattering was obtained from the numerical work of Bruenn (1985) as was the cross-section for neutrino pair annihilation. The scattering is assumed to be isotropic and for a constant neutrino energy, a constant composition, and assuming that density $\propto r^{-3}$, the change in r between interactions is $\Delta r = r_o\sqrt{r_o^2/(r_o^2 - 2\lambda_o r_o cos\theta - \lambda_o^2 sin^2\theta)} - r_o$, where λ_o is the mean free path computed at r_o and θ is the angle between the scattering direction and the radius. If the particle crosses the shock, a change in composition and density will occur. The neutrinosphere, or radius at which a neutrino will escape to infinity, depends on E_ν and θ. We define this radius to be $r_{esc} = (1 + cos\theta)\lambda_{esc}$, where λ_{esc} is the mean free path at r_{esc}.

Energy changes result from first-order Fermi shock acceleration for neutrinos that scatter across the shock (Kazanas and Ellison 1981), second-order Fermi acceleration for neutrinos that scatter off the hot nucleons, and inelastic electron scattering. In second order scattering, it is possible for a neutrino to gain or lose energy depending on whether it scatters off approaching or receding nucleons. Approaching nucleons will present a larger cross-section (the interaction cross-section increases with E_ν^2) and more collisions producing energy gains will occur. The energy exchange is averaged over angle and the loss of energy due to the recoil of the nucleon is included.

The neutrinos are produced and injected into the simulation throughout the shock heated material and the production is maintained from core bounce until the shock reaches some arbitrary maximum radius ($\sim$ 150 km).

3. RESULTS. The emitted energy of neutrinos will be affected by the energy exchange processes and absorption occuring during transport. Electron scattering and inelastic scattering off nucleons tend to equilibrate the neutrinos with the matter. However, the emitted spectrum is modified from a Fermi-Dirac distribution because high energy ν_e's are absorbed more readily than low energy ones and first-order Fermi shock acceleration will tend to accelerate neutrinos. In addition, different energy neutrinos escape from different depths with different characteristic temperatures and surface areas.

In the figures we show the emitted spectra for the shock associated neutrinos (solid lines) along with Fermi-Dirac distributions (chemical potential set equal to zero) with the same total and average energies (dotted lines). Tau and muon neutrinos are treated identically in this work and the emitted spectrum labeled (ν_τ) is the *sum of the four muon and tau neutrino types* (ν_μ, ν_τ, $\bar{\nu}_\mu$, $\bar{\nu}_\tau$).

In general, the emitted spectra develop a peak above the Fermi-Dirac distributions and fall below the Fermi-Dirac distributions at both high and low energies. The different neutrino types do, however, behave quite differently. The tau and muon neutrinos develop a pronounced peak at about 20 MeV which is absent in the electron neutrinos. First and second order Fermi acceleration and electron scattering, all tend to accelerate low energy neutrinos, however, few ν_τ's reach energies much above 50 MeV due to losses by nucleon recoil and electron scattering. Electron scattering tends to shift both high and low energy ν_τ's towards ~ 20 MeV depending on the electron degeneracy.

The situation is quite different for ν_e's. Here the strong absorption by nucleons and nuclei dominates and only low energy ν_e's are emitted. If a ν_e gains energy, its scattering mean free path decreases and it is more likely to be absorbed before diffusing out. The high energy component seen in ν_τ's is absent for ν_e's.

The total number and total energy of the emitted neutrino spectra, along with the kT of the matching Fermi-Dirac distributions, are:

(a) ν_e: 1.8×10^{55} ν's, 2.6×10^{50} erg, $kT \simeq 2.9$ MeV

(b) sum of ν_μ, $\overline{\nu}_\mu$, ν_τ, $\overline{\nu}_\tau$: 2.5×10^{55} ν's, 8.8×10^{50} erg, $kT \simeq 6.8$ MeV.

Acknowledgements. The authors wish to thank Demos Kazanas and David Eichler for much help. Partial support came from NSF Grant AST 86-11939.

Bethe, H.A., Applegate, J.H., and Brown, G.E., *Ap.J., 259,* 343, 1980.
Bethe, H.A., and Wilson, J.R., *Ap.J., 295,* 14, 1985.
Bruenn, S.W., *Ap.J.(Suppl.), 58,* 771, 1985.
Burrows, A., and Lattimer, J.M., *Ap.J.Lett., 299,* L19, 1985.
Kazanas, D., and Ellison, D.C., *17th ICRC (Paris), 7,* 176, 1981.
Shapiro, S.L., and Teukolsky, S.A., *Black Holes, White Dwarfs, and Neutron Stars,* New York: John Wiley & Sons, 1983.
Tubbs, D.L., and Schramm, D.N., *Ap.J., 201,* 467, 1975.

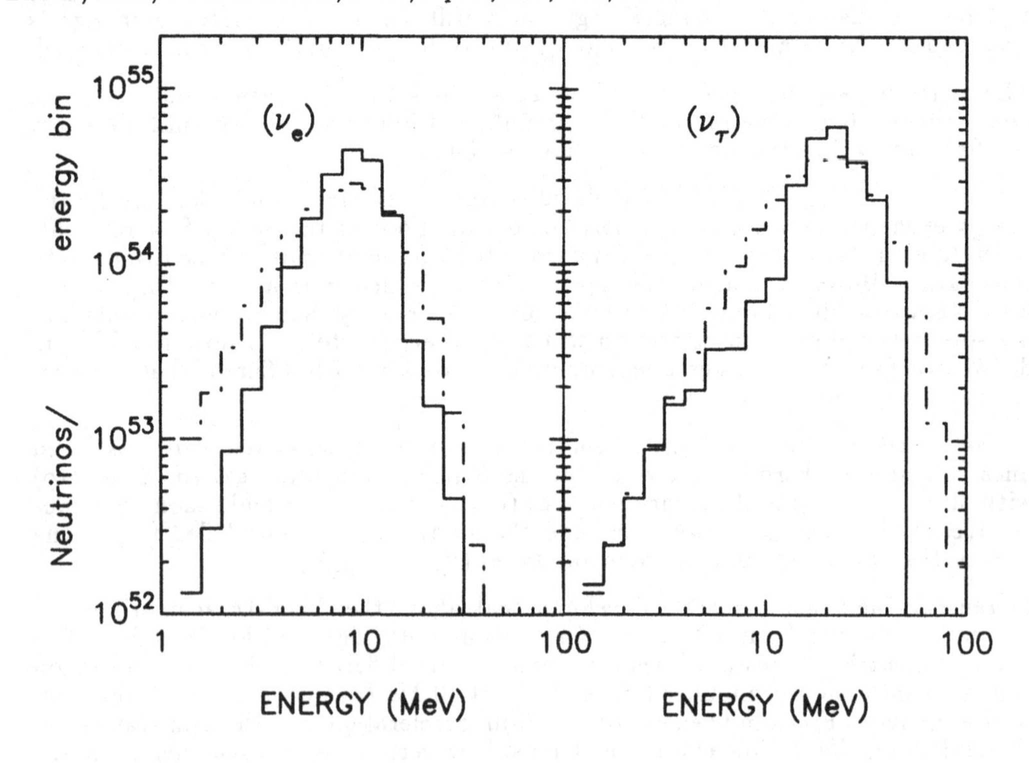

NEUTRINO MASSES FROM SN1987A

J. Franklin
Temple University, Philadelphia, PA 19122

Abstract. We investigate the hypothesis that all of the neutrinos from SN1987A were emitted simultaneously and that their energy-time distribution results from time of flight differences for massive neutrinos.

The Kamiokande (Hirata et al. 1987) and IMB (Bionta et al. 1987) detectors observed a short burst of 20 neutrinos from the SN1987A supernova. Their energy-time distribution is plotted in Fig. 1 with the energies determined by assuming that all the observed events were actually caused by electron antineutrinos interacting with protons in the detectors. We have adjusted each detector's initial time by using time of flight for the antineutrinos as described later.

Fig. 1. Antineutrino energies as a function of arrival times for the Kamiokande (solid circles) and IMB (open circles) detectors. (a) Events for t<1.5 sec. (b) events for t>1.5 sec. The curves represent Eq.(2) for antineutrinos of mass 4.3 eV (solid curve), 19.4 eV (dashed curve), and 31 eV (dash-dot curve).

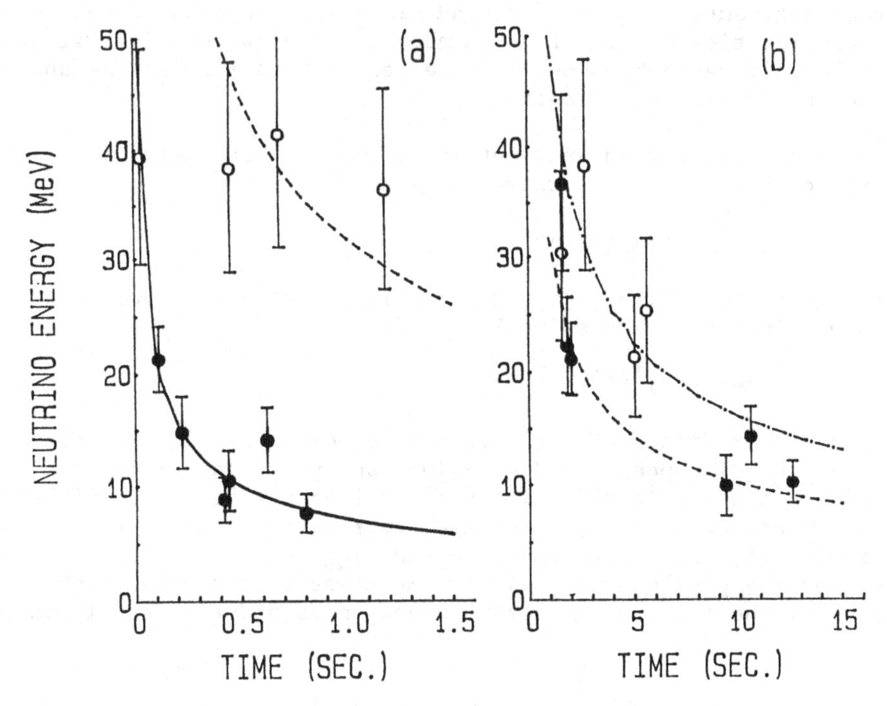

Figure 1 has been split into two time scales because the first ten events occurred in such a brief interval. A striking feature of Fig. 1 is the time distribution for the Kamiokande events of three short pulses, of increasing duration. The first pulse includes a spread of energies, but the second and third pulses seem to be correlated in energy, with a high energy pulse followed in about 7 seconds by a lower energy pulse. The IMB events have a different time structure with a high energy pulse of 6 antineutrinos in 2.7 seconds followed in 2.3 seconds by a lower energy pulse of 2 antineutrinos within 0.6 seconds. The assumption of massless (or very light) antineutrinos would require this time pulse structure, differing for the two detectors, to be an unusual and unexplained feature of the supernova explosion.

Another visible feature in Figs 1a and 1b is a general downward trend in energies as time increases within two different sets of events, with a distinct jump in energy at around half a second. The two sets seem to be quite similar distributions, but on different time scales. We will see below that these features of the time-energy correlation can be understood naturally as resulting from time of flight differences for massive antineutrinos.

The hypothesis that all 20 observed antineutrinos were produced essentially simultaneously in the supernova has been made by several groups (Kim 1987; Cheng _et al_. 1987; Evans _et al_. 1987; Franklin 1987), all of whom arrived at somewhat similar conclusions for the neutrino masses. This assumption is at odds with "standard" theories of supernova collapse, but such theories (that suggest continuous emission of thermal antineutrinos over a period of several seconds) cannot explain the peculiarities of Fig. 1 noted above. Here, we consider simultaneous emission as a hypothesis to be tested by a statistical analysis of the energy-time distribution.

The arrival time for an antineutrino of mass m and energy E traveling a distance L from the supernova is given by

$$t=t_o+(L/2)(m/E)^2, \tag{1}$$

where t_o is the arrival time of a massless particle traveling at the speed of light c=1. We solve Eq.(1) for E,

$$E=m(L/2)^{1/2}/(t-t_o)^{1/2}, \tag{2}$$

and perform χ^2 fits of Eq. (2) to the 20 points of Fig. 1. For SN1987A, the distance $L=(5.3\pm0.5)X10^{12}$ seconds. The error on L does not affect the χ^2 fit since it shifts all points equally. Since the two detectors were not synchronized, there are two separate t_o's which we take as $-t_K$ and $-t_{IMB}$, where t_K and t_{IMB} represent the period of time from the possible arrival of a massless particle (or light signal) until the first antineutrino event in each detector. These two arrival times, now synchronized, are used, along with antineutrino masses, as free parameters for the fits.

It is clear from Fig. 1 that a single mass could not fit both sets of points and χ^2/df (degrees of freedom) is 48/17 for this hypothesis. The fit for two different antineutrino masses, m_a and m_b, has a reasonable χ^2/df=20/16 with m_a=4.3±0.8 eV, m_b=21.5±2.0 eV, $t_K=0.11^{+0.08}_{-0.03}$ sec., and $t_{IMB}=0.03^{+0.03}_{-0.01}$ sec. A much better fit is obtained, however, with three antineutrino masses. Then, χ^2/df=9/15 with m_b=19.4±2.4 eV and m_c=31±4 eV. The light mass m_a and the initial times are the same as for the two mass fit because these are determined primarily by the (a) events. This three mass fit is plotted in Fig. 1, where t_K and t_{IMB} have been used to offset events from each detector separately so that t=0 now corresponds to the arrival time of a light signal from the supernova collapse. Either two masses or three masses gives a reasonable fit to Fig. 1 and smaller error bars would be required to decide whether two masses or three are required by the data.

The plotted curves on Fig. 1 show how two (or three) antineutrino masses account for two separate fall-offs in energy with time delay. An attempt to reproduce this behavior with two fall-offs in temperature (one time delayed) would have much more spread in the energies (if the antineutrinos had a thermal distribution) than is evident in Fig 1. The gaps observed in the time distribution are also seen to be correlated on Fig. 1 to corresponding (definitely non-thermal) gaps in the energy distribution. (Franklin 1987)

What is the meaning of two or three antineutrino masses? Neutrino mixing among the known families of this amount seems ruled out by laboratory experiments. If the antineutrino has only one mass of 4.3 eV, corresponding to the events on the solid curve, then an unexplained energy-time correlation remains for the other points. The masses found here would probably require that the electron antineutrino (and probably other antineutrinos and neutrinos) have three states of differing masses without much mixing. This may not be a welcome situation, but colored quarks were also unwelcome twenty years ago. The basic question is whether it is better to "smooth over" the remarkable structure in Fig. 1 or to ask why?

References

Bionta, R.M. et al. (1987). Phys. Rev. Lett., 58, 1494.
Chen, Z-M. et al. (1987). Chinese Physics Letters, Vol. 4.
Evans, D. et al. (1987). Durham University preprint no. DTP/87/12.
Franklin, J. (1987). Temple University preprint no. TUHE-87-52.
Hirata, K. et al. (1987). Phys. Rev. Lett., 58, 1490.
Kim, C.W. (1987). Johns Hopkins University preprint no. JHU-HEP8705.

SUPERNOVA NEUTRINOS AND THEIR OSCILLATIONS

T.K. Kuo
Physics Dept., Purdue University, West Lafayette, IN 47907

J.T. Pantaleone
Physics Dept., Purdue University, West Lafayette, IN 47907

The recent observations of neutrinos from a supernova (Hirata et al. 1987; Bionta et al. 1987) have many implications for astrophysics and particle physics. Besides containing information on the supernova, the signal depends on the properties of neutrinos. In order to interpret the recent observations, the uncertainties in supernova dynamics must be disentangled from the effects of neutrino propagation. In this talk we will concentrate on the mixing of neutrino fluxes from neutrino oscillations, both in vacuum and in matter as discussed by Kuo & Pantaleone (1987) and also Arafune et al. (1987); Notzold (1987) and Rosen (1987).

If neutrinos are massive then the flavor and mass eigenstates are, in general, different. The two bases are related by a unitary transformation. For quarks this unitary matrix is the Cabbibo, Kobayashi- Maskawa mixing matrix. Since the propagation basis and the flavor basis are different the flavors will mix during propagation. For neutrinos from the supernova there is a simple expression describing these mixing effects. Using unitarity and the observation that, since the neutrino energies are well below the muon mass, the muon and tau neutrinos interact only via neutral currents, we can derive the following relation between the produced ($F_\alpha{}^0$) and detected (F_β) neutrino fluxes.

$$F_e = F_e{}^0 - [1 - P(\nu_e \to \nu_e)] \cdot [F_e{}^0 - F_x{}^0]$$

$$F_\mu + F_\tau = 2F_x{}^0 + [1 - P(\nu_e \to \nu_e)] \cdot [F_e{}^0 - F_x{}^0] \qquad (1)$$

The subscripts denote the flavor of the neutrino, e for ν_e and x for ν_μ or ν_τ. $P(\nu_e \to \nu_e)$ represents the probability for a ν_e produced in the supernova to remain a ν_e as observed on the Earth. For antineutrinos, eqs. (1) hold with all $\nu \to \overline{\nu}$.

Neutrino mixing is not sensitive to masses but to mass^2 differences. The oscillation wavelength is $4\pi E/(m_i^2-m_j^2)$ so the supernova signal is sensitive to vacuum oscillations with $(m_i^2-m_j^2) > 10^{-20}$ eV^2 (this is twenty orders of magnitude below what purely terrestial experiments at reactors and accelerators are sensitive to). Such small mass^2 differences do not affect the timing of detected events but the number of events and their average energy is altered. As a measure of mixing effects let us define R to be the ratio of the number of events detected with mixing to the number of events detected without mixing. Then for $\bar{\nu}_e$ from thermal emission we find R to be

$$R^t = 1 + [1-P(\bar{\nu}_e \to \bar{\nu}_e)]\cdot B^t$$

$$B^t = 0.20 \;\; \text{Kamiokande}\;; \;\; 4.5 \;\; \text{IMB} \qquad (2)$$

Where we have chosen typical supernova parameters for the temperatures $(2T_e=T_\mu=T_\tau=6\text{MeV})$ and luminosities $(L_e=2L_\mu=2L_\tau)$ of the neutrino flux. The strong sensitivity of the IMB detector to mixing is because the detection threshold is much larger than the flux temperatures.

In principle the detected neutrino signal can tell us something about neutrino mixing, through $P(\bar{\nu}_e \to \bar{\nu}_e)$, and about supernova dynamics through F_e^0. In practice neither goal can be achieved with the present data. The number of events is too small and the models of the supernova neutrino fluxes are, at present, too crude to reliably extract information on the oscillation probability. Since the oscillation probability is sensitive to a wide range of parameters which can not be tested in terrestial experiments, one can not reliably deduce a temperature and flux for the $\bar{\nu}_e$ from the present data.

The first two events in Kamiokande are highly directional, suggesting that they are not $\bar{\nu}_e$ events but instead come from ν_e-electron scattering. The probability that these events are due to $\bar{\nu}_e$ is about $[(20°\cdot\pi/180°)^2\pi/4\pi]^2 \approx 1/1000$. It is thus tempting to identify these events as coming from the neutronization of the supernova core. The expected number of neutronization events is very model dependent but is typically about .05 to .3 events. Mixing will decrease the expected number of events since the cross section for ν_μ or ν_τ scattering off of electrons is about 1/7 that for ν_e. With full mixing, the probability of getting two events is reduced by about $(1/7)^2$.

Figure 1 shows the reduction in the expected number of neutronization events from matter enhanced mixing (Mikheyev & Smirnov 1985; Wolfenstein 1979). We use three flavors of neutrinos, typical see-saw constraints, and a neutronization flux temperature of 3 MeV. Also shown is the range of parameters that solve the solar neutrino problem. We see that most of the parameters that solve the solar neutrino problem also give a maximum suppression of the supernova neutronization signal. The one exception is where the diagonal contours for the sun and supernova overlap (the e-τ, nonadiabatic contours). These contours are especially model dependent and cannot be said to suppress the supernova neutronization signal.

In conclusion, if one or two neutronization events are contained in the Kamiokande sample then the e-τ, nonadiabatic solution to the solar neutrino problem is favored over other solutions via matter induced mixing. The observed $\bar{\nu}_e$ events are sensitive to vacuum mixing. According to eq. (2), mixing always increases the observed $\bar{\nu}_e$ flux. For IMB this increase ranges from about 50% when one uses the quark mixing angles for the neutrino mixing angles, to about a factor of 5, for maximal mixing. However the small number of events and uncertainty of the supernova models make it difficult to quantitatively probe the mixing. Conversely, because the mixing depends on a range of neutrino parameters never before probed, it is not possible to unambiguously extract temperature and flux information from the data.

Arafune, J. et al. (1987). Kyoto preprint RIFP-699.
Bionta, R.M. et al. (1987). Phys. Rev. Lett., 58, 1494.
Hirata, K. et al. (1987). Phys. Rev. Lett., 58, 1490.
Kuo, T.K. & Pantaleone, J.T. (1987). Purdue preprint TH-87-08.

Mikheyev, S.P. & Smirnov, A.Yu. (1985). Sov. J. Nucl. Phys., 42, 913.
Notzold, D. (1987). Phys. Lett., B196, 315.
Rosen, S.P. (1987). Los Alamos preprint 87-1296.
Wolfenstein, L (1978). Phys. Rev., D17, 2369.

Fig. 1 A contour plot of R, the ratio of the number of events with oscillations to those without for directional events from neutronization, for the Kamiokande detector. Here we use three flavors and "typical" constraints on the relevant neutrino vacuum parameters: $m_1=0$, $(m_3/m_2) = 250$ and $(|U_{e2}|^2/|U_{e3}|^2) = 100$. The dashed lines show the (3σ) solutions to the solar neutrino problem. The shaded regions are excluded by reactor oscillation experiments where we have taken $|U_{\mu 3}| \approx |U_{e2}|$.

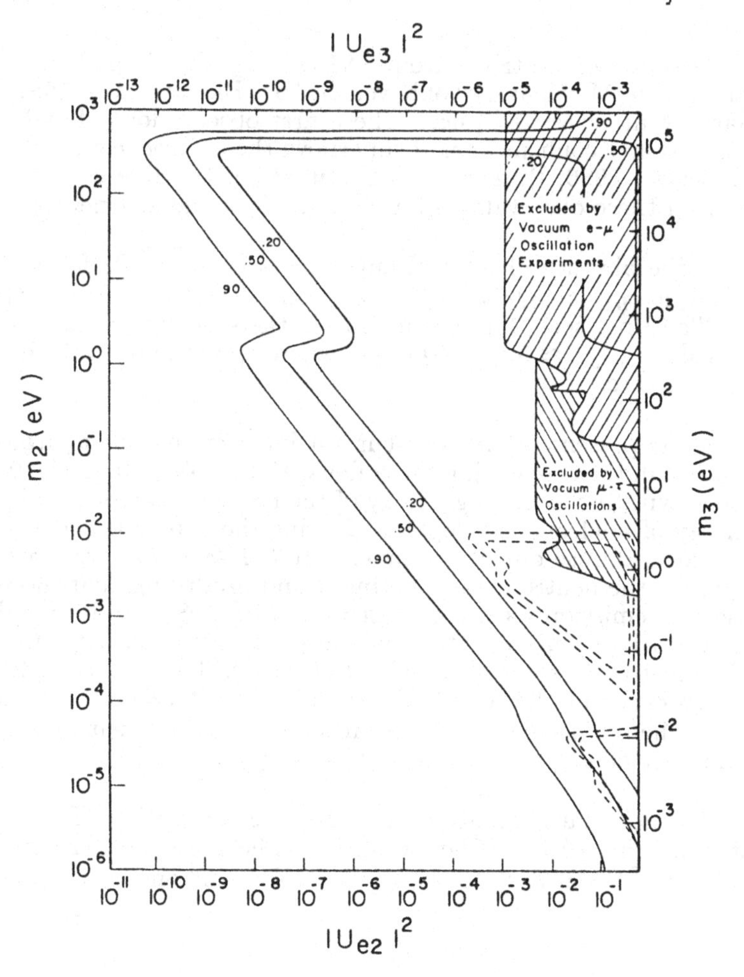

NEUTRINOS FROM SN 1987A AND COOLING OF THE NASCENT NEUTRON STAR

D. Q. Lamb
Dept. Astronomy and Astrophysics and Enrico Fermi Institute
U. Chicago, 5640 S. Ellis Avenue, Chicago, IL 60637, U.S.A.

Fulvio Melia
Dept. Physics and Astronomy
Northwestern University, Evanston, IL 60208, U.S.A.

Thomas J. Loredo
Dept. Astronomy and Astrophysics and Enrico Fermi Institute
U. Chicago, 5640 S. Ellis Avenue, Chicago, IL 60637, U.S.A.

Detection of neutrinos from SN1987A by the Kamiokande II (Hirata *et al.* 1987) and Irvine-Michigan-Brookhaven (IMB) (Bionata *et al.* 1987) detectors was a landmark event in astrophysics. These first observations of stellar collapse have been a spectacular confirmation of supernova theory (see, e.g., Wilson, Mayle, Woosley, and Weaver 1986; Burrows and Lattimer 1986). Here we concentrate on the implications of these observations for cooling of the nascent neutron star.

In the absence of neutral currents, the time scale for ν_e losses due to free streaming is $\tau_{\rm free} \approx R/c < 10^{-3}$ sec, where R is the radius of the nascent neutron star. The time scale for ν_e emission is therefore governed by the stellar collapse time scale $\tau_{\rm collapse} \approx 5\tau_{\rm ff} \approx 10$ msec, where $\tau_{\rm ff}$ is the free fall time scale for the collapsing iron core.

In the presence of neutral current interactions, the ν_e from neutronization are trapped in the core during the collapse (Sato 1975; Mazurek 1976; Arnett 1977), and the gravitational binding energy of the nascent neutron star is converted into Fermi energy of nucleons and leptons. During the subsequent deleptonization of the core, these ν_e diffuse out of the inner core ($M \approx 0.8 M_\odot$) and are absorbed in the outer part of the neutron star, heating it and producing neutrino pairs of all flavors via thermal emission (see, e.g., Burrows *et al.* 1981). Additional neutrino pairs of all flavors are produced by thermal emission behind the accretion shock at the edge of the inner core (see, e.g., Mayle *et al.* 1987). The time scale for neutrino cooling is thus governed by the time scale for the ν_e to diffuse out of the inner core, $\tau_{\rm diff} \approx \sigma R^2 n/c$, where σ is the cross section for the relevant neutrino interaction, and n is the number density of the relevant target particles.

The dominant diffusion processes are absorption of ν_e on protons and coherent scattering of ν_e off heavy nuclei. The cross section for the former is $\sigma = 1.5\sigma_0(\epsilon_\nu/m_e c^2)^2 = 9.77 \times 10^{-42}(\epsilon_\nu/10 \text{ MeV})^2$ cm^2 and $n = Y_e\rho/m_p$, where

$\sigma_0 = 1.7\times10^{-44}$ cm^2 and Y_e is the electron (proton) number fraction. The cross section for the latter is $\sigma = \sigma_0(N_n^2/24)(\epsilon_\nu/m_ec^2)^2 f(\epsilon_\nu, A) = 2.71\times10^{-39}(N_n/100)^2(\epsilon_\nu/10\ \mathrm{MeV})^2 f(\epsilon_\nu, A)$ and $n \approx \rho/Am_p$, where N_n and A are the neutron number and atomic weight of the dominant nucleus, and $f(\epsilon_\nu, A)$ is the form factor for coherent scattering off heavy nuclei (Burrows *et al.* 1981). The form factor $f(\epsilon_\nu, A)$ depends strongly on ϵ_ν, with $f \to 1$ when $\epsilon_\nu \to 0$ and $f \to 0$ when $\epsilon_\nu \to \infty$.

The Fermi energy ϵ_{F,ν_e} of the ν_e in the inner core can be estimated from the virial theorem, $n_\nu\epsilon_{F,\nu_e} \approx 3GM_G^2/5R$, giving $\epsilon_{F,\nu_e} \approx (3GM_G^2/5R)/(Y_eM_G/3m_p) \approx 88(M_G/1.4M_\odot)(R/100\ \mathrm{km})^{-1}(Y_e/0.4)$ MeV. In the absence of coherent scattering off heavy nuclei, the ν_e diffusion time is $\tau_{\mathrm{diff,abs}} \approx 5.0(R/15\ \mathrm{km})^2(\rho/\rho_0)(Y_e/0.4)(\epsilon_\nu/100\ \mathrm{MeV})^2$ sec, where we have scaled the density to the nuclear density ρ_0. In the presence of coherent scattering, the ν_e diffusion time is $\tau_{\mathrm{diff,coh}} \approx 41(N_n/200)^2(A/320)^{-1}(\epsilon_\nu/100\ \mathrm{MeV})^2(R/15\ \mathrm{km})^2(\rho/\rho_0)(f/1)$ sec.

In the absence of neutral currents, the ν_e escape with roughly the Fermi energy of the captured electrons (up to 40 MeV). In the presence of neutral currents, the energy of the escaping $\bar{\nu_e}$ is typically that for which the $\bar{\nu_e}$ mean free path is comparable to the size of the core R, i.e. the energy is defined implicitly by $\rho\sigma R/m_p \approx 1$. Substituting $\epsilon_{\bar\nu} \approx 3.15T$ and taking the scattering cross section off neutrons to be $\sigma = \sigma_0(\epsilon_\nu/m_ec^2)^2/2$, gives $T_{\bar\nu} \approx (m_n/5R\rho\sigma_0)^{1/5}(T/m_ec^2)^{3/5}m_ec^2 \approx 3.2$ MeV (Schramm 1987), assuming $R = 15$ km and that the stellar collapse follows an adiabat from a density $\rho = 10^{10}$ gm cm^{-3} and temperature $T = 1$ MeV.

We construct the likelihood function $\mathcal{L}(N, T_0, \tau)$ for detecting neutrinos with energies E_i at times t_i (i=1, 19), using the published values of the efficiencies, threshholds, and energy dispersions of the Kamioka and IMB detectors and taking into account Poisson statistics. Here N is the total number of emitted $\bar{\nu_e}$ and T_0 is their apparent initial temperature. We assume a Fermi-Dirac spectrum with zero chemical potential and an apparent $\bar{\nu_e}$ energy flux $F_{\bar\nu}(t) = F_0\exp(-t/\tau)$ [i.e., an apparent $\bar{\nu_e}$ temperature $T(t) = T_0\exp(-t/4\tau)$], where τ is the apparent cooling time scale of the neutron star. We neglect ν_e scattering events (Bahcall *et al.* 1987). We combine the Kamioka and IMB data sets, and (in this study) assume zero time lag between the first event in each detector.

Maximizing $\mathcal{L}(N, T_0, \tau)$ gives best fit values $(N, T_0, \tau) = (4.5\times10^{57}$, 4.0 MeV, 4.7 sec). These values confirm those found by Spergel *et al.* (1987) and Bludman and Schinder (1987). The 68% and 95% likelihood regions for three degrees of freedom are "banana"-shaped volumes, lying in a (N, T_0)-plane in (N, T_0, τ)-space. Thus T_0 and N are strongly correlated, with the smallest number N corresponding to the highest temperature T_0, as noted by Spergel *et al.* (1987). Figures 1 and 2 show cross sections of the 68% and 95% likelihood regions in the (T_0, τ)-plane for the best fit value of N and in the (T_0, N)-plane for the best fit value of τ. As conservative error estimates, we take the total ranges of parameter values lying in the 95% likelihood volume. This gives $(N, T_0, \tau) = (4.5^{+11.1}_{-3.3}\times10^{57}, 4.0^{+1.6}_{-1.0}\ \mathrm{MeV}, 4.7^{+5.2}_{-2.5}\ \mathrm{sec})$.

The number N implies many more ν_e than are expected from neutronization. Correspondingly, the energy $3.15T_0 \approx 12$ MeV is much smaller than the ν_e energy of up to 40 MeV and the time scale $\tau = 4.7$ sec is much longer than

Fig. 1.–The best fit values $(T_0, \tau) = (4.0\ \text{MeV}, 4.7\ \text{sec})$, indicated by a cross, and the cross sections of the 68% and 95% likelihood volumes in the (T_0, τ)-plane at the best fit value $N = 4.5 \times 10^{57}$.

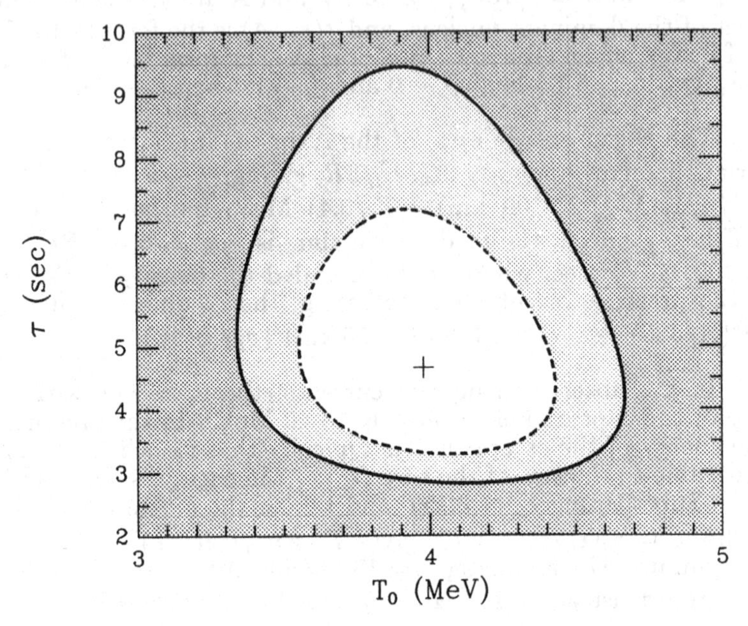

Fig. 2.–The best fit values $(T_0, N) = (4.0\ \text{MeV}, 4.5 \times 10^{57})$, indicated by a cross, and the cross sections of the 68% and 95% likelihood volumes in the (T_0, N)-plane at the best fit value $\tau = 4.7$ sec.

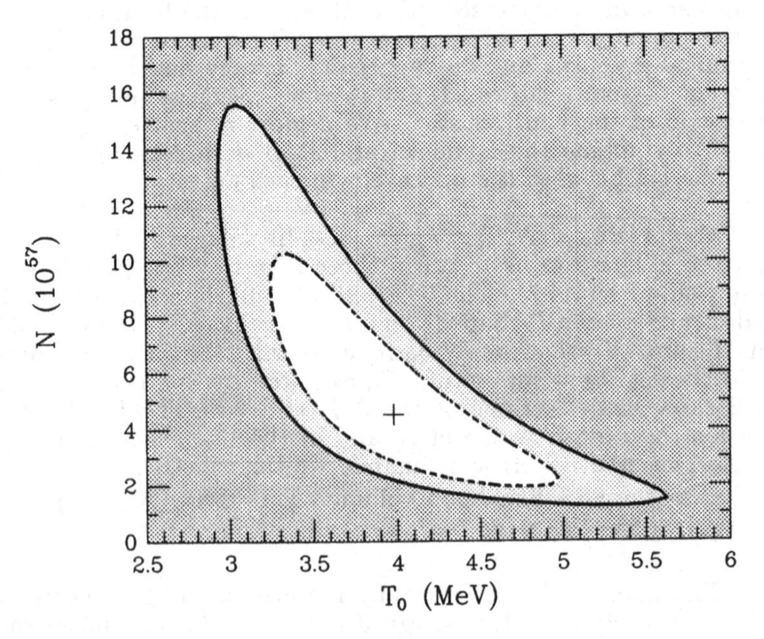

$\tau_{\text{collapse}} \approx 10^{-2}$ sec expected in the absence of neutral currents. In contrast, all three values are in good agreement with those estimated when neutral currents are present, confirming their importance in supernovae. These results verify the basic picture of neutrino trapping, which leads to core bounce at nuclear matter density ρ_0 due to the degeneracy pressure of neutrons (Lamb *et al.* 1978, Bethe *et al.* 1979), rather than free escape and bounce at densities $\rho < 0.1\rho_0$ due to the thermal pressure of nucleons. The time scale $\tau \ll \tau_{\text{diff,coh}} \approx 41$ sec for $f = 1$. Only if $\epsilon_{F,\nu_e} \gtrsim 100$ MeV is $f \approx 0.1$ at the edge of the inner core, leading to $\tau_{\text{diff,coh}} \approx \tau_{\text{diff,abs}} \approx 5$ sec. These results therefore also verify the value of ϵ_{F,ν_e} estimated above from the virial theorem.

Each point in (N, T_0, τ)-space can be projected onto the (R_{obs}, E_b)-plane, where $R_{\text{obs}} = R\,(1 - 2GM_G/Rc^2)^{-1/2}$ and E_b are the apparent radius and the binding energy of the newly formed neutron star. The needed relations are

$$R_{\text{obs}} = \left[\frac{(hc)^3}{4\pi^2 c}\,\frac{3N}{4F_2(0)\tau T_0^3}\right]^{1/2}, \tag{1}$$

and

$$E_b = 4\pi R_{\text{obs}}^2\,\frac{\pi c}{(hc)^3}\,6F_3(0)T_0^4\tau = \frac{3NT_0F_3(0)}{8F_2(0)}, \tag{2}$$

where $F_n(0) = \int_0^\infty [x^n/(\exp(x)+1)]dx = 1.803,\ 5.682$ for $n = 2$ and 3, and we have assumed 3 flavors of neutrinos (6 species).

Figure 3 shows the best fit values $(R_{\text{obs}}, E_b) = (32.1$ km, 4.1×10^{53} erg) and the 68% and 95% likelihood volumes *projected* onto the (R_{obs}, E_b)-plane. Also shown in Figure 3 are (R_{obs}, E_b)-curves for a representative set of neutron star models. The projected 95% likelihood region overlaps the gravitational mass ranges $M_G = 1.0$-1.1, 1.0-1.25, 1.15-1.5, 1.1-1.45, and 1.35-1.95 $M_\odot$ for models based on the P(Λ), P(n), BJI, BJV, and PS(tensor) equations of state (see Arnett and Bowers 1977, and references therein), in remarkable agreement with the expected gravitational core mass $M_{\text{core}} \approx 1.4\ M_\odot$. Comparison of the projected 95% confidence region with these neutron star models provides an upper bound on neutrino flavors of ≈ 7, while consideration of more extreme equations of state gives a more conservative upper bound of ≈ 12.

In conclusion, the $\bar{\nu}_e$ number N, apparent temperature T_0, and cooling time scale τ measured by the Kamioka and IMB detectors, and the inferred neutron star apparent radius R_{obs} and binding energy E_b all provide striking verification of current supernova theory.

We gratefully acknowledge discussions with John Bahcall, Adam Burrows, Jim Lattimer, Dave Schramm, Dave Spergel, Terry Walker, and especially Dave Arnett, Sid Bludman, Paul Schinder, and Michael Turner. This work is supported in part by NASA Grant NAGW-830.

Fig. 3.–Comparison of the best fit values and the 68% and 95% likelihood volumes projected onto the $(R_{\rm obs}, E_b)$-plane, and the $(R_{\rm obs}, E_b)$-curves for neutron star models based on a representative set of equations of state.

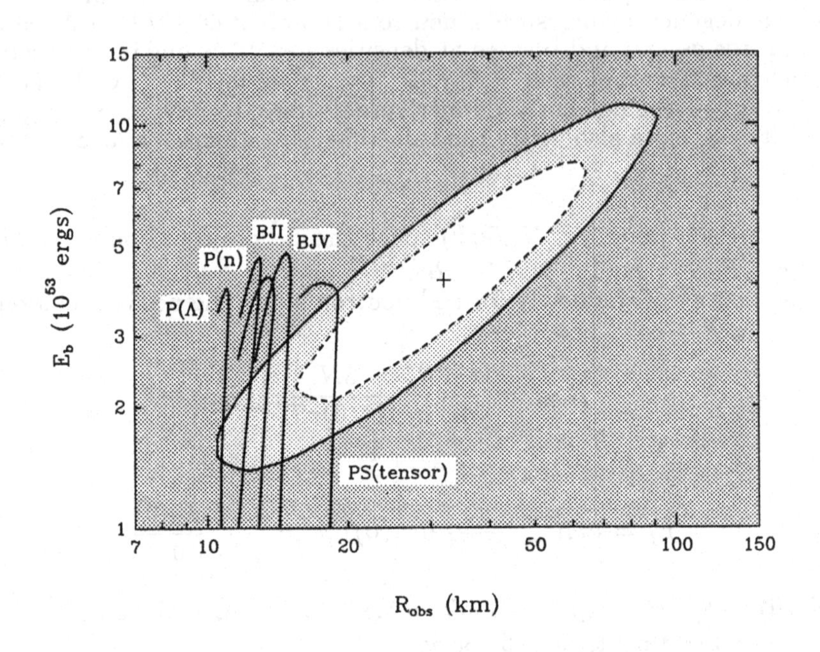

REFERENCES

Arnett, W. D. 1977, *Ap. J.*, **218**, 815.

Arnett, W. D., and Bowers, R. L. 1977, *Ap. J. Suppl.*, **33**, 415.

Bahcall, J. N., Piran, T., Press, W. H., and Spergel, D. N. 1987, *Nature*, **327**, 682.

Bethe, H. A., Brown, G. E., Applegate, J., and Lattimer, J. M. 1979, *Nucl. Phys.*, **A324**, 487.

Bionta, R. M., *et al.* 1987, *Phys. Rev. Lett.*, **58**, 1494.

Bludman, S. A. and Schinder, P. J. 1987, *Ap. J.*, in press.

Burrows, A. and Lattimer, J. M. 1986, *Ap. J.*, **307**, 178.

Burrows, A., Mazurek, T. J., Lattimer, J. M. 1981, *Ap. J.*, **251**, 325.

Hirata, K., *et al.* 1987, *Phys. Rev. Lett.*, **58**, 1490.

Lamb, D. Q., Lattimer, J. M., Pethick, C. J., and Ravenhall, D. G. 1978, *Phys. Rev. Lett.*, **41**, 1623.

Mayle R., Wilson, J. R., Schramm, D. N. 1987, *Ap. J.*, in press.

Mazurek, T. J. 1976, *Ap. J.*, **207**, 187.

Sato, K. 1975, *Progr. Theoret. Phys.*, **53**, 595.

Schramm, D. N. 1987, *Comments on Nucl. and Part. Phys.*, in press.

Spergel, D. N., Piran, T., Loeb, A., Goodman, J., and Bahcall, J. N. 1987, *Science*, **237**, 1471.

Wilson, J. R., Mayle, R., Woosley, S. E., and Weaver, T. E. 1986, *Ann. N. Y. Acad. Sci.*, **470**, 267.

NEUTRINO ENERGETICS OF SN 1987a

James M. Lattimer
Dept. of Earth and Space Sciences, SUNY, Stony Brook, NY

Amos Yahil
Dept. of Earth and Space Sciences, SUNY, Stony Brook, NY

Abstract. The total energy and average temperature of neutrinos emitted from SN 1987a are estimated from the Kamioka and IMB observations. Special care has been taken to include the detector response function, which we find improves the agreement between the two detections. We determine the average neutrino temperature is 3–5 MeV and the total energy emitted in $\bar{\nu}_e$'s to be 3–5$\times 10^{52}$ ergs. One sigma errors for the temperature and total energy are large, due primarily to counting statistics. We also present Monte Carlo simulations which strongly suggest that the gaps which appear in the Kamioka events are due to low number counting statistics rather than to any pulsing or bursting behavior in the source, as has been suggested in the literature.

It has long been appreciated that neutrino, not photon, emission dominates the last phases of the evolution of massive stars. The core collapse and subsequent supernova are accompanied by the most energetic neutrino emission of all, and a neutron star, or black hole, is formed. This prediction was dramatically confirmed by the simultaneous observation of a neutrino burst by the Kamioka (Hirata et al. 1987) and IMB (Bionta et al. 1987) collaborations. The 8 IMB neutrino events, of average energy 34 MeV, spread over 5.6 s, and the 11 Kamioka events (17 MeV, 12.5 s) are in amazing agreement with the standard predictions of supernova theory (Burrows and Lattimer 1986, Bruenn 1987, Mayle and Wilson 1987). We show these energies and counts imply a (redshifted) emission temperature of 4–5 MeV and total emitted energy of 2–3$\times 10^{53}$. This energy equals the binding energy of a 1–1.5 $M_\odot$ neutron star. The multisecond timescale indicates that the neutrinos did indeed diffuse out of a cooling neutron star.

Two general methods we utilize to estimate the neutrino temperature, T, are the moment method, which only uses the average detected energy, and the maximum-likelihood method, which employs the individual energies and should give more reliable results. In fact, the results of both methods are nearly identical. We assume that the time averaged spectrum is Fermi-Dirac, with zero chemical potential. This is a good assumption for the case of a cooling neutron star (Myra et al. 1987, this volume). The data are too meagre to allow more spectral parameters than T to be fitted. The published data consist of detection times, estimated electron energies, and electron angles with respect to the supernova. It is not possible, except in a statistical sense, to determine which ν events are due to scattering and which to absorption. We treat all events as absorptions since they are expected to dominate the signal. We take the incident neutrino energy to be the detected electron energy plus the neutron–proton mass defect, independently of the observed angle. We use updated detector efficiencies (private communications). For details of the schemes, see Lattimer (1987) and Lattimer and Yahil (1987).

The derived temperatures and total $\bar{\nu}_e$ energy for each detector are given in the Table. In both moment and maximum-likelihood approaches, T≃2.8(4.6) MeV for Kamioka (IMB) are found. Similarly, the total $\bar{\nu}_e$ energies are $30(60)\times10^{51}$ ergs. Although the experiments yield somewhat different results, this is not statistically significant. Figure 1 illustrates the sensitivity of the total energy to the observed ν energies and also shows that including detector response is not a small effect. In the maximum-likelihood approach, it is also possible to combine the two data sets. These estimates (see Table) lie between the separate determinations, with smaller error bars.

Table. $\bar{\nu}_e$ temperatures (MeV) and total $\bar{\nu}_e$ energies (10^{51}ergs)

	$T_{Kamioka}$	T_{IMB}	$E_{Kamioka}$	E_{IMB}
Response not included				
Moment	2.81±0.6	4.62±0.75	61(-30/+50)	29(-16/+26)
Max-Likelihood	2.85±0.4	4.60±0.8	59(-28/+34)	27(-10/+26)
" combined	3.8±0.4		43(-18/+20)	
Response included				
Moment	2.75±0.5	4.0±0.7	69(-38/+55)	57(-28/+100)
Max-Likelihood	2.80±0.4	4.2±1.1	63(-31/+40)	45(-34/+120)
" combined	3.7±0.4		48(-18/+21)	
Relative counts	5.0(+1.25/-1.0)		26(-16/+19)	

Figure 1. Total $\bar{\nu}_e$ energy per detected neutrino as a function of the average detected energy. K refers to Kamioka, I to IMB. Solid curves include detector response; dashed curves do not. Vertical lines are observed values and one sigma deviations.

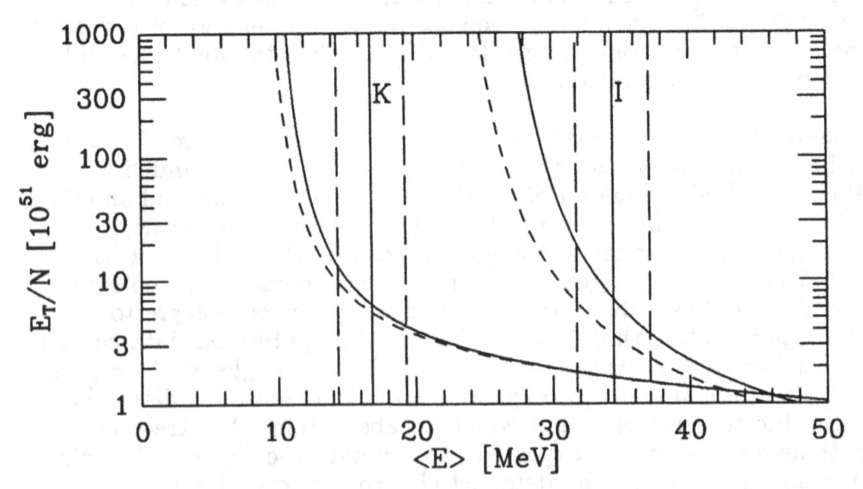

Besides estimating temperatures from the observed neutrino energies, it is possible to infer T from the ratio of counts in the detectors. Because of the higher threshold of the IMB detector, T is sensitive to this ratio. Somewhat higher temperatures and lower total energies are inferred from this approach (see Table and Figure 2). Nevertheless, the error bars are large enough that all our estimates are self-consistent. The simple interpretation of the signal as a thermal spectrum from a cooling protoneutron star seems justified.

Figure 2. Expected Kamioka/(Kamioka+IMB) count fraction vs. T. The solid curve includes detector response, the dashed curve does not. Horizontal lines are observations and one sigma errors.

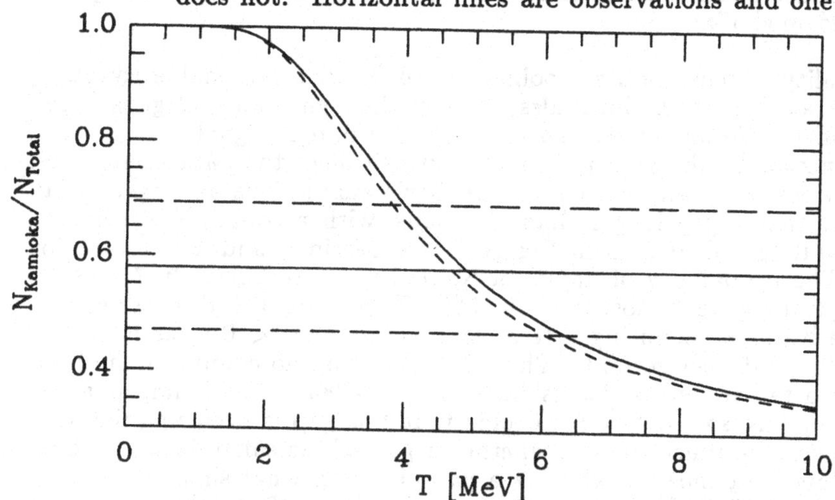

Figure 3. Probabilites that gaps with duration longer than t seconds and with n trailing neutrino events occur in randomly generated data sets. Assumed parent signal accumulations are from theory (solid curves) or Kamioka data (dashed curves).

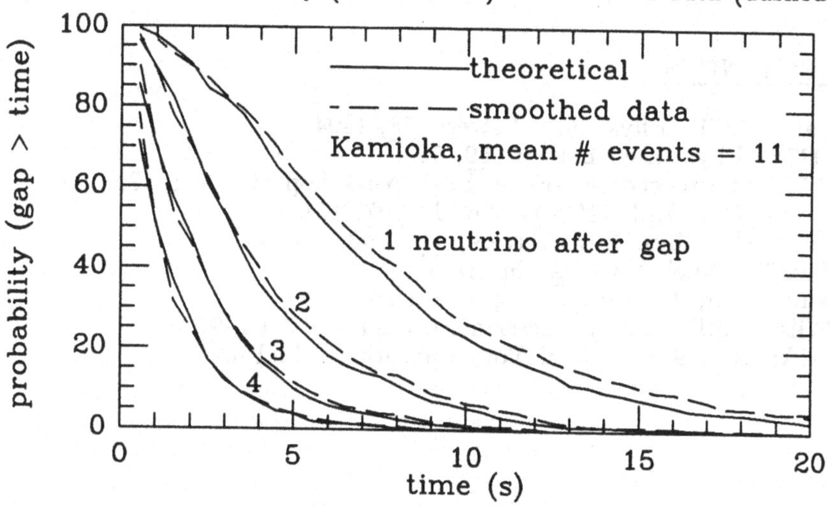

We note that the inferred $\bar{\nu}_e$ energy can be used to estimate the mass of the neutron star that might have formed. According to detailed models of neutrino emission (Mayle and Wilson 1987, Bruenn 1987), each neutrino species carries a nearly equal share, about 1/6, of the total. Thus, the total binding energy of the neutron star is about $2\text{–}3\times10^{53}$ ergs. This translates into neutron star gravitational masses of 1.2–1.5 $M_{\odot}$ (Burrows 1987, Sato and Suzuki 1987).

Many authors argue that the bunched structure of the Kamioka data is inconsistent with a simple cooling model and propose a variety of explanations. These efforts persist despite the fact that the last few IMB events fill in the largest Kamioka gap. We feel compelled to show that the observed bunching is a simple consequence of small-number statistics, and not an effect that requires more complicated modelling.

Based on the ability of neutron star cooling models to give reasonable counts, average energies, and emission timescales, we test the significance of gaps with a theoretical model. We also derive an experimental parent signal accumulation function by drawing a smooth curve through the data themselves. With these functions, one can perform Monte Carlo simulations and ask: what is the probability that a gap longer than t seconds, with n trailing neutrinos, occurs? Our results are displayed in Figure 3 (see Lattimer and Yahil 1987 for more details). The probability of having a Kamioka-like gap, namely 7 s in duration with 3 trailing neutrinos, is about 5%. To understand this value, approximate the Kamioka counting rate at late times ($3\ s<t<10\ s$) as a constant, $\simeq 3/10$ counts per second. The odds of having no counts in the first seven seconds of a ten second period is then $(0.7)^7 \simeq 0.085$. This is improbable, but not infinitessimally so. Before we decide that this gap is even marginally significant, however, we must enquire whether a gap of the same duration, but having two trailing neutrinos, or whether a gap of a somewhat shorter duration, would have elicited the same preprint activity. We think so, and the higher probability of these examples means that the observed gap is not very significant. Moreover, the presence of the late-time IMB events also justifies the conclusion that small-number statistics are at play.

This work was supported in part by USDOE Grant DE-FG02-87ER40317.A000.

REFERENCES

Bionta, R.M. et al. (1987). Phys. Rev. Letters, 58, 1494.
Bruenn, S.A. (1987). Phys. Rev. Letters, 59, 938.
Burrows, A. (1987). In Proceedings of the ESO Workshop on SN 1987a.
Burrows, A. and Lattimer, J.M. (1986). Ap. J., 307, 178.
Hirata, K. et al. (1987). Phys. Rev. Letters, 58, 1490.
Lattimer, J.M. (1987). Nucl. Phys. A, in press.
Lattimer, J.M. and Yahil, A. (1987). In preparation.
Mayle, R. and Wilson, J.R. (1987). Livermore preprint UCRL-97355.
Sato, K. and Suzuki, H. (1987). U. of Tokyo preprint UTAP-52.

NEUTRINO EMISSION FROM COOLING NEUTRON STARS

E.S. Myra
Dept. of Earth and Space Sciences, SUNY, Stony Brook, NY, USA
J.M. Lattimer
Dept. of Earth and Space Sciences, SUNY, Stony Brook, NY, USA
A. Yahil
Dept. of Earth and Space Sciences, SUNY, Stony Brook, NY, USA

Abstract. Most interpretations of the neutrinos collected from SN 1987*a* assume the emitted spectrum was in a Fermi-Dirac distribution with zero chemical potential. To investigate this assumption, we examine the results of numerical simulations by Bruenn (1987) and Mayle *et al.* (1987). We also examine models with a power-law atmosphere employing multi-group neutrino transport. We find many of the numerical spectra are best fit by thermal distributions having small positive chemical potentials. Fortunately the effect of such corrections is not appreciable enough to require extensive re-interpretation of the SN 1987*a* neutrino data.

Introduction

It has been suggested by Domogastsky *et al.* (1978) that instantaneous neutrino spectra emitted by a supernova deviate from a zero-chemical-potential Fermi-Dirac distribution. In this paper we examine this question by developing an analytic model of neutrino emission in order to find the nature of the expected deviation. We then compare this prediction with the results of detailed numerical calculations.

To obtain the form of the emitted spectrum, we construct a simple analytic model. In this model, neutrinos of any given energy remain in thermal equilibrium with the local matter until they encounter a region where their optical depth reaches unity. We stipulate that they subsequently escape with no further interaction. The matter density and temperature profiles are assumed to be of the form $\rho \propto r^{-\alpha}$ and $T \propto r^{-\beta}$, where r is the distance from the center of the star. Since the neutrino atmosphere is scattering dominated, the inverse neutrino mean free path has the form $1/\lambda \propto \rho\epsilon^2$, where ϵ is the energy of the neutrino. Because of the strong energy dependence of the mean free path, neutrinos of different energies depart from regions of the star that are at different radii and temperatures.

At the surface of last scattering,

$$\int_{r_i}^{\infty} \frac{1}{\lambda}\, dr = 1, \tag{1}$$

where r_i is radius at which neutrinos of energy ϵ_i are emitted. With the given matter profile this yields,

$$r_i \propto \epsilon_i^{2/(\alpha-1)}, \tag{2}$$

with corresponding temperature at each r_i being

$$T_i \propto \epsilon_i^{-2\beta/(\alpha-1)}. \tag{3}$$

These expressions can be used to obtain the form of the spectral luminosity profile

$$L(\epsilon_i) \propto r_i^2 \frac{\epsilon_i^3}{\exp[\epsilon_i/T_i]+1} \propto \frac{\epsilon_i^{4/(\alpha-1)+3}}{\exp[K\epsilon_i^{2\beta/(\alpha-1)+1}]+1}, \tag{4}$$

where K is a function of α and β. For clarity, consider the specific example $\alpha = 3$ and $\beta = 1$. Then

$$L(\epsilon_i) \propto \frac{\epsilon_i^5}{\exp[K\epsilon_i^2]+1}. \tag{5}$$

It is evident that such a spectrum is "pinched" relative to a thermal distribution. At low energies the factor ϵ^5 depletes the low energy neutrinos, while the ϵ^2 in the argument of the exponential depletes the high energy tail. This result holds qualitatively for all $\alpha > 1$ and $\beta > 0$.

Fits to Numerical Models

Having predicted the shape of the emitted spectrum, we next examine numerical models to see the magnitude of the effect. The models we consider are by Bruenn (1987) and Mayle *et al.* (1987). These calculations employ

Figure 1. Emergent $\bar{\nu}_e$ spectrum from the 2-$M_\odot$ iron core calculation of Bruenn (1987) at $t = 0.47$ s (solid line). The dashed line shows a fit to a Fermi-Dirac distribution with $\eta_\nu = 0$ for which $T_{\rm fit} = 4.0$ MeV. The inset shows Bruenn's spectrum is well fit by a $\eta_\nu = 3.4$ distribution for which $T_{\rm fit} = 3.0$ MeV.

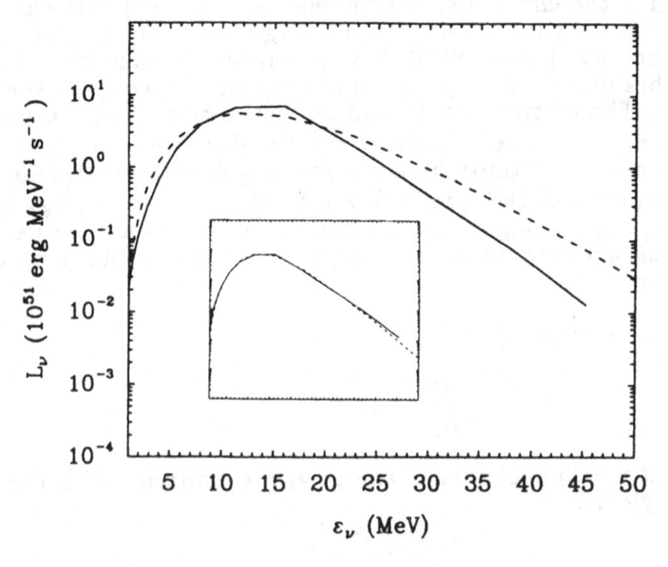

multi-group neutrino transport and therefore make detailed predictions for the neutrino distributions. A temperature fit to their distributions is made by use of the relation

$$\frac{L_\nu}{N_\nu} = T_{\text{fit}}\frac{F_3(\eta_\nu)}{F_2(\eta_\nu)}, \tag{6}$$

where L_ν and N_ν are the spectral energy and number luminosities respectively, and the F's are the standard Fermi integrals. We also constrain the total energy luminosities of the calculated models to be equal to the fitted luminosities. With this prescription, an effective radius of the neutrinosphere can be inferred.

Figure 1 shows that an electron anti-neutrino spectrum from Bruenn (1987) is well fit by a positive neutrino chemical potential. For $\eta_\nu = 0$, the radius of the neutrinosphere $R_{\nu,\text{fit}}$ is 83 km while for $\eta_\nu = 3.4$, $R_{\nu,\text{fit}}$ shrinks to 35 km. At later times in his calculation ($\simeq$ 1 s), Bruenn's spectra are more nearly $\eta = 0$ distributions. However, the fitted temperatures and especially neutrinosphere radii show rather erratic behavior. (The fitted radius at 0.75 s is 120 km!)

A similar analysis can be performed using the calculations of Mayle *et al.* (1987). The results are also similar. Figure 2 shows the $\eta = 0$ and $\eta = 3.6$ fits for which the values of $R_{\nu,\text{fit}}$ are 43 and 18 km respectively. Unlike the calculations of Bruenn, at later times Mayle *et al.* show a persistence of the high energy depletion, though to a lesser degree than evident at 0.47 s. (One should note that these "late" times of $\sim$ 1 s are small compared to the 10-s timescale for neutrino cooling in neutron stars (Burrows & Lattimer 1986)).

Figure 2. Emergent $\bar{\nu}_e$ spectrum from Model 25C of Mayle *et al.* (1987) at $t = 0.47$ s (solid line) from calculations using their Boltzmann equation solver. The dashed line shows a fit to a Fermi-Dirac distribution with $\eta_\nu = 0$ for which $T_{\text{fit}} = 3.7$ MeV. The inset shows the fit with $\eta_\nu = 3.6$ and $T_{\text{fit}} = 2.8$ MeV.

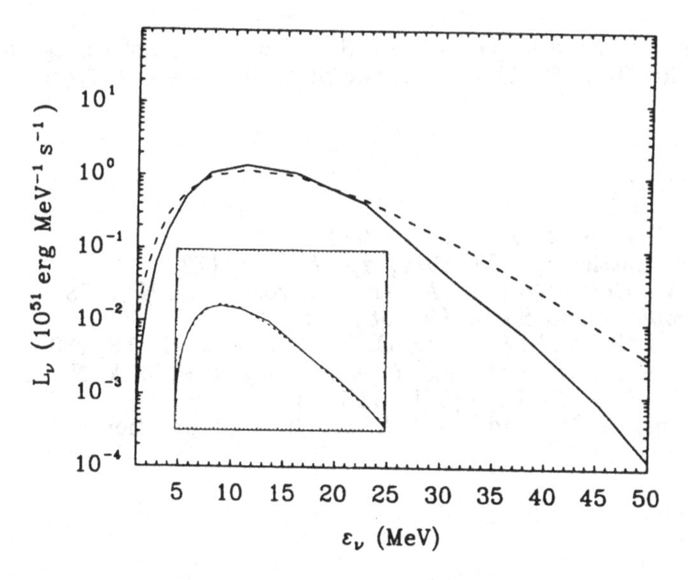

To study the effect further, we have constructed a static matter profile and, by fixing the matter composition, obtain steady state solutions of the emergent electron anti-neutrino luminosity. To model the neutrino transport, we use a multi-group flux-limited diffusion code similar to that described in Myra *et al.* (1987a). We include antineutrino production by positron capture on free protons (the dominant process), by the pair annihilation process, and by the plasma process. Preliminary results show that the power-law atmosphere yields the same low and high energy depletion shown in the detailed simulations examined above. For the details of these calculations, see Myra *et al.* (1987b).

Discussion and Conclusions.

We have demonstrated that with the current models for neutrino atmospheres in proto-neutron stars, the assumption of a Fermi-Dirac emission spectrum is not necessarily justified. However, because of the paucity of data and the relatively high energy threshold for the detectors, this does not greatly affect the results inferred from SN 1987*a* (Myra *et al.* 1987b).

Spectra which deviate from $\eta_\nu = 0$ distributions are well fit by Fermi distributions with $\eta_\nu > 0$. The radii of the neutrinospheres given by these fits, as well as the positive chemical potentials obtained for anti-neutrinos, imply that the results must be seen as *merely* fits and the emission spectra are actually non-thermal in nature. Hence, one can derive only limited information about the physical conditions under which such neutrinos were emitted.

While we have concentrated in this paper on the form of the instantaneous spectrum, it is also important to investigate the effects of time-integrating the neutrino signal. It is easy to show that in an atmosphere in which the temperature is changing, a time integrated spectrum will not be Fermi-Dirac, even if the instantaneous spectrum is. Combining this effect with a *non*-thermal instantaneous spectrum could have interesting consequences and merits further study.

This research was supported in part by the U.S. Department of Energy under Grant No. DE-FG02-87ER40317.A000 at the State University of New York at Stony Brook.

References.

Bruenn, S. W. 1987, *Phys. Rev. Lett.*, **59**, 938.
Burrows, A. S. and Lattimer, J. M. 1986, *Ap. J.*, **307**, 178.
Domogastsky, G. V., Eramzhyan, R. A., and Nadyozhin, D. K. 1978, *Astrophys. and Space Sci.*, **58**, 273.
Mayle, R., Wilson, J. R., and Schramm, D. N. 1987, *Ap. J.*, **318**, 288.
Myra, E. S., Bludman, S. A., Hoffman, Y., Lichtenstadt, I., Sack, N., and Van Riper, K. A. 1987a, *Ap. J.*, **318**, 744.
Myra, E. S., Lattimer, J. M., and Yahil, A. 1987b, in preparation.

STATISTICAL ANALYSIS OF THE TIME STRUCTURE OF THE NEUTRINOS FROM SN 1987A

Paul J. Schinder
Physics Department,University of Pennsylvania, Philadelphia, PA 19104,USA

Sidney A. Bludman
Physics Department,University of Pennsylvania, Philadelphia, PA 19104, USA

INTRODUCTION

In the excitement following the Kamiokande (Hirata *et al.* 1987) and IMB (Bionta *et al.* 1987) observations of neutrino bursts from Supernova 1987A, there has been strong temptation to overinterpret the scant neutrino data: 11 events over 12.4 seconds with energy 7.5-36 MeV from the 2.1 kton Kamiokande II detector plus 8 events over 5.6 seconds with energy 20-40 MeV from the 5 kton IMB detector. We present here the results of our statistical analysis of the entire data sample (11 + 8 events over 12.4 seconds) for consistency and for dynamical evidence for cooling of the nascent neutron star. We attempt to fit the data with three simple cooling models, *power law cooling*, $T = T_s\,(at/n+1)^{-n}$, for various values of n, *exponential cooling*, $T = T_s\,e^{-at}$ (the limit of power law cooling as $n \rightarrow \infty$) and *loaded sphere cooling*, $T = T_s \lim_{n\rightarrow\infty} \frac{1}{n}\sum_{j=0}^{n} e^{-j^2 at}$. We assume that all of the observed events were $\bar{\nu}_e$ and ignore the weak angular dependence of the cross section $\sigma(\bar{\nu}_e + p \rightarrow n + e^+)$. (A more complete discussion of our techniques and the results presented below may be found in Bludman & Schinder (1988)).

A complete statistical analysis must determine three things: (1) Best fit values for the parameters N_o, T_s, a, where N_o is the total number of electron antineutrinos emitted over all time; (2) The statistical uncertainties in these parameters; (3) The goodness of fit between the data and our assumed cooling curve. Our *best fits* are obtained by maximum joint likelihood analysis, including the Poisson probability of the temporal gaps between neutrino events. The long gaps between the millisecond or microsecond events cannot be ignored: these non-events contain as much information as the reported events. With these best fit parameters, approximately 750 Monte Carlo simulations of events at the two detectors are then run to determine the probable *error bounds* on these parameters. *Goodness-of-fit* is checked graphically with "Kolmogornov-Smirnov by eye" comparisons between the best fit curves and the data histograms, and by scatter plots of the parameters determined from the Monte Carlo synthetic data.

We also search for periodicity in the sparse neutrino data. There is some theoretical expectation (Mayle 1986; Wilson 1985) for oscillations in the $\bar{\nu}_e$ luminosity with periods ≈ .1 sec. Harwit *et al.* (1987) suggest that there is an 8.91 ms period to be found in the IMB and Kamiokande data. We search for periodicity by using both *power spectrum* (Fourier) analysis and *minimum residual* analysis.

Figure 1: Comparison of our best fit model with the observations of Kamiokande and IMB.

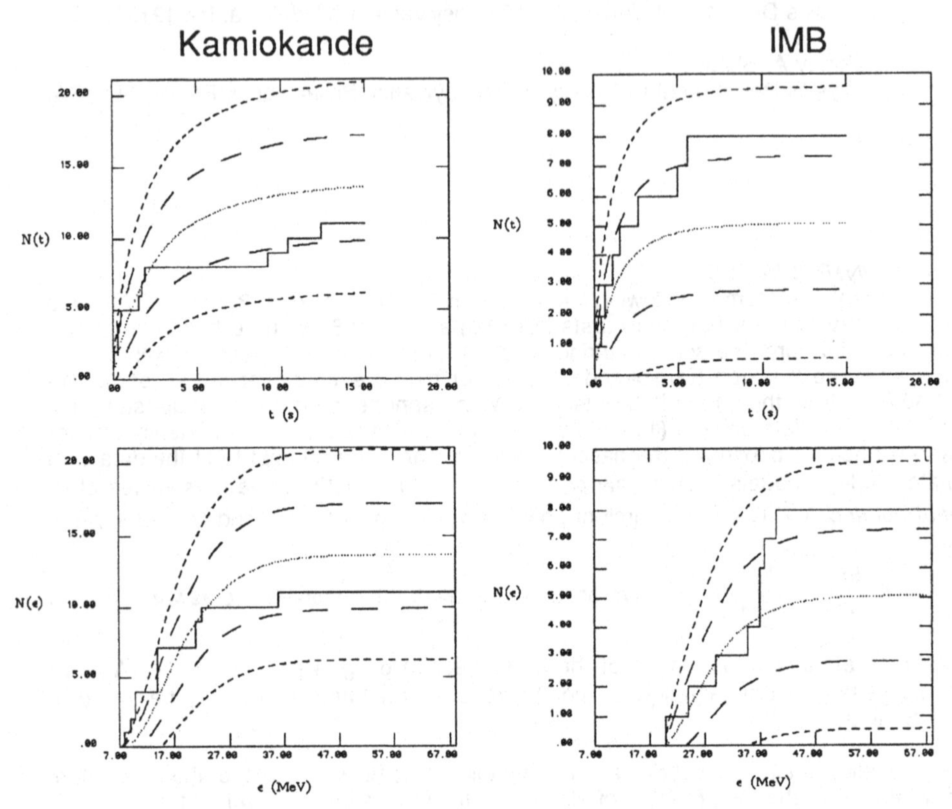

Results

Our best fit turns out to be a n=0.4 power law with T_s = 4.2±0.5 MeV, a = 0.14±0.05 s^{-1}, releasing $N_0 = (1.3\pm0.4)\times10^{58}\ \bar{\nu}_e$ and $E_0 = (9.2\pm3.3)\times10^{52}$ erg (but only about 70% in the first 15 s). However, the difference between our various cooling models when fit to the observations is not statistically significant. Figure 1 shows that the observations (solid step-like line) agrees within $\approx 1\sigma$ with the best fit n=0.4 power law (central dotted line) in both energy (top two figures) and angle (bottom two figures). The long dashed and dashed lines surrounding the central curve are the 1 and 2 σ bands, respectively.

Our search for periodicity lead us to the conclusion that there is *no* periodicity in any frequency range. We analyzed the Kamiokande data and IMB data separately, focusing our attention on the Kamiokande (larger) data set, since the relative timing between the Kamiokande and IMB data is not known. Our best fit cooling model implies that the offset time t_{off} = 0.1±0.5 s with IMB leading, but does not allow us to precisely state what t_{off} is. We compared the power spectrum and minimum residual analysis with the analysis of 1000 random data sets of 11 neutrinos each. Based on this comparison, we conclude that the Kamiokande data contains no periodicity on any timescale.

Prospects for Galactic core collapse

To conclude, we briefly discuss our simulations of a *Galactic* core collapse. We emphasise that neutrino bursts of the type detected by Kamiokande and IMB are the signature of *core collapse*. The supernova associated with the core collapse is an energetically superficial (0.0001% of the energy is emitted in photons) phenomena which may not be a necessary product of a core collapse. This gives us hope that the rate of neutrino bursts associated with Galactic core collapse may be appreciably greater than the Type II supernova rate in the Galaxy. In our simulation, we used our best fit power law and moved it to 10kpc. We added a 0.5 s heating phase with a linear ramp up to the maximum temperature. We induced a 10% sinusoidal oscillation in T during the cooling phase, lasting for either 15, 5, or 1 s. We "captured" about 330 $\bar{\nu}_e$ in our simulated Kamiokande tank, and about 130 in IMB. In addition to being able to determine the parameters T_s, a, and N_0 with much greater precision, we are able to detect the 0.5 s heating phase, measuring the heating time to within 50%. We need about 50 cycles (5 s) to reliably detect the 10% temperature oscillation. These results give us hope that with the data measured in a Galactic core collapse, we may be able to use experimental data to confirm or rule out specific models of core collapse.

References

Bionta, R. M. *et al.* (1987), *Phys. Rev. Lett.* **58**, 1494.

Bludman, S. A. & Schinder, P. J. (1988), *Ap. J.*, in press.

Harwit, M. *et al.* (1987), preprint.

Hirata, K. *et al.* (1987), *Phys. Rev. Lett.* **58**, 1490.

Mayle, R. (1986), Ph.D. Thesis, U. of California, Berkeley.

Wilson, J. (1985), in *Numerical Astrophysics*, eds. J. M. Centrella, J. M. LeBlanc, and R. L. Bowers, pp. 422-434, Boston: Jones and Bartlett Publishers.

NEUTRINO PROPERTIES FROM OBSERVATIONS OF SN 1987A

A. Dar
Department of Physics and Space Research Institute, Technion, Haifa, Israel.

Abstract. Observations of the neutrino burst from SN 1987A by KAMIOKANDE II and IMB water Cherenkov detectors yield strong constraints on the lifetime, mass, mixings, electric charge and magnetic moment of the electron neutrino and on the number of flavours of light neutrinos.

Table I compares constraints on the properties of the neutrinos that were obtained from experiments with terrestrial neutrino sources and constraints that were derived by us and by other authors from observations of the neutrino burst from SN 1987A. Question marks indicate that the conclusions are based on assumptions that may be disputed.

(a) Neutrino Life Time. The idea of unstable neutrino was proposed many years ago by Bahcall et al. (1972) and by Pakvasa and Tennakone (1972) as a solution to the solar neutrino problem (Bahcall et al. 1982) and has been recently revived (Bahcall et al. 1986) in the light of theoretical models for ν-instability (Gelmini and Valle, 1984, Valle et al. 1987). The observation of $\bar{\nu}_e$'s and (?) ν_e's with $E \sim 15$ MeV from SN 1987A (Hirata et al. 1987, Bionta et al. 1987, Aglietta et al. 1987, Pomansky 1987) has been widely perceived ruling out ν-decay as an explanation to the solar neutrino problem (Bahcall et al. 1987a). In the theoretically preferred regime of small flavour-mixing and small mass differences the observation of the ν-burst from SN 1987A indeed implies that $\gamma\tau > 5\times10^{12}$ sec for ν_e's. However, it has been pointed out recently (Frieman et al. 1987, Raghavan et al. 1987) that if $\bar{\nu}_e$'s and ν_e's are unstable but admixed significantly with other neutrinos which are stable, then they could have generated the observed neutrino signals from SN 1987A. (This however may require a neutrino luminosity which is too large compared with the binding energy release expected from a type II supernova like SN 1987A).

(b) Neutrino Mass: The basic idea was discussed first by Zatsepin (1968) who pointed out that if neutrinos had a finite mass, the higher energy neutrinos from a supernova explosion would arrive before the more slowly moving, lower energy neutrinos. The difference in time of flight from SN 1987A to Earth (distance D) of two neutrinos of energies E_1 and E_2, respectively, is given by $\Delta t \cong (D/2c)(m_\nu c^2)^2(E_1^{-2}-E_2^{-2})$. A

TABLE I. Comparison Between Electron Neutrino Properties As Determined From Terrestrial Neutrinos And From SN 1987A Neutrinos

Property	Terrestrial Experiments	SN 1987 A
Life Time ($\gamma\tau_\nu$ in sec)	Atmospheric ν's: $>4\times10^{-2}$ Solar ν's : $\gtrsim5\times10^{2}$	$>5\times10^{12}$ (if no mixing)
Mass (m_{ν_e} in eV/c^2)	<18 (Zurich) 34 (Lubimov	$<3.4\pm1$? <15 ?
Mixing with ν_L (m_{ν_L} < MeV)	$\Delta m^2 \gtrsim 0.1 eV^2$, $\sin^2 2\Theta > 0.1$ } excluded	Practically Excluded? (See Fig.)
Mixing with ν_h (m_{ν_h} > 1 MeV)	Left Narrow Window	Ruled Out For m_{ν_h} < 100 MeV
Electric Charge (q/e)	$<10^{-13}$	$<2\times10^{-17}$
Magnetic Moment (μ/μ_B)	$<10^{-9}$	$<10^{-13}$ $<2\times10^{-15}$?
ν Flavours	$\lesssim4$ (Cosmology)	$\lesssim5\pm1$

finite mass of the neutrino will cause, accordingly, particles of different energies to arrive at different times. This dispersion in arrival times can be used to extract a neutrino mass if one makes assumptions on the emission time. For instance many independent authors (see e.g. Abott et al. 1987 and references therein) have assumed that the neutrino luminosity ($\mathcal{L}_\nu \sim R^2T^4$) can be fitted by a smoothly varying function of time, and from a statistical analysis deduced that $m_{\nu_e} \lesssim 15$ eV. However, some authors (e.g. Arfune and Fukugita 1987) pointed out that the first two events observed by KAMIOKANDE II are most probably ν_ee scattering events generated by electron neutrinos from the neutronization burst. (These events were both observed in the forward direction with respect to the LMC-Earth axis). A neutronization pulse should precede almost all of the neutrino emission, although it is expected to be weak in the water Cherenkov detectors. But the expectations are based on a theory which has not incorporated angular moment effects during the collapse, neither neutrino transport via convection , nor neutrino magnetic moment effects (if the neutrino has a magnetic moment), etc., which may lead to an enhanced neutronization burst (limited only by lepton number conservation). The 4 eV upper limit on electron neutrino mass was obtained (Dar and Dado 1987a) by requiring that events 3, 4, 5, in the KAMIOKANDE detector were generated by $\bar{\nu}_e$'s (as evident from the large angles of the recoiling electrons) which were emitted in the thermal burst i.e. after the ν_e's which generated events 1 and 2.

The validity of the various assumptions used by different authors in extracting limits on m_{ν_e} will be tested only when much larger neutrino signals from galactic supernovae explosions will be detected. Moreover, larger signals from galactic supernovae explosions and the different sensitivity of some future underground neutrino detectors to different neutrino flavours (e.g. D_2O detectors) will yield very significant bounds on the ν_μ and ν_τ masses (present bounds are $m_{\nu_\mu} \leq 250$ KeV and $m_{\nu_\tau} \leq 56$ MeV).

(c) Mixing of ν_e With Other Generations: If the first two events which were generated by neutrinos from SN 1987A in the KAMIOKANDE II detector were induced by ν_e's from the neutronization burst, they indicate that the ν_e's from the neutronization burst did not suffer flavour flip due to the MSW effect (Mikheyev and Smirnov 1986, Wolfenstein 1978) in the progenitor envelope, because their inferred number is about the maximum number allowed by lepton number conservation for a complete neutronization of an iron core with $M_c < 2M_\odot$. Since the range of densities encountered in the progenitor envelope include the range of densities encountered in the sun, the absence of any significant MSW conversion of ν_e's into another flavour in the progenitor envelope therefore indicates that the MSW effect is not responsible for the solar neutrino problem (Dar and Dado 1987a, Arafune et al. 1987, Nötzold 1987). In fact, the absence of any significant evidence for the MSW effect in the progenitor envelope excludes such a large range of ν_e mixing parameters that practically excludes ν_e mixing with light ν's.

(d) Mixing of ν_e With ν_i, 1 MeV$<M_{\nu_i}<$100 MeV: Mixing of the ν_e with ν_i induces muon like decay $\nu_i \to \nu_e e^+ e^-$, if $m_{\nu_i} > 2m_e$. Experiments searching for such decays of ν_τ's and of higher neutrino generations at beam dumps and at meson decay facilities have led to m dependent lower bounds on the lifetime τ_{ν_i}, while the abundance of elements synthesized in the early universe and the observed electromagnetic background radiations set upper bounds of τ_{ν_i} (for a recent summary see e.g. Dar et al. 1987). These lower and upper bounds left a small open window for the existence of unstable neutrinos with $m_i \sim 50$ MeV and $\tau_i \sim 10^3$ sec. However, ν_i that are produced thermally ($e^+e^- \to \nu_i \bar{\nu}_i$, $\nu_L \bar{\nu}_L \to \nu_i \bar{\nu}_i$) in the hot core of the photoneutron star and decay outside the progenitor star produce e^+e^- pairs, which are accompanied by internal and external bremsstrahlung photons and by photons produced by pair annihilation. From the fact that no enhancement of the γ-ray background from the direction of SN 1987A has been detected by the SMM Gamma Ray Spectrometer since core collapse on Feb. 23,316 through April 9 (Chupp et al. 1987) Dar and Dado (1987b) have derived a new lower bound on τ_i which closed the window for the existence of ν_i with $1 \leq m_i \leq 100$ MeV that are mixed with ν_e.

(e) Neutrino Electric Charge: Barbiellini and Cocconi (1987) pointed out that if neutrinos have an electric charge q they are deflected by galactic and intergalactic magnetic fields. Thus their path lengths and flight times from SN 1987A to Earth depend on their energy. Assuming an intergalactic field $B=10^{-3}\mu G$ over a 50 kpc path, and a galactic magnetic field $B=1\mu G$ over a 10 kpc path, they derived from the energy spread and the dispersion in arrival times of the neutrinos from SN 1987 that q/e is constrained to be smaller than $2x10^{-15}$ and $2x10^{-17}$, respectively.

(f) Neutrino Magnetic Moment: An anomalous magnetic moment $\eta=\mu/\mu_B\simeq10^{-10}$ has been suggested by Okun et al. (1986) as an explanation to the suppressed counting rate in the solar neutrino experiment and its possible anticorrelation with the sun spot cycle. However, Goldman et al. (1987) have argued that if $\eta>2x10^{-15}$ then a magnetic field of the order $B\sim10^{12}$ gauss believed to be present near the surface of the proto-neutron star will induce $\nu_L\leftrightarrow\nu_R$ precesion. Thus the observed numbers of ν_L's is only half of the number of ν's emitted by SN 1987A, i.e. the inferred binding energy is twice that carried by ν_L's. However, the binding energy estimated by the KAMIOKANDE group (Hirata et al. 1987) is already about the maximum gravitational binding energy expected to be released by a neutron star, which excludes the possibility that $\eta>2x10^{-15}$.

(g) The Number of Light Neutrino Flavours: SN 1987A could radiate energy not only via the known neutrinos but also via other light neutrinos with $m_i\lesssim T_i$ where $T_i\sim10$ is the temperature at the neutrino-sphere for emission of μ and τ like interacting neutrinos. In this case the energy which was radiated in the form of $\bar{\nu}_e$'s and which was estimated by KAMIOKANDE II from their events to be $W_{\bar{\nu}_e}\sim8x10^{52}$ ergs (Hirata et al. 1987) should be multiplied by N_f the number of light neutrino flavours to obtain the total binding energy that SN 1987A has released. If this total binding energy is smaller than $4x10^{53}$ ergs, as expected from the best available equation of states of nuclear matter, then $N_f\lesssim5$.

References

Abott, et al. (1987), Preprint, BUHEP-87-24.
Aglietta M., et al. (1987), Europhysics Lett. 3, 1315.
Arafune, J. and M. Fukugita, (1987), Phys. Rev. Lett. 59, 367.
Arafune, J. et al. (1987), Phys. Rev. Lett. 59, 1864.
Bahcall, J.N. et al. (1972), Phys. Rev. Lett. 28, 316.
Bahcall, J.N. et al. (1982), Rev. Mod. Phys. 54, 767.
Bahcall, J.N. et al. (1986), Phys. Lett. 181B, 369.
Bahcall, J.N. et al. (1987a), Nature 326, 125.
Barbiellini, G. and Cocconi G. (1987), Nature 329, 21.
Bionta, R.M. et al. (1987), Phys. Rev. Lett. 58, 1494.
Chupp, E.L. et al. (1987), Phys. Rev. Lett. 58, 2146.

Dar, A. et al. (1987), Phys. Rev. Lett. 58, 3246.
Dar, A. and Dado, S.(1987a),Preprint Technion-PH-000.
Dar, A. and Dado, S.(1987b),Phys. Rev. Lett. 59, 2368.
Frieman, J.A. et al. (1987), Preprint SLAC-PUB 4261.
Gelmini, G. and Valle, J.W.F. (1984), Phys. Lett. 142B, 181.
Goldman, I. et al. (1987), Preprint TAUP 1543-87.
Hirata, K. et al. (1987), Phys. Rev. Lett. 58, 1490.
Mikheyev, S.P. and Smirnov A. Yu (1986), Nuovo Cimento C9, 17.
Notzold, D. (1987), Preprint MPI-PAE/PTh 09/87.
Okun L.B. et al. (1986), Preprint ITEP 82-86.
Pakvasa S. and Tennakone (1972), Phys. Rev. Lett. 28, 1415.
Pomansky, A. (1987), Proceedings of the XXIII Rencontre De Moriond (Ed. Tran Thanh Van).
Raghavan, R.S. et al. (1987), Preprint.
Spergel, D. and Bahcall, J.N. (1987), IAS Preprint.
Valle, J.W.F., et al. (1987), Phys. Lett. 131B, 83.
Wolfenstein, L. (1978), Phys. Rev. D17, 2369.
Zatsepin, G.I. (1968), JETP Lett. 8, 205.

SN 1987A and COMPANION

Costas Papaliolios, Margarita Karovska
Peter Nisenson, and Clive Standley

Harvard-Smithsonian Center for Astrophysics
Cambridge, MA 02138

INTRODUCTION

When Bob Kirshner called to announce SN 1987A and to ask if we were going to look at it with our speckle system the first thought that came to mind was to measure the size of the expanding shell. The best way to do this is with an interferometer, such as the one we had used last summer at CERGA, where it would be easy to measure the expected diameter of a star as bright as SN 1987A. Unfortunately the supernova, because of its southern location, cannot be seen from CERGA which is located in France. We remembered, however, the interferometric work in progress by John Davis at Sydney, Australia and his recently reported measurement of the diameter of Sirius (Davis & Tango 1986) to a precision of 80 microarcseconds. We called Davis the next day but he was not confident that he could make the needed observations for a variety of reasons. His problems were not related to any limitations of the techniques he had been using, but involved the fact that the interferometer he used was built to look specifically at Sirius and thus the location of SN 1987A would require extensive modifications to his system. The more we talked with Davis the more convinced we became that perhaps we were in a better position to make these important measurements than was Davis. Our system had been used successfully on another project on the 4-meter at the Cerro Tololo Inter-American Observatory (CTIO) two years ago and could again be used there without any modifications. If any modifications were needed to make observations specific to the supernova, they would be simple and could be done quickly. On the basis of the expansion velocities measured by many other observers, and the reasonably well known distance to the Large Magellanic Cloud (LMC), which presumably contained the supernova, we concluded that the calculated size of the shell might be within our measurement capabilities after a month's time had elapsed.

The Director of CTIO, Bob Williams gave us two nights on the 4-meter, March 25 and April 2. These dates were just about a month after the supernova exploded. Our three objectives were: 1) to measure the diameter of the expanding shell, 2) to measure the position of the supernova relative to Sanduleak -69 202 (believed at that time to still be present) and, 3) to look for light echoes from the interaction of dust with the intense expanding radiation front.

OBSERVATIONS

Following the technique used by Davis (Davis & Tango 1986) we tried to determine the shell diameter by measuring, with precision, the loss of fringe contrast at the higher spatial frequencies relative to the measured fringe contrast of an unresolved star. In this way we can go beyond the 25 mas resolution expected of a 4-meter telescope. Because of the maximum photon count rate that our camera can detect (10^5 per sec), and the brightness of the supernova, it was necessary to reduce the light through the telescope. The first method we tried consisted of using only the light from a set of seven small apertures close to the outer edge of the primary mirror. This reduces the count rate but maintains the spatial resolution needed for the experiment. The second method was just to reduce the light uniformly over the full telescope aperture with neutral density filters. This second method of data acquisition is identical to what we have been doing for many years while the first method was a new technique for us and did not provide data we could use.

To determine the position of the supernova relative to the Sanduleak star we recorded data in our usual way but with a 400 nm filter which would reduce the large magnitude difference between the supernova and Sanduleak -69 202.

In the search for light echoes we recorded supernova data again in our usual mode of operation except for the addition of neutral density filters that were needed to make the count rates manageable. We used several different 10 nm wide spectral filters (in H alpha and in the continuum around it) to look at the supernova and at a few nearby stars which serve as unresolved references.

RESULTS

Analysis of the data taken through the seven small apertures proved troublesome for a variety of reasons, so this data was temporarily put aside. The data taken with neutral density filters did, however, allow us to determine the shell diameters and will be discussed later.

Analysis of the data taken at 400 nm, where the expected magnitude difference between the supernova and the Sanduleak star is diminished, revealed no evidence of the Sanduleak star, though by that time others had concluded (Kirshner et al 1987) that Sanduleak -69 202 was the supernova precursor and was no longer present.

We then turned to our third objective, to examine the surroundings of the supernova for the presence of reflected light. To our surprise, we saw fringes in the power spectrum, a sure sign of a binary companion (Karovska et al 1986, Nisenson et al 1985). We have seen this many times before with other stellar systems, sometimes under much less favorable conditions (Papaliolios et al 1985, Nisenson et al 1985, Karovska et al 1986). Since we also measure phase in addition to power, we were able to reconstruct the full image from the data thus giving the position of the companion relative to the supernova (Karovska et al 1987a, Nisenson et al 1987).

The supernova has a faint companion at the position indicated in the table below (with the magnitude difference given in the last column).

DATE	WAVELENGTH	SEPARATION	P.A.	MAG DIFF
3/25	656 nm	59 ± 8 mas	194° ± 2	2.7 ± 0.2
4/2	"	"	"	"
"	533	52 ± 7	"	3.0 ± 0.2
"	450	"	"	3.5-4.0
"	400	-----	---	>4.0

NEW RESULTS ON SHELL DIAMETER

We calculated the power spectra of each of 30,000 5msec exposures (only about 3 minutes of data) and added them to give a power spectrum that has a good signal out to the telescope diffraction limit. In order to improve the signal-to-noise ratio we then formed an azimuthal average (this assumes circular symmetry which is in fact consistent with the unaveraged data). This was done to the supernova data as well as to the data of an unresolved star. This averaged supernova power spectrum was then divided point by point by the similar power spectrum of the reference star, whose data were recorded close in time and sky position. This procedure is essentially the same one used by Davis (1986) and effectively cancels out the telescope abberations, the atmospheric transfer function, and the telescope transfer function. The only thing left that makes the spectrum of the supernova fall off more rapidly with frequency than that of the unresolved star is the finite angular size of the supernova.

The shape of the power spectrum was fit by the appropriate Bessel function that is expected for a uniform disk. The only free parameter is the angular diameter of the disk. The fit was done over almost the full spatial frequency range of the data, from DC out to nearly the telescope diffraction limit. The very lowest frequencies were excluded because of seeing fluctuations while the very highest frequencies were excluded because of poor signal-to-noise ratios there. The diameter of the supernova at several wavelengths was obtained by this method. This technique is a very straightforward extension of that used by Davis to measure the diameter of Sirius (Davis & Tango 1986). The results of this experiment (Karovska et al 1987b) are tabulated below.

DATE	WAVELENGTH	DIAMETER
4/2	450 nm	12 ± 4 mas
"	533	11 ± "
"	656	1 ± "
6/1	450	23 ± "
"	533	18 ± "
"	656	8 ± "
"	775	15 ± "

CONCLUDING REMARKS

A companion to SN 1987A was observed one month after the birth of the supernova but was not visible on two later observing trips. The observing conditions, however, were far from optimal on those later dates and the lack of a subsequent detection should not detract from the initial observations in which the companion was clearly present.

The diameter measurements are quite different from the approximate 4 mas diameter of the photosphere determined photometrically by assuming the supernova to be radiating as a blackbody. This discrepancy should not be surprising since even a casual glance at the supernova spectrum shows that the blackbody assumption is extreme.

Clearly more data is essential to track the development of the supernova and to verify, where possible, both of the very informative results given here.

REFERENCES

Davis, J. & Tango, W.J. (1986). New determination of the angular diameter of Sirius. Nature 323, 234-5.

Karovska, M., Nisenson, P., & Noyes, R. (1986). On the Alpha Orionis Triple System. Ap.J. 308, 260-9.

Karovska, M., Nisenson, P., Noyes, R., & Papaliolios, C. (1987a). IAU Circular #4382

Karovska, M., Nisenson, P., Papaliolios, C., & Standley, C. (1987b). IAU Circular # 4457

Kirshner, R.P., Sonneborn, G., Crenshaw, D.M., & Nassiopoulos, G.E. (1987). Ultraviolet Observations of SN 1987A. Ap.J. 320, 602-8.

Nisenson, P., Stachnik, R.V., Karovska, M., & Noyes, R. (1985). A New Optical Source Associated with T Tauri. Ap.J. (Letters) 297, L17-20.

Nisenson, P., Papaliolios, C., Karovska, M., & Noyes, R. (1987). Detection of a Very Bright Source Close to the LMC Supernova SN 1987A. Ap.J. (Letters) 320, L15-8.

Papaliolios, C., Nisenson, P., & Ebstein, S. (1985). Speckle imaging with the PAPA detector. Applied Optics 24, 287-92.

SUPERNOVAE LIGHT ECHOES

B.E. Schaefer
Code 661
NASA/Goddard Space Flight Center
Greenbelt, MD, 20771

DISCUSSION

The sudden brilliance of a supernova (SN) eruption will be reflected on surrounding dust grains to create a phantom nebula. If the SN is far away, the phantom nebula will be unresolved from the SN itself, and will appear as light added to the light curve and spectrum. For nearby SN like SN 1987A, the echo can be resolved as being separate from the SN itself.

The dust responsible for the echo can be either in the interstellar medium (ISM) up to many parsecs away from the explosion or it can be in a circumstellar shell previously ejected by the progenitor star. The effects from echoes will be significant only if the optical depth of the dust is significant. SN 1987A, type Ia SN, and some type II SN do not have thick circumstellar shells, and hence can show echo effects only from reflections off the intervening ISM. Many type II SN have thick circumstellar shells (as demonstrated by the strong correlation between the observed extinction and the _shape_ of the light curve) and this will provide a significant echo.

I have performed Monte Carlo calculations for echoes from circumstellar shells (Schaefer 1987a). I find that echo effects will often _dominate_ the light curve and spectrum of SN at late times. For the case of type II SN, the decay in brightness at late times will be governed by the size of the circumstellar shell, while the brightness will be determined by the optical depth of the dust shell. Since the size and optical depth should vary continuously, the late time behaviour of type II SN should show a continuum in both brightness and decay rate. Indeed, for 12 type II SN with good light curves, I find that the decay rate varies by an order of magnitude (contrary to conventional wisdom) and the brightness at late times is not bimodally distributed (contrary to the conventional classification scheme of types IIL and IIP).

I have also calculated the brightness of echoes off the ISM for a variety of situations (Schaefer 1987b and see Schaefer 1988 for similar calculations and observations regarding novae). For any reasonable distribution of the magnitude or so of extincting dust to SN1987A, the phantom nebula should be quite bright and easily resolvable. The visibility of the nebula will be best when the SN has faded in brightness. The echo will be visible in small telescopes for many decades. The SN 1986G should also show a bright echo because the famous dust lane in Cen-A provides 3.6 mag of extinction to the SN. In fact, I find that this SN is anomalously bright by 2.7 magnitudes over any reasonable extrapolation of the light curve 400 days past maximum (Schaefer 1987c).

PREDICTIONS

I predict the following: (1) SN 1987A will have a bright echo which will last for decades and perhaps be visible in binoculars. (2) The brightness of SN 1986G will decay at a remarkably slow rate. (3) The late time spectrum of SN 1986G will be that of the SN at maximum. (4) The Hubble Space Telescope can resolve light echoes off the ISM for nearby SN. (5) For type II SN which are optically bright at late times, the infrared echo should be exceptionally bright. (6) These same SN should have a late time spectrum similar to the spectrum at maximum. (7) Since the radioactive tail presumably does not vary greatly in brightness, the strength of gamma-ray lines should not correlate with the late time optical brightness. (8) The echo of a nova explosion off a pre-existing dust shell should be resolvable by the Hubble Space Telescope for bright naked eye novae.

REFERENCES

Schaefer, B.E. (1987a). Light Echoes: Type II Supernovae. Ap. J. (Letters), 323, L51-54.

Schaefer, B.E. (1987b). Light Echoes: Supernovae 1987A and 1986G. Ap. J. (Letters), 323, L47-50.

Schaefer, B.E. (1987c). SN1986G in NGC 5128 (Cen-A). IAU Circulars no. 4421.

Schaefer, B.E. (1988). Light Echoes: Novae. Ap. J., 327, in press.

A REAL LIGHT ECHO: NOVA PERSEI 1901

J.E. Felten
Code 685, NASA Goddard Space Flight Center,
Greenbelt, MD 20771, USA

There has been much discussion of possible "light echoes" from Supernova (SN) 1987A. I thought you might like to see how a real light echo looked. Nova Persei 1901 (GK Per) was first seen in February. In August a surprisingly large nebulosity was discovered around it. This was soon observed to be "expanding" rapidly. Here I reproduce, as Figures 1a-1b, two of the five beautiful "diagrams" by Ritchey (1902), based on his photographs. These show a large outer (broken) ring, an inner and brighter ring, and numerous bright interior features. Many of the features appear to be moving. The outer ring increased its radius at $\approx$11 arcmin year^{-1}. The presently accepted distance to this nova is D $\approx$ 400-600 pc, so the apparent projected velocity of expansion was $\approx$ 4c-6c. This was the first observed "superluminal" cosmic source!

I give below a few references on the remarkable history of observation and theory of this nebula. Kapteyn (1901) perceived quickly that this was a light echo, but didn't realize that it could be superluminal, though the idea was in the air (Hinks 1902). Kapteyn assumed that the motions were "luminal" and thereby derived D $\approx$ 90 pc. His idea gained wide acceptance. By 1916 it was realized that this distance was too small, but still no one put two and two together. Finally Couderc (1939), in a classic but little-known paper, elucidated the superluminal kinematics, analyzed Ritchey's diagrams, and showed that theory and observation were in good agreement. Although the light echo of 1901-02 had faded with a time scale of 6 months or so, Perrine (1903), at the late date of November 1902, had obtained (by a 34-hour exposure!) a spectrum of the brightest patch of nebulosity (closest to the nova, in the southwest quadrant). He found it to be similar to that of the nova at maximum light in February 1901. This confirmed the reflection-echo theory. Perrine (1902) had also looked for polarization in this patch and found none; Katz & Jackson (1987) conclude that polarization would have been expected. This remains a puzzle. For more history, see their paper and Couderc's. The five diagrams by Ritchey have been used as a sequence in an instructional film, "Nonrecurrent Wavefronts" (1962), by Franklin Miller, Jr.

Of course there is no real motion of matter. The thin spherical light front from the nova sweeps outward, illuminating successive parts of blobs and swatches of dust, wherever they lie. (Note that the "motions" on Figure 1 are not, in general, radial.) Nova Persei 1901 was an idealized case of a light echo, because the optical outburst by the nova was very brief (duration $\approx$ 15 days)-- essentially a delta function in time. Therefore all patches of dust visible by reflection on a given photo lie in a thin paraboloidal sheet (the "kinematically accessible paraboloid") having a common time delay. This paraboloid sweeps outward with time. The large outer ring must be a more or less planar sheet of dust swatches lying well in the foreground ($\approx$ 15 pc in front, according to Couderc) and intersecting the paraboloid in a (projected) circle. The inner ring is another sheet, also in the foreground but closer to the nova. Bright features even closer to the nova lie more or less at right angles to the observer. Wherever there is enough dust, we see the paraboloid. Kapteyn (1901) writes in German that a time series of photos "will betray to us little by little the whole structure of the cloud, just as the biologist learns the structure of an organism from a series of cross sections...except that here the cross sections are bounded not by parallel planes but by paraboloids of revolution."

Portions of Couderc's results have been revived or re-derived several times in recent years, in various contexts: possible echoes from ancient supernovae (van den Bergh 1965); superluminal motions of radio sources (Rees 1966, Lynden-Bell 1977); light curves of classical supernovae (Morrison & Sartori 1969); infrared echoes of supernovae from heating of circumstellar dust (Dwek 1983); and luminous arcs in clusters of galaxies (Katz 1987, Milgrom 1987, Katz & Jackson 1987, Braun & Milgrom 1987). Emulating Perrine's spectroscopy, Walborn & Liller (1977) and López & Meaburn (1986) have shown that dust in the Carina Nebula is echoing the variable η Carinae, but the "limit" D $>$ 2 kpc in the latter paper is a kinematical error which mirrors Kapteyn's.

In the context of SN 1987A, Dopita et al. (1987), Hillebrandt et al. (1987), and Felten et al. (1987) have treated the "mystery spot" as possibly some kind of echo; Chevalier & Fransson (1987) and Dwek (1987, and this volume) have discussed the expected infrared echo; and Schaefer (1987, and this volume) has predicted a general visual echo from the surroundings, possibly bright enough to be seen in binoculars. The general angular dimensions will be smaller than in Figure 1, being inverse with the distance; thus we'll be dealing with angular sizes of seconds rather than minutes. The SN echo may be geometrically more complicated than the one shown here, because of the greater duration of the SN; in effect, we could see a superposition of several photos similar to those shown here. On the other hand, if the distribution of dust around the SN is simpler (e.g. a thin spherical shell swept up by the fast stellar wind of the blue supergiant), maybe we'll get lucky and see a simpler pattern. Let's hope we can see it and understand it!

REFERENCES

Braun, E. & Milgrom, M. (1987). Light reprocessing in the echo model of luminous arcs. Astrophys. J., submitted.

Chevalier, R.A. & Fransson, C. (1987). Circumstellar matter and the nature of the SN1987A progenitor star. Nature,328, 44-5.

Couderc, P. (1939). Les aureoles lumineuses des novae. Ann. d'Astrophys.,2, 271-302.

Dopita, M.A., Meatheringham, S.J., Nulsen, P. & Wood, P. (1987). The ionization effects of shock breakout in SN 1987A. Astrophys. J. (Lett.),322, L85-9.

Dwek, E. (1983). The infrared echo of a Type II supernova with a circumstellar shell: applications to SN 1979c and SN 1980k. Astrophys. J.,274, 175-83.

_____. (1988). Will dust black out SN 1987A? Astrophys. J.,329, in press.

Felten, J.E., Dwek, E. & Viegas-Aldrovandi, S.M. (1987). Companion to Supernova 1987A: reflection or fluorescence by an interstellar cloud? Astrophys. J. (Lett.), submitted.

Hillebrandt, W., Höflich, P., Schmidt, H.U. & Truran, J.W. (1987). On the interaction of the UV burst of Supernova 1987A with a nearby cloud: a possible explanation of the speckle images. Astr. Astrophys., in press.

Hinks, A.R. (1902). The movements in the nebula surrounding Nova Persei. Astrophys. J.,16, 198-202.

Kapteyn, J.C. (1901). Ueber die Bewegung der Nebel in der Umgebung von Nova Persei. Astr. Nachr.,157, 201-4.

Katz, J.I. (1987). Arcs, light echoes, and supergalaxies. Astr. Astrophys.,182, L19-20.

Katz, J.I. & Jackson, S. (1988). Predictions for arc polarization. Astr. Astrophys., in press.

López, J.A. & Meaburn, J. (1986). The broad Balmer profiles from η Carinae scattered by dust over the Car II region. Rev. Mex. Astr. Astrofis.,13, 27-32.

Lynden-Bell, D. (1977). Hubble's constant determined from super-luminal radio sources. Nature,270, 396-9.

Milgrom, M. (1987). The light-echo model for luminous arcs. Astr. Astrophys.,182, L21-4.

Morrison, P. & Sartori, L. (1969). The light of the supernova outburst. Astrophys. J.,158, 541-70.

Perrine, C.D. (1902). Photographs and measures of the nebula surrounding Nova Persei; observations of the nebulosity about Nova Persei for polarization effects. Astrophys. J., 16, 249-62.

_____. (1903). The spectrum of the nebulosity surrounding Nova Persei. Astrophys. J.,17, 310-4.

Rees, M.J. (1966). Appearance of relativistically expanding radio sources. Nature,211, 468-70.

Ritchey, G.W. (1902). Nebulosity about Nova Persei: recent photographs. Astrophys. J.,15, 129-31.

Schaefer, B.E. (1987). Light echoes: Supernovae 1987A and 1986G. Astrophys. J. (Lett.),323, L47-9.

van den Bergh, S. (1965). Light echos from ancient supernovae. Publ. Astr. Soc. Pac.,77, 269-71.

Walborn, N.R. & Liller, M.H. (1977). The earliest spectroscopic observations of Eta Carinae and its interaction with the Carina Nebula. Astrophys. J.,211, 181-3.

Figure 1. The field of Nova Persei (Ritchey 1902). Nova at center; scale in minutes of arc. (a) September 20, 1901; (b) November 13, 1901.

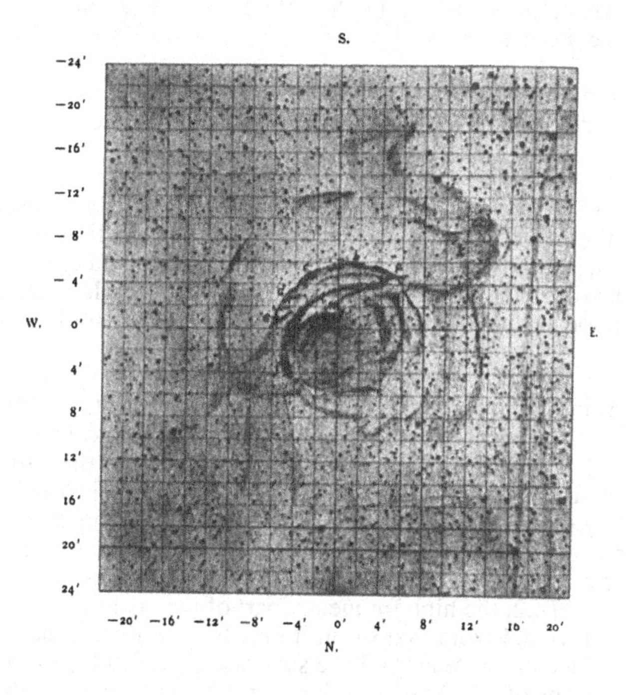

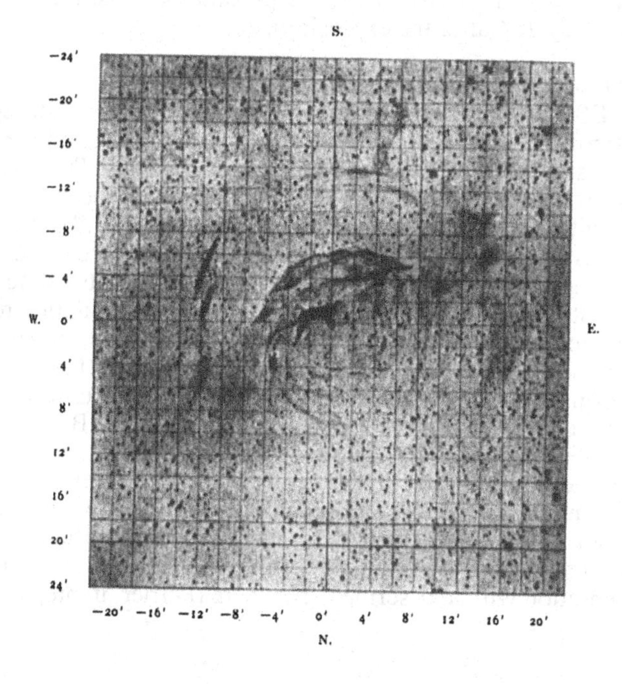

IR SPECKLE-INTERFEROMETRY OF SN 1987A

A.A. Chalabaev
CNRS, Observatoire de Haute-Provence, F-04870, Saint Michel l'Observatoire, France

C. Perrier and J.M. Mariotti
Observatoire de Lyon, F-69230, Saint Genis Laval, France

Abstract. The speckle observations in the near IR, carried out on August 6, 1987 at the ESO 3.6m telescope, yielded the detection of a weak oscillation superimposed on the visibility of the unresolved ejecta. A ring halo with the diameter 420±80 mas appears as the most plausible image. Its interpretaion as due to an IR light echo is briefly discussed.

1 Introduction

Due to the relative proximity of SN1987a, one has an exceptional opportunity to study the angular structure of the supernova and of possible related phenomena in its vicinity. The first near IR speckle-interferometry of the supernova, carried out in L and M bands on May 8 and 9 at the ESO 3.6 m telescope in Chile (Chalabaev *et al.*, 1987), yielded the visibilities of an unresolved supernova. The observations continued on June 7-22, 1987 (Perrier, 1987). The visibilities in the K and M bands were also those of an unresolved source as well as the visibilities in the L band until at least June 8. On June 17-22, the high frequency part of the visibilities in L, both in NS and EW directions, deviated from that of an unresolved source, indicating the presence of a partially resolved structure around the supernova (probably, an extended halo). The preliminary account of the May and June results was given at the ESO Workshop on SN1987a (Perrier *et al.*, 1987). Here we present the most recent results obtained on August 6, 1987 (day 164 after the explosion).

2 Observing procedure, data reduction and data.

We used the ESO slit scanning IR specklegraph working in the 2-5 μm spectral range (Perrier, 1986). The observing procedure consists of scans on the object and the nearby sky recorded at a speed high enough to "freeze" speckles. The maximum effective frequency of measurements is strongly seeing dependent with the cut-off set by the telescope diffraction limit which is 7.9, 4.8 and 3.7 $arcsec^{-1}$ in K (2.2 μm), L (3.6 μm) and M (4.6 μm) bands respectively. Owing to distant references and to the brightness of the SN, the uncertainty on the measured visibility points comes from systematic errors due to variations of the atmospheric and telescope modulation transfer function (MTF) with the zenith distance rather than from the speckle, detector and background noise. To account for variations of the MTF, we used four different reference stars. As additional controls, observations "reference-to-reference" and measurements of well known double stars were performed each night. Besides of the mentioned sources of possible troubles, it was also realized that due to large passbands of the filters and the extremely red colour of the SN relative to reference stars, the telescope MTF's on the SN and on a reference star are not identical (different λ_{eff} of the measurements). This colour effect can affect the visibility and introduce spurious features. In August, its importance was about 1-2% of the SN flux. In more detail, the colour effect and its compensation will be described elsewhere (Perrier, in preparation).

The visibilities of SN1987a in K, L and M filters, measured in the NS direction are shown in Fig. 1. We also measured the visibility in K filter in the EW direction, which closely ressembled that in the NS direction. All visibilities display a constant term of about 95% and a weak superimposed oscillation, having the first minimum at about 1.5 $arcsec^{-1}$. The constant term certainly comes from the expanding supernova ejecta. Indeed, taking 10^4 $km \cdot s^{-1}$ as un upper limit on the mean expansion velocity of the ejecta, we estimate its diameter on August 6 to be less than 36 milliarcsec (mas) which is still small enough to neglect the corresponding inflection of the power Fourier transform of the ejecta image. There are three possible images which can account for the oscillation. The case 1 consists of two weak spots, one lying to the N (or S) from the SN and another to the E (or W), both at equal separation from the SN. Best fits to the visibilities yield the following values for ϕ_s, the separation from the SN, and for ε_s, the contribution of each source to the total flux : K(NS and EW), $\phi_s = 330 \pm 80$ mas, ε_s = 2.6% (3 Jy); L(NS), $\phi_s = 350 \pm 50$ mas, ε_s =2.7% (2.9 Jy); M(NS), $\phi_s = 380 \pm 100$ mas, ε_s = 1.8% (2 Jy). The weighted mean separation value is $\phi_s = 350 \pm 40$ mas. The case 2 consists of one secondary spot lying on an intermediate direction (NE, NW, SE, or SW) relative to the SN, separated from it by $\sqrt{2} \cdot \phi_s$ and contributing $2\varepsilon_s$ to the total flux, where ϕ_s and ε_s have the values found in the former case. Finally, the third possible image is a ring with the angular diameter $\phi_r = 420 \pm 80$ mas, contributing 3.4% (3.5 Jy) in K, 3% (3.2 Jy) in L and 2.2% (2.2 Jy) in M. The fits to the visibilities by a ring are somewhat worse that those by a secondary source. A clumpy ring would give a better agreement. The values of ϕ and ε, given above, must be taken with a caution since the results are preliminary and further data reduction might slightly modify them.

Fig. 1 The visibilities of SN1987a in K, L and M bands in the North-South direction, measured on August 6, 1987.

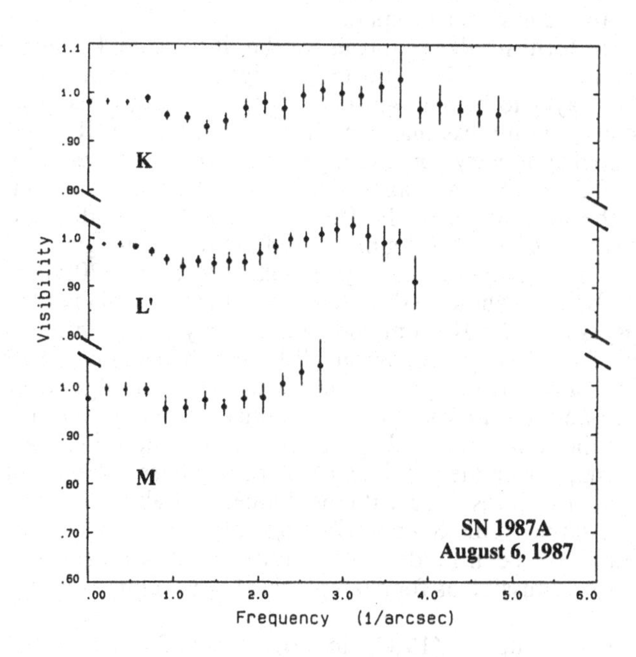

3 *Discussion*

The full understanding of the visibilities will require the knowledge of their time evolution as well as an analysis of spectroscopy and spectrophotometry in the IR. Here we present only a simple tentative interpretaion. The projected velocity of the phenomenon is the least in the case of the ring and equal to 0.4c. Therefore, whatever is the real image, the velocity scale points to a light echo as the most plausible explanation.The echo arises when the light pulse of the explosion heats pre-existing dust grains in the SN vicinity (cf. Graham *et al.*, 1983; Dwek, 1983). The picture, proposed for the environment of SN1987a (Chevalier and Fransson, 1987), consists of a low density cavity of the radius R_c, created by the hot and fast wind of the blue supergiant (BSG) in the larger unperturbed "relict" shell of gas and dust, formed during the red supergiant (RSG) stage of the progenitor. In this model the IR spot(s) could be understood as a dust clump(s) in the low-density cavity. The distance from the spot(s) to the SN can be estimated from the following relation (Couderc, 1939) : $R = 0.5(ct+p^2/ct)$, where t is the time interval after the explosion, and p is the impact parameter. One gets the distance $R \approx 3\cdot10^{17}$ cm for the two spots case and $R \approx 4.2\cdot10^{17}$ cm for the one spot case. The dust density necessary to account for the observed flux may be roughly estimated from the expression given by Dwek (1983, expression 17) which shall be completed by the dilution factor W. Taking Q=1, s=1, λ_o=0.2 μm, T_{ev}=1000°K, t_{SN}=10 days and t_1=12 days (for the meaning of the parameters see the article by Dwek), one gets F_L (Jy) $\approx 90W\tau_D$. Given that the contribution of the spot(s) to the total flux is small compared to the noise, their size, ΔR, cannot be derived from the visibilities. Nevertheless, taking $\Delta R<R$ seems to be a generous upper limit. Then, $\tau_D>0.1$ (case 1) and $\tau_D>0.2$ (case 2). The dust density, n_d, is related to τ_D by the following expression : $\tau_D \approx \pi a^2 Q n_d \Delta R$, where a is the radius of the dust particles, and Q=1. Assuming a=0.1μm, one has the following estimates : $n_d>1\cdot10^{-9}$ cm^{-3} (two spots) and $n_d>1.5\cdot10^{-9}$ cm^{-3} (one spot). It is difficult to understand how such dense clumps could escape any destruction by the BSG wind while the material all around has been swept out.

One can perform similar estimates for the ring image. In this case, the dust would lie at the distance $R \approx 2.5\cdot10^{17}$ cm ($\approx$100 light-days) from the supernova. Assuming the emitting layer to be 10 light-days thick, the dust density is $n_d=4\cdot10^{-10}$ cm^{-3}. This value is considerably less than what is required for the IR spot(s). The ring image also offers an axial symmetry consistent with the expected spherical symmetry of the BSG wind cavity. Tentatively, we suggest to identify the ring emission with the IR echo expected from the inner edge of the RSG dust shell (Chevalier and Fransson, 1987). Taking carbon dust grains (ρ=2 g·cm^{-1}) and assuming the gas-to-dust ratio of 100, the density of the associated gas can be estimated to be n_H=200 cm^{-3}.The RSG mass loss rate would be then about $4\cdot10^{-6} M_\odot yr^{-1}$ which is a plausible value. The radius of the cavity R_c is then $2.5\cdot10^{17}$ cm, indicating a very short life time of the BSG. Taking the velocity of the cavity propagion 50 km·s^{-1} (Chevalier, 1987), one gets t(BSG)=6400 yr. If the identification of the ring with the inner edge of the RSG dust shell is correct, then one would expect the ring emission being continously present around the SN during the time interval $2R_c/c$ after the explosion, i.e. about 200 days, and in particular, in June, during the previous ESO speckle observations. At that epoch, the ring would have been well resolved with the diameter of about 600 mas. However, the echo was weaker relative to the SN contributing only 1.5% to the total flux and the weak oscillation could have been hidden in the noise. Thus, we cannot discard other interpretations of the ring emission as for example due to dust clumps in the low-density cavity.

Recently, Catchpole (1987) identified the 2.3 μm CO band in the supernova spectrum; there are other so far unidentified features e.g. at 3.11 μm and at

3.34 μm (Bouchet *et al.*, 1987) which colud also be due to a gas emission. One may suggest that a part of the extended emission, giving rise to the oscillation on the visibilities, is due to a light echo from gas, associated with the dust. The derived dust density estimates should be then decreased.

The whole set of the speckle observations allows one to conclude that from May to August the contribution of the unresolved emission in the near IR was always at least 95% with the exception of L band during the June 17-22 period when the lower limit can be set to 80%. This implies that if there is an IR "excess" emission in the spectrum of the supernova, exceeding those limits, it is formed either in the ejecta, or in its interior, or in the outer region immediately close to the ejecta.

4 Summary and conclusions

The speckle observations in the near IR, carried out on August 6, 1987 at the 3.6m telescope of the European Southern Observatory, yielded the detection of a weak oscillation superimposed on the visibility of the unresolved supernova ejecta. This oscillation of about 5% amplitude is present in all visibilities, obtained through three filters (K, L and M) in the North-South direction, as well as on the visibility in the East-West direction, obtained in the K filter. It clearly indicates the presence of a source of emission in the SN outer environment. Three possible images can fit the oscillation : firstly, two IR spots, to the North (or South) and to the East (or West) from the SN, both at the separation of 350±40 mas; second, one IR spot, to the NE (or NW, or SE, or SW) at the separation of 495±60 mas; third, a ring of diameter 420±80 mas around the SN. The velocity scale of the phenomenon is comparable to the speed of light, thus pointing to a light echo explanation. Estimates of the dust density necessary to account for the measured flux and spherical symmetry favors the ring as the most physically plausible image fitting the visibilities. It requires the dust density $n_d=4\cdot10^{-10}$ cm^{-3} at the distance $2.5\cdot10^{17}$ cm. Tentatively, we suggest to identify the ring with the light echo from the inner edge of the dust shell, formed during the red supergiant stage of the progenitor. However, we cannot discard other interpretations of the ring emission as for example due to dust clumps in the low-density cavity. The solution of the "RSG shell or dust clumps" dilemma awaits further IR speckle observations.

Acknowledgements. We are grateful to the ESO Director-General Prof. L. Woltjer for allocating the observing time at the La Silla 3.6m telescope. We wish also to warmly thank the La Silla staff, in particular P. Bouchet for communicating the IR spectrophotometry of the supernova prior to publication and D. Hofstadt, J. Roucher and L. Baudet for their efficient help during the observations.

References

Bouchet, P. *et al.*, 1987, in the *ESO Workshop on SN1987a* (ed. I.J. Danziger), Garching, July 6-8, 1987
Catchpole, R., 1987, *IAU Circ.* 4457
Chalabaev, A.A., Perrier, C., and Bouchet, P., 1987, *IAU Circ.* 4389
Chevalier, R.A., 1987, in the *ESO Workshop on SN1987a*
Chevalier, R.A., and Fransson, C., 1987, *Nature*, **328**, 44
Couderc, P., 1939, *Ann. d'Astrophys.*, **2**, 271
Dwek, E., 1983, *Ap.J.*, **274**, 175
Graham, J.R. *et al.*, 1987, *Nature*, **304**, 709
Perrier, C., 1986, *Messenger,* No. 45, p.29
Perrier, C., 1987, *IAU Circ. 4417*
Perrier, C., Chalabaev, A.A., Mariotti, J.M., and Bouchet, P., 1987, in the *ESO Workshop on SN1987a*

INFRARED OPPORTUNITIES FOR SUPERNOVA 1987A

Eli Dwek
Laboratory for Astronomy and Solar Physics
NASA/Goddard Space Flight Center, Greenbelt, MD 20771, U.S.A.

1 ABSTRACT

SN 1987a, the recent supernova that exploded in the Large Magellanic Clouds (LMC), is the first supernova observable with the naked eye in the last 383 years. Its proximity to earth (D=170,000 light years) offers astronomers a once–in–a–lifetime opportunity to conduct the most detailed multi–wavelength analysis of the supernova phenomenon and its effect on the ambient medium. Infrared observations of SN 1987a are an integral part of such analysis with the objectives of: 1) obtaining the time history of the temperature, density and velocity structure of the expanding material; 2) determining the abundances of the elements in the SN ejecta; 3) determining the presence of dust in the ejecta, and conducting a detailed study of its composition and process of formation; 4) studying the formation and excitation of molecules in the ejecta; 5) analyzing the medium around the SN, and determining the amount of mass lost by the progenitor star during the post main sequence phase of its evolution; 6) studying the interaction of the expanding SN blast wave with the ambient dusty medium; and 7) examining the possibility that a cometary cloud may be present around the supernova.

2 INFRARED PROCESSES IN SN 1987A

Until the photometric observations of SN 1979c by Merrill (1980), there were no infrared (IR) data on a type II supernovae (SN). These observations were followed by observations of SN 1980k (Dwek *et al.* 1983) and SN 1982g (Graham *et al.* 1983). In all these supernovae the IR exhibits a similar characteristic evolution, depicted in Figure 1 for SN 1980k. The figure clearly illustrates the existence of two distinct phases in the evolution of the spectrum. An initial phase during which the IR emission

Figure 1: The evolution of the spectrum of SN 1980k (Dwek *et al.* 1983) was characterized by the appearance of an infrared excess above the photospheric continuum. This infrared excess was first detected on May 31, 215 days after the first visual detection of the supernova.

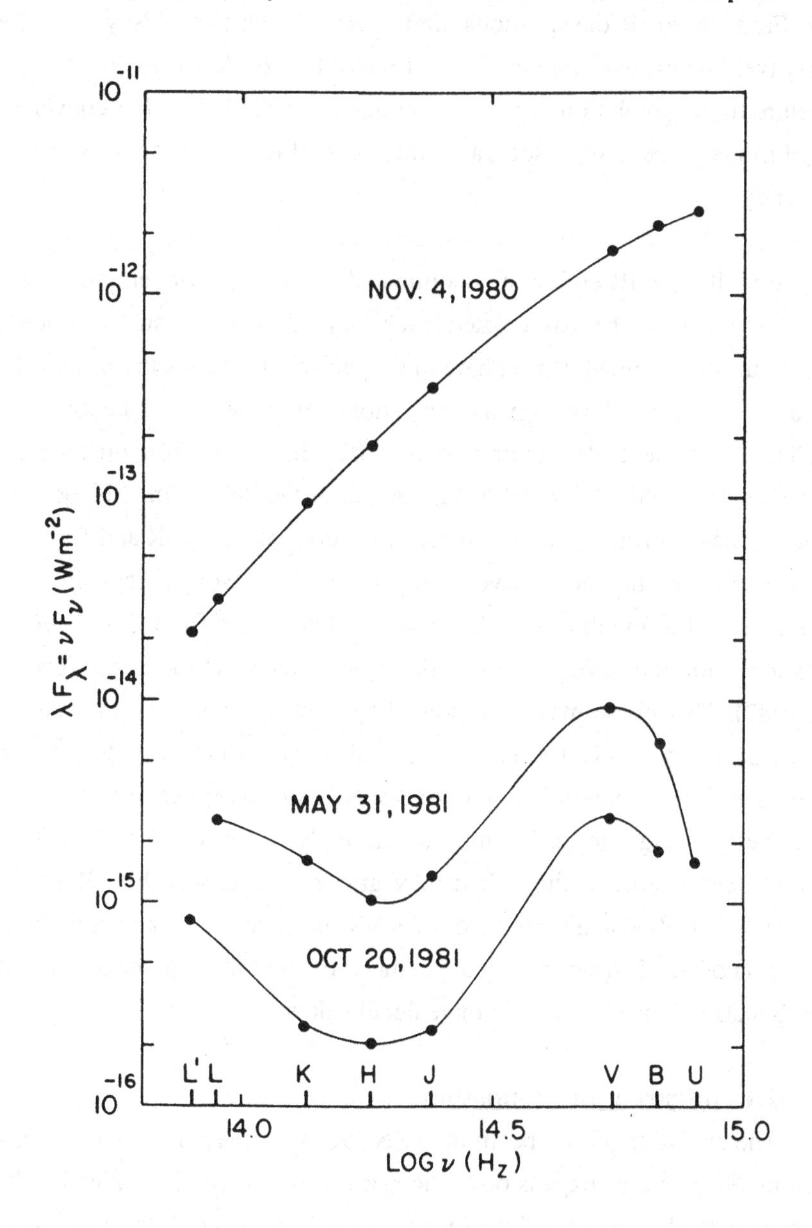

originated from the photosphere of the expanding ejecta was followed by a second phase (first observed in SN 1980k ~ 7 months after the explosion) characterized by the appearance of a distinct thermal component that dominated the emission at the longer wavelengths. From these IR observations, and those of novae (e.g. Ney & Hatfield 1978) and evolved supernova remnants (see review by Dwek 1987a), one can put together a picture for the evolution of the IR emission from SN 1987a. For convenience here, this evolutionary scenario is separated into several distinct processes that may occur concurrently.

Initially, the IR emission is dominated by atmospheric emission, which will exhibit an excess above the extrapolated blackbody emission at visual wavelengths, probably the result of extended atmosphere and opacity effects. As the photospheric emission declines, a secondary thermal emission component will become more pronounced. This component, also referred to as an IR echo, arises from dust present in a circumstellar shell heated by the UV–visual output of the SN. As the SN ejecta cool they will enter a phase during which the spectrum is dominated by IR and far–IR fine structure lines excited by the radioactive energy source (Fransson & Chevalier 1987). Below temperatures of a few thousand degrees, molecular lines from CO or SiO may appear in the spectrum, indicating the formation of molecules in the ejecta (Danziger 1987; Feast 1987). This phase may be followed by dust formation, giving rise to IR emission from dust radiatively heated by the radioactive decay of ^{56}Co, or by an underlying pulsar. The expanding SN blast wave may sweep up the dust in the circumstellar shell, giving rise to IR emission from shock heated dust. Finally, if a cometary cloud existed around the SN, it may give rise to observable IR emission (Mumma 1987). The UV–visual output from the SN may evaporate and subsequently irradiate a fraction of the dust trapped in the comets. Each of these processes and their observational effects will be discussed in more detail below.

2.1 Infrared line emission

The evolution of the spectrum of SN 1987a is determined by the decaying gamma ray luminosity and the expansion of the ejecta (Fransson & Chevalier 1987). As the expansion ensues, the cooling of the ejecta will shift from cooling in UV–optical lines to cooling by IR and far–IR line emission. Table 1 lists important IR transition lines for the most common elements. The expected IR fluxes in several prominent IR lines at day 300 after the explosion are given by Fransson & Chevalier (1987) and Chevalier

(1987). The expanding SN ejecta will eventually cool to temperatures below which molecules may be stable. The first molecule to form in the ejecta is CO, which has a binding energy of 11.3 eV. Other molecules which may form are SiO and CS, which have binding energies of 8.3 and 7.4 eV, respectively. The fundamental vibrational frequencies in these three molecules are 2170 cm^{-1} (4.6 μm), 1242 cm^{-1} (8.0 μm), and 1285 cm^{-1} (7.8 μm), respectively. Table 2 summarizes the IR lines from abundant molecular species that may form in the SN ejecta. As of the writing of this report, the CO vibration line at 4.6 μm, and its 2.3 μm overtone have been detected in the spectrum of SN 1987a (e.g. McGregor and Hyland 1987, Danziger 1987, Feast 1987). In at least two novae, the dust formation epoch was preceded by the appearance of an excess of IR emission at 5 μm, which is now widely interpreted as emission from CO that formed in the nova ejecta (Gehrz *et al.* 1980). If the CO lines detected from SN 1987a are emerging from the outflowing material, their appearance may be signalling the imminent onset of dust formation.

Table 1

Infrared fine structure lines from atomic and ionized species in the supernova ejecta

Specie	λ(μm)	Specie	λ(μm)
C II	– 157.7	S I	– 25.3
N II	– 121.9		– 56.3
O I	– 63.2	S III	– 33.5
	– 145.5	Fe I	– 24.0
O III	– 51.8		– 34.7
	– 88.4	Fe II	– 26.0
Al I	– 89.2		– 35.4
Si I	– 68.5	Fe III	– 22.9
	– 129.7		– 33.0

Table 2
IR lines from molecular species that may form in the SN ejecta

Molecule	λ(μm)
CO	4.6, 2.3
SiO	8.0, 4.0
CS	7.8, 3.9

2.2 Infrared emission from supernova condensates

Grains may form in the expanding SN ejecta as soon as they can survive the heating by the ambient radiation field. The formation of dust will have immediate implications on other wavelength observations, as the dust may obscure the UV and visual output of any pulsar that may have formed in the explosion (Gehrz & Ney 1987; Dwek 1987b). That dust can form in SN is suggested by two lines of evidence: 1) dust formation has been observed to take place in novae (e.g. Ney & Hatfield 1978) and Wolf–Rayet stars (Williams, Hucht, and Thé 1987) under conditions that are similar to those encountered in the ejecta of SN 1987a (Dwek 1987b); and 2) the presence of isotopic anomalies in meteorites suggests that dust particles with isotopic and elemental composition that are only encountered in SN ejecta were present in the early solar system (Clayton 1982). Depending on the value of the condensation temperature, dust formation can commence as early as 3 to 8 months after the explosion. A mass of 0.01 $M_\odot$ of dust in the ejecta will obstruct the UV–visual output of the pulsar for ≈ 5– 20 yrs, depending on dust composition. The supernova will be brightest in the infrared between the onset of dust formation (between 3 to 8 months after the explosion) and the time of maximum γ-ray transparency (~ day 600). If the dust shell is optically thick, the infrared luminosity initially will be ~ 10^{41} erg s^{-1}, declining to less than ~ 10^{39} erg s^{-1} by day 600. Thereafter, all the IR luminosity will be provided by the UV–visual output of the pulsar. If the pulsar has an initial luminosity and energy distribution like that of the Crab, then typical IR luminosities will be ~ $5x10^{38}$ erg s^{-1}.

Dust forming in the ejecta of a supernova may have a composition that is distinct from the dust in the general interstellar medium. For example, some elements may condense in a sulphur–rich atmosphere (Clayton & Ramadurai 1977), giving rise to the possibility that MgS, CaS, and FeS may condense in the ejecta. Table 3 lists some well known IR features commonly attributed to interstellar dust particles. The detection of these features will provide clues on the dust excitation mechanisms, and on the location and composition of the dust–forming layers in the ejecta.

Table 3

IR features from solid particles that may condense in the SN ejecta

Particle	λ(μm)	Particle	λ(μm)
H_2O ice	– 3.1	MgS	– ~ 36
Silicates	– 9.7, 20.0	CaS	– ~ 40
SiC	– 11.3	FeS_2	– ~ 24, 29
PAH's	– 3.3, 3.4, 6.2,		– 38
	7.7, 8.6, 11.3	SiS_2	– ~ 22

2.3 Infrared emission from circumstellar dust (IR echoes)

The infrared light curves of various Type II supernova have exhibited a dramatic rise at infrared wavelengths that has been attributed to thermal emission from circumstellar dust heated by the UV–visual output of the exploding star (Bode & Evans 1980; Dwek 1983; Graham *et al.* 1983). The dust presumably formed in the outflowing matter ejected during the red giant phase of the evolution of the progenitor star. The evolution of the IR emission, also known as an infrared echo , is completely determined by the UV–visual lightcurve, and by the density and structure of the circumstellar shell. If the SN lightcurve is known, analysis of the IR echo can determine the structure and density of the shell, from which the mass lost from the progenitor star can be inferred (Dwek 1983). In addition, the observations can determine the composition of the dust, and hence infer the C/O elemental abundance in the ejected shell. This quantity is of special interest since it will provide valuable information on nuclear and convective processes that determine the composition of the envelope of the presupernova star (Dwek 1985).

The IR spectrum of SN 1987a clearly exhibits an excess of IR emission above the extrapolated photospheric emission. A significant fraction of this IR excess may, however, be due to photospheric line and continuum emission. In fact, the most difficult problem in modeling an echo is the separation of the circumstellar component from the atmospheric or line emission components. Additional difficulties in modeling an echo arise from uncertainties in the composition and distribution of dust around the supernova. Prospects for the presence of an IR echo increased since the discovery of very narrow UV lines (Kirshner 1987; Panagia 1987) which indicate the existence of a circimstellar shell around SN 1987a. The spectrum suggests an overabundance of ^{14}N in the shell, presumably the result of CNO processing. If the shell is dusty, then it could give rise to an observable IR echo. Preliminary analysis of the UV data (Panagia 1987) suggests a minimum shell distance of $\sim 2 \times 10^{17}$ cm and an electron density of 10^4 cm^{-3}.

To examine the presence of an echo in the spectrum of SN 1987a, I calculated the IR flux expected from a spherically–symmetric circumstellar dust shell centered on the supernova with a radius R_s, *arbitrarily* chosen to be 0.1 pc . The dust–to–gas mass ratio in the shell was taken to be 0.25×10^{-2} (i.e. its value was depleted by a factor of 4 relative to its galactic value). The dust radiative absorption coefficient was taken to be equal to Q_0 at $\lambda < \lambda_0 = 2\pi a$, and equal to $Q_0(\lambda_0/\lambda^{-1})$ at $\lambda > \lambda_0$, where $a = 0.1$ μm is the grain radius. The density structure of the shell was assumed to follow an r^{-2} law. Given the shell geometry and radiative output of the SN, the only adjustable parameter of the model is τ_V, the visual optical depth of the shell, which for a given value of Q_0 determines its density and hence its mass. The value of τ_V was chosen to give the best overall fit to the available IR data taken from: Bouchet *et al.* 1987, day 16; Gregory & Elias 1987, day 49; Aitken 1987, day 54; Rank *et al.* 1987, day 57; and Danziger 1987, days 113, 131, 160, and 200. The SN photosphere and luminosity for these days was characterized by a single–temperature blackbody given by Catchpole *et al.* (1987). The value of τ_V, which gave the best fit to the IR observations, was 0.16. For a value of Q_O=1, the resulting shell density was ~ 3×10^3 cm^{-3}, yielding a shell mass of ~ 12 $M_\odot$. Since the shell mass and density are inversely proportional to Q_O, smaller masses, more compatible with current ideas on the mass loss history from the progenitor star; (i.e. Woosley 1987) can be obtained by inceasing Q_O by the desired factor.

Figure 2a–h compares the result of the model to the observations. The shaded area in the figure depicts the contribution of the IR echo to the total observed flux. The figure shows that most of the IR emission above 3.4 μm, with the exception oi the 5 μm CO emission, can be attributed to an IR echo. In general, the fit of the model the observations appears to be quite good, considering the simplicity of the model and the arbitrary choice of shell radius. However, detailed inspection of the figure shows that the predicted IR emission is significantly higher than the observed fluxes on days 57 and 200. In fact, the observed IR emission on day 57 is more compatible with no IR echo at all. Furthermore, the 10 and 20 μm fluxes decreased between days 160 and 200 contrary to the general predictions of an echo model. If there is an echo, then clearly the model presented here needs to be significantly modified to fit the detailed time variations in the observed IR fluxes. These rmodifications include spatial deviations from spherical symmetry, and time variability in the amount of circumstellar dust, expected if the dust has a cometary origin. Furthermore, additional IR observations at wavelengths longer than ~ 20 μm are needed to constrain the radial distribution of dust around the supernova.

2.4 Infrared emission from shock–heated dust

IR observation of supernova remnants obtained with the Infared Astronomical Satellite (IRAS; Dwek 1987a and references therein) show that at gas temperatures above ~ 10^6 K, thermal emission from dust swept up by an expanding SN blast wave dominates the cooling of the shocked X–ray emitting gas. In young remnants

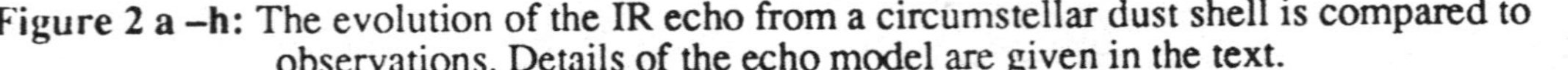

Figure 2 a –h: The evolution of the IR echo from a circumstellar dust shell is compared to observations. Details of the echo model are given in the text.

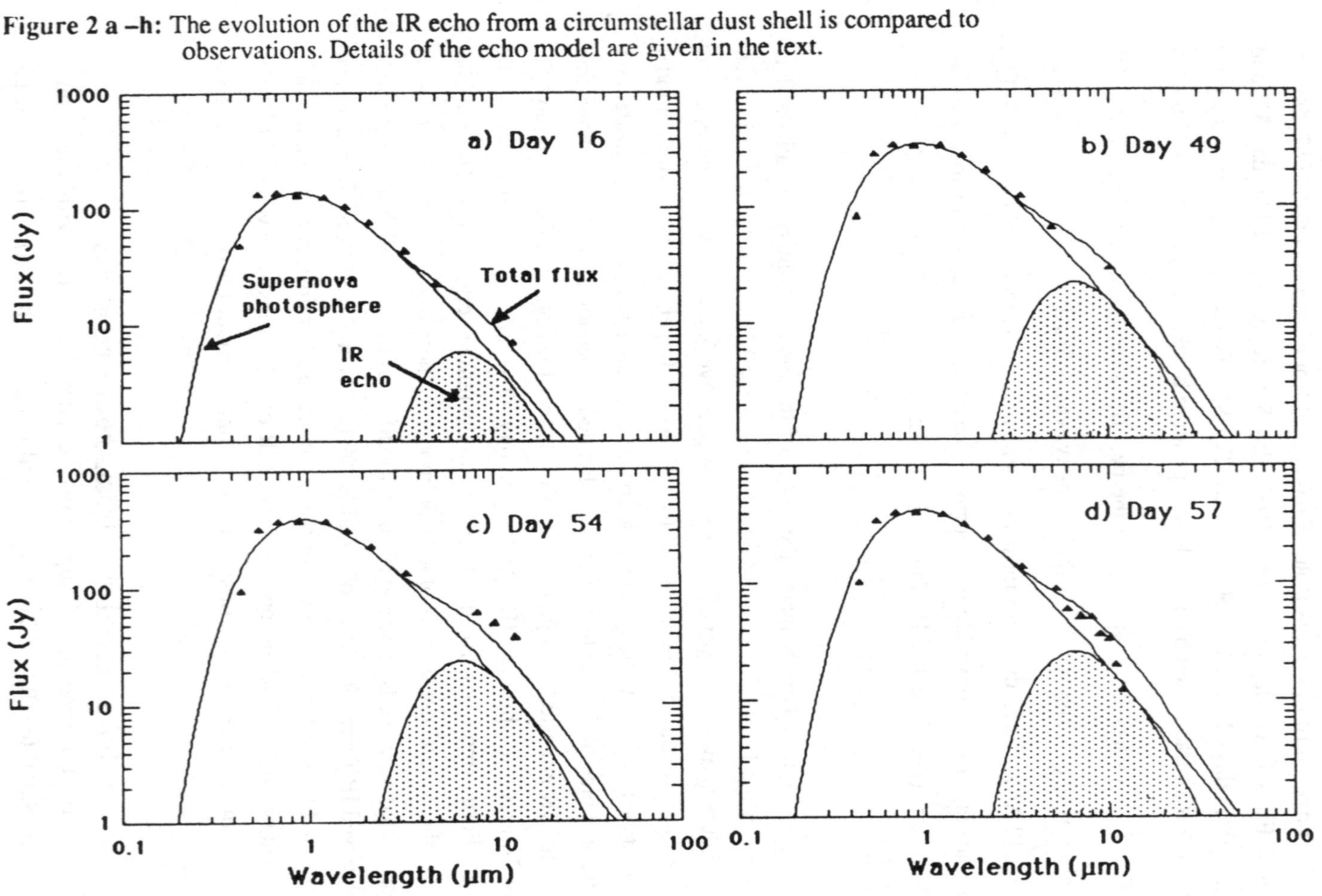

Figure 2: continued

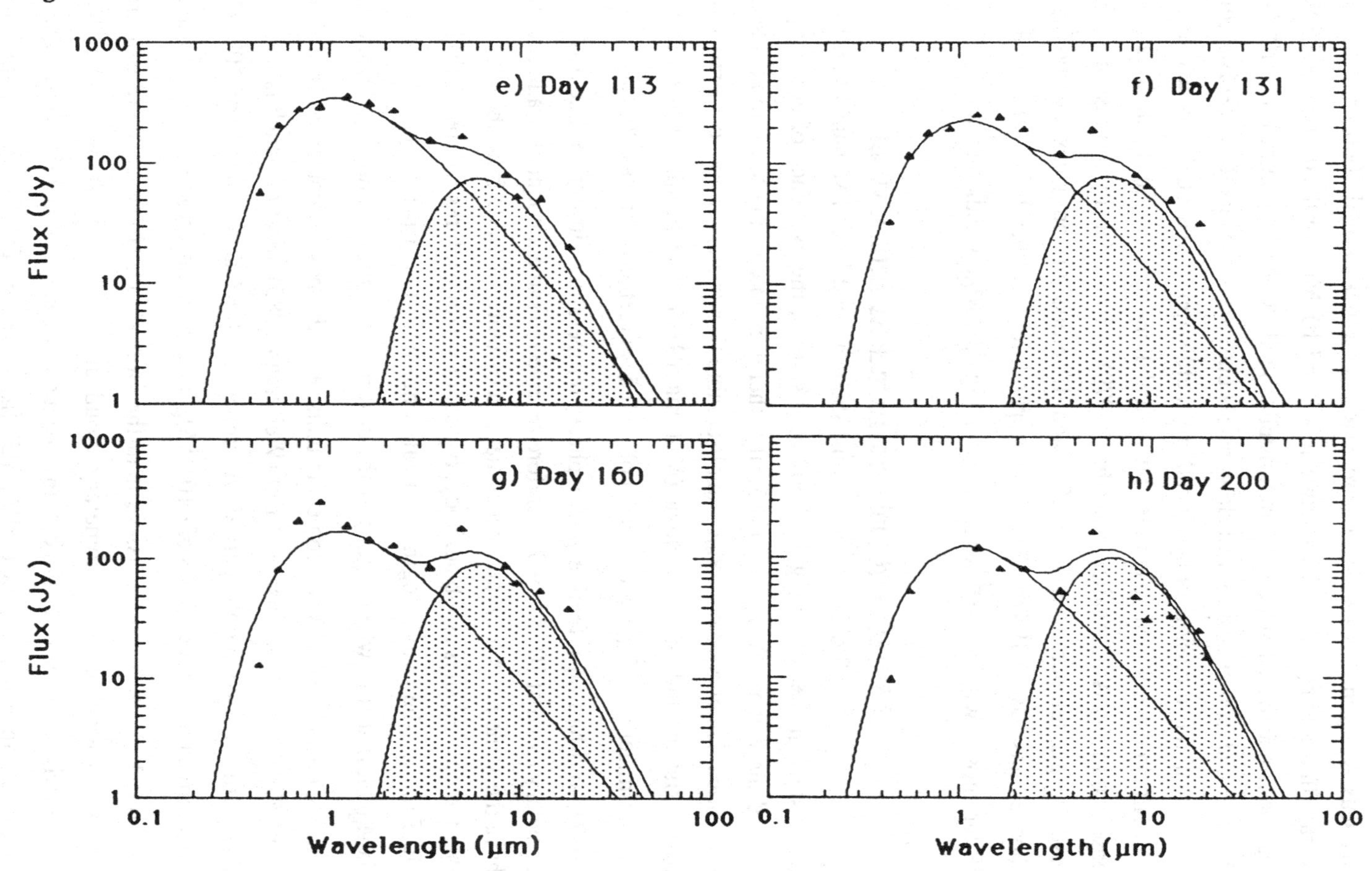

like Cas A, Tycho, or Kepler, where shock velocities exceed 1000 km/sec, the IR to X–ray flux ratio reaches values between 300 and 10. Most of the dust around SN 1987a may be concentrated in a dense circumstellar shell. A density contrast of ~ 10^4 between the shell and its inner cavity will slow the SN blast wave to velocities of ~ 100 km/sec so that optical line emission, rather than thermal emission from dust, dominates the cooling. The shocked shell will resemble the optical line–emitting quasi–stationary floculli observed in Cas A (Chevalier & Kirshner 1979). A lower density contrast of ~ 1000 will result in gas temperatures > 10^6 K, where IR emission dominates the cooling of the shocked gas (Dwek 1987c). The dust temperatures will then be ~ 200 K, and the peak IR luminosity will be ~ 10^{40} erg/sec, assuming 0.01 $M_\odot$ of radiating dust.

2.5 Infrared emission from a cometary cloud

IRAS observations of dust shells around Vega (Aumann *et al.* 1984) and other stars in the solar neighborhood may suggest the existence of cometary clouds around these stars (Weissman 1985). It is therefore interesting to examine whether the supernova lightcurve can produce an observable IR signature from a cometary cloud that may exist around the supernova (Mumma 1987). The luminosity of the SN was approximately 10^8 $L_\odot$ over a period of ≈ 60 days (centered on day 80; Catchpole *et al.* 1987), and had an effective photospheric temperature similar to that of the sun. A cometary cloud at 10^4 AU will therefore receive the same amount of irradiation and will have the same mass loss rate as a comet at a distance of 1 AU from the sun. Assuming that the dust radiates as a blackbody, typical dust temperatures will be ~ 300 K, and the IR flux at the wavelength of peak emission (17 μm) is given by F(Jy) ~ $5 \text{x} M_d(M_\oplus)/a(\mu m)$, where $a(\mu m)$ is the dust radius in microns ($a(\mu m) \approx 1$), and M_d is the mass of dust released by the comets in units of earth–mass. A comet can lose about 0.01 – 0.1 % of its mass in dust per orbit (Weissman 1985), most of which is lost around 1 AU within a two–month period. A cometary cloud at a distance of 10^4 AU will therefore have to have a mass of ~10^{3-4} $M_\oplus$ to give an observable 20 μm IR flux, which is about 100 times more massive than the Oort cloud. However, there is a priori no reason to assume that a cometary cloud around SN 1987a will have the same characteristics as the Oort cloud. It may be more massive and at a significantly smaller or larger distance (compared to 10^4 AU) from the SN. Finally, the IR signature of a cometary cloud may be distinguishable from that of a circumstellar dust shell, since the former will have a distinct IR evolution determined by the combined effects of finite light travel times and the temporal increase in the amount of dust released by the cometary nuclei.

Table 4

Investigation teams and instrument characteristics

PI/Institute	Instrument	λ Range	Sensitivity*	λ/Δλ
April 1987 Campaign				
1) F. C. Witteborn NASA/ARC	Grating Spectrometer	4 – 14 μm	3 Jy @10μm	120,50
2) H. Larson U. of Arizona	Michelson Interferometer	1 – 3.5 μm		~ 2000
November 1987 Campaign				
1) E. F. Erickson NASA/ARC	Echelle Spectrometer	20 – 90 μm	$\sim 5 \times 10^{-19}$ Wcm^{-2}	~ 500
2) F. C. Witteborn NASA/ARC	Grating Spectrometer	4 – 14 μm	3 Jy @10 μm	120,50
3) S. H. Moseley NASA/GSFC	Grating Spectrometer	20 – 65 μm	3 Jy @20 μm	~ 30
4) P. M. Harvey Univ. of Texas	Photometers	50 , 100 μm	1 Jy @100 μm	~ 2

* Calculated for S/N ratio of 3, and a 1 hour integration time.

3 SUMMARY

As of this writing, several expeditions are conducting infrared observations with the Kuiper Airborne Observatory (KAO) from New Zealand. Table 4 lists the various observing teams and their instrumental capabilities). These observations, and their ground–based counterparts, constitute the most detailed infrared studies ever undertaken of a supernova. They will provide astronomers with an exciting new database to examine various processes inherent to the SN phenomenon, including: the excitation of elements in the ejecta, the formation of molecules and dust, and the interaction of the supernova with its surrounding medium. The determination of the elemental abundances in the ejecta and the detection of dust in the supernova will provide new information on the evolution of dust and gas in the interstellar medium. An infrared echo from the supernova will produce important clues for determining the pre–supernova evolution of the progenitor star. And finally, the existence of a cometary cloud around the supernova will give us new clues about the preponderance of cometary clouds around other stellar systems.

REFERENCES

Aitken, D. K. 1987, *IAU Circular*, No. 4374.
Aumann, H. H. *et al.* 1984, *Ap. J. (Letters)*, **278**, L23.
Bode, and Evans, 1980, *M. N. R. A. S.*, **193**, 21p.
Bouchet, P. *et al.* 1987, *Astr. Ap.*, **177**, L9.
Catchpole, R. 1987, *IAU Circular*, No. 4457.
Chevalier, R. A., and Kirshner, R. P. 1979, *Ap. J.*, **233**, 154.
Chevalier, R. A. 1987, this volume.
Clayton, D. D., and Ramadurai, S. 1977, *Nature*, **265**, 427.
Clayton, D. D. 1982, *Q. J. R. A. S.*, **23**, 174.
Danziger, I. J. 1987, this volume.
Dwek, E. *et al.* 1983, *Ap. J.*, **274**, 168.
Dwek, E. 1983, *Ap. J.*, **274**, 175.
Dwek, E. 1985, *Ap. J.*, **297**, 719.
Dwek, E. 1987a, in "The Interaction of Supernova Remnants with the

Interstellar Medium", IAU Colloq. # 101.
Dwek, E. 1987b, submitted to *Ap. J. (Letters).*
Dwek, E. 1987c, *Ap. J.* , **322**, 812.
Feast , M. W. 1987, this volume.
Fransson, C., and Chevalier, R. A. 1987, *Ap. J. (Letters)*, in press.
Gehrz, R. D. *et al.* 1980, *Ap. J.*, **237**, 855.
Gehrz, R. D., and Ney, E. P. 1987, *Publ. Natl. Acad. Sci. (USA)*, in press.
Graham, J. et al. 1983, *Nature*, **304**, 709.
Gregory, B. and Elias, J. 1987, *IAU Circular*, No. 4368.
Kirshner, R. P. 1987, this volume.
Larson, H. P. et al. 1987, *IAU Circular*, No. 4370.
McGregor, P. J., and Hyland, A. R. 1987, *IAU Circular*, No. 4468.
Merrill, K. M. 1980, *IAU Circular* , No. 3444.
Mumma, M. J. 1987, private communications
Ney, E. P., and Hatfield, B. F. 1978, *Ap. J. (Letters)*, **219**, L111.
Panagia, N. 1987, this volume.
Rank, D. M. *et al.* 1987, *Ap. J. (Letters)*, submitted.
Weissman, P. R. 1985, in *Protostars and Planets II*, eds.D. C. Black and M. S. Mathews, pp.895. Tucson: The University of Arizona Press.
Williams, P. M., Hucht, K. A., and Thé, P. S. 1987, *Astr. Ap.*, **182**, 91.
Woosley, S. E. 1987, this volume.

THE UV INTERSTELLAR SPECTRUM AND ENVIRONMENT OF SN 1987A.

F.C. Bruhweiler
Department of Physics, Catholic University of America, Washington, DC 20064, U.S.A

Abstract. The unusually strong C IV and Si IV features in the UV spectrum of SN 1987A, plus X-ray and optical data obtained prior to this explosive event suggest that SN 1987A exploded in a large low density cavity or superbubble. This cavity was carved out of the interstellar medium by the stellar winds and previous supernovae due to the large concentration of massive stars near SN 1987A. These stars, cavity, and associated nebulosity, as evidenced by extended [O III], form an H II region complex with a diameter of approximately 1 Kpc.

Introduction

The first few days after the initial discovery of Supernova 1987A, before it faded at ultraviolet wavelengths, provided an ideal opportunity to probe the interstellar gas toward this object using the International Ultraviolet Explorer (IUE) satellite. By investigating the line-of-sight toward SN 1987A, we could perhaps for the first time determine the pre-existing conditions of the ambient interstellar medium (ISM) about a supernova. Besides determining the conditions in the local ISM about the supernova, The initial UV brightness of the supernova provided us IUE data for the galactic halo and the Large Magellanic Cloud (LMC) gas of unequaled quality that would be very important addition to previous IUE studies of this gas (see the review by Savage 1986.) However, in what we present here, we will concentrate on what the IUE results and other supporting data can tell us about the pre-existing interstellar environment around SN 1987A.

The high-resolution IUE spectra of SN 1987A show a wide range of interstellar species. (See the poster paper by Blades et al. in these proceedings and de Boer et al. 1987.) These interstellar features indicate a large number of velocity components (or velocity systems) representing numerous clouds along the line-of-sight. Fitting multiple Gaussian profiles to the interstellar features reveals up to 12 systems for the strongest lines. these components cover a heliocentric velocity range from below 0 to over 300 km s^{-1}.

In Figure 1, we present the seven component Gaussian fit to the interstellar absorption of Mg I 2852 A. One can easily separate the observed interstellar components within three distinct velocity ranges;

Figure 1. Mg I in SN 1987A. The interstellar data for Mg I 2852 A along with the Gaussian fit to the profile are shown. The instrumental profile of the IUE is approximately 20 km s^{-1}. The IUE data show up to 12 components for the strongest lines. Optical data reveal up to 24 for Ca II (Vidal-Madjar et al. 1987).

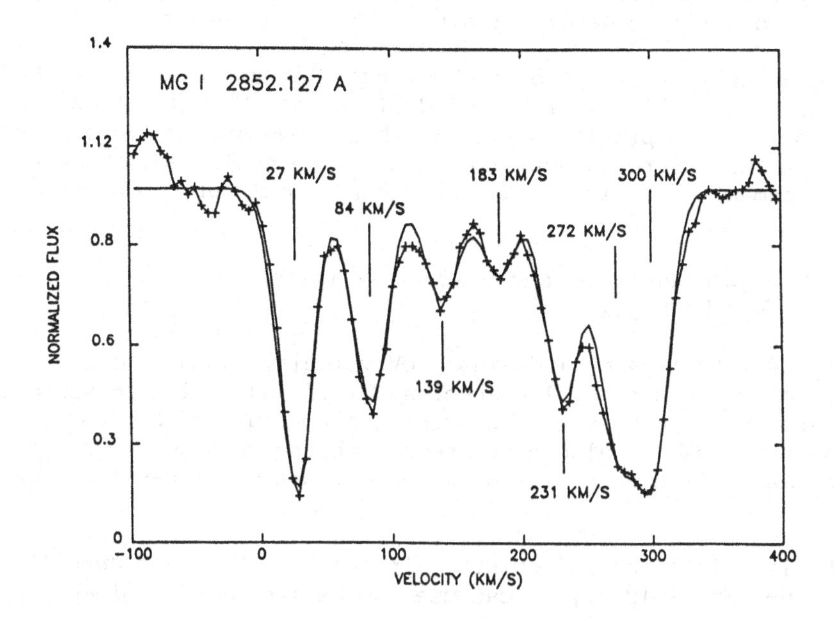

Figure 2. C IV in SN 1987A. The interstellar feature of C IV 1550 A along with the Gaussian fit are displayed. the axes are the same as in Figure 1.

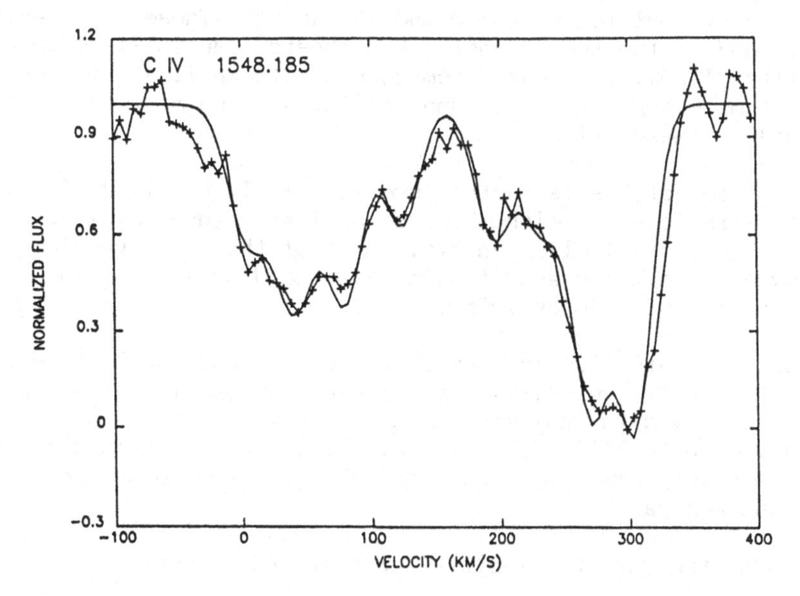

a.) the strong features between 0 to 100 km s^{-1}, b.) the weak features between 100 to 200 km s^{-1}, possibly arising from diffuse, low column density gas, and c.) the very strong features originating in the LMC at 200 to 300 km s^{-1}. The gas in velocity range (a) may be all galactic in origin. However, the placement of the component near 60 km s^{-1} along the line-of-sight can still be debated (Savage 1986).

Perhaps the most notable aspect of the interstellar spectrum of SN 1987A is the unusually strong lines of C IV and Si IV, at LMC velocities (see Fig. 2). One possible explanation is that these ions are produced in a "UV flash" from the supernova explosion. But as we shall see, the pre-existing environment of SN 1987A was more than capable of producing these ions.

Column Densities and Depletions in the Gas Toward the LMC

We have used our fits to each individual UV velocity components to derive ionic column densities. The species, Si II and Fe II, provided in most cases very reliable curves-of-growth with b-values of 2.5 to 3 km s^{-1}. These b-values are also consistent with the results for other species, such as Mn II and Zn II, for which estimates for b-values can be derived.

Since Zn exhibits little or no depletion in the ISM (York and Jura 1983; van Steenberg and Shull 1987), we can use the derived Zn II column densities to estimate H I column densities.

As is shown in Table 1, the bulk of the inferred H I column density is at three velocities. There is to be a major component near 20 km s^{-1} which is associated with local gas in our own galaxy. Two other major components are at velocities near 260 and 280 km s^{-1}. These components seem to correspond to the two extensive H I sheets seen in 21-cm data (McGee and Milton 1966). The weaker components with Zn II at or near the IUE detection limits suggest H I column densities at no more than a few times 10^{19} cm^{-2} at these velocities.

The column densities of the refractory species Si II, Fe II, Mn II, and Al II seen at ultraviolet wavelengths and Ca II at visual wavelengths (Vidal-Madjar et al. 1987) allow an evaluation of the relative elemental depletions compared to those seen toward galactic disk stars (van Steenberg and Shull 1987; Hobbs 1975).

Both UV and optical data clearly show much lower gaseous depletions for components in the entire velocity range from approximately 46 to 215 km s^{-1}. Moreover, the UV data indicate that the weak, likely low density gas in the range 100 to 215 km s^{-1} may have very low or negligible depletions. These very low depletions could be interpreted as the signature of shocked gas.

The depletions in the gas at velocities greater than 215 km s^{-1} are much

TABLE 1

GAS PHASE DEPLETION ESTIMATES

(Depletions Relative to Zn)

Comp.	B	C	D	EFG	HI	JK	*DISK
approx. velocity	20	46	70	130-205	230-270	290-320	
log N(Zn)	13.08	12.02	12.08	11.90	13.33	13.29	
log N(H)	20.63	19.57	19.63	19.45	20.88	20.84	
D (Mn)	-0.44		-0.64		-1.53	-0.12:	-0.96
D (Si)	-0.95		-0.75	-0.42	-1.50:	-1.67:	-0.78
D (Fe)	-0.69	-1.53	-1.02			-2.21:	-1.80
D (Al)	-1.00	-0.59	-1.37	+0.3	-1.51:	-1.34:	-D>2.00
D (Ca)		low?	-2.76	-2.22	-3.75		-3.09

Depletion:

$$D(A) = \log (N(A)/N(Zn))_{ISM} - \log (N(A)/N(Zn))_{\odot}$$

*Disk star depletions from IUE studies of Van Steenberg & Shull (1987)

Figure 3. SN 1987A and the surrounding associations or star clouds are shown. The figure is adopted from Lucke and Hodge (1970). The outline of the diffuse [O III] emission as seen in the optical narrow bandpass imagery is denoted by a dashed surface. The approximate location of SN 1987 A is also indicated. The stars inside the [O III] envelope represent 25% (over 700 stars) of the luminous stars in the LMC. Of these, 225 luminous stars are in star cloud No. 96, nearby to SN 1987A.

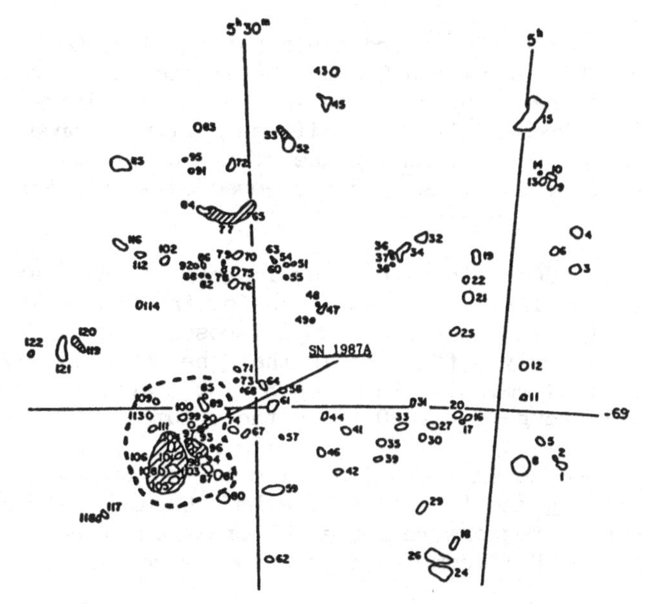

more uncertain. Extreme line blending complicates any attempts to disentangle the interstellar features and obtain usable column densities. In addition, since the abundance mix of the LMC is not well known, the relative depletions given in Table 1 for components H, I, J, and K were derived using solar abundance ratios.

The Origin of the Strong Interstellar Features of C IV and Si IV

If we combine the IUE data with other data at other wavelengths obtained before the occurrence of SN 1987A, we realize that that we do not need a UV flash to produce the observed C IV and Si IV.

Our analysis of narrow bandpass optical imagery taken in the light of forbidden [O III] 5007 A clearly shows a large extended region of faint [O III] emission with a size roughly one degree, or 1 Kpc, in diameter. This region encompasses not only SN 1987A, but also the 30 Doradus complex. This imagery outlines a very large H II complex, of which 30 Doradus is only a minor part. More importantly, O III requires ionization intermediate to Si IV and C IV. Thus, the large region giving rise to the observed [O III] most certainly will produce Si IV and most likely, C IV as well.

This large H II complex is also quite similar to another H II region in our galaxy, in which very strong C IV and Si IV interstellar features are seen. Walborn, Heckathorn, and Hesser (1981), in an IUE study of a number of lines-of-sight toward stars in the Eta Carina Nebula, find strong, but highly variable, interstellar C IV and Si IV in stars separated by only a small angular distance. Likewise, previous studies of interstellar gas in the LMC (Savage and de Boer 1981; Savage 1986) show similar characteristics.

Soft X-ray imagery from the IPC aboard the Einstein X-ray Observatory (Seward et al. 1979; Chlebowski and Seward 1984) also showed that the Eta Carina Nebula contained many discrete sources plus a large diffuse X-ray emitting region. Presumably, this diffuse region is produced in a hot, low density cavity carved from the gas of the nebula by the stellar winds and supernovae from the massive stars embedded in the Eta Carina H II region.

When we examine similar deep Einstein X-ray imagery of the regions around 30 Doradus and SN 1987A, we also, like for the Eta Carina Nebula, find a large number of discrete sources superimposed on a diffuse, extended component. The only difference is that the LMC H II complex delineated by the [O III] emission in which SN 1987A appears to be embedded or behind, is roughly a 1000 times the volume of Eta Carina.

What is the physical mechanism that gives rise to the strong highly ionized species including the [O III] in the LMC? Usually, strong [O III] in galactic H II regions reflects the presence of hot photoionizing stars. Our H II complex in the LMC is no exception. If we look at the results of Lucke and Hodge (1970), we find that there are at least 700 stars with M_V brighter than -4 within the region defined by the [O III] emission (see Fig. 3). These bright stars are massive and are the stars that produce both stellar winds and supernova. Given the concentration of massive stars in this region, the probability of a supernova occurring in this region is quite high. Therefore, it should not be a surprise that SN 1987A occurred in this concentration of stars.

The momentum and kinetic energy of the stellar winds, not to mention that of previous supernovae, from the concentration of massive stars around SN 1987A will have a definite effect upon the interstellar environment. We find, if we apply the results of Bruhweiler et al. (1980), that the stellar winds alone acting only for $3x10^6$ years will, in such a short time, carve out a hot ($T \sim 10^{6-7}$ K), low density ($n \sim 10^{-3}$ cm^{-3}) cavity of at least 200 parsecs radius. If we take longer evolutionary times and consider also previous supernova explosions, the radius of this cavity or "superbubble" would be much larger.

The Inferred Interstellar Environment for SN 1987A

Based upon the observational data and the available energetics provided by the large number of massive stars in the LMC near where SN 1987A occurred, we conclude the SN 1987A occurred either inside or behind a large superbubble complex. The C IV and Si IV (and O III as well) are formed in the H II regions of the clouds or filaments interior to and at the perimeter of the cavity. The high ionization is largely due to photoionization by the strong EUV radiation field from the many O stars in the superbubble complex. However, X-ray ionization from the numerous discrete sources and thermal Bremsstrahlung from the cavity's diffuse hot gas may also play an important role.

I suggest that most of the gas representing the two high H I column density sheets that we see in the IUE spectrum of SN 1987A (components H,I,J,and K) and at 21-cm are produced in the low velocity expanding shell system around the superbubble. Perhaps the low density, wispy gas representing the intermediate velocity range (b) may be high velocity, shocked filaments from this large superbubble complex. However, we cannot yet eliminate other possible origins for this intermediate velocity gas.

Implications

The question is, "Is the interstellar environment for SN 1987 A typical for supernova events?" If so, what does it imply about the visibility of supernova remnants (SNRs) at later times for such events? If such remnants expand into low density cavities, it means that the SNRs would quickly have very low emission measure. Thus, many supernova may quickly fade leaving no detectable remnant. Then, we should ask if the supernova rate deduced from the presently detectable SNRs is not much higher than observations suggest.

References

Bruhweiler, F., Gull, T., Kafatos, M., and Sofia, S. 1980, Ap.J.(Letters), 238, L27.

Chlebowski, T., Seward, F.D., Swank, J., and Szymkowiak, A. 1984, Ap.J., 281, 665.

De Boer, K.S., Grewing, M., Wamsteker, W., Gry, C., and Panagia, N. 1987, Astron. Astrophs., 177, L37.

Dupree, A. et al. 1987, Ap.J.(Letters), in press.

Hobbs, L.M. 1974, Ap.J., 191, 381.

Lucke, P.B. and Hodge, P.W. 1970, A.J., 75, 171.

McGee, M.X. and Milton, J.A. 1966, Australian Jnl. of Phys., 19, 345.

Savage, B. 1986, IUE European Symposium ESA-SP 263, 259.

Savage, B. and de Boer, K.S. 1981, Ap.J., 243, 460.

Seward, F., Forman, W., Giacconi, R., Griffiths, R., Harnden, F.R., Jones, C., and Pye, J. 1979, Ap.J.(Letters), 234, L55.

Van Steenberg, M.E. and Shull, J.M. 1987, preprint.

Vidal-Madjar, A., Andreani, P., Cristiani, S., Ferlet, R., Lanz, T., and Vladilo, G. 1987, Astron. Astrophys., 177, L17.

Walborn, N.R., Heckathorn, J.N., and Hesser, J.E. 1984, Ap.J., 276, 524.

THE INTERSTELLAR SPECTRUM OF SN 1987A IN THE ULTRAVIOLET

J.C. Blades, J.M. Wheatley, N. Panagia
Space Telescope Science Institute, Baltimore, USA
M. Grewing
Astronomisches Institut, Tuebingen, FRG
M. Pettini
Anglo-Australian Observatory, Epping, Australia
W. Wamsteker
ESA IUE Observatory, Madrid, Spain

Abstract

We have compiled a high resolution ultraviolet atlas of the interstellar spectrum toward SN 1987A. The atlas consists of summed IUE spectra covering the range 1250 to 3200 Å. We identify ~ 200 absorption components from 12 interstellar species that cover a wide range of ionization and show complex velocity structure, namely: C I, II, IV; O I; Mg I, II; Al II, III; Si II, IV; S II; Cl I; Cr II; Mn II; Fe II; Ni II and Zn II. Fine structure lines of C I*, C I**, C II*, Si II* are also seen. In addition, a number of weak absorption lines remain unidentified.

Observations and Data Reduction

The IUE observations that comprise our interstellar absorption altas were obtained over the three day period, February 25–27, 1987, as part of the ESA Target of Opportunity Program for observing bright supernovae (Wamsteker *et al.* , 1987). A wide range of exposure times was needed to accomodate both the limited dynamic range of the IUE cameras and the intrinsic flux distribution of the supernova—which varied rapidly during the course of the observations.

The ultraviolet flux decreased very quickly with time. In particular, the average flux in the SWP range decreased at a rate of 0.87 dex/day and that in the LWP range at a rate of 0.49 dex/day during the first three days (Panagia *et al.* , 1987) In addition, as the supernova faded in the UV, its spectrum became contaminated by the light of two "neighbour" stars (Walborn *et al.* , 1987), also included in the field of view of the IUE large aperture which was used for the high-resolution observations. We have no significant contamination in our spectral atlas because in the region below ~1670 Å we used data taken only on February 25, with data longward coming from both February 25 and 26. At these early dates, contamination from the neighbouring stars was negligible.

The supernova's rapid decline in the ultraviolet severely limited the useful period for obtaining high dispersion measurements. Nevertheless, we have been able to assemble a set of high resolution images that have allowed us to produce a high quality interstellar atlas towards SN1987A. Over virtually the entire wavelength range, the atlas represents the sum of two or more well-exposed spectra.

To produce the atlas we worked from the standard IUESIPS data output, starting from the non-ripple corrected, net extraction. Our procedure was as follows. First we overplotted common echelle orders for those regions where we had two or more adequate exposures; data flags showing reseau positions, known blemishes and saturated pixels were superposed. These comparisons allowed us to select the most appropriate images for the atlas. Additionally, they allowed us to search for weak absorption lines, by looking for consistency from spectrum to spectrum, and to recognise additional blemishes.

After producing a summed spectrum for each echelle order, we normalized by fitting a high-order spline to the continuum, using software developed by Dr. K. Horne (ST ScI). The normalization does not have any deleterious effects on sharp absorption lines, but does remove any broad absorption features wider than ~ 5 Å, including those intrinsic to the supernova. Finally, the normalized spectra were joined together at the positions where the flux equalled that in the adjacent orders.

The velocity structure that we see in these ultraviolet interstellar profiles is difficult to disentangle at the resolution of IUE. Higher resolution studies of the optical species Ca II and Na I by Vidal-Madjar *et al.* (1987) show the presence of *at least* 24 components covering the velocity range 0–300 km/sec. In Figure 1 we compare a few of the ultraviolet profiles with the Ca II K line from Vidal-Madjar *et al.* Based on these 3km/s resolution optical observations, we expect seven major components at the 30km/s IUE resolution, with heliocentric velocities of 9–24, 65, 126, 165, 216, 250–265 and 280 km/sec. At the higher optical resolution, we see that many of the ultraviolet components are actually blends of two or more subcomponents.

A number of lines, especially in the SWP wavelength region, remain unidentified, although other workers have not reported unidentified lines towards Magellanic Stars (de Boer *et al.* 1985). An important component of our analysis will be in the verification and interpretation of these unidentified lines.

In a series of papers that we are preparing, we shall publish our analysis of the velocity structure and give detailed abundance determinations using the spectral atlas and a multi-cloud curve of growth analysis.

REFERENCES

de Boer, K. S., Fitzpatrick, E. L., and Savage, B. D., 1985, *M.N.R.A.S.,* **217**, 115.

Panagia, N., Gilmozzi, R., Clavel, J., Barylak, M., Gonzalez, Riestra, R., Lloyd, C., Sanz Fernandez de Cordoba, L. and Wamsteker, W. 1987, *Astron. Astrophys.,* **177**, L25.

Vidal-Madjar, A., Andreani, P., Cristiani, S., Ferlet, R., Lanz, T., Vladilo, G., 1987, *Astron. Astrophys.,* **177**, L17.

Walborn, N. R., Lasker, B. M., Laidler, V. G. and Chu, Y.-H., 1987, *Ap. J. (Letters),* **321**, L41.

Wamsteker, W., Panagia, N., Barylak, M., Cassatella, A., Clavel, J., Gilmozzi, R., Gry, C., Lloyd, M., van Santvoort, J., and Talavera, A., 1987, *Astron. Astrophys.,* **177**, L21.

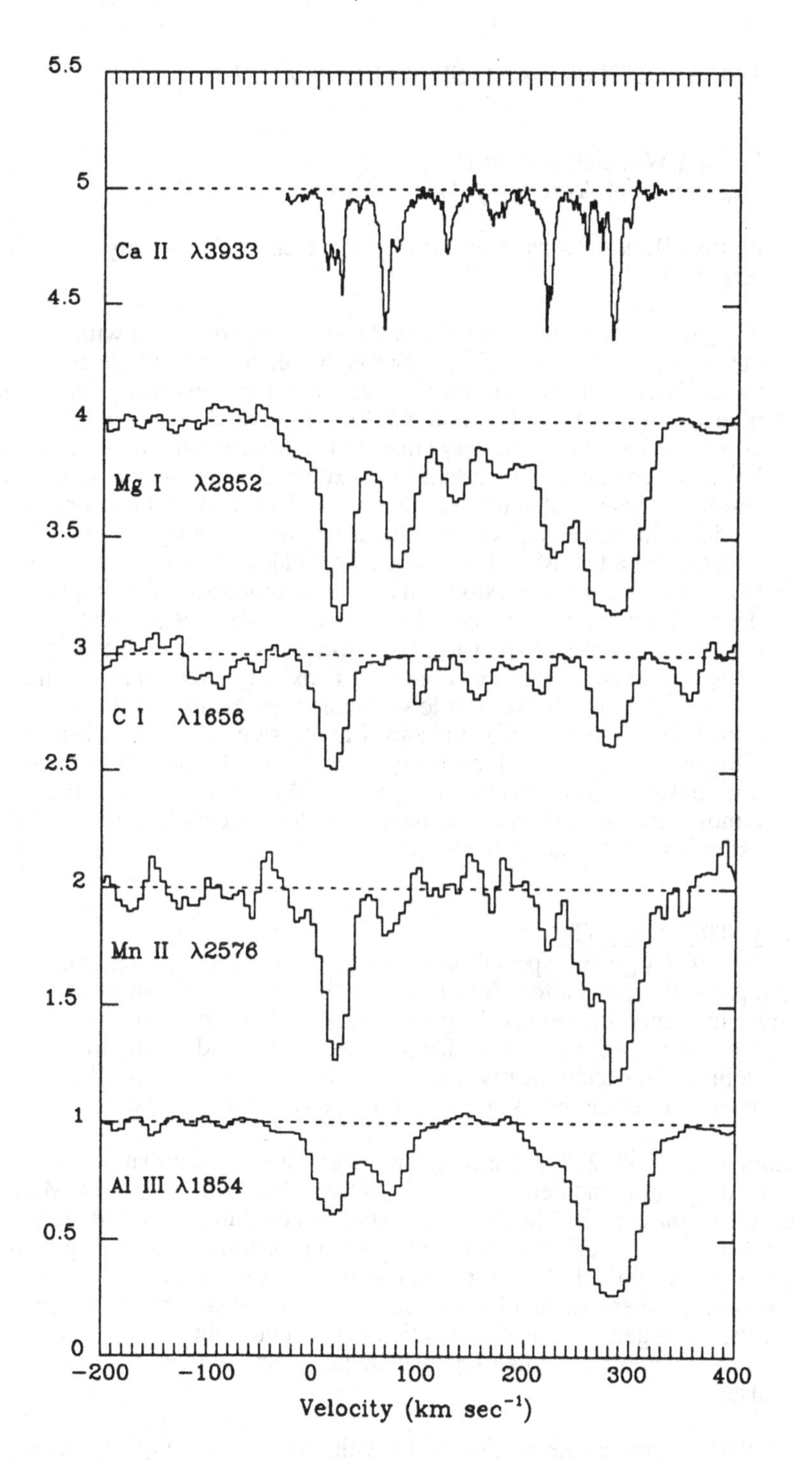

Figure 1. Selected interstellar velocity profiles from our ultraviolet atlas compared with the Ca II K line profile from Vidal-Madjar *et al.* (1987).

THE STRUCTURE AND SPECTRUM OF SN 1987A

J. Craig Wheeler, Robert P. Harkness
Department of Astronomy, University of Texas at Austin

Zalman Barkat, Department of Physics, Hebrew University of Jerusalem

Abstract. Consideration of the stellar structure consistent with observations of SK -69° 202 suggests that there are multiple solutions to the stellar structure equations that pertain to the question of red versus blue supergiant evolution, and that for normal hydrogen abundance only very large (~ 10 $M_\odot$) or very small (~ 0.1 $M_\odot$) envelopes are allowed. For increased helium abundance the excluded range of envelope masses shrinks. This result and other constraints based on observations and model light curves suggest that the progenitor had a helium-rich (Y ~ 0.5) envelope of a few $M_\odot$. LTE atmosphere calculations corresponding to two days after the explosion constrain the luminosity, density, and density gradient in the ejecta. These calculations give a natural explanation of the "Type I-like" UV spectrum in terms of resonance scattering of Mg II and Fe II lines. A density gradient of approximately $\rho \propto r^{-11}$ is favored over shallower density gradients. At this particular epoch, the absence of Ca II H and K absorption sets a lower limit to the temperature and hence luminosity. The derived luminosity is consistent with observations corresponding to m - M $\gtrsim$ 18.5 and Av $\gtrsim$ 0.6. The atmospheres are scattering dominated with implications for distance estimates by the Baade method .

I. INTRODUCTION

SN 1987A gives a special opportunity to learn about the structure and evolution of the particular progenitor. It also has much to teach us about other supernova by seeking similarities and differences. The superb spectral observations are a crucial diagnostic and represent a proving ground for the developing field of supernova atmosphere modeling. This technique will help us to learn more about SN 1987A and other supernovae, and to refine the use of supernovae as distance indicators.

The identification of SK -69° 202 as the progenitor star allows an estimate of the luminosity of the progenitor and hence estimates of the helium core mass (~ 6 $M_\odot$) and initial main sequence mass (~ 20 $M_\odot$) assuming standard evolution and core mass-luminosity relations. Dynamical models based on such progenitor structures give a reasonable reproduction of the light curve (see contributions by Nomoto, Woosley, and Arnett in this volume). These calculations do not directly constrain the mass and composition of the hydrogen-rich envelope which in turn determine the expansion of the core and the rate of uncovering of inner heavy elements, some of them radioactive, and eventually a pulsar.

There are hints that an appreciable portion, but not all, the outer envelope has been lost, and that the remaining several solar masses is enriched in processed material in a manner that does not conform to standard evolution calculations. Several contributions to these proceedings address aspects of this potential anomaly. Walborn reported that the classification spectra of SK -69° 202 may be consistent with enhanced He and N. Kirshner and Panagia discussed the narrow N lines in UV spectra indicative of a circumstellar nebula enhanced in N, and hence, presumably, a similarly altered progenitor envelope. Williams (1987) has identified Ba in the ejecta and Lucy estimated that enhancements by factors of order 30 might be necessary, subject to caveats of the radiative transfer process. Woosley and Nomoto discussed the notion that helium enhancements might help to explain the early evolution of the light curve.

In this paper we will discuss the progenitor structure and the question of the helium abundance of the envelope from another perspective. In addition, we will present atmosphere models for the spectrum at two days which account for the unique optical and UV spectra. These models also constrain the structure of the envelope, give a distance independent lower limit to the luminosity and give some new perspective on the use of the Baade method to estimate supernova distances.

II. PROGENITOR STRUCTURE

Barkat and Wheeler (1988; see also Wheeler, Harkness, and Barkat, 1988) have discussed aspects of the structure of the progenitor star by assuming external properties (L, T_{eff}) consistent with SK -69° 202 and then using an efficient program which numerically integrates envelope structure with the assumption of hydrostatic and thermal equilibrium. These calculations are used to make a parameter study of structures which are consistent with the observations.

One important aspect of this study is to show that multiple solutions to the stellar structure equations can exist despite the specification of all the seemingly important structural parameters. Figure 1 illustrates this phenomenon by presenting the core radius as a function of effective temperature at a fixed luminosity of 55,000 $L_\odot$ for a model with total mass 15 $M_\odot$ and a helium core mass of 4 $M_\odot$. Note that for a *fixed* core radius $\lesssim 8 \times 10^{10}$ cm, there are *three* solutions with different core and envelope structure and different effective temperature. One of the solutions corresponds to the Hyashi track, but two are blue. The question of which blue solution might be attained by a given evolutionary or structural code is not obvious. Small differences in the core radius, or luminosity, caused by physical assumptions or numerical procedures can lead to situations in which there are *no* blue solutions. The implicit existence of a critical point in the curve of Figure 1 may help to explain why some researchers find evolution to the red, some to the blue, and some swinging back and forth. Clearly, care must be exercised in constructing progenitor models.

An important question is the amount of mass loss the progenitor underwent and hence the mass of the remaining envelope. We have approached the problem by constructing envelopes of varying mass and examining them for self-consistency. For a "standard" composition (X = .75, Y = .245, Z = .005) we find that envelopes of arbitrary mass are not self-consistent. For models with L ~ 100,000 $L_\odot$ corresponding to core masses ~ 6 $M_\odot$ and initial main sequence mass ~ 20 $M_\odot$, we find that large mass envelopes, $M_{env} \gtrsim 12\ M_\odot$, remain cool, with temperature less than the hydrogen ignition temperature, ~ 4-5 x 10^7 K, all the way down to the helium core. The composition switches to helium at that point, a previous hydrogen burning shell having presumably become extinct, and the integration can be extended self-consistently. For a smaller envelope mass, the temperature climbs steeply inward and exceeds the hydrogen ignition temperature well outside the core mass consistent with the adopted luminosity. That envelope structure is thus self-inconsistent. Another range in allowed masses is found for masses so small ,$\lesssim$ 0.1 M$_\odot$, that despite the steep temperature gradient in the envelope, the core is encountered before the hydrogen ignition temperature is reached.

As the helium abundance is increased, the temperature gradient becomes shallower for a given envelope mass and the excluded range of envelope masses shrinks. For Y $\gtrsim$ 0.5 no envelope mass can be excluded on this basis. Figure 2 gives the excluded total and hence envelope masses as a function of Y for stars with cores of 6 M$_\odot$. For normal Y ~ 0.25, the allowed envelope masses are either very large or very small. The large envelopes are in potential conflict with the suggestions of He, N, and Ba excesses. The small envelope masses are excluded in general because they yield light curves or kinematics that deviate badly from observations.

With reasonable kinetic energy ~ 10^{51} ergs, a small mass hydrogen envelope is ejected with excessive velocity. In addition, the core expands too quickly in the absence of a tamping mantle and the trapped heat of radioactivity is released too quickly, causing the light curve to brighten too rapidly to too bright a peak. This is illustrated in Figure 3 which shows a light curve corresponding to a 6 M$_\odot$ core with an envelope of 0.1 M$_\odot$ expanding with 6.6 x 10^{50} ergs of kinetic energy and 0.07 M$_\odot$ of ^{56}Ni. The minimum hydrogen velocity is 6500 km s^{-1}, whereas hydrogen is observed to move as slowly as ~ 2000 km s^{-1} (Phillips, these proceedings). The other light curve corresponds to the very low kinetic energy of 10^{50} ergs. Now the hydrogen expands too slowly. The minimum velocity for hydrogen in the model is 2100 km s^{-1}, but very little moves at speeds of $\gtrsim$ 20,000 km s^{-1} as observed. Despite this, the light curve still rises too quickly to

Figure 1 - The radius of the helium core in units of 10^{10} cm is given as a function of effective temperature for a model with total mass of 15 M$_\odot$, a helium core mass of 4 M$_\odot$, and a luminosity of 55,000 L$_\odot$. Note that for fixed total mass, core mass, core radius, and luminosity, there can be three solutions with different effective temperature and different density structure.

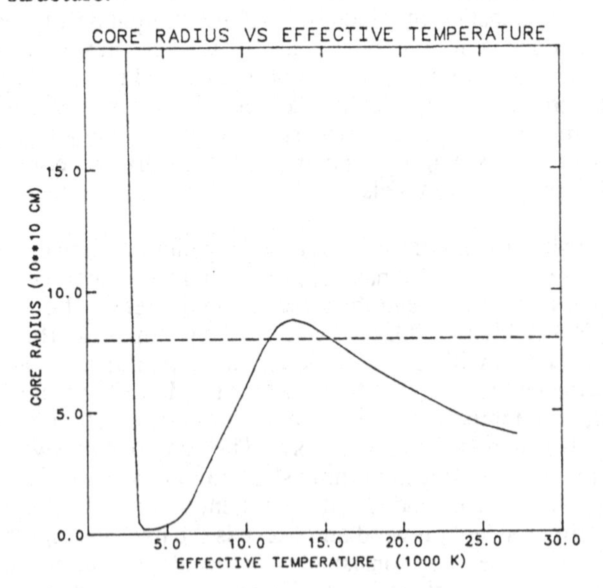

Figure 2 - The range of total stellar mass that can be excluded on structural grounds is given as a function of the helium abundance in the envelope for models with helium core mass of 6 $M_⊙$ and luminosity L = 110,000 $L_⊙$.

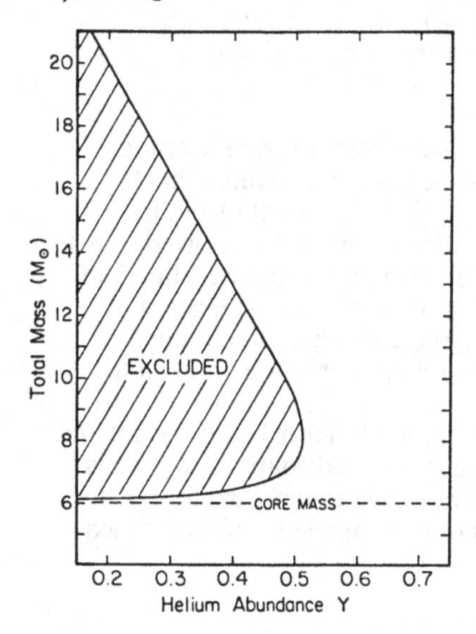

Figure 3 - Bolometric light curves are given for models with a helium core of 6 $M_⊙$ and a hydrogen envelope (Y=.25) of 0.1 $M_⊙$. The solid curve represents an injected kinetic energy of 6.6 x 10^{50} ergs, the dashed curve 1x10^{50} ergs, and the dot-dash curve 1x10^{50} ergs, but Y~1. Note all three rise much faster than the observed light curve taken from Catchpole, et al. (1987).

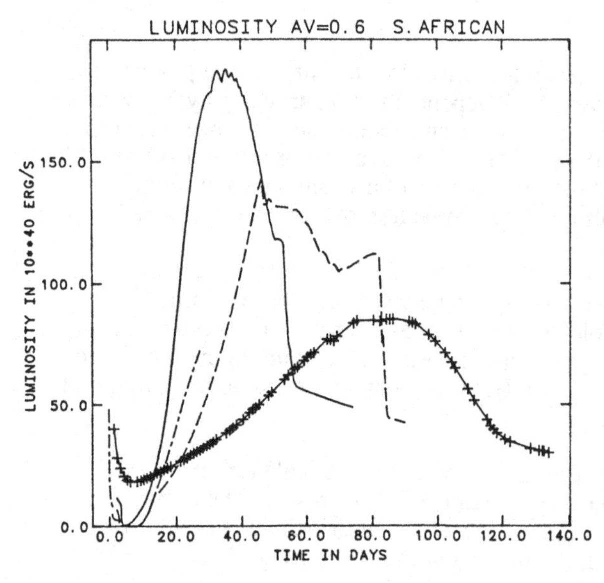

maximum. Both of these models are too dim at around 10 days. The third curve in Figure 3 corresponds to the lowest energy model with the envelope enriched to nearly 100 percent helium. The luminosity at 10 days is still too low, but it rises even faster toward the peak. This type of low-mass envelope model is clearly severely deficient. Models of this type proposed by Wood and Faulkner (1987) and by Maeder (1987), while based on astrophysically reasonable evolution with strong mass loss, can nevertheless be ruled out.

Models which match the light curves, kinematics, and other constraints thus seem to require at least several solar masses of envelope to slow the core by a reverse shock (Woosley, Pinto and Ensman 1987, Nomoto et al. 1987). The constraint of Figure 2 suggests that if helium is enhanced, envelopes of only moderate mass are allowed. Such a helium enhancement is consistent with the hints of composition peculiarities, but does not correspond to any "standard" model of evolution. Rather there is a suggestion of some "anomalous" mixing. Figure 2 does not exclude large mass, helium enriched envelopes, but the evolution to produce such a result would be even more anomalous.

The particular results of Figure 2 are based on the assumption of thermal equilibrium and constant composition in the envelopes. There may be quantitative deviations if there is a helium gradient or a breakdown of strict thermal equilibrium due to rapid evolution, but it is not clear that these are dominant effects. To the best of our knowledge all published models for SN 1987A are consistent with Figure 2.

III. ATMOSPHERE MODELS

Atmosphere models have been constructed with an LTE atmosphere code (Harkness 1985, 1986). The current discussion will be confined to the first few days when the hot photosphere is expected to be nearly in LTE. After that time, the Balmer lines, particularly Hα, depart strongly from LTE and a more general approach will be needed. The early phase is an important one since it was during this time that the ultraviolet faded very rapidly due to line blanketing and evolved from an "unprecedented" spectrum to that "like a Type I."

We concentrate here on atmosphere models with power low density profile $\rho \propto r^{-n}$ in order to explore constraints on the atmosphere independent of a particular hydrodynamic model. Branch (1987) and Dopita et al. (1987) have suggested that the power law index is $\gtrsim 11$ for the first 5 days, then ~ 5 subsequently. We have explored n = 5, 11 and 7, the latter being the value expected from the propagation of a strong shock through a polytrope of index 3. The velocity profile is taken to be homologous and the metal abundances are $Z = 1/4\ Z_{\odot}$.

We consider an epoch corresponding to two days to be specific, corresponding to the earliest optical spectrum from Cerro Tololo. At this phase there is little sign of Ca II H and K or of the IR triplet, and this emerges as an important constraint on the atmosphere models. To avoid the strong signatures of Ca II, the atmosphere must be hot enough at all significant densities to ionize Ca II.

At two days, no models with n = 5 are adequate. For any reasonable luminosity the atmospheres are extended and cool and very strong Ca II lines result. The derived colors are too blue and there is no discernable sign of Hγ. Similar criticisms, not as strong, apply to models with n = 7. Of this set, the models with n = 11 are preferred. This atmosphere is steep enough that conditions can be found for which the Ca is all hot enough to substantially ionize away all Ca II.

For a given density gradient, the normalization of the density is determined by the requirement that the absorption minima of the Balmer lines fall at the correct minima. For the models with n = 11, the normalization density that gives the proper Doppler shift is about 2×10^{-14} gm cm^{-3} at a velocity of 20,000 km sec^{-1}, a convenient fiducial velocity near the photospheric velocity, which is about 17,000 km sec^{-1} at two days. Dynamic models should thus reproduce this slope and density if they are to be considered successful.

Even with a relatively steep density gradient, the Ca II features are too strong unless the temperature of the atmosphere is high enough to ionize it away. Raising the temperature results in an increased luminosity, and hence the requirement that the Ca II not be a strong feature at 2 days provides an intrinsic, distance independent, lower limit to the luminosity of a given atmosphere model. For n = 11, the lower limit is about 5×10^{41} erg s^{-1}. This is illustrated in Figure 4 which shows atmosphere models with n = 11, and density at 20,000 km s^{-1} of 1.5×10^{-14} gm cm^{-3}. Three luminosities are shown with only the brightest having sufficiently small Ca II features to correspond to the observations. A somewhat steeper atmosphere would allow a smaller minimum luminosity. While this analysis has not resulted in any critical new insights to SN 1987A, it does serve to illustrate the potential of the method to provide intrinsic information concerning the luminosity as well as the structure of the supernova atmosphere.

Figure 4 also shows the ultraviolet portion of the models and the UV spectrum at two days as presented by Sonneborn in these proceedings. The UV flux and spectrum were changing rapidly at this epoch, but at this phase bore a noticeable resemblance to Type I supernovae, unlike any other Type II supernova. The atmosphere calculations show that in both Type Ia and Type Ib supernovae the UV deficit and characteristic spectra are formed primarily by resonance line scattering by overlapping Fe II lines (Wheeler and Harkness 1986, Wheeler, Harkness, and Cappellaro 1987). A characteristic feature of these types of events is a peak at about 2000 Å which is not emission, but due to a natural lack of Fe II scattering lines in that wavelength range. Figure 4 shows that same characteristic, but Doppler shifted to about 1800 Å in the high velocity atmosphere. Another important feature is the Mg II λ2797 line that causes the severe drop between about 2600 and 2500 Å.

These basic atmosphere models thus give a reasonable account of the "unprecedented" UV spectrum at this epoch. Other Type II do not display a similar UV deficit. This is presumably because they explode within a denser circumstellar environment which can affect both the UV emission directly and the nature of the supernova atmosphere. If the atmosphere steepens in a shock resulting from the collision with a circumstellar nebula, the whole atmosphere can be so hot as to ionize Fe II and it will then radiate nearly as a black body at a single temperature.

These atmospheres are strongly scattering dominated. On the short wavelength side of the Balmer discontinuity where there the bound-free opacity is strongest, the absorptive opacity is still only about 10% of the scattering opacity at unity scattering optical depth, and on the long wavelength side, it is much less. This must affect distance estimates to supernovae using the Baade method in which the color temperature is assumed to represent the same surface as that which represents the minimum of the absorption in a line formed at the photosphere. In a scattering atmosphere, the color temperature represents a deeper, hotter layer. As a crude measure of this effect based on the models of Figure 4, we can make an estimate of the color temperature from the calculated unreddened B-V color. The middle theoretical spectrum of Figure 4, for which the actual computed luminosity is 3.9×10^{41} erg s^{-1}, gives B-V = 0.115 (corresponding to B-V = .303 with $A_V = 0.6$; the observed value is closer to B-V ~ 0.2, Menzies et al. 1987). This corresponds to a black body temperature of about 9,500 K. In the Baade method, the photospheric radius is estimated by multiplying the velocity corresponding to an observed absorption minimum by the time since the explosion. At this early time, Hα suffices for this since the hot atmosphere is highly ionized and the optical depth in Hα is not anomalously large as it is at later epochs. The surface of unity optical depth in electron scattering corresponds closely with the Hα minimum in the theoretical model at 16,600 km s^{-1}. At two days, this velocity gives a kinematic radius of 2.86×10^{14} cm.

Figure 4 - Three model atmospheres are given corresponding to a density profile of $\rho \propto r^{-11}$, $\rho = 1.5 \times 10^{-14}$ gm cm^{-3} at 20,000 km s^{-1} at two days past explosion. The three curves correspond to luminosities of 4.9, 3.9 and 3.2×10^{41} erg s^{-1}, from top to bottom, respectively. The observed bolometric luminosity at this epoch is about 4×10^{41} erg s^{-1} (Menzies et al. 1987). The lower curves are optical spectra from Cerro Tololo provided by Mark Phillips and UV spectra provided by Bob Kirshner. Note the diminishing Ca II features with increasing luminosity, and the strong UV deficit.

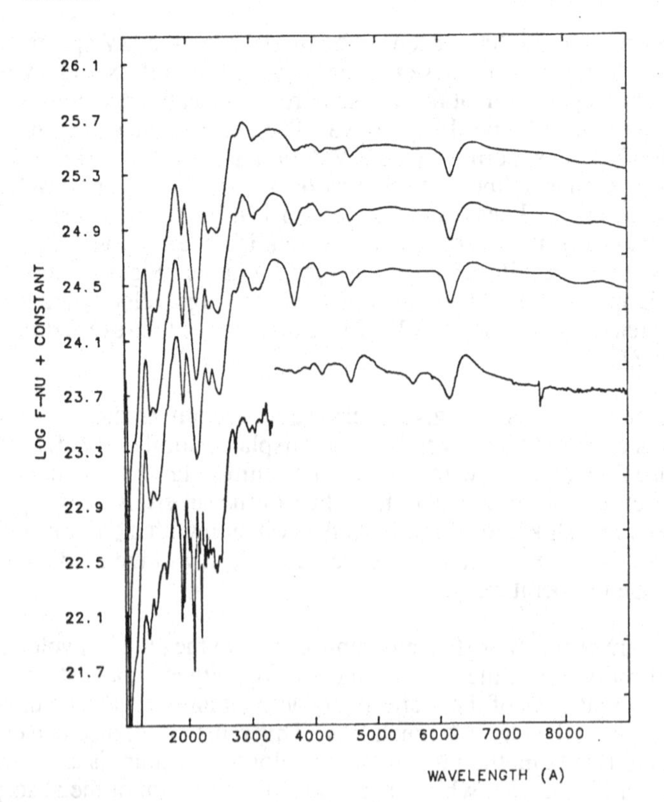

The resulting bolometric "Baade" luminosity, using the derived temperature and radius is about 4.8 x 10^{41} erg s^{-1} corresponding to a 23 percent error in luminosity and an 11 percent error in distance. For a model with density profile index n = 7, a similar exercise gives errors in luminosity and distance of 78 percent and 33 percent, respectively. Use of the B-V color to estimate the bolometric luminosity is crude, and not expected to be reliable, given the steep UV deficit. Rather, the models can be analyzed in terms of a monochromatic color temperature, and the ratio of the flux predicted by the "Baade" method to that actually produced by the model computed as a function of wavelength. The Baade method will presumably be more accurate in the red. The current models may also be overly sensitive to the inner boundary condition for wavelengths longward of the Balmer discontinuity where scattering is particularly dominant. This question is under investigation.

IV. CONCLUSIONS

Our study of the structure of the progenitor of SN 1987A has pointed to a star with an envelope of moderate mass which is enriched in helium, in keeping with other emerging lines of evidence. This is a particularly intriguing conclusion because it suggests that the progenitor did not undergo an evolution in strict accord with standard models, but rather was anomalously well mixed during its quasi-static evolution. Further study of SN1987A might thus teach us some important lessons concerning the nature of stellar evolution.

The atmosphere models show that LTE is a reasonably good approximation at least in the first few days. These models provide a satisfactory explanation of the UV spectrum which was so surprising at first glance. A rather steep slope of the density gradient is indicated, consistent with estimates based on the rate of recession of the photosphere, and with some dynamical models (e.g. Arnett 1987 and these proceedings). The dynamical models must also give the correct absolute density at a given epoch in order to reproduce the observed Doppler shift of the lines. This will in practice give yet another constraint on the energy of the explosion for models of a given mass.

At two days, a particularly interesting constraint emerges from the failure to see strong Ca II lines. This gives a lower limit on the luminosity for a given model, and suggests, for instance, that the distance modulus to the LMC should not be much less that 18.5, nor should the extinction be much less than $A_V = 0.6$. Beyond this particular detail, this analysis illustrates the potential power of the atmosphere models to define new intrinsic constraints on the structure and distance of supernovae. The atmosphere models are scattering dominated and will yield new insights on the traditional Baade method of estimating supernova fluxes and distances.

This research is supported in part by NSF grant 8413301 and by the R.A. Welch Foundation. We are grateful to Mark Philips and Bob Kirshner for providing the optical and UV spectral data in numerical form, and to Stan Woosley and Ken Nomoto for providing structural models with which we could compare.

REFERENCES

Arnett, W. D. 1987, *Ap. J.*, in press.
Barkat, Z., and Wheeler, J. C. 1988, in preparation.
Branch, D. 1987, *Ap. J. (Letters)*, **320**, L23.
Catchpole, R. M. et al. 1987, *M.N.R.A.S.*, in press.

Dopita, M. A., Achilleos, N., Dawe, J. A., Flynn, C., and Meatheringham, S. J. 1987, preprint.
Harkness, R. P. 1985, in *Supernovae as Distance Indicators*, ed. N. Bartel (Berlin: Springer-Verlag), p. 183.
Harkness, R. P. 1986, in *Radiation Hydrodynamics in Stars and Compact Objects*, ed. D. Mihalas and K.-H.A. Winkler (Berlin: Springer-Verlag), p. 166.
Maeder, A. 1987, *Proceedings of the ESO Conference on SN 1987A*, in press.
Menzies, J. W., et al. 1987, *M.N.R.A.S.*, **227**, 39p.
Nomoto, K., Shigeyama, T., and Hashimoto, M. 1987, in *Proceedings of the ESO Conference on SN 1987*, in press.
Wheeler, J. C. and Harkness, R. P. 1986, in *Proceedings of the NATO Advanced Study Workshop on Distances of Galaxies and Deviations from the Hubble Flow*, eds. B. M. Madore and R. B. Tully (Dordrecht: Reidel), p. 45.
Wheeler, J. C., Harkness, R. P., and Cappellaro, E. 1987, in *Proceedings of the 13th Texas Symposium on Relativistic Astrophysics*, ed. M. Ulmer (World Scientific; New York) p. 402.
Wheeler, J. C., Harkness, R. P., and Barkat, Z. 1988, in *IAU Colloquium 108*, in press.
Williams, R. E., 1987, *Ap. J. (Letters)*, **320**, L117.
Wood, P. R. and Faulkner, D. J. 1987, preprint.
Woosley, S. E., Pinto, P. A., and Ensman, L. 1987, *Ap. J.*, in press.

SUPERNOVA 1987A: CONSTRAINTS ON THE THEORETICAL MODEL

Ken'ichi Nomoto and Toshikazu Shigeyama

Department of Earth Science and Astronomy, University of Tokyo, Meguro-ku, Tokyo 153

Abstract. Hydrodynamical models for SN 1987A are compared with the observations. From the expansion velocity and the light curve, constraints on the explosion energy E, the mass of the hydrogen-rich envelope M_{env}, and the mass and distribution of ^{56}Ni are obtained. Models and observations are in reasonable agreement for $E/M_{env} = 1.5 \pm 0.5 \times 10^{50}$ erg/$M_\odot$, $M_{env} > 3$ $M_\odot$, and $M_{Ni} \sim 0.07$ $M_\odot$. The best fit among the calculated models is obtained for $M_{env} = 6.7$ $M_\odot$ and $E = 1.0 \times 10^{51}$ erg. Mixing of ^{56}Ni in the core and moderate helium enrichment in the envelope are suggested.

1 INTRODUCTION

The supernova 1987A in the LMC is providing us with valuable information on the chemical and dynamical structure of the exploding massive star. Current theory of massive star evolution, nucleosynthesis, and supernova explosion contains three major uncertainties, namely, 1) the mechanism that tranforms collapse into explosion, 2) mass loss mechanism, and 3) convection (material mixing, in general). These lead to uncertainties of explosion energy E, mass and distribution of ^{56}Ni, mass of the hydrogen-rich envelope M_{env}. We derive some constraints for these quantities by comparing our hydrodynamical models of the supernova explosion and the optical light curve with the observations. These constraints are useful for deeper understanding of the above processes.

Recent X-ray observations of SN 1987A by the Ginga (Dotani et al. 1987) and Kvant (Sunyaev et al. 1987) provides another constraints on the hydrodynamical model and mass loss process. If the hard component of the X-rays is due to Compton degradation of γ-rays from ^{56}Co (McCray et al. 1987), its detection is much earlier than the prediction and may give information on the material mixing in the supernova interior (M. Itoh et al. 1987; see Kumagai et al. 1987 and references therein). Thermal X-rays from the collision between the supernova ejecta and the circumstellar material have been predicted and modeled by H. Itoh et al. (1987). If the soft component corresponds to this prediction, useful quantities on the mass loss history and circumstellar material can be obtained (Masai et al. 1987; see Nomoto et al. 1987b). Refined hydrodynamical model will be further tested by γ-ray observations.

2 PROGENITOR'S CORE AND ENVELOPE

The progenitor of SN 1987A is very likely Sk-69 202. Its luminosity of about 1.3×10^5 $L_\odot$ (Woosley et al. 1987; Hillebrandt et al. 1987) corresponds to the presupernova luminosity of the 6 $M_\odot$ helium star (Hashimoto and Nomoto 1987; see Nomoto and Hashimoto 1986, 1987a,b and Nomoto et al. 1987a). Its main-sequence mass is estimated to be $M_{ms} = 17$

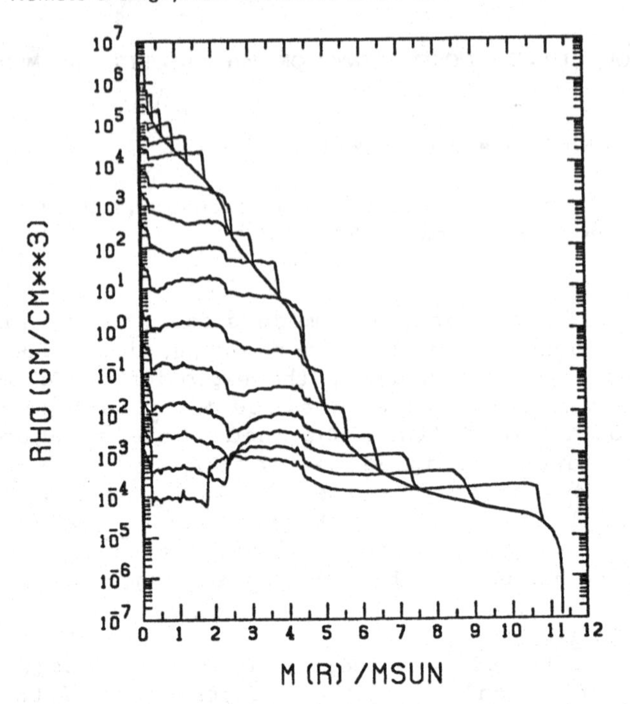

Figure 1: Change in the density distribution during the propagation of a shock wave and a reverse shock.

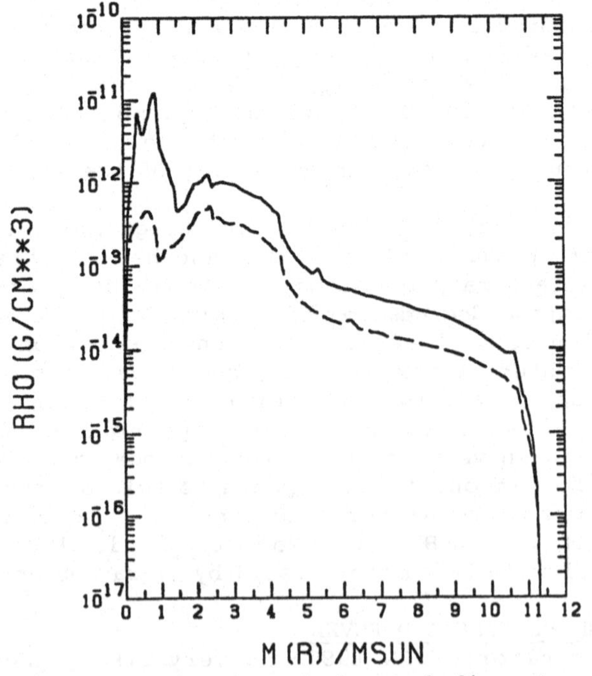

Figure 2: Density distribution of the exploding star at t = 116 d. Shown are Model 11E1 (solid): $M = 11.3\ M_{\odot}$, $M_{env} = 6.7\ M_{\odot}$, $E = 1 \times 10^{51}$ erg and Model 11E2 (dashed): $E = 2 \times 10^{51}$ erg.

- 20 $M_\odot$ which depends on the convective overshooting during hydrogen burning (Maeder 1987b).

We constructed the initial model for the explosion calculation from the above 6 $M_\odot$ helium star with a hydrostatic hydrogen-rich envelope. The inner 1.4 $M_\odot$ was replaced with a point mass neutron star. The composition structure of the ejecta is: the heavy element layer of 2.4 $M_\odot$, helium-rich layer of 2.2 $M_\odot$, and the hydrogen-rich envelope of M_{env}. Thus the total mass of the ejecta is $M = 4.6\ M_\odot + M_{env}$.

The mass, M_{env}, and the composition of the hydrogen-rich envelope depends on mass loss and convective mixing during the presupernova evolution. Recent UV observations have shown that the UV emission lines originated from circumstellar materials where the ratios of N/C and N/O are much larger than the solar values (Kirshner 1987a). This implies that the progenitor had evolved once to a red supergiant stage, lost a significant fraction of its hydrogen-rich envelope containing the CNO processed material, and then contracted to the blue supergiant size.

The surface composition changes during mass loss (Maeder 1987a). At a certain stage of mass loss, the surface abundances change from the original ones. Helium and nitrogen are overabundant at almost the same stages. M_{env} and helium abundance at the explosion depend on the mass loss rate. The large N/C ratio in the circumstellar matter suggests that the progenitor had lost a large fraction of its mass and its envelope was somewhat helium-rich at the explosion. Interestingly the early light curve and the photospheric velocity depends somewhat on the surface abundance and thus provides information on the abundance.

3 HYDRODYNAMICAL MODELS

In the hydrodynamical calculation of the supernova explosion, an energy E was deposited instantaneously in the central region of the core to generate a strong shock wave. The subsequent propagation of the shock wave, the expansion of the star, and the optical light curve were calculated for models 11E1Y4, 11E1Y6, 11E1.5, 11E2, 7E1Y4, where 11 and 7 denote the ejecta mass of $M = 11.3\ M_\odot$ ($M_{env} = 6.7\ M_\odot$) and 7.0 $M_\odot$ ($M_{env} = 2.4\ M_\odot$), respectively; E1, E1.5, and E2 denote explosion energy of $E = 1$, 1.5, and 2×10^{51} erg, respectively; Y4 denotes a surface composition $X = 0.59$, $Y = 0.40$, and $Z = 0.01$, while Y6 stands for $X = 0.39$ and $Y = 0.6$. For all models the initial radius of the progenitor is assumed to be $R_0 = 3 \times 10^{12}$ cm (see Nomoto et al. 1987a, Shigeyama et al. 1988 for details).

3.1 Mixing

The shock wave generated at the inner edge of the ejecta propagates through the hydrogen-rich envelope. The expansion of the inner core is decelerated by the low-density envelope and a reverse shock is produced (Fig. 1). This forms a density inversion (Fig. 2). Because of Rayleigh-Taylor instability, the core material will be mixed during early stages (see also Woosley et al. 1988). Since the time scale and the extent of mixing are uncertain, we calculated several models 11E1Y6 assuming mixing up to $M_r/M_\odot = 0.07$ (no mixing), 3.4, 4.6 (outer edge of helium layer), and 6.0. For the mixed layer, we assume uniform composition. Rayleigh-Taylor instability may also form clumpy medium and

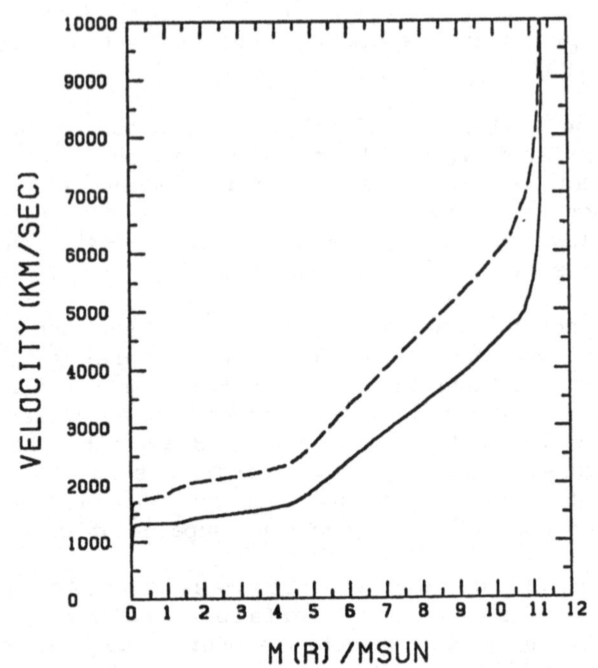

Figure 3: Velocity distribution for Models 11E1 (solid) and 11E2 (dashed) at t = 116 d.

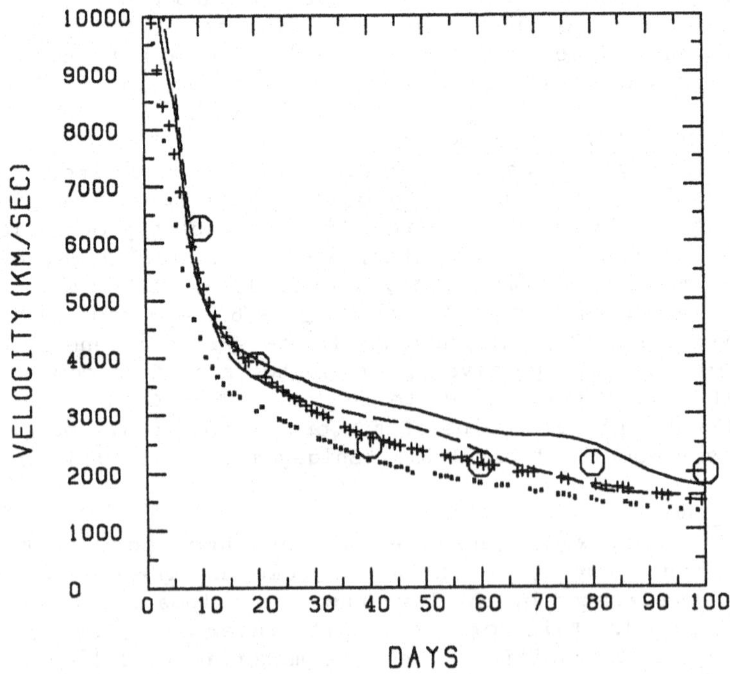

Figure 4: Changes in the expansion velocity of the material at the photosphere for Models 11E1Y6 (solid: $E = 1 \times 10^{51}$ erg, $M = 11.3\ M_{\odot}$, $R_0 = 3 \times 10^{12}$ cm, helium abundance in the envelope Y = 0.6) and 11E1Y4 (dashed: Y = 0.4). Observed values are r_{ph}/t obtained at SAAO (+ mark; Catchpole et al. 1987) and CTIO (small x mark; Hamuy et al. 1987). Radial velocity measurements for the absorption minimum of Fe II 5018, 5169 lines are plotted by open circles (Phillips 1987).

could cause a leak of γ- and X-rays; this might be crucial for the optical and X-ray light curves (see §5.1 and Kumagai et al. 1987).

3.2 Shock propagation time

The shock wave arrives at the original surface of the star at t_{prop}, which is approximated for different values of the initial radius R_0 and the ejected mass M, and for different explosion energies E as (Shigeyama et al. 1987)

$$t_{prop} \sim 2 \text{ hr } (R_0/3 \times 10^{12} \text{ cm}) \, [(M/10 \, M_\odot)/(E/1 \times 10^{51} \text{ erg})]^{1/2}. \quad (1)$$

The condition $t_{prop} < 3$ hr (Kamiokande time is used) rules out all the models with too large R_0 and too low E/M (Shigeyama et al. 1987). A progenitor radius larger than 4.5×10^{12} cm is ruled out for Model 11E1 simply from this condition. If $R_0 = 3.5 \times 10^{12}$ cm, E/M should be larger than 0.6×10^{50} erg/$M_\odot$.

3.3 Velocity profile

After the shock wave reaches the surface, the star starts to expand and soon the expansion becomes homologous as $v \propto r$. In Fig. 3, the velocity distribution for the homologous expansion is shown for 11E1 and 11E2. The velocity gradient with respect to the enclosed mass, M_r, is very steep near the surface, while it is almost flat in the helium layer and the heavy element core (see Table I). This is because the core material is decelerated and forms a dense shell due to the reverse shock when the expanding core hits the hydrogen-rich envelope. The expansion velocities of the helium and heavy element layers are so small that the kinetic energy of these layers is only 10 % of the total kinetic energy for 11E1.

Expansion velocity has been measured and give important constraints on the above hydrodynamical models (Woosley 1988). The velocity at the bottom of H-rich envelope should be lower than the lowest velocity of the hydrogen line which is 2100 km s^{-1} from Bγline (Phillips 1987) and 2200 km s^{-1} from Pβline (Dopita et al. 1987). Table I shows that models with $E/M_{env} \gtrsim 2.2 \times 10^{50}$ erg/$M_\odot$ are ruled out.

Table I. Expansion velocities for $M_{env} = 6.7 \, M_\odot$

Model	Explosion energy	Bottom of H-rich envelope	Bottom of helium layer	Bottom of oxygen layer
11E1	1.0×10^{51} erg	1700 km s^{-1}	1500 km s^{-1}	1300 km s^{-1}
11E1.5	1.5×10^{51} erg	2100 km s^{-1}	1800 km s^{-1}	1500 km s^{-1}
11E2	2.0×10^{51} erg	2400 km s^{-1}	2100 km s^{-1}	1700 km s^{-1}

3.4 Photospheric velocity

As the star expands, the photosphere moves inward in M_r. Because of the steep velocity gradient near the surface, the velocity of the material at the photosphere decreases as seen in Fig. 4 for Models 11E1Y4 and 11E1Y6. This may be compared with $v_{ph} = r_{ph}/t$ where r_{ph} is

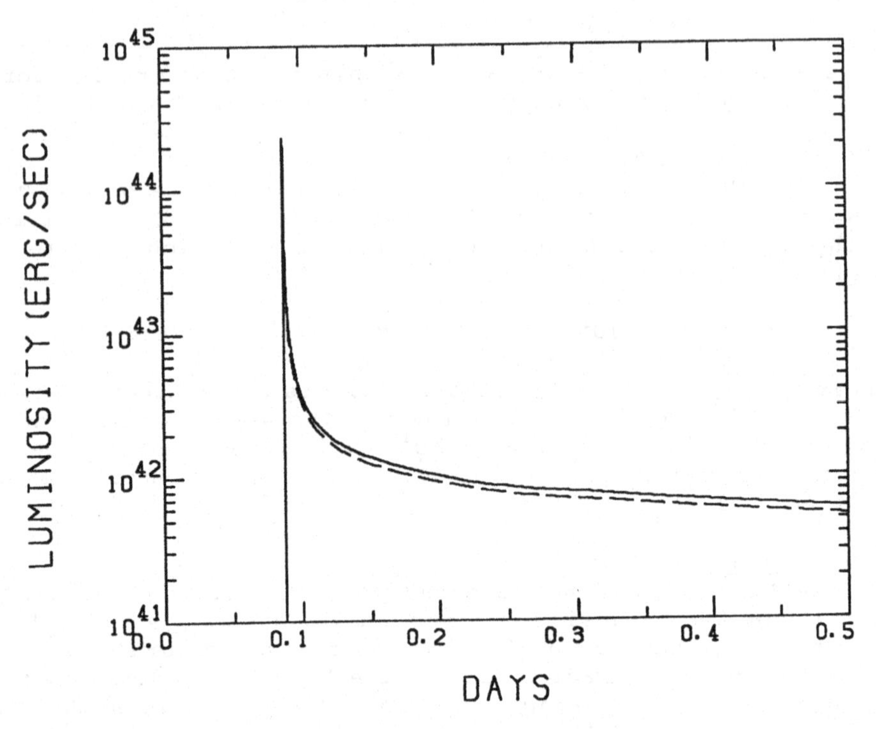

Figure 5: Change in the bolometric luminosity near the shock break out at the surface for 11E1Y6 (solid) and 11E1Y4 (dashed).

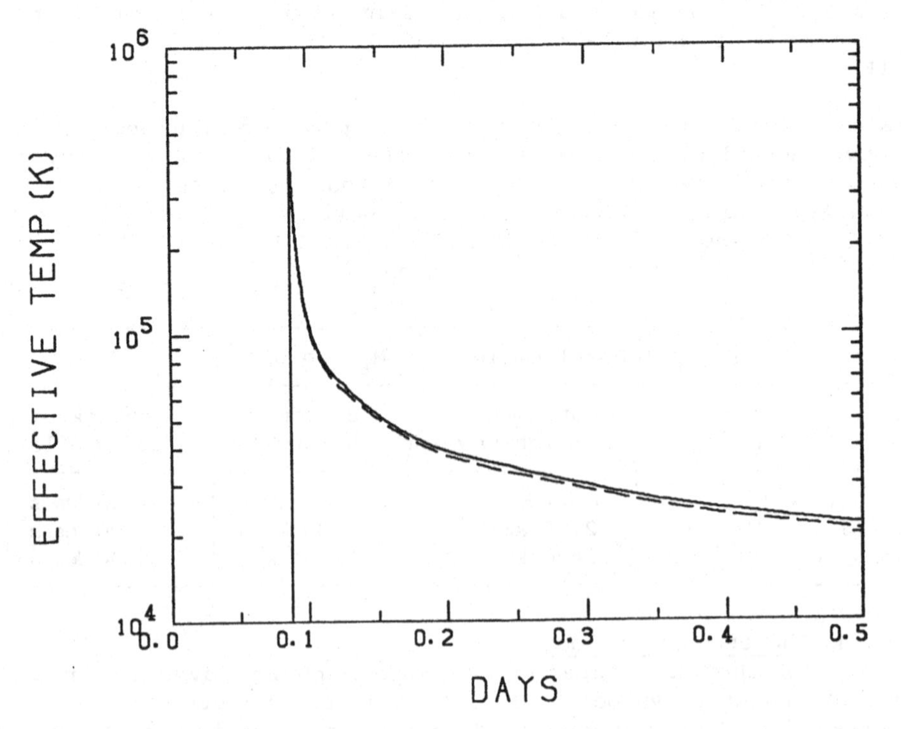

Figure 6: Same as Fig. 5 but for the effective temperature.

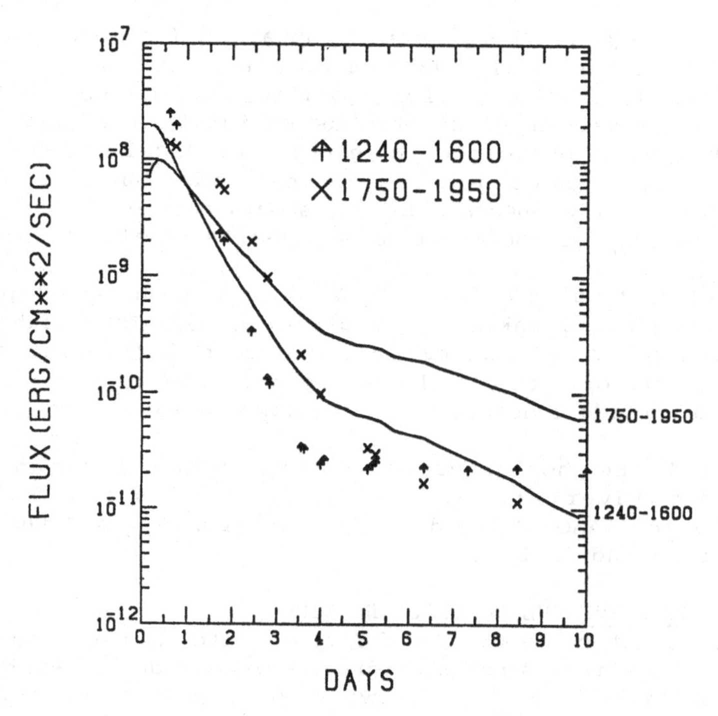

Figure 7: UV light curves for 11E1Y6 compared with the IUE SWR data.

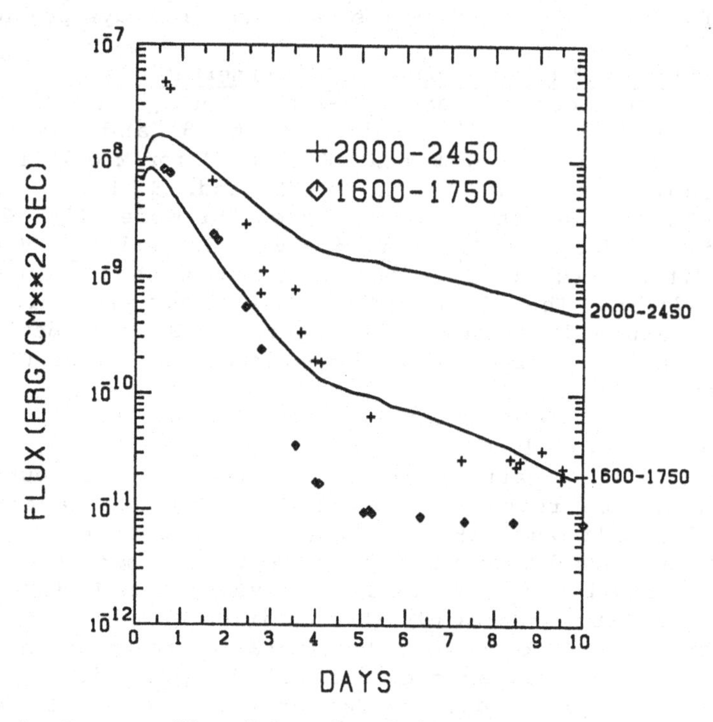

Figure 8: Same as Fig. 7 but for different energy bands.

the radius of the black body surface obtained from the photometric observations (Menzies et al. 1987; Catchpole et al. 1987; Hamuy et al. 1987). Approximate photospheric velocities are also obtained from the radial velocities measured for the absorption minima of Fe II lines (Phillips 1987). The photospheric velocity of Model 11E1Y4 is in good agreement with SAAO data and Fe II line velocities but slightly larger than CTIO values. This suggests E/M_{env} should not be so different from $\sim$ 1.5 x 10^{50} erg/$M_{\odot}$ for the model to be consistent with observations.

It is interesting to note that v_{ph} depends on Y in the envelope because the helium-line opacity makes r_{ph} larger (Fig. 4). This might indicate that the envelope is not extremely helium-rich, although uncertainties are involved in the opacity and the determination of r_{ph}. A more careful analysis could provide constraints on the composition of the envelope.

In Model 11E1Y4, the photosphere entered the helium layer and then the heavy element layer around t = 90 d (70 d) and 110 d (80 d), respectively, for SAAO (CTIO) data as seen from v_{ph} and the velocity profile in Fig. 5 and Table I.

4 LIGHT CURVE DUE TO SHOCK HEATING.

The light curve of SN 1987A show quite unique features, which give important constraints on the hydrodynamical model of explosion. In the theoretical model, the light curve is powered by two energy sources. 1) The shock wave initially establish the radiation field with energy of roughly a half of explosion energy E; e.g., 4.4 x 10^{50} erg for 11E1. The early light curve up to t $\sim$ 25 d can be accounted for by diffusive release of this energy. 2) Afterwards radioactive decays provide energy.

4.1 UV flash and comparison with IUE observations

When the shock wave arrives at the photosphere, the bolometric luminosity reaches 2 x 10^{44} erg s^{-1} (Fig. 5) and the effective temperature becomes as high as 5 x 10^5 K (Fig. 6) for Model 11E1. Hence, most of the radiation is emitted in the UV band. In Figs. 7 and 8, the black body UV light curves are constructed for model 11E1Y6 taking a distance modulus of 18.5, A_v = 0.6, E(B-V) = 0.2 and the UV reddening curve from Fitzpatrick (1985), and compared with IUE observations (Kirshner 1987a,b). The black body UV light curves decline as the surface temperature decreases. The observed IUE curves decline faster than that, which is ascribed to the line blanketing effect (Fransson et al. 1987; Lucy 1987).

4.2 Early optical light

Two unique features are: 1) It took only 3 hours for the visual magnitude to reach 6.4 magnitude after the neutrino burst (McNaught 1987; Zoltowski 1987). 2) At the subsequent plateau, the optical light was much dimmer than typical Type II supernovae. The steep rise in luminosity in 3 hr requires relatively large E/M and small R_0 (Shigeyama et al. 1987). Certainly the condition of t_{prop} < 3 hr (Eq. 1) should be satisfied. In order for the optical flare-up of the supernova to be seen at t = 3 hr, the ejected gas and the radiation field should have expanded rapidly so that the temperature becomes lower and the radius of the photosphere becomes larger. Therefore, the expansion

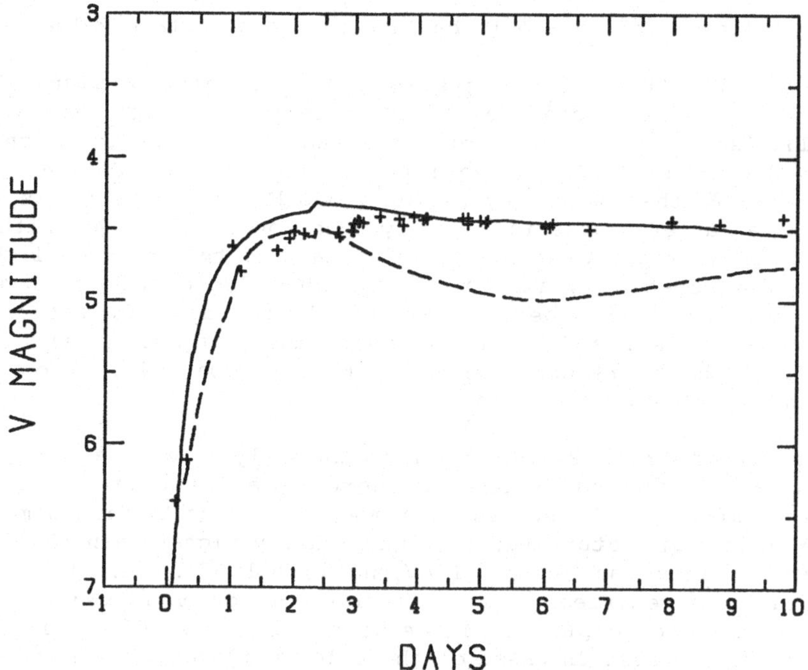

Figure 9: The calculated visual magnitude for Models 11E1Y6 (solid) and 11E1Y4 (dashed). Observed data are taken from ESO (Cristiani et al. 1987), CTIO (Blanco et al. 1987), and SAAO (Menzies et al. 1987) except for the two early points measured on films (McNaught 1987; Zoltowski 1987).

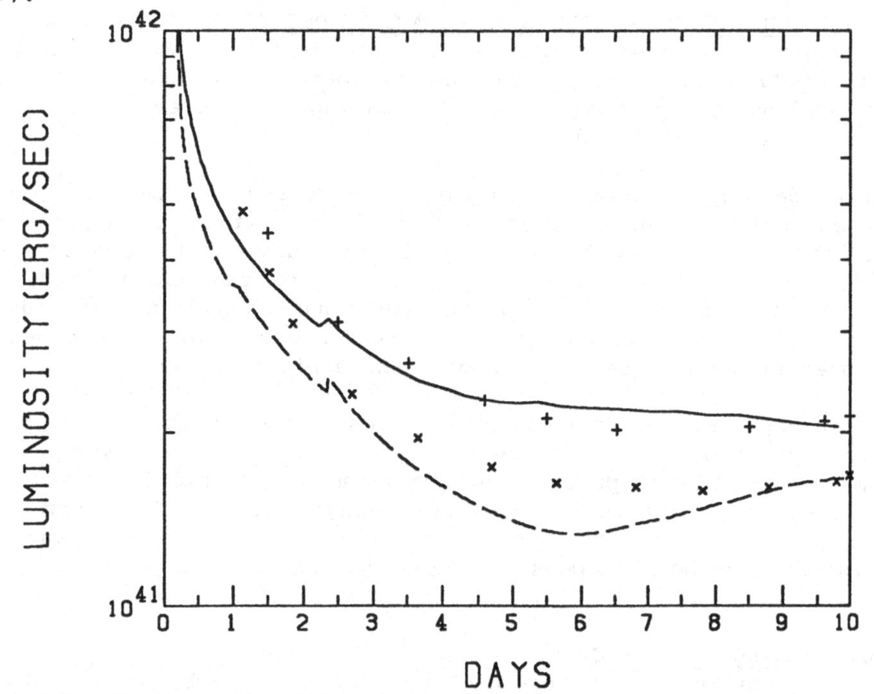

Figure 10: Same as Fig. 9 but for the bolometric luminosity. The solid and dashed lines are Models 11E1Y6 and 11E1Y4, respectively. Observations at SAAO (Menzies et al. 1987; Catchpole et al. 1987) and CTIO (Hamuy et al. 1987) are indicated by + and small x, respectively.

velocity and thus E/M_{env} should be larger than a certain value.

Figures 9 - 11 show the changes in the calculated V magnitude, the bolometric luminosity, and the effective temperature for 11E1Y4 (dashed) and 11E1Y6 (solid). In order for the visual magnitude to reach 6.4 mag in time, $E = 1 \times 10^{51}$ erg is required for $M = 11.3\ M_{\odot}$. For the envelope with smaller Y, the luminosity is lower and hence a larger E is required because of a larger scattering opacity. Moreover, the model with Y = 0.6 is in better agreement with the overall shape of the observed light curve than Y = 0.4 especially at t > 3 d. Such a dependence on Y deserves further careful study, because it would provide information on the presupernova mass loss. Such an enhancement of helium abundance is consistent with the restriction from the envelope solution obtained by Barkat and Wheeler (1987).

Although the theoretical models are generally in good agreement with observations at the early phase, there remain uncertainties in the theoretical models. The supernova atmosphere is scattering dominated so that the color temperature may be significantly higher than the effective temperature (Shigeyama et al. 1987; Hoflich 1987), i.e., diluted black body radiation is emitted. The bolometric correction is sensitive to the color temperature because it is as high as $4 - 6 \times 10^4$ K. If we apply the bolometric correction based on the color temperature, the theoretical visual luminosity is lower than in Fig. 9 so that larger E/M and smaller R_0 are required as derived in Shigeyama et al. (1987). More careful calculations would be required to examine the question regarding the Mt. Blanc time and the early light curve (Wampler et al. 1987; Arnett 1987b).

5 LIGHT CURVE POWERED BY RADIOACTIVE DECAYS

After these early stages, the observations show the increase in the bolometric luminosity. The energy source that continually heats up the expanding star is certainly the decaying ^{56}Co as evident from the light curve tail after 120 d.

The theoretical light curve with radioactive decays of ^{56}Ni and ^{56}Co was calculated assuming the production of 0.07 $M_{\odot}$ ^{56}Ni. Figures 12 - 13 compare the calculated bolometric light curve with observations (Catchpole et al. 1987; Hamuy et al. 1987) for several models. The light curve shape is sensitive to the hydrodynamics and thus is a useful tool to infer 1) the distribution of the heat source ^{56}Ni, 2) the mass of the hydrogen-rich envelope M_{env}, and 3) the explosion energy E.

As the star expands, the photosphere becomes deeper as the recombination front proceeds through the hydrogen-rich envelope deeper in mass. At the same time a heat wave is propagating out from the interior. At a certain stage, energy flux due to radioactive decays exceeds that from shock heating. The dates when the radioactivity starts to dominate and when the luminosity reaches its peak depend on the above three factors 1) - 3) as follows.

5.1 Mixing of Ni - Co

Rayleigh-Taylor instability is likely to lead to mixing of ^{56}Ni into outer layers (§3.1). Figure 12 shows how the light curve

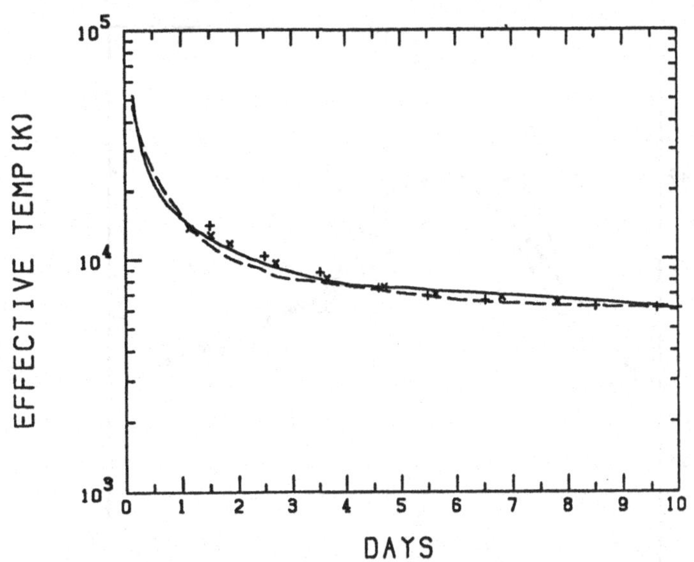

Figure 11: Same as Fig. 10 but for the effective temperature.

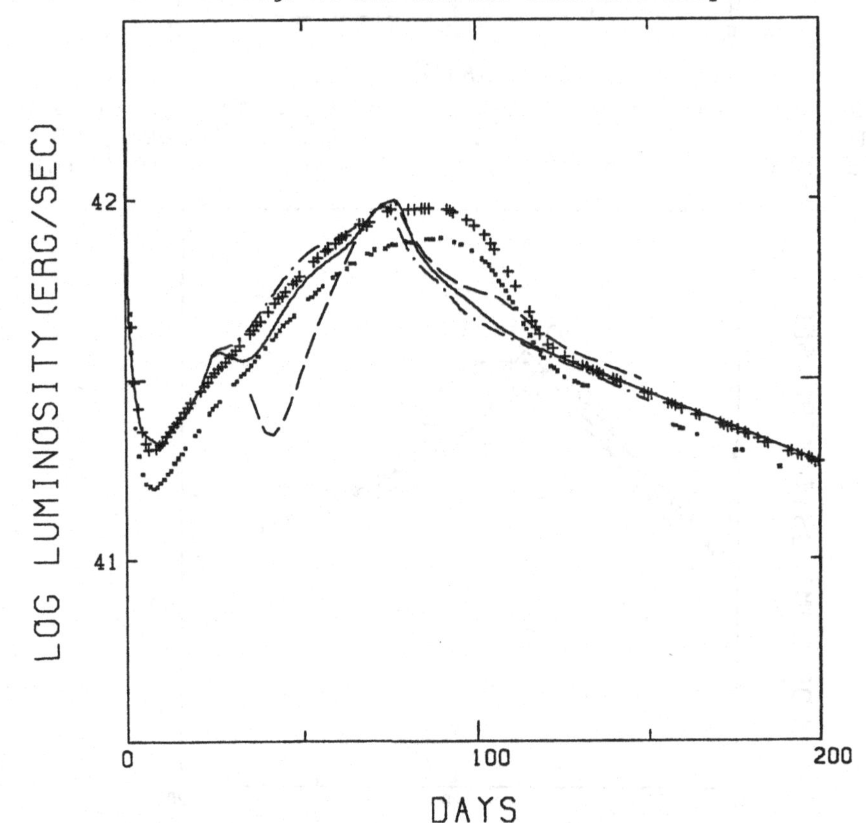

Figure 12: Bolometric light curve for several cases of mixing (11E1Y6). Production of 0.07 $M_\odot$ ^{56}Ni is assumed. For the dashed curve ^{56}Ni is assumed to be confined in the innermost layer while the solid curve assumes that ^{56}Ni is mixed uniformly up to M_r = 4.6 $M_\odot$ (solid) and 6.0 $M_\odot$ (dash-dotted). Observed points are taken from SAAO (+; Catchpole et al. 1987) and CTIO (x; Hamuy et al. 1987).

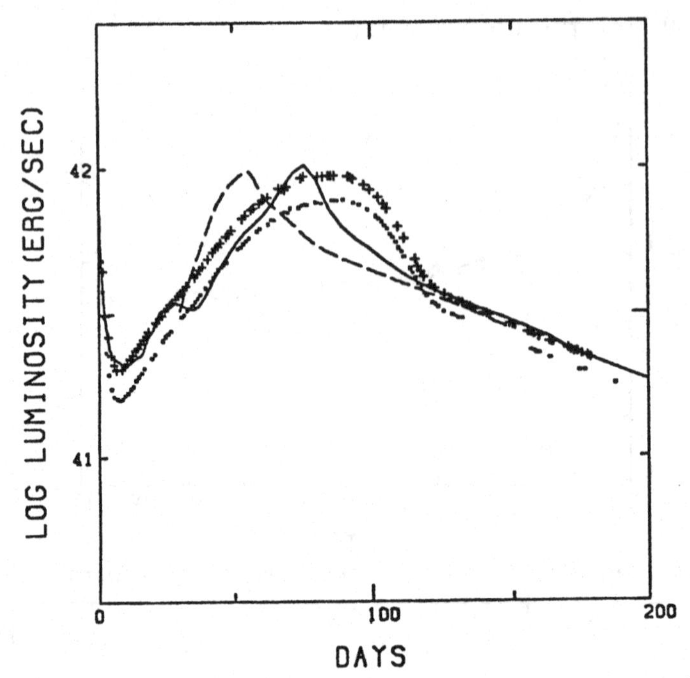

Figure 13: Dependence of the light curve on M_{env}. The solid line is Model 11E1Y6 ($M = 11.3\ M_{\odot}$, $M_{env} = 6.7\ M_{\odot}$; mixing of ^{56}Ni at $M_r < 3.4\ M_{\odot}$) and the dashed is 7E1Y6 ($M = 7\ M_{\odot}$, $M_{env} = 2.4\ M_{\odot}$, $E = 1 \times 10^{51}$ erg).

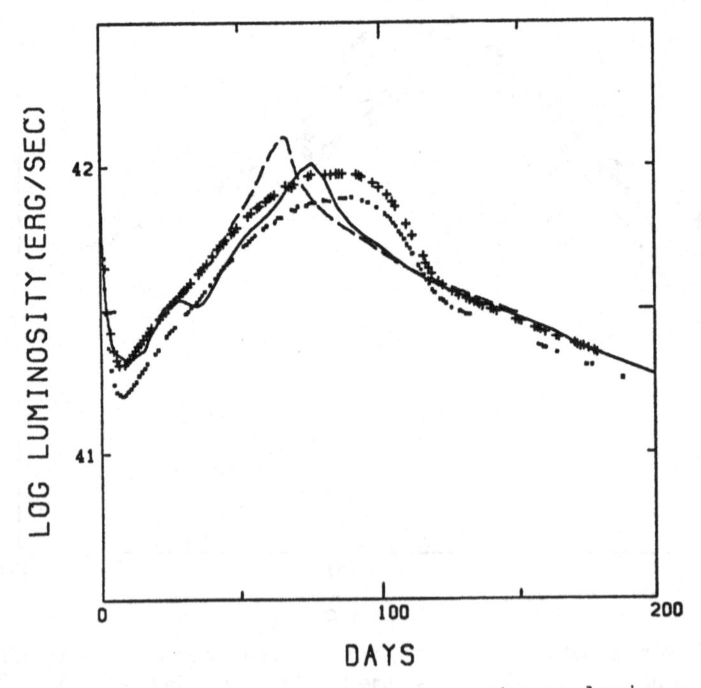

Figure 14: Dependence of the light curve on the explosion energy E. The solid line is the same as in Fig. 13 and the dashed is for Model 11E1.5Y6 ($E = 1.5 \times 10^{51}$ erg).

depends on the distribution of ^{56}Ni by comparing the cases with mixing of ^{56}Ni up to $M_{mix}/M_\odot$ = 0.07 (no mixing; dashed line), 4.6 (outer edge of helium layer; solid), and 6.0 (dashed-dotted) for 11E1Y6. If we assume that ^{56}Ni is confined in the innermost layer of the ejecta, the increase in luminosity due to radioactive heating is delayed to t = 42 d, a dip appears in the curve, and the light curve shape in the rising part is too steep as compared with the observations. On the other hand, if ^{56}Ni is mixed into outer layers, heat is transported to the envelope earlier. As a result the optical light increases earlier, i.e., t = 33 d and 26 d for M_{mix} = 4.6 and 6.0 $M_\odot$, respectively, and the light curve shape is less steep, being in better agreement with the observations.

This may suggest that the heat source had actually been mixed into outer layers and its effect began to dominate the light curve from t $\sim$ 26 d. From the observational side, Phillips (1987) noted that the color changes and kinks started from t = 25 d may indicate the appearance of heat flux due to radioactive decays. More pronounced effects are seen in X- and γ-rays (see Kumagai et al. in these proceedings).

5.2 Envelope mass and Explosion energy

For Model 7E1 with a hydrogen-rich envelope of 2.6 $M_\odot$, a peak luminosity of $\sim 10^{42}$ erg s^{-1} is reached too early at about t = 50 d (Fig. 13), because the expansion velocities of the helium and heavy element layers are larger than in 11E1 and the photosphere approaches the helium layer earlier. Therefore M_{env} should be larger than $\sim$3 $M_\odot$. Probably we need M_{env} = 5 - 8 $M_\odot$ for a good agreement with the observations. This indicates the importance of the deceleration of the expanding core by the hydrogen-rich envelope (Woosley et al. 1988). On the other hand, M_{env} should be lower than a certain value because the nitrogen-rich layer had been lost from the star before the explosion. The important question is if there exists an allowable range of M_{env} to satisfy both conditions.

For Model 11E1.5Y4 (E = 1.5 x 10^{51} erg), the luminosity peak of 1.3 x 10^{42} erg s^{-1} is reached at t = 63 d which is a little too bright and too early compared with the observation (Fig. 14). The cases with E = 1.0 - 1.3 x 10^{51} erg show a better fit to the observations. This constraint is consistent with the condition obtained from the photospheric velocity.

5.3 Peak and tail

For Model 11E1, the bolometric luminosity reaches a peak value of L_{pk} = 1 x 10^{42} at t = 77 d. After the peak, the luminosity decreases more rapidly than SN 1987A which formed a broad peak. Finally the light curve enters the radioactive tail (Fig. 12).

It should be noted that the observed peak luminosity is higher than the energy generation rate due to Co-decay which gives L = 5.4 x 10^{41} erg s^{-1} at t = 70 d for M_{Co} = 0.07 $M_\odot$. This implies that previously deposited energy from Co-decay is also radiated away during the broad peak of the light curve (Woosley et al. 1988). How this additional energy is radiated away is rather sensitive to the dynamical behavior and to the opacity. The difficulty to reproduce the broad peak may be related to the fact that the total photon energy emitted from the supernova is smaller than 10^{49} erg, less than 1 % of the kinetic energy of the expansion (Catchpole et al. 1987; Hamuy et al. 1987). The optical flux may easily be affected by slight changes in the hydrodynamics. Also the flux-limited diffusion approximation used in the calculation needs to be improved. Contributions due to oxygen recombination to the light curve (Schaeffer et al. 1987) were found to be negligible in our calculation.

Despite this difficulty, both the peak luminosity and the tail can be consistently accounted for by the radioactive decay of ^{56}Co. The amount of ^{56}Ni initially produced is estimated to be 0.07 $M_{\odot}$.

5.4 Deviation from the exponential decline

As the optical depth of the expanding star becomes thin, more X- and γ-rays escape from the supernova. Eventually its effect will appear in the optical light curve which is currently following the exponential decline due to the ^{56}Co decay. X-rays have been observed by Ginga (Dotani et al. 1987) and Kvant (Sunyaev et al. 1987) much earlier than predicted. This may be interpreted by the mixing of ^{56}Co into outer layers (M. Itoh et al. 1987). Such a mixing will certainly affect the optical light curve as well as the γ - and X-ray light curves. Figure 15 shows the change in the luminosity above 5 keV for two cases of ^{56}Co mixing, i.e., $M_{mix}/M_{\odot}$ = 4.6 (dash-dotted) and 6.0 (dashed) (taken from the calculation by Kumagai et al. 1987). The peak of γ-ray flux will be reached at t ∿ 300 d and 350 d for the latter and former cases, respectively. The resulting deviation of the optical bolometric light curve from the exponential decline will be significant starting from t ∿ 300 d (Fig. 15; also Ebisuzaki and Shibazaki 1988).

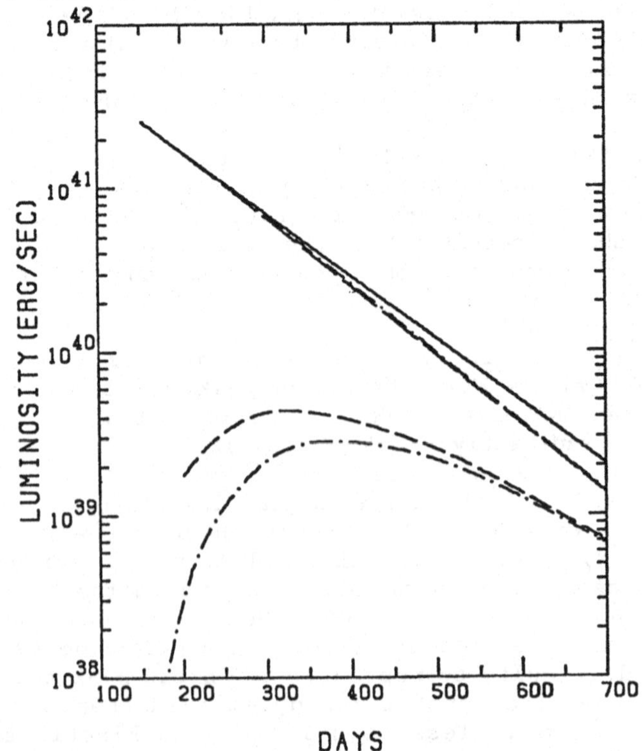

Figure 15: Change in the X- and gamma-ray luminosity above 5 keV for M_{mix} = 6.0 $M_{\odot}$ (dashed) and 4.6 $M_{\odot}$ (dash-dotted). The associated changes in the optical (bolometric) luminosity are compared with the energy generation rate due to ^{56}Co decay (solid line).

6 CONCLUDING REMARKS

Comparison between the hydrodynamical models and the observations imposes several interesting constraints on M_{env} and E, i.e., on the presupernova mass loss history and the explosion mechanism. These are summarized as follows:

1) The very early light curve gives a lower bound of E/M_{env} (§ 4.1), while the later light curve and the slowest hydrogen velocity give an upper bound of E/M_{env}. Reasonable agreement between the model and observations is obtained for $E/M_{env} = 1.5 \pm 0.5 \times 10^{50}$ erg/$M_{\odot}$ (see also Woosley 1988).

2) The pre-maximum light curve sets a lower bound of M_{env} which is about 3 $M_{\odot}$. The upper bound of M_{env} may be obtained from the observed nitrogen to carbon ratio in the circumstellar shell observed with IUE. However, N/C ratio at the surface depends on convection, etc., and the observed ratio involves a large uncertainty. At the moment, therefore, it does not give a clear upper limit to M_{env}. Kirshner (1987a) and Dopita (1988) also obtained a relatively large M_{env}. Such a large M_{env} is inconsistent with the evolutionary models by Maeder (1987b) and Wood and Faulkner (1987) where M_{env} is smaller than 1 $M_{\odot}$.

The suggested explosion energy with $E/M_{env} = 1.5 \pm 0.5 \times 10^{50}$ erg/$M_{\odot}$ could be obtained by both the prompt (Baron et al. 1987) and the delayed explosion mechanism (Wilson and Mayle 1987). The mass cut that divides the neutron star and the ejecta may be at $\sim$1.6 $M_{\odot}$ irrespectively of the explosion mechanism, because a very steep density jump appears at $M_r \sim$ 1.6 $M_{\odot}$ in the 6 $M_{\odot}$ helium core model (Nomoto et al 1987a).

We would like to thank Drs. R. Catchpole and R. Williams for sending us data taken at SAAO and CTIO, respectively, prior to publication and Dr. R. Kirshner for IUE data. A part of this work is based on the collaboration with Dr. M. Hashimoto, Mr. M. Itoh, and Ms. S. Kumagai. We are grateful them for their contribution. This work has been supported in part by the Grant-in-Aid for Scientific Research (62540183) of the Ministry of Education, Science, and Culture in Japan and by the Space Data Analysis Center, Institute of Space and Astronautical Sciences.

References

Arnett, W.D. (1987a). Astrophys. J., **319**, 136.
Arnett, W.D. (1987b). in Danziger, p. 373.
Barkat, Z., Wheeler, J.C. (1987). preprint.
Baron, E., Bethe, H., Brown, G.E., Cooperstein, J., Kahana, S. (1987).
Blanco, V.M. et al. (1987). Astrophys. J., **320**, 589.
Catchpole, R. et al. (1987). M.N.R.A.S., in press.
Cristiani, S. et al. (1987). Astron. Astrophys., **177**, L5.
Danziger, I.J. (ed.) (1987). Proc. of ESO workshop on SN 1987A (ESO).
Danziger, I.J. et al. (1987). Astron. Astrophys., **177**, L13.
Dopita, M. (1988). Nature, submitted.
Dopita, M. et al. (1987). preprint.
Dotani, T., Ginga team (1987). Nature, **330**, 230.
Ebisuzaki, T., Shibazaki, N. (1988). Astrophys. J. Lett., submitted.
Fitzpatrick, E. (1985). Astrophys. J., **299**, 219.

Fransson, C. et al. (1987). Astron. Astrophys., **177**, L33.
Hamuy, M., Suntzeff, N.B., Gonzalez, R., Martin, G. (1987). Astron. J., **94**, in press.
Hashimoto, M., Nomoto K. (1987). in preparation.
Hillebrandt, W., Hoflich, P., Truran, J.W., Weiss, A. (1987). Nature, **327**, 597.
Hoflich, P. (1987). in IAU Colloq. 108, Atmospheric Diagnostics of Stellar Evolution, ed. K. Nomoto (Springer), in press.
Huebner, W., Merts, A., Magee, M., Jr., Argo, M. (1977). Los Alamos Sci. Lab. Rept. No. LA6760M; Opacity table for Nomoto mixture.
Itoh, H., Hayakawa, S., Masai, K., Nomoto, K. (1987). Publ. Astron. Soc. Japan, **39**, 529.
Itoh, M., Kumagai, S., Shigeyama, T., Nomoto, K., Nishimura, J. (1987). Nature, **330**, 233.
Kirshner, R.P. (1987a). in Danziger, p. 121.
Kirshner, R. P. (1987b). private communication.
Kirshner, R.P., Sonneborn, G., Crenshaw, D.M., Nassiopoulos, G.E. (1987). Astrophys. J., **320**, 602.
Kumagai, S., Itoh, M., Shigeyama, T., Nomoto, K., Nishimura, J. (1987). in these proceedings.
Lucy, L.B. (1987). Astron. Astrophys., **182**, L13.
Maeder, A. (1987a). Astron. Astrophys., **173**, 247.
Maeder, A. (1987b). in Danziger, p. 251.
Masai, K., Hayakawa, S., Itoh, H., Nomoto, K. (1987). Nature, **330**, 235.
McCray, R., Shull, J.M., Sutherland, P. (1987). AP. J. Lett., **317**, L73.
McNaught, R.H. (1987). IAU Circ. No. 4389.
Menzies, J.M. et al. (1987). M.N.R.A.S., **227**, 39P.
Nomoto, K., Hashimoto, M. (1986). Prog. Part. Nucl. Phys., **17**, 267.
Nomoto, K., Hashimoto, M. (1987a). in Proc. Japan-France Seminar on Chemical Evolution of Galaxies with Active Star Formation, Sci. Rep. Tohoku Univ., **7**, 259.
Nomoto, K., Hashimoto, M. (1987b). in Proc. Bethe Conference on Supernovae, ed. G.E. Brown, Physics Report, in press.
Nomoto, K., Shigeyama, T., Hashimoto, M. (1987a). in Danziger, p. 325.
Nomoto, K., Shigeyama, T., Hayakawa, S., Itoh, H., Masai, K. (1987b). in these proceedings.
Phillips, M.M. (1987). in these proceedings.
Schaeffer, R., Casse, M., Mochkovitch, R., Cahen, S. (1987). Astron. Astrophys., **184**, L1.
Shigeyama, T., Nomoto, K., Hashimoto, M. (1988). Astron. Ap., submitted.
Shigeyama, T., Nomoto, K., Hashimoto, M., Sugimoto, D. (1987). Nature, **328**, 320.
Sunyaev, R. et al. (1987). Nature, **330**, 227.
Wampler, E.J., Truran, J.W., Lucy, L.B., Hoflich P., Hillebrandt, W. (1987). Astron. Astrophys., **182**, L51.
Wilson J.R., Mayle, R. (1987). in Proc. Bethe Conference on Supernovae, ed. G.E. Brown, Physics Report, in press.
Wood, P.R., Faulkner, D.J. (1987). in Proc. Astron. Soc. Australia, **7**.
Woosley, S.E. (1988). Astrophys. J., submitted.
Woosley, S.E., Pinto, P.A., Ensman, L. (1988). Astrophys. J., **324**.
Woosley, S.E., Pinto, P.A., Martin, P.G., Weaver, T.A. (1987). Astrophys. J., **318**, 664.
Żoltowski, F. (1987). IAU Circ. No. 4389.

Supernova 1987a: A Model and Its Predictions

S. E. Woosley
Board of Studies in Astronomy and Astrophysics
Lick Observatory, University of California at Santa Cruz
Santa Cruz CA 95064

Abstract. *Attention is focused upon a single model that is most nearly consistent with all observations thus far, and its predictions regarding future x-ray and γ-ray emission. The model is a 20 $M_\odot$ star, presumed to have lost 4 $M_\odot$ during its helium burning stage as a red supergiant. The star explodes with a kinetic energy of* 1.45×10^{51} *erg and leaves behind a neutron star of gravitational mass 1.4 $M_\odot$. Better agreement with the optical light curve and x-ray spectrum is achieved if mixing has occurred during the explosion so that a small fraction of ^{56}Ni exists at velocities as great as 3000 km s^{-1}. The assumed mixing does not imply complete homogenization within the mixed region but a gradient of heavy elements. The hard x-ray flux of this model turns on at a detectable level in mid-August and remains nearly constant for 200 days thereafter. Peak γ-line flux at 847 keV is* 6×10^{-4} *cm*$^{-2}$ s^{-1} *at day 450.*

I. Introduction

During the 8 months since its light arrived many models have been put forward for Supernova 1987a (henceforth 1987a). Early on, especially when the nature of the progenitor star was in doubt, many of these models were quite creative, even marvelous. Some were close to correct and have changed very little; others have evolved or fallen by the wayside. This is not the place for an historical accounting. However, by now there has been a convergence so that several groups support very similar scenarios for the evolution and explosion of Sk -69 202. Detailed descriptions have been provided by Woosley, Pinto, and Ensman (1987); Woosley (1988ab);

Arnett (1987 and this volume); Truran (this volume); Nomoto and Shigeyama (this volume); and Nomoto, Shigeyama, and Hasimoto (1987). X-ray and γ-line fluxes have also been calculated for some of these models by a number of groups, notably Pinto and Woosley (1988); Ebisuzaki and Shibazaki (1988); Shibazaki and Ebisuzaki (1988); Itoh *et al.* (1987); Xu *et al.* (1987); and Shull and Xu (this volume). Details can be read in these papers.

Here for sake of brevity and clarity, and to avoid duplication of other workshop papers, I present a single model and explore its consequences. There is now general agreement, at least among the papers cited above, that 1987a was the explosion of Sk -69 202, a star within a few solar masses of 20 $M_\odot$ when it lived on the main sequence and which had a helium core near 6 $M_\odot$ at the time it died. At death, its luminosity was near 100,000 $L_\odot$, its temperature near 16,000 K, and its radius near 50 $R_\odot$(Figure 1). Two critical unknowns are the explosion energy and the mass of the hydrogen envelope on the 6 $M_\odot$ helium core at the time it exploded, *i.e.*, how much mass loss occurred. The two are related. Observations can be fit by a lower energy explosion if the envelope mass is less. Woosley (1988a) concludes that the explosion energy and envelope mass were related by $E_{expl} = 8 \times 10^{50}(M_{env}/5\ M_\odot)$ erg so that, for a reasonable range of envelope masses, the explosion energy was within a factor of two of 10^{51} erg (Figure 2). All agree that 0.07 $M_\odot$ of radioactive ^{56}Ni was produced in the explosion. Some disagree upon the exact path taken in the Hertzsprung-Russell diagram, but observations (see *e.g.*, papers by Kirshner and by Walborn, this volume) strongly suggest Sk -69 202 spent a portion of its evolution, presumably most of its helium burning lifetime, as a *red* supergiant, evolving back to the blue just in time to explode (Figure 1; the blueward evolution would have commenced about 20,000 years ago). Many, but not all, model builders attribute the blue nature of the progenitor to the low metallicity of the LMC, but mass loss and uncertain convective algorithms are also possibilities. There are multiple solutions to the stellar structure equations for the envelope of a supergiant star and it is hard to say just which perturbation broke the symmetry in Sk -69 202.

The particular model to be discussed is a 20 $M_\odot$ star that lost 4 $M_\odot$ while a red supergiant and ended its life as a 16 $M_\odot$ star having a 6 $M_\odot$ helium core. The helium core employed was extracted from a previous 20 $M_\odot$ presupernova model (Woosley and Weaver 1986) and capped by a 10 $M_\odot$ envelope constructed separately in thermal and hydrostatic equilibrium with a radius and luminosity appropriate to Sk -69 202. Explosion was simulated in the hybrid configuration by removing the collapsing iron core (1.4 $M_\odot$ with an entropy jump at 1.55 $M_\odot$; Woosley 1988a; Woosley and Weaver 1987) and replacing it with a piston of specified trajectory. Rapid motion of the piston initiated a shock wave that ejected all exterior matter. After a time, when the expansion had become homologous, the total kinetic energy, hereafter referred to as the "explosion energy", could be sampled. For the case considered this energy was 1.45 $\times 10^{51}$ erg. Other quantities of interest for this

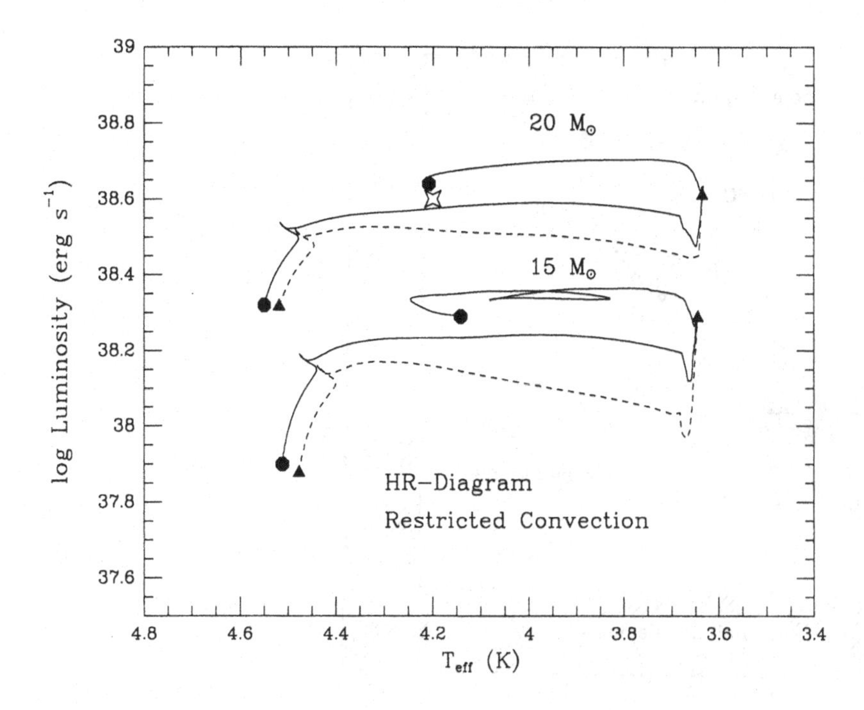

Figure 1. HR-diagram for 15 and 20 $M_{\odot}$ stars having solar (dashed line) and 1/4 solar (solid line) metallicity evolved in a calculation (Woosley 1988a) that employed the LeDoux criterion for convection. The four-pointed star indicates the best estimated properties of Sk -69 202.

Model (*aka* Model 10H) include the time when the shock broke through the surface of the star – 6400 s; the velocity of the slowest moving hydrogen in the unmixed configuration – 1700 km s^{-1}; the amount of ^{56}Ni explosively synthesized – 0.07 $M_{\odot}$ by *fiat*; the time when the recombination traversed the hydrogen envelope and the light curve became radioactive powered – 40 days; and the column depth to the edge of the Ni/Co shell when the supernova was 10^6 s old and expanding homologously – 7.1×10^4 g cm^{-2}.

This model, unmodified, does not give a good fit to the intermediate light curve (20 to 40 days) nor does it present a detectable hard x-ray flux during August when MIR (Sunyaev *et al.* 1987) and GINGA (Dotani *et al.* (1987) first observed the downscattered photons from ^{56}Co decay. Thus the model was arbitrarily and artificially *mixed*. That is the density-velocity profile of Model 10H was retained but the composition was smeared. Specifically a numerical operation was performed wherein the composition extending 2 $M_{\odot}$ out from a given zone boundary was completely homogenized. Marching through the star zone by zone this mixing operation was carried out until the surface of the star was encountered. This procedure retains, in an approximate fashion, the elemental stratification of the original model

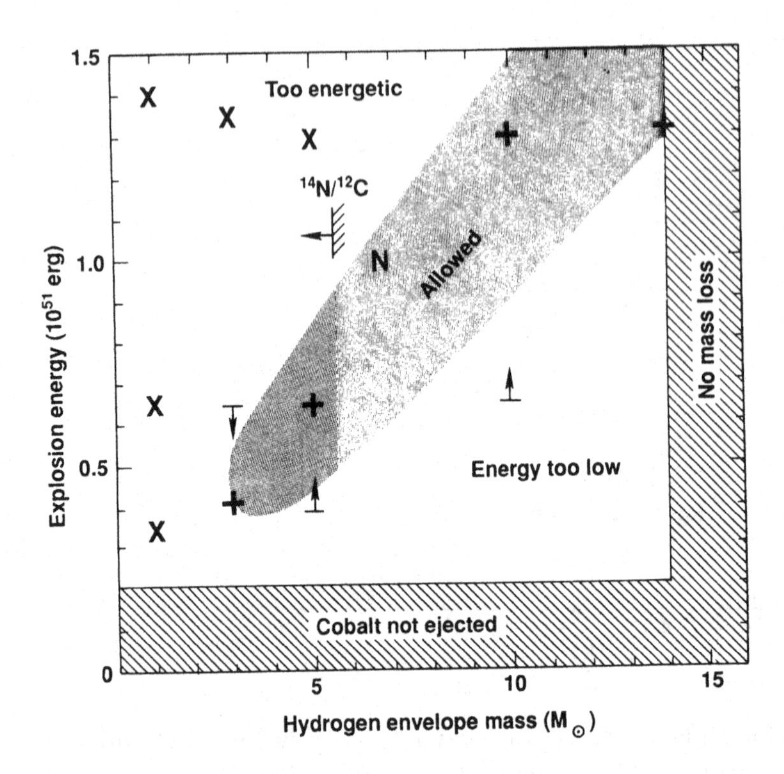

Figure 2. Allowed values of explosion energy and hydrogen envelope mass are broadly delineated for 1987a. Based upon the explosion of a 6 $M_\odot$ core (main sequnce mass 20 $M_\odot$), the atmosphere can be no greater than 14 $M_\odot$. Symbols "X" denote a model that can be excluded on the basis of one or more observational constraints (Woosley 1988a), chiefly too great a velocity for hydrogen at the base of the envelope or a light curve that brightens too early or stays too faint; "+" indicates a moderately successful model; arrows indicate lower and upper bounds provided by three of the models; and "N" a successful model recently published by Nomoto *et al* (1987). Explosion energies below 3×10^{50} erg lead to reimplosion of the core and loss of all ^{56}Co.

(Figure 3) and should be superior to the complete homogenization of a given region employed by Nomoto and Shigeyama and by Arnett (this volume). Clearly a theoretical problem of high priority is the multi-dimensional modelling of the Rayleigh-Taylor instabilities expected to give rise to the mixing during and after the explosion (Woosley 1988a).

Figure 4 shows the asymptotic velocity structure of this model. The abundance of ^{56}Ni at 3000 km s^{-1} (6 $M_\odot$ into the hydrogen envelope) is 2% of its central value. If mixing is due to the decay of ^{56}Ni and its daughter ^{56}Co, the total energy, 9.4×10^{16} erg g^{-1}, is unlikely to propel material to greater velocity. Because of the tamping effect of overlying material and the fact that much of this energy is lost to adiabatic expansion, most of the mass 56 material would not be mixed even this far out.

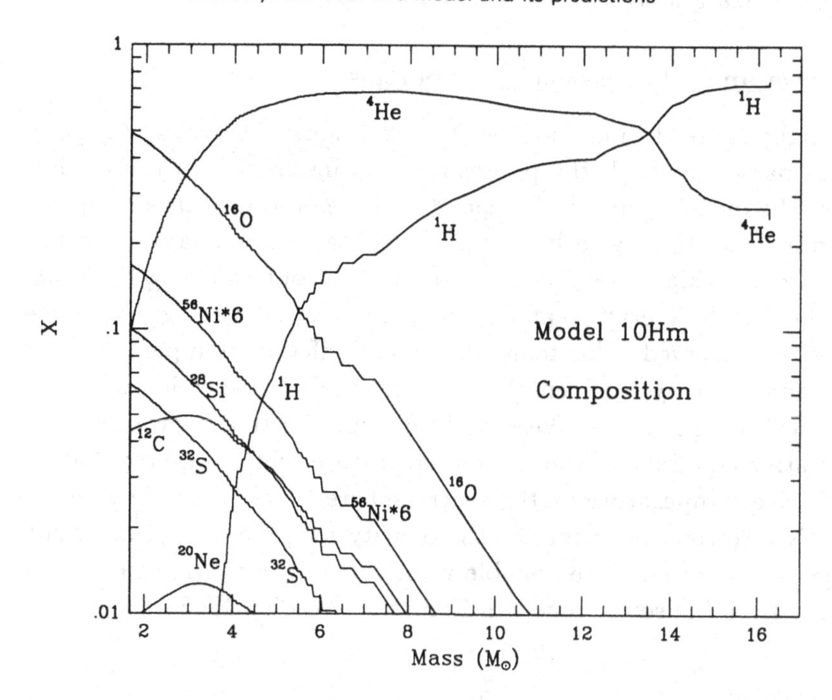

Fig. 3 - Composition of major elements in mixed Model 10HM.

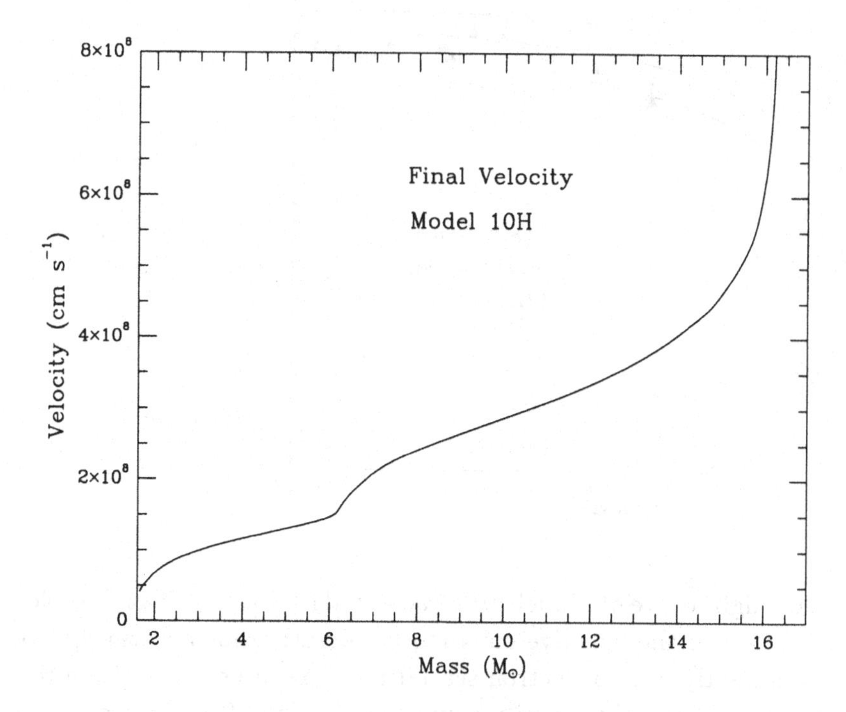

Fig. 4 - Final velocity in Model 10HM.

II. The Light Curve and Photospheric Properties

The light curve during the first 2 days (Figure 5), when only 0.01 $M_{\odot}$ of material has passed through the photosphere, is insensitive to the mixing operation and, once the model is specified, depends only upon the surface composition and opacity employed. At very early times, when observational data regarding the color temperature is lacking, one must be concerned about equating the color temperature and effective temperature at the surface of last scattering. At times later than one day the observed color temperature and effective temperature are in good agreement and one may confidently employ a simple bolometric correction to obtain a theoretical V-light curve. Very early on this is not the case. For an atmosphere in radiative equilibrium the color temperature will be approximately $\beta^{1/8}$ times the effective temperature at the scattersphere where β is the ratio of total opacity (mostly electron scattering) to the opacity for processes that do not preserve the energy of a photon. A reasonable value for β at early times is in the range 0.1 to 1 which gives the set of curves in Figure 5 (Woosley 1988b).

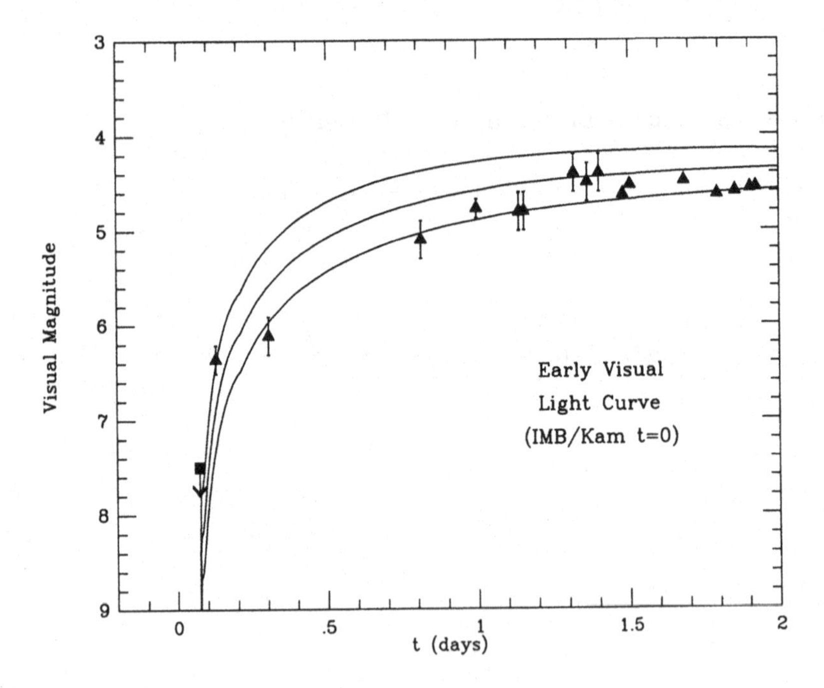

Figure 5. Early visual light curve of Model 10HM adjusted for the fact that the color temperature does not equal the effective emission temperature for an atmosphere whose opacity is dominantly due to electron scattering. The three curves from top to botton have the non-conservative opacity equal to 1, 0.3, and 0.1 of the electron scattering opacity.

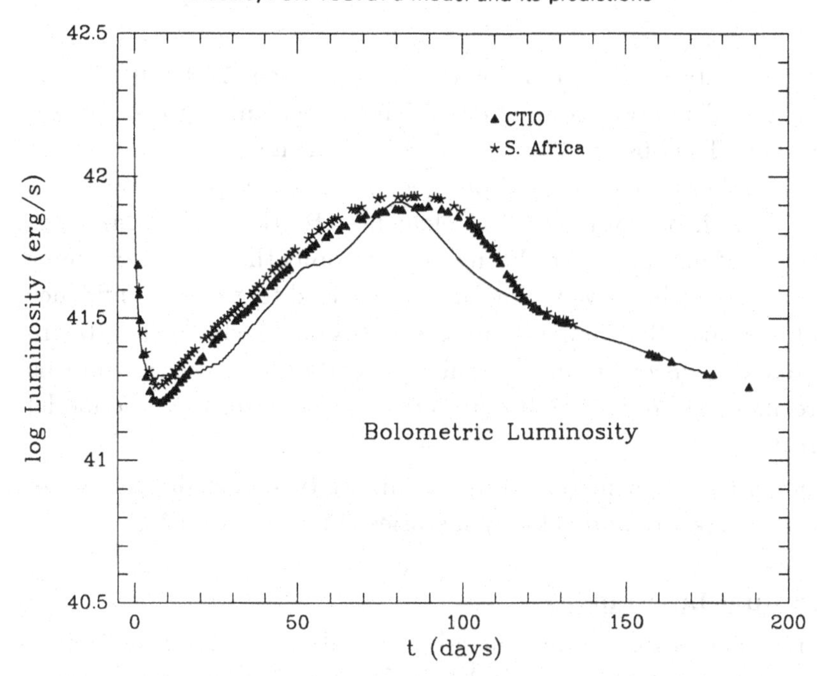

Fig. 6 - Comparison of the observed bolometric light curve to the mixed Model 10HM. Data are from Menzies *et al.* (1987); Catchpole *et al.* (1987); and Hamuy *et al.* (1987).

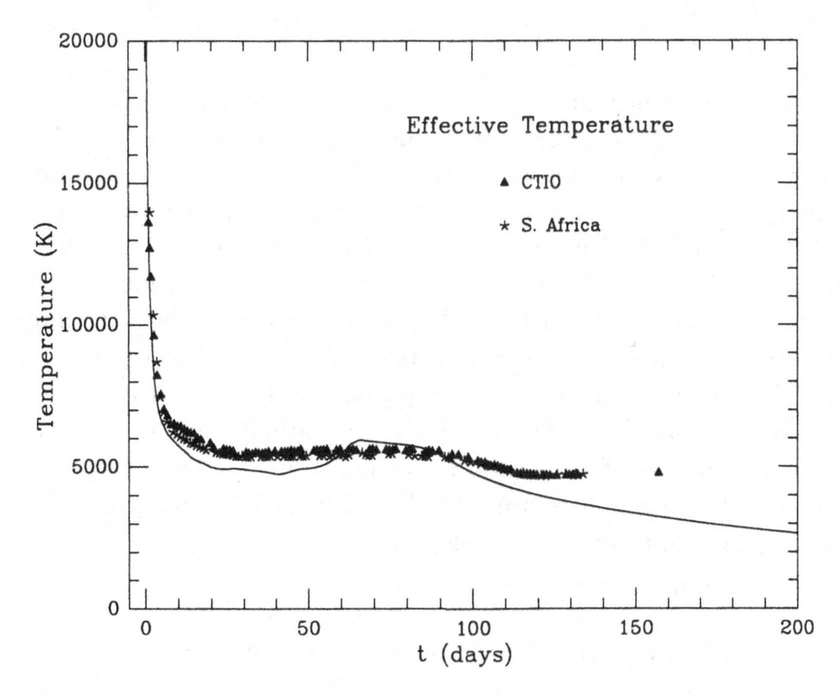

Fig. 7 - Photospheric temperature of Model 10HM compared to observations by Catchpole *et al.* (1987) and Hamuy *et al.* (1987).

Figure 6 shows the bolometric light curve for the mixed Model 10H (10HM) for the first 200 days compared to observational data. The agreement is good though, admittedly, not excellent. The observed peak is broader and smoother. A better fit might be obtained by further adjustments of parameters at our disposal, especially the opacity history for each of several critical elements (H, He, and O) and the extent and completeness of mixing. It has been my philosophy that such parameter variation would be an exercise in curve fitting that might lead to greater confidence in the particulars of the model (flux-limited radiative diffusion, blackbody spectrum, single temperature, one dimension) than is justified. Clearly the observers win this one, but overall agreement in bolometric flux to 50% demonstrates that the model is qualitatively correct.

Confidence in the model is further increased by examining two other diagnostics of the photosphere, its temperature and velocity histories (Figures 7 and 8).

III. X-ray and Gamma-Line Fluxes

The only species which will emit γ-line radiation at a level that might possibly be detected in the next two years is ^{56}Co. The flux from a mass, M_{56}, of ^{56}Co in solar masses located *in the middle of the supernova*, (and assuming a LMC distance of 50 kpc) is

$$F = 0.602 \left(\frac{M_{56}}{0.1 M_\odot}\right) \exp\left(-t/113.6\,\mathrm{d} - \kappa_\gamma\,\phi_o (10^6\mathrm{s})/t)^2\right)\ \mathrm{cm^{-2}\ s^{-1}} \qquad (1)$$

where t is the elapsed time since the explosion, ϕ_o is the column depth to the edge of the centrally located ^{56}Co at age 10^6 s, and κ_γ is the opacity to 1 MeV γ-rays. Here F is the flux of some line, such as 847 keV, through which all decays proceed and homologous expansion has been assumed. An appropriate value of κ_γ is 0.06 cm^2 g^{-1}. For Model 10H (unmixed), ϕ_o is 7.1×10^4 g cm^{-2}. Actually, even in the unmixed model, the ^{56}Co is not centrally located which means that ϕ_o needs to be increased owing to the extra column depth to the polar extremity of the radioactive core (as viewed along a line passing through the center and equator equator). This angle averaging results (approximately) in multiplying ϕ_o by a quantity $[(1+\alpha)/(1-\alpha)]^{1/2} \approx \sqrt{2}$, where α is the ratio of the radius of the cobalt containing region to that of the region providing substantial γ-ray opacity, about 1/3. Hence though the radial column depth of 10H is 7.1×10^4 g cm^{-2}, a value closer to 10^5 is more appropriate to employ in eq. (1).

This flux will have a maximum at time

$$t_{max} = (2\tau_{Co}\,\kappa_\gamma\,\phi_o\,t_o^2)^{1/3} = 263\,(\phi_o/10^4)^{1/3}\ \mathrm{days} \qquad (2)$$

of

$$F_{max} = 0.602 \left(\frac{M_{56}}{0.10\,M_\odot}\right) \exp\left(-0.161\,\phi_o^{1/3}\right)\ \mathrm{cm^{-2}\ s^{-1}}. \qquad (3)$$

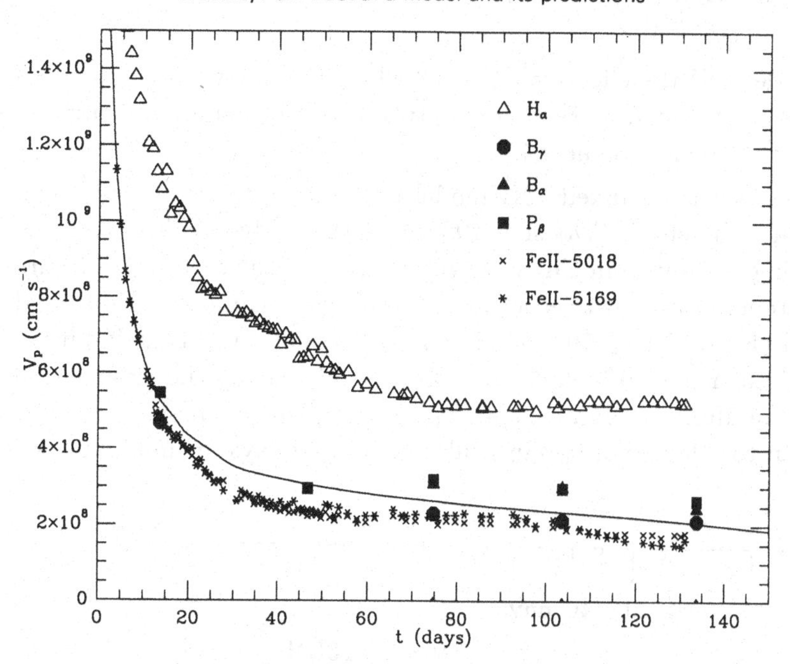

Fig. 8 - Photospheric velocity of Model 10HM compared to observations by Elias and Gregory (1988) and Phillips *et al.* (1987).

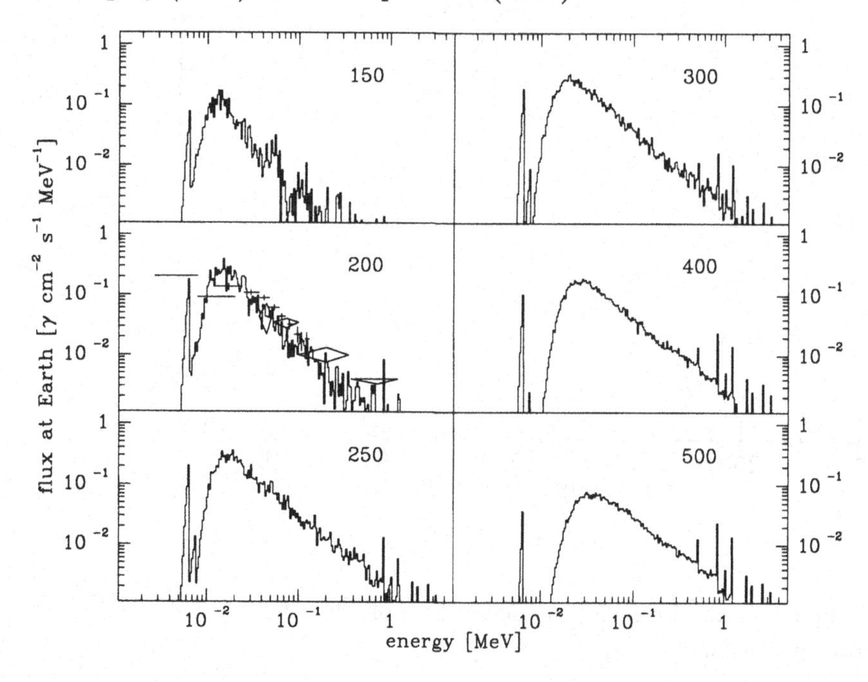

Fig. 9 - X-ray spectra at several times for Model 10HM evaluated at the dates shown from Monte Carlo calculations by Pinto and Woosley (1988). Horizonal error bars show the GINGA sensitivity. Other data points are from MIR (Sunyaev *et al.* 1987; see also papers by Trumper and by Skinner, this volume).

For the unmixed Model 10H the effective ϕ_o is 10^5 which implies a peak γ-line flux of 2.4×10^{-4} cm^{-2} s^{-1} at day 570. The γ-ray optical depth at maximum emission is 1.4 $(\phi_o/10^4 \text{ g cm}^{-2})^{1/3}$, which is near 3.

In fact, as noted before, the unmixed 10H model does not provide either a good match to the optical light curve (Woosley 1988a) or the early break out of the hard x-rays. Assuming mixing as in Figure 3, the simple formulae (eqs. 1 – 3) are no longer appropriate and one must resort to Monte Carlo techniques. Pinto and Woosley (1988) find (Figs. 9 and 10) for Model 10HM that the 847 keV line of ^{56}Co decay will peak near 6×10^{-4} cm^{-2} s^{-1} on day 450. Given the exponential sensitivity of the γ-line flux to column depth and mixing this estimate should be regarded as uncertain to a factor of two in peak flux and 50 days in time of peak.

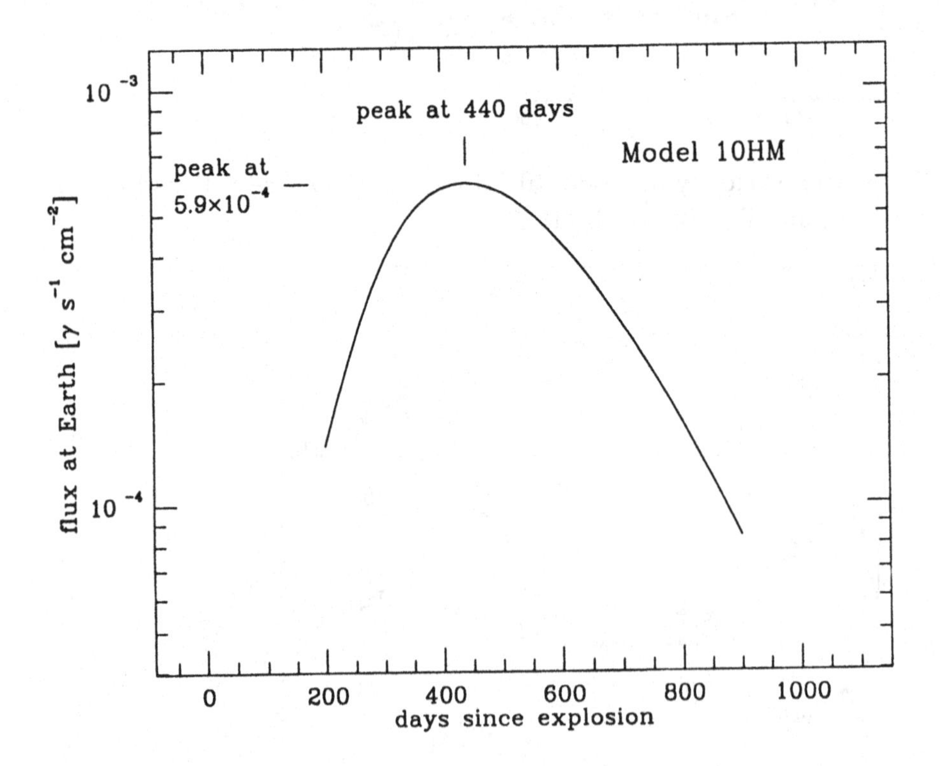

Figure 10. Light curve in the 847 keV decay line of ^{56}Co for Model 10HM (from Pinto and Woosley 1988).

This work has been supported by the National Science Foundation (AST-84-18185).

Bibliography

Arnett, W. D. 1987, *Ap. J.*, **319**, 136.

Catchpole, R. M., Menzies, J. W., Monk, A. S., Wargau, W. F. and 16 others, preprint, South African Astronomical Observatory, *MNRAS*, in press.

Dotani, T., and 36 others 1987, *Nature*, **330**, 230.

Ebisuzaki, T., and Shibazaki, N. 1988, *Ap. J. Lettr.*, in press.

Elias, and Gregory 1988, in preparation for *Ap. J.*

Hamuy, M., Suntzeff, N. B., Gonzalez, R., and Martin, G. 1987, NOAO Preprint No. 102, *Astro. J.*, **94**, in press.

Itoh, M., Kumagai, S., Shigeyama, T., Nomoto, K., and Nishimura, J. 1987, *Nature*, **330**, 233.

Menzies, J. W., Catchpole, R. M., van Vuuren, G., Winkler, H. and 12 others 1987, *MNRAS*, **227**, 39P.

Nomoto, K., Shigeyama, T., and Hashimoto, K. 1987, in *Proc. ESO Workshop. on SN1987a*, ed. J. Danziger, in press.

Phillips, M. M., Heathcote, S. R., Hamuy, M., and Navarrete, M. 1987, CTIO preprint submitted to *Astron. J.*

Pinto, P. A., and Woosley, S. E. 1988, *Ap. J.*, in press.

Shibazaki, N., and Ebisuzaki, T. 1988, *Ap. J. Lettr.*, in press.

Shigeyama, T., Nomoto, K., Hashimoto, K., and Sugimoto, D. 1987, *Nature*, **328**, 320.

Sunyaev, R. and 33 others 1987, *Nature*, **330**, 227.

Woosley, S. E. 1988a, *Ap. J.*, in press.

Woosley, S. E. 1988b, in *Proc. IAU Colloq. No. 108: Atmospheric Diagnostics of Stellar Evolution*, ed. K. Nomoto, (D. Reidel: Dordrecht), in press.

Woosley, S. E., and Weaver, T. A. 1986, in *Radiation Hydrodynamics in Stars and Compact Objects*, ed. D. Mihalas and K.-H. A. Winkler, (Springer Verlag: Berlin), p. 91.

Woosley, S. E., Pinto, P. A., and Ensman, L. E. 1988, *Ap. J.*, **324**, in press.

Xu, Y., Sutherland, P., McCray, R., and Ross, R. R. 1987, submitted to *Ap. J. Lettr.*

SN1987a: CIRCUMSTELLAR AND INTERSTELLAR INTERACTION

R. A. Chevalier
University of Virginia, Charlottesville, Virginia, U.S.A.

Summary. Both observational and theoretical evidence points to the progenitor of SN1987a being Sk-69 202, a B3I star. Such a star is expected to have a wind with a mass loss rate of a few times 10^{-6} $M_\odot$ yr^{-1} and a velocity of 550 km s^{-1}. The outer density profile of the supernova is likely to have a power law form, so the initial interaction of the supernova with the wind can be described by similarity solutions. The expected thermal X-ray emission from the interaction region is below the upper limits initially set by the Ginga satellite. However, radio emission from the shocked region is expected and provides a model for the prompt radio emission that was observed. The radio turn-on is due to the wind outside the shock wave becoming optically thin to free-free absorption. The later turn-on and higher luminosities of other radio supernovae can be attributed to the denser winds expected around red supergiant stars. It is plausible that Sk-69 202 was a red supergiant star in a previous evolutionary phase so denser gas at some distance from the supernova is likely. The fast blue supergiant wind tends to sweep the red supergiant wind into a shell. Ultraviolet emission lines from highly ionized atoms have been observed and are likely to be from ions in the shell that have been ionized by the initial burst of radiation from the supernova. Initial calculations of the line emission suggest a shell radius of (2-3) x 10^{18} cm. If the shell contains dust, the infrared echo from a shell at this radius is expected to be weak. Current infrared observations imply that a dusty shell with a gas mass of $1M_\odot$ must be at a radius of 10^{18} cm or larger; there is not yet conclusive evidence for the presence of an infrared echo. The absence of absorption effects and the thermal spectrum of the soft (< 10 keV) X-ray component observed by the Ginga satellite suggests an origin in circumstellar interaction. The observed time variability then requires shock interaction with clumps embedded in the blue supergiant wind. These may be the result of a Rayleigh-Taylor instability of the wind-driven shell. Finally, the detection of light echoes formed by dust scattering in interstellar clouds near SN1987a is promising. The echoes may appear as luminous overlapping annular regions with a size scale of order one arcmin. These topics are discussed in more detail in Chevalier (1987).

Reference

Chevalier, R. A. (1987). In ESO Workshop on the SN1987a ed. I. J. Danziger, pp. 481-494, Garching: ESO.

THEORETICAL MODELS OF SUPERNOVA 1987A

W. David Arnett
Enrico Fermi Institute, University of Chicago, 5640 S. Ellis Ave., Chicago IL 60637, USA

Abstract. Theoretical models of the early and late light curve and other observations of Supernova 1987a are compared with the data. The hydrostatic evolution of the presupernova is calculated with an extensive and carefully chosen network. The explosion mechanism is modeled from a "long term" (2 minute) hydrodynamic calculation of core collapse. The early ($t \lesssim 20$days) luminosity, effective temperature and fluid velocity at the photosphere are well reproduced for the initial radius ($R \approx 3 \times 10^{12}$ cm) and an explosion energy density of $E/M \approx 0.75 \times 10^{17}$ erg/g. The rapidly rising luminosity is consistent with the Jones limit and the Kamiokande/IMB neutrino detection, but difficult to reconcile with the Mont Blanc event. The action of the supernova shock on the outer layers of the presupernova star generate a density structure which has a (relatively) uniform core and a mantle which falls as a power-law in radius. The power-law region dominates the formation of the spectra. Later ($t \gtrsim 20$ days) evolution requires the presence of 0.7 ± 0.1 $M_\odot$ of ^{56}Ni to be synthesized and ejected by the explosion, which is consistent with the models. The decay of this material to ^{56}Co and subsequently to ^{56}Fe causes a pronounced Rayleigh-Taylor instability in the inner regions of ejecta; this modifies the light curve and predictions of x-ray and γ-ray escape. The accuracy of the exponential decay of the light curve for $t \gtrsim 120$ days places severe limits ($L \lesssim 3 \times 10^{39}$ ergs/s) on the luminosity of the newly formed pulsar.

BEFORE THE EXPLOSION

The evolution of the star prior to explosion is dominated by the nuclear processes in the interior which attempt to balance cooling by neutrino emission. However, a crucial factor in the observed light at early times is a relatively superficial feature: its initial radius. The idea of blue supergiants exploding is not new (e. g., Arnett 1977), but there are a number of factors involved in the star being blue when it explodes. There is no reliable theoretical argument as to which effect (if any) is dominant. Mass loss, mixing, abundances and opacity are probably all involved. The models discussed here are the simplest, having no mass loss, and were chosen for convenience. There is no indication from the data that this choice is less appropriate than any other.

The nuclear reaction network used for C, Ne and O burning is shown in Figure 1. It contains 60 species and was designed on the basis of the results of Arnett & Thielemann (1985) and Thielemann & Arnett (1985). Silicon burning was examined both with a larger network (100 species) and with a typical nucleus approximation.

Fiqure 1. The Nuclear Reaction Network Used .

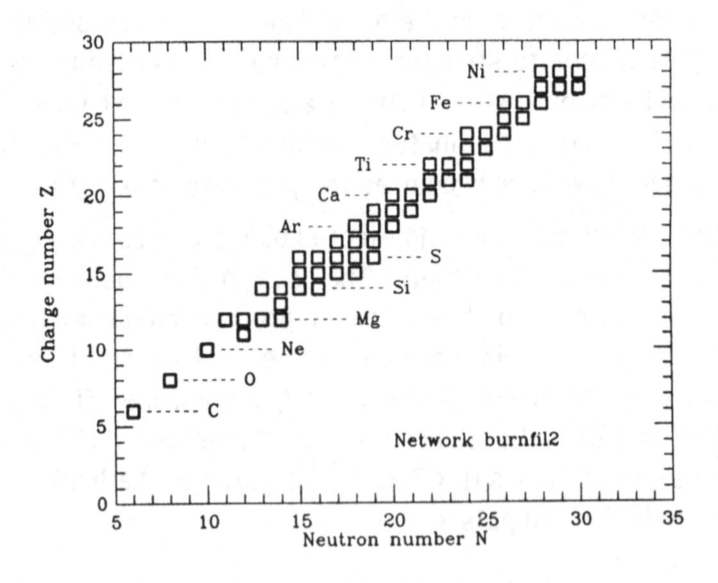

The entire distribution of abundance for a 20 $M_{\odot}$ star with 1/4 solar abundances ("LMC") is shown in Figure 2. In Figure 3 shows in more detail the complex abundance pattern which develops in the inner 3 solar masses. At the time shown the Si flash is occurring at the center of the

Figure 2. Abundance Structure of a 20 $M_\odot$ Star.

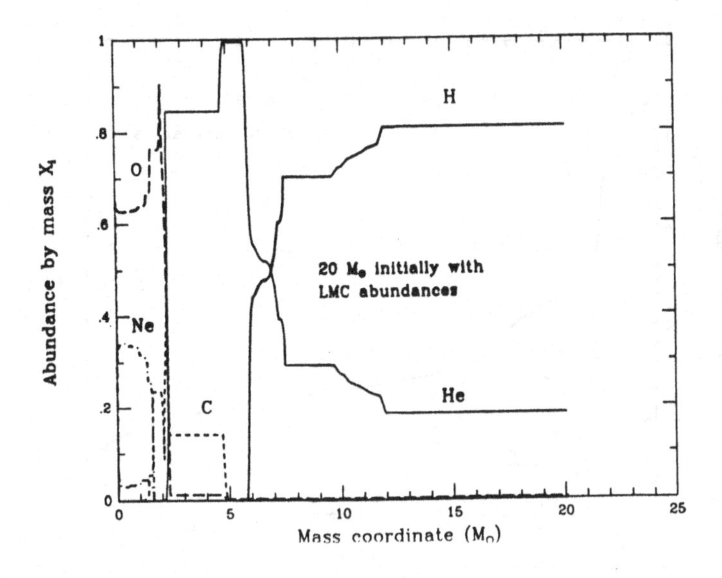

Figure 3. Detailed Abundance in the Inner 3 $M_\odot$.

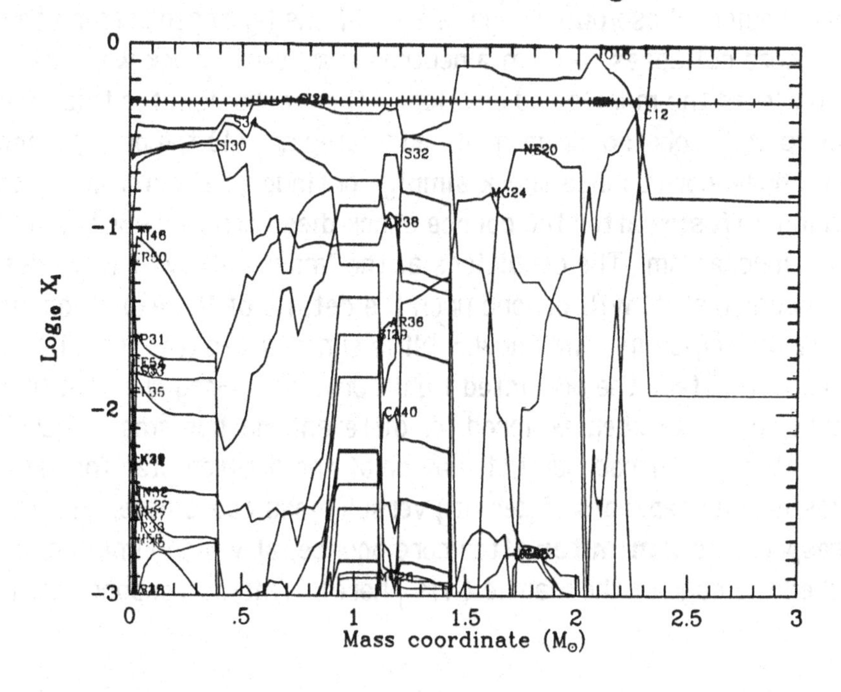

star. Because of electron capture during oxygen burning, $Y_e \rightarrow 0.48$ and this results in ^{30}Si and ^{34}S in high abundance in the core. The star develops a

Figure 4. Typical Density Structure for a Presupernova.

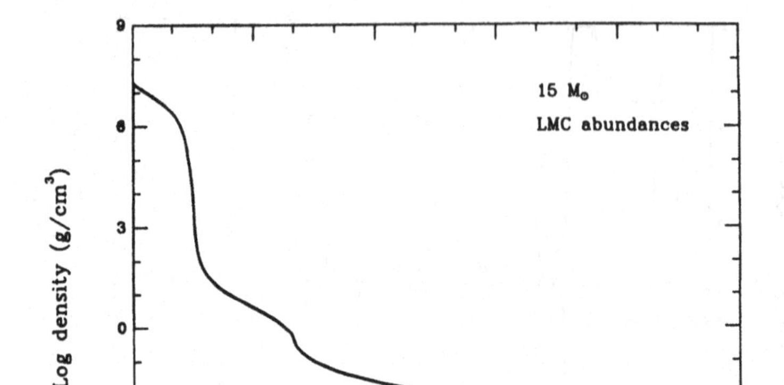

pronounced core-envelope structure in density, with an intermediate mantle region whose outer edge is that of the hydrogen burning shell. Most of the core collapses to form a neutron star, with shock wave generated which ejects the mantle and envelope. The mechanism for this "mantle" shock is still debated; prominent candidates are (1) the prompt mechanism in which the core bounce shock simply continues, (2) outer core heating by neutrino diffusion after the bounce shock dies (Arnett 1966), and (3) the Wilson mechanism. The conditions at the "mass cut" (which divides ejecta from neutron star) will depend upon the details of the explosion mechanism. This is the region in which new ^{56}Ni is synthesized (the O shell), and its nature will affect the predicted light curve and γ-ray spectra; these features have not been explored for different mechanisms. Figure 5 illustrates the hydrodynamic behavior of the neutron star formation, and explosion, by snapshots of density versus mass coordinate. The last time corresponds to 2 **minutes** after core bounce, at which time the shock has reached the edge of the mantle (4 $M_\odot$ here). This corresponds to a "prompt shock" mechanism; these characteristics were used to instigate the

Figure 5. The Formation of a Neutron Star.

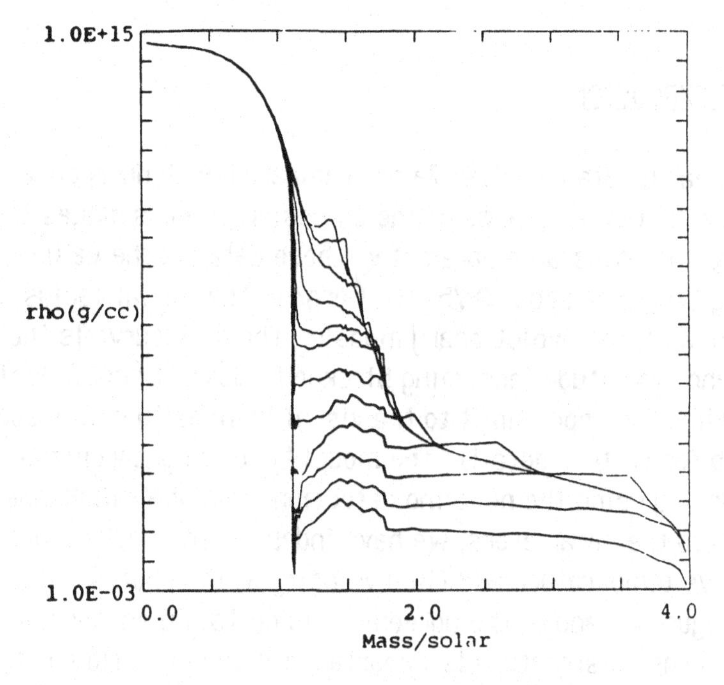

Figure 6. The Early Light Curve of SN1987a.

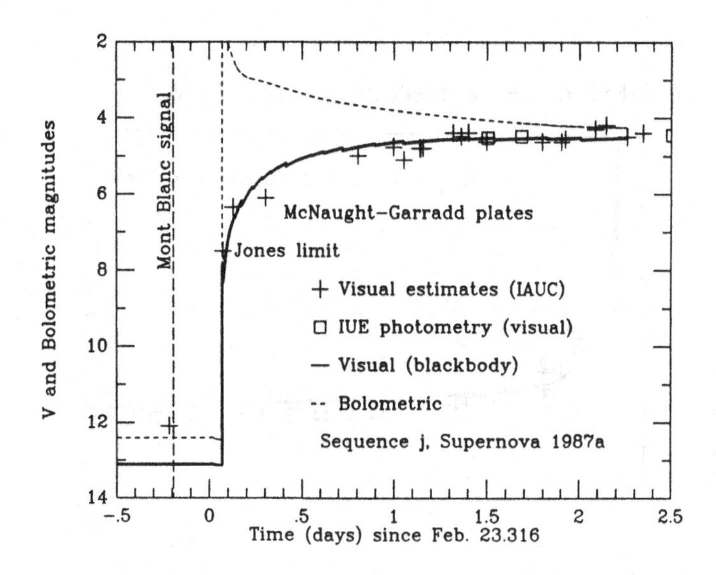

explosions which model the light curve. Note the steep density gradient (15 powers of 10) at the edge of the neutron star; this feature seems to be unresolved in other calculations.

THE EXPLOSION

The early data on SN1987a are compared with hydrodynamic models in Figure 6. Our ignorance of the explosion process makes the explosion energy an adjustable parameter; these data fix the value of the specific energy E/M to be about 0.75×10^{17} erg/g. The initial radius is in good agreement with the evolutionary models. The dark curve is the predicted V-band magnitude (assuming blackbody flux); it agrees well with the data, including the upper limit to the visual luminosity due to Jones (see Arnett 1987a,b for further detail). The model time is synchronized with the observations by using the neutrino detection time of Kamiokande and IMB. Having fixed the parameters, we have independent checks from the data on effective temperature and fluid velocity at the photosphere. These are shown in Figures 7 and 8; the agreement is certainly encouraging.

The density structure is important for the evolution of the light curve and for eventual x-ray and γ-ray escape. Two hours after the shock

Figure 7. Effective Temperature of SN1987a.

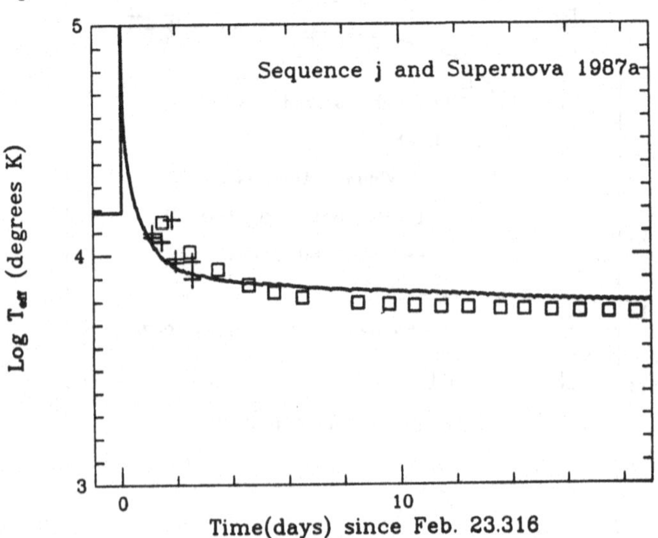

Figure 8. Velocities in SN1987a.

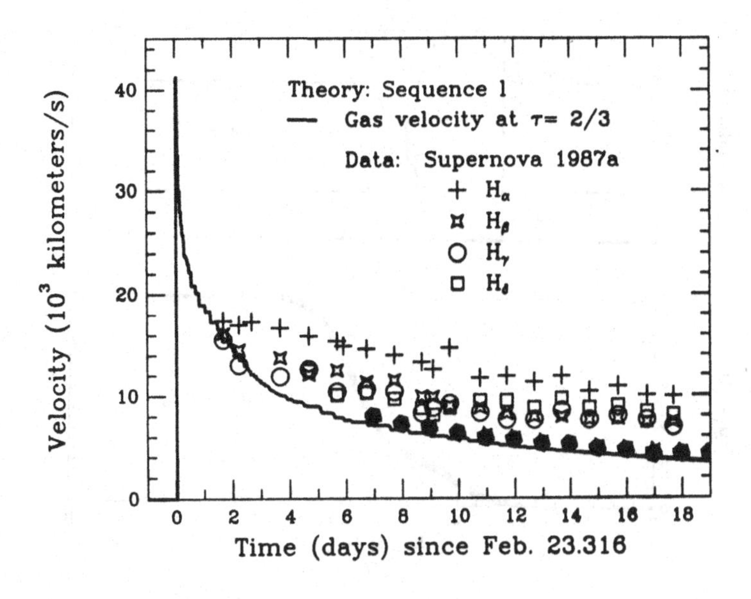

Figure 9. Snapshots of Density Structure.

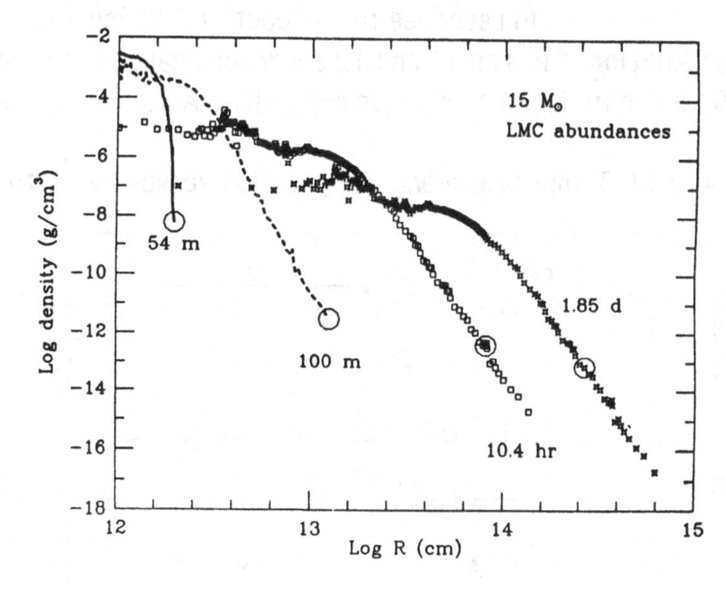

hits the surface of the star, the shape of the density structure is fixed, and "coasts" until modified much later by snowplowing into matter around the presupernova. Most of the mass is moving at several thousand km/s, with

Figure 10. Mass and Composition versus Expansion Velocity.

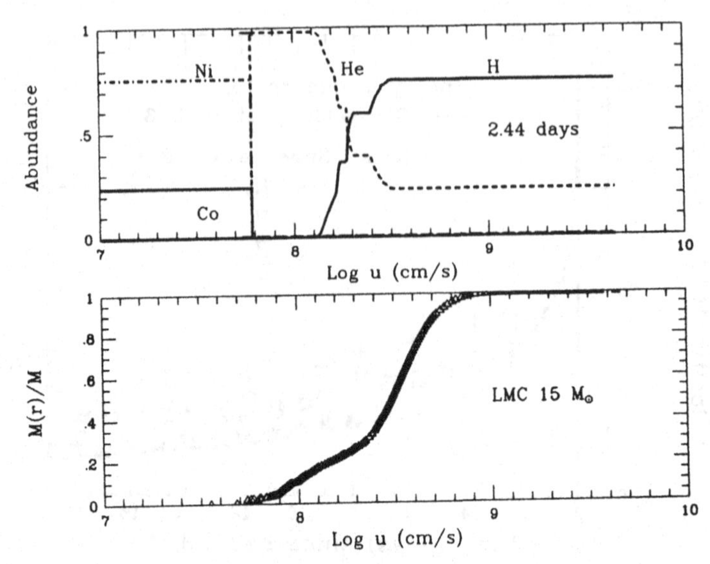

a small amount moving at much higher velocity (see Figure 10). The temperature T still varies in response to radioactive heating and cooling by expansion and radiation. Figures 11 and 12 show the change in thermal structure at 9.26 and at 19.75 days after explosion. A step in T moves to

Figure 11. Temperature and Composition versus Velocity (9.26d).

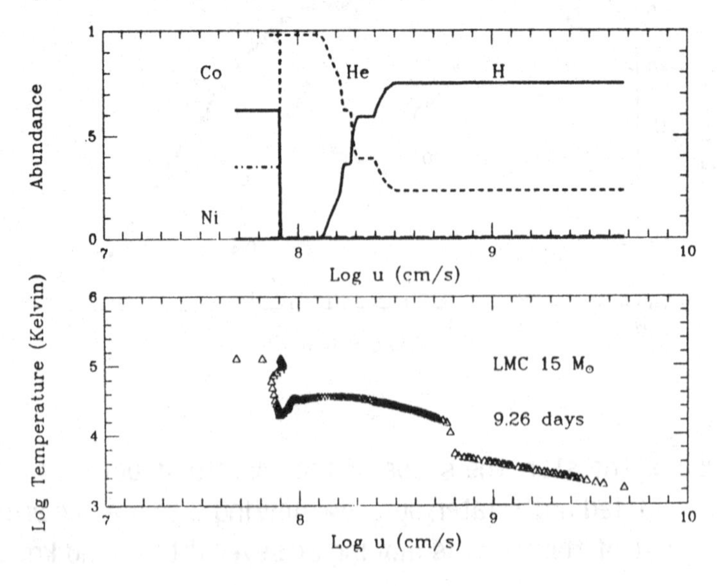

Figure 11. Temperature and Composition versus Velocity (19.75d).

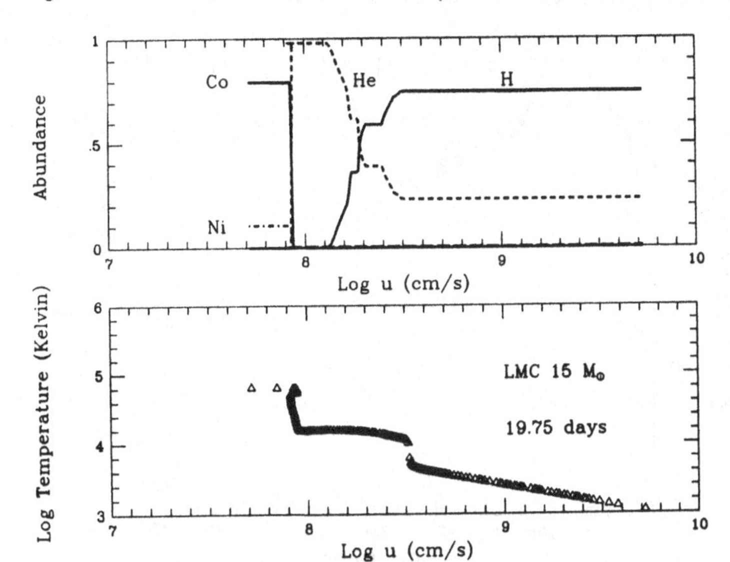

lower velocites as the recombination wave lowers opacity and the photosphere recedes in mass and radius. A heating wave from radioactive decay heats the center and moves outward. Actually the matter at the second (inner) step in T is Rayleigh-Taylor unstable, and will penetrate overlying, slower matter. This spreads the source of heat and enhances the prospects for x-ray and γ-ray observations.

To examine the later part of the light curve, we (1) generalize the analytic methods of Arnett (1982) to include recombination effects in heating and opacity, (2) make use of the unchanging density structure and the relatively uniform core (see Figure 9), (3) use the parameters which worked for the early part of the light curve discussed above, and (4) smoothed the unrealistically sharp interfaces of composition and heat sources which are built into spherically symmetric hydrocodes, but are destroyed as discussed above. Figure 13 compares the resulting light curve to the bolometric luminosity inferred from observations of the South African Astronomical Observatory (crosses) and Cerro Tololo International Observatory (pluses). The model had 0.75 $M_\odot$ of ^{56}Ni. After 20 days, the agreement is excellent; the early portion is affected by the power law

Figure 13. Light Curve at Late Times

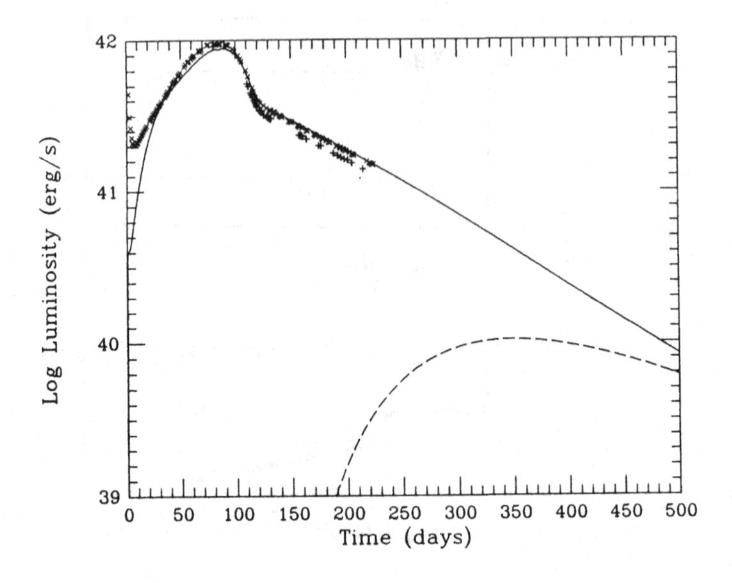

Figure 14. Late Light Curve with L(pulsar) = 10^{40}erg/s.

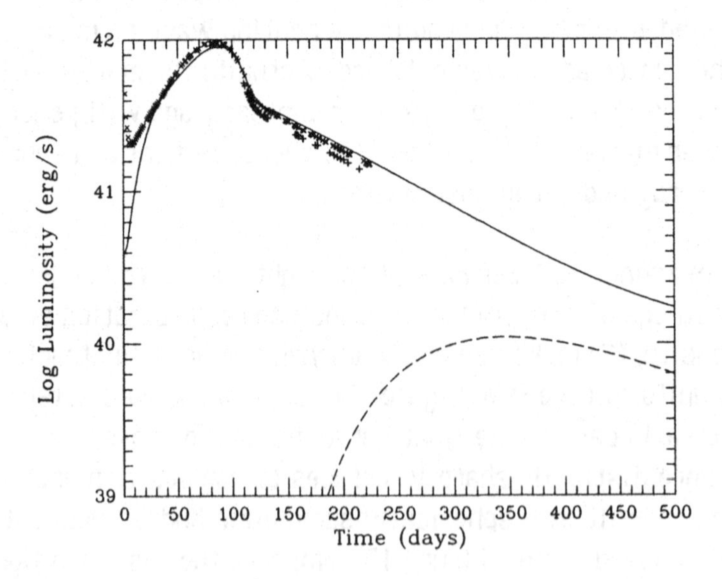

density structure and transient effects which are ignored by the analytic method. The dashed line shows the luminosity of escaping γ-rays. Note that although qualitatively correct, this uniform density model tends to overestimate the escape of γ-rays; this can be corrected and will be discussed in a later publication.

Figure 15. Late Light Curve with L(pulsar) = 10^{41}erg/s.

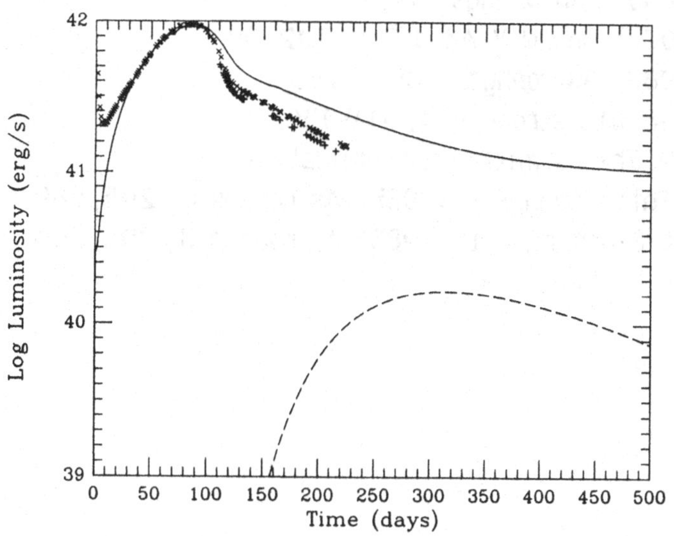

The exponential decline is in spectacular agreement with the release of radioactive energy by ^{56}Co. We may use this to place a stringent limit on pulsar luminosity trapped by the dense expanding shell. Figure 14 shows the light curve as before, but with a pulsar of 10^{40} ergs/s. The slope of the luminosity disagrees with the observations. Figure 15 shows the result for a pulsar luminosity of 10^{41} erg/s. Such a limit on the pulsar energy input suggests that the pulsar may have been born "slowly" rotating in the sense that centrifugal force was small compared to pressure gradients and gravity; this helps explain why spherically symmetric computations of the collapse and neutrino emission are in agreement with observations.

I wish to thank the many people who have provided helpful discussions and preprints of their work, and especially Robin Catchpole of SAAO and N. Suntzeff of CTIO for providing last minute data on bolometric luminosity.

REFERENCES

Arnett, W. D. (1966). Can. J. Phys., **44**, 2553.
Arnett, w. D. (1977). Ann. N. Y. Acad. Sci., **302**, 90.
Arnett, W. D. (1982). Astrophys. J., **253**, 785.
Arnett, W. D. (1987a). Astrophys. J., **319**, 136.
Arnett, W. D. (1987b). submitted for publication.
Arnett, W. D. & Thielemann, F. K. (1985). Astrophys. J., **295**, 589.
Thielemann, F. K. & Arnett, W. D. (1985). Astrophys. J., **295**, 604.

EVOLUTION OF THE STELLAR PROGENITOR OF SUPERNOVA 1987A

James W. Truran and Achim Weiss
Department of Astronomy, University of Illinois
Urbana, Illinois 61801, USA

Abstract. We address the question as to how the presence of a blue supergiant progenitor for Supernova 1987A in the Large Magellanic Cloud can best be understood in the context of stellar evolution theory. Proposed explanations for this observed behavior which will be discussed include: (i) it is a consequence of the lower metallicity characteristic of the Large Magellanic Cloud; (ii) it is a consequence of a high rate of mass loss, which enables the progenitor to evolve to the red and then return to the blue; and (iii) it is attributable to the effects of convection on the evolution of supergiants. Our own calculations, performed and with the choice of the Schwarzschild criterion for convection, predict that the evolution of a star of initial main sequence mass ~ 20 $M_\odot$ and initial metal composition $Z = 0.25\ Z_\odot$, in the absence of appreciable mass loss, is quite consistent with the occurrence of a blue supergiant progenitor: the B3 Ia supergiant Sanduleak -69 202. Some critical observational contraints on supergiant evolution are identified and discussed.

1 INTRODUCTION

The recent occurrence of Supernova 1987A in the Large Magellanic Cloud is providing an unprecedented opportunity for astrophysical theorists to test their models of massive star evolution, gravitational collapse, supernova explosions, explosive nucleosynthesis, and the evolution of supernova light curves. Stellar evolution theorists, have already encountered some surprises. In particular, the unambiguous identification (West et al. 1987; Panagia et al. 1987) of a blue supergiant progenitor for Supernova 1987A – the B3 Ia super-giant Sanduleak -69 202 – has raised questions concerning our understanding of the details of the later stages of evolution of massive stars. Our aim in this paper will be to explore some possible implications of this identification for models of the evolution of typical massive star progenitors of Type II supernovae.

We first require an estimate of the allowed range of luminosities and associated stellar masses for the supernova progenitor. The fact that

Supernova 1987A is situated in the Large Magellanic Cloud, at a relatively well determined distance, allows an accurate estimate to be obtained. The identification of Sanduleak -69 202 as of spectral type B3 Ia implies a surface effective temperature of approximately 15,000 K. With the choices of an apparent visual magnitude $m_v = 12.24$ mag (Isserstedt 1975), a distance modulus m-M = 18.5 for the LMC, and a reddening correction $A_v = 0.6$ mag, a photometric analysis of the several components of the Sanduleak -69 202 system by West *et al.* (1987) yielded an absolute visual magnitude $M_v = -6.8$ mag for the B3 Ia star. The further assumption of a bolometric corrections of 1.3 yields an absolute bolometric magnitude for the supernova progenitor $M_{bol} = -8.1$ mag and a corresponding luminosity of 5×10^{33} erg s^{-1} or 1.3×10^5 $L_\odot$.

An estimate of the mass of the progenitor may be obtained from a survey of published models of massive star evolution. As we will learn from our subsequent discussion, it is necessary to give careful attention to results obtained for various assumptions concerning initial metallicity, the rate of mass loss, and the treatment of convection. The models of Brunish and Truran (1982a,b) indicate that a 15 $M_\odot$ star will reach the onset of carbon burning at a luminosity $\approx 3.1 \times 10^{38}$ erg s^{-1} ($\approx 8 \times 10^4$ $L_\odot$) quite independent of initial metallicity and the presence of a moderate rate of stellar mass loss. Weaver, Zimmerman, and Woosley (1978) found a 15 $M_\odot$ star of solar metallicity to have a luminosity of 3.7×10^{38} erg s^{-1}, while the 20 $M_\odot$ model discussed by Wilson et al. (1986) had a luminosity of 5.7×10^{38} erg s^{-1}. Similar results are reflected in the evolutionary tracks for massive stars undergoing mass loss, as described by Chiosi and Maeder (1986). We will therefore proceed on the assumption that the progenitor of Supernova 1987A had a mass ≈ 20 $M_\odot$.

The general opinion of stellar evolution theorists was that stars of this initial main sequence mass should evolve to red super-giants prior to the ignition of carbon in the central regions and thus should subsequently explode as red supergiants (Chiosi and Maeder 1986). The inclusion of mass loss effects encourages evolution to the red, unless the mass loss rates are so high that the hydrogen envelope of the star is almost completely removed by the end of helium burning. Such high mass loss rates are observed in the case of Wolf-Rayet stars, which typically have much higher initial masses than SK -69 202. We note, however, that the calculations of Lamb, Iben, and Howard (1976), for stars of solar metallicity, indicated that significant redward evolution need not be experienced for stars of mass ~ 25 $M_\odot$.

In the next section we will review several possible means by which a star of ~ 20 $M_\odot$ may evolve to a blue supergiant progenitor, including our own calculations (Hillebrandt et al. 1987), which are guided by the fact that stars in the LMC might have a lower metal content than those in our own galaxy. We then identify and discuss observational constraints which can serve to guide theoretical models of the evolution of supergiant stars.

2 CALCULATIONS

Due to the presence of semiconvective regions in stars of 15-40 $M_{\odot}$, the results of evolutionary calculations are extremely uncertain. For Population I stars, this fact has long been known and has been reviewed by Chiosi (1978). In general, different physical assumptions about the treatment of convection (e.g. Stothers and Chin 1975,1976a), different input data such as opacity tables (e.g. Stothers and Chin 1976b; Chiosi 1978), and even different numerical schemes yield sometimes strikingly different results. However, for Population I stars, the models generally reach their final configuration, at the onset of core collapse, as red supergiants, no matter what the differences in the proceeding evolution might have been. A possible exception to this is the evolution in the presence of extremely high mass loss rates.

The fact that SK -69 202 was indeed a blue supergiant before exploding as SN 1987A raises the following question: for what plausible assumptions can a star of ~ 20 $M_{\odot}$ evolve in such a way that it is a B3 supergiant at the end of its evolution? Possible physical explanations include a low metal content consistent with the LMC, or the occurrence of a high rate of mass loss. More technical possible explanations which reflect our insufficient knowledge of basic physical processes include different assumptions concerning semiconvection, overshooting, opacity data or nuclear reactions rates. The very existence of cool supergiants in the LMC, furthermore, makes it desirable, though in our opinion not essential, for a model first to evolve to the red and later to return to the blue.

Our own approach is based on the assumption that the metal content of SK -69 202 is comparable to that found in H II regions of the LMC. This is the only assumption we make. All other input physics for our models is based upon standard procedures. We do not include mass loss, or overshooting, or semiconvection. For the opacities, we use the most recent tables available (Hübner et al. 1977), and for the nuclear reactions, the standard rates (including an increased $^{12}C(\alpha,\gamma)^{16}O$ rate). For more details see Truran and Weiss (1987). The results of our calculations for models of 15 and 20 $M_{\odot}$ and different metal contents have been presented in Hillebrandt et al. (1987) and are shown in Figure 1. Clearly, in both cases, a moderately low metal abundance compared to the sun results in a blue supernova progenitor. We thereby confirm earlier results by Brunish and Truran (1982b), and Stothers and Chin (1975), indicating that a low metallicity favors a "bluish" evolution. However, we fail to obtain any red supergiant evolution whatsoever. This fact appears to be in contradiction with the occurrence of cool stars in the LMC. It therefore would be very interesting to confirm indications that SK -69 202 actually has been a red supergiant at earlier phases. We return to this matter in the next section.

Brunish and Truran (1982a,b), furthermore, have investigated the influence of moderate mass loss on the evolution of stars of both populations. Although their final models at the end of the carbon

Figure 1. Evolution for 15 $M_\odot$ and 20 $M_\odot$ stellar models to the onset of carbon burning for different metallicities.

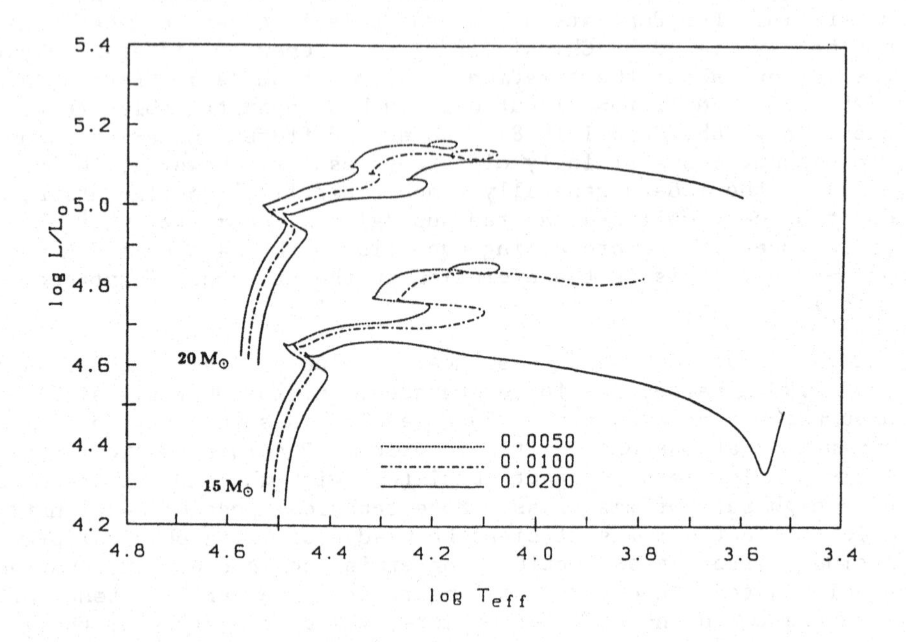

Figure 2. Evolution of a possible progenitor for SN 1987A with high mass loss (adapted from Maeder 1987).

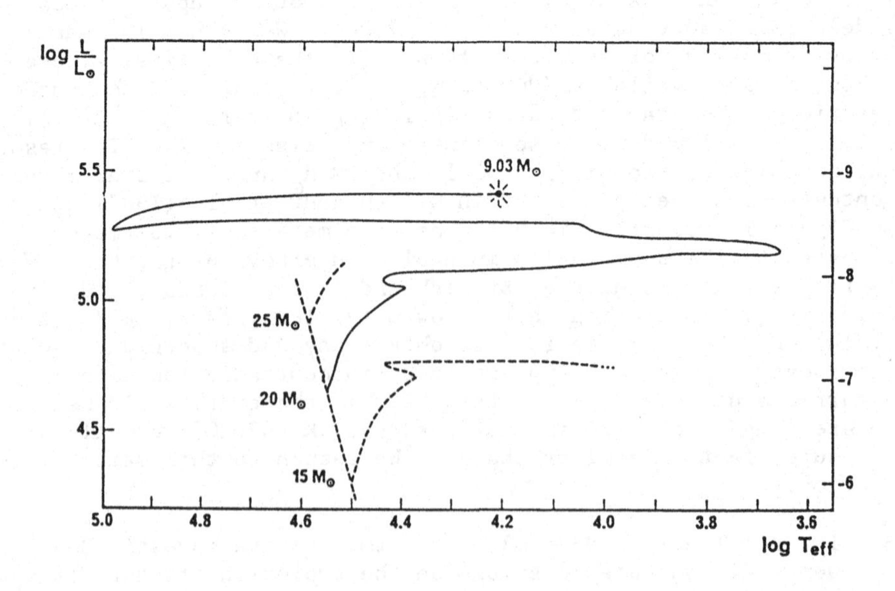

burning phase, after suffering the loss of approximately 10% of their initial mass, have a tendency to be slightly cooler than the models without mass loss, the basic results are unchanged. Maeder (1981) finds the same weak influence on Population I models becoming red supernova progenitors when utilizing moderate mass loss. However, when he used a higher mass loss rate, models losing 50% or more of their initial mass were found to reach the end of their evolution as blue supergiants. We reproduce Maeder's (1987) recent result for the evolution of the progenitor of SN 1987A in Figure 2, which confirms that heavy mass loss yields a blue supergiant. However, the total amount of hydrogen left at the time of core collapse is only $\sim 3\ M_\odot$ (the star has lost mass from 20 $M_\odot$ to 9 $M_\odot$). This value may be too small to reconcile with models of the light curve and spectral evolution of SN 1987A, as discussed elsewhere in these proceedings.

An alternative approach known to us to obtain a blue progenitor that previously has been a red supergiant is based upon a treatment of semiconvection different from our standard (Schwarzschild) approach. Using some particular assumptions concerning the mixing timescale, Woosley, Pinto, and Ensman (1987) succeed in obtaining the desired results. Baraffe and El Eid (1987) presently are investigating evolution with the Ledoux criterion for convection. Their results agree with our calculations for a 20 $M_\odot$ star, which are shown in Figure 3. We apply the same criterion and also assume that mixing only takes place in the region where $\nabla_{rad} > \nabla_{Ledoux} = \nabla_{ad} + \beta/4\text{-}3\beta \cdot d\ln\mu/d\ln P$. This corresponds to the extreme case of assuming that mixing in semiconvective regions, i.e. where $\nabla_{ad} < \nabla_{rad} < \nabla_{Ledoux}$, is infinitely slow. The other extreme, instantaneous mixing as in the fully convective regions, corresponds to the application of Schwarzschild's criterion, i.e. $\nabla_{rad} > \nabla_{ad}$. In fact, Langer, El Eid, and Fricke (1985), using a diffusion approximation for semiconvection, mentioned that in most cases the mixing is very fast, so that usually the Schwarzschild case is almost recovered. Evolution with the Schwarzschild criterion is shown in Figure 4.

The significant difference between the results of evolution with the two criterion is evident, although the calculations shown in Figure 3 have not yet been carried through carbon burning. It seems, however, that the model will finish as a red supergiant, since the blueward loop during helium burning will be completed in the red before carbon burning begins. Internally, the model calculated with the Ledoux criterion shows a shallow gradient in composition in the envelope, while the one for the Schwarzschild criterion shows several discreet steps. As a preliminary remark we state that our extreme treatment of semiconvection does not result in a realistic model for SK -69 202. However, an intermediate treatment such as described by Woosley, Pinto, and Ensman (1987) might be more successful.

3 SOME CONSTRAINTS ON SUPERGIANT EVOLUTION

The supergiant phase of evolution of the progenitor of Supernova 1987A provides an important test of our understanding of

Figure 3. Evolution through helium burning with the choice of the Ledoux criterion.

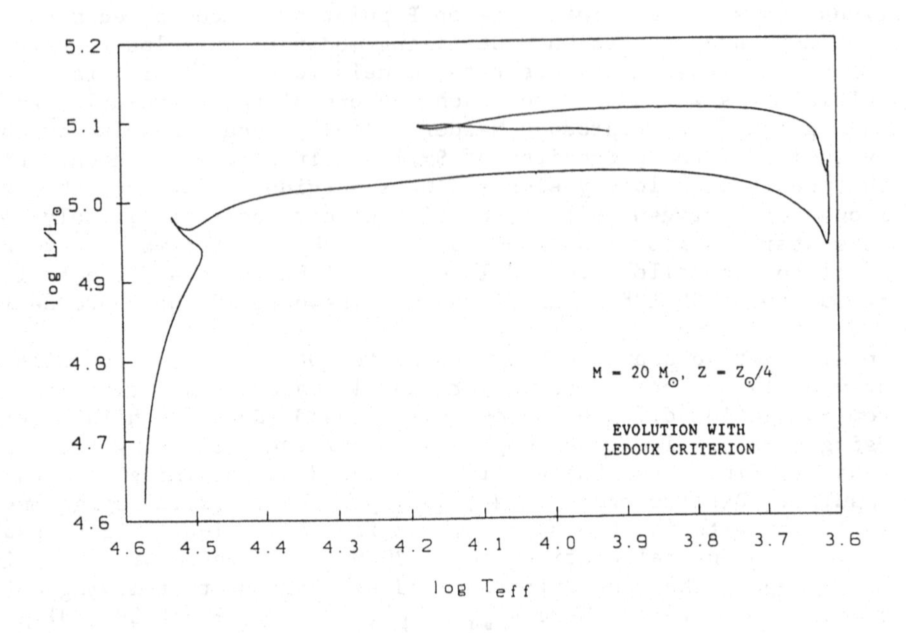

Figure 4. Evolution through carbon burning with the choice of the Schwarzschild criterion (numbers give ages in 10^6 years).

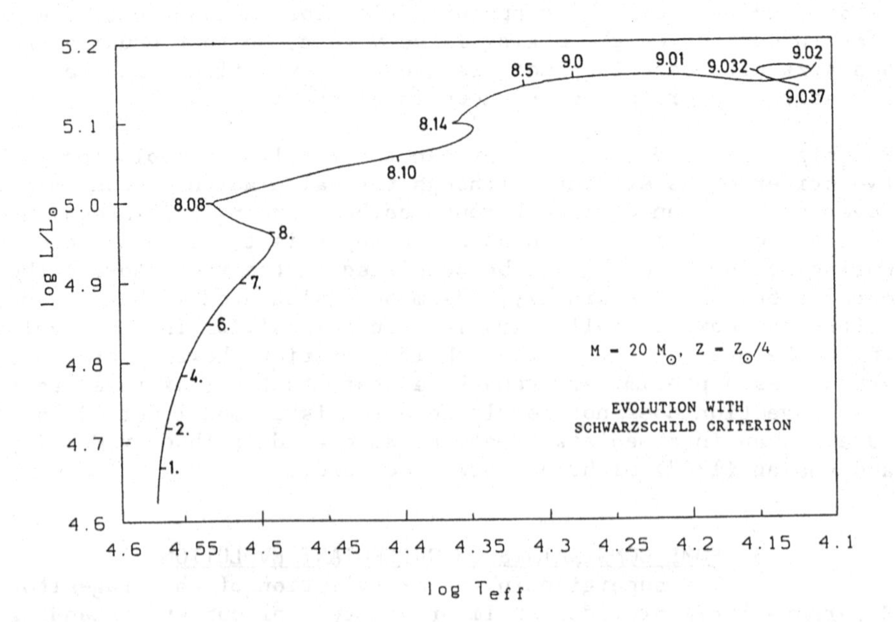

stellar evolution theory. A critical question here concerns the extent of evolution experienced by the supernova progenitor as a red or blue supergiant. Was SK -69 202 ever a red supergiant? While it was apparently a blue supergiant immediately prior to the outburst, its earlier evolution in the supergiant region of the Hertzsprung-Russell diagram is a matter of discussion and debate. There are some general constraints on the character of this evolution which we believe may help guide us to an understanding of the nature and evolutionary history of Sanduleak -69 202.

The question quite naturally arises here as to whether our stellar evolution calculations, and those of other researchers (Arnett 1987; Brunish and Truran 1982b; Hillebrandt et al. 1987), for metal deficient stars imply a significant distortion of the Hertzsprung-Russell diagrams for low metallicity populations in general and for the Large Magellanic Cloud in particular. Humphreys and Davidson (1979) have argued that massive star evolution has resulted in very similar supergiant populations in the Galaxy and the LMC, in spite of the small differences in the heavy element abundance levels. Of particular concern might be the fact that the observed M_{bol} verses log T_e H-R diagrams for both the LMC (Humphreys and Davidson 1979) and the SMC (Humphreys 1983) reveal the presence of a significant population of red supergiants at luminosities consistent with stars in the mass range ~ 15-20 $M_\odot$. For the case of the LMC, we believe that this observed behavior might be simply a consequence of a spread in metallicity of a factor of a few, since our numerical results reveal that there exists a strong dependence of the extent of redward evolution on metallicity Z for values in the range $0.1\ Z_\odot \leq Z \leq Z_\odot$. There also exists some uncertainty as to whether the H II region abundances (Dufour 1984) for oxygen and a few other elements accurately reflect the level of abundances of heavier metals like iron. A somewhat broader spread in metallicity might be necessary to allow an understanding of the H-R diagrams for the SMC, which is known to have a lower Z than the LMC. This could be a problem for our models. However, we believe the observations of both the LMC and the SMC suggest a relatively larger population of supergiants in the temperature range T_e ~ 10,000-20,000 K, compatible with the results of our calculations (Brunish, Gallagher, and Truran 1986).

The alternative interpretation of these data is generally based on calculations of massive star evolution at high mass loss rates. Evolutionary tracks for massive stars of solar composition undergoing mass loss, reviewed by Chiosi and Maeder (1986), reveal a more rapid evolution to the red subsequently followed by quite significant blueward excursions. For such models, the breadth of the blue supergiant region (and its relative population) is generally attributed to the presence of these blue loops. The occurrence of an immediate blue supergiant progenitor of Supernova 1987A is not easy to understand in this context, since mass loss effects are not expected to be so significant for stars of mass ~ 20 $M_\odot$. However, Maeder (1987) has shown that, for the choice of an appropriate high rate of supergiant phase mass loss, the evolutionary models can reproduce both the observed distributions of red and blue supergiants and the occurrence of a blue supernova progenitor (see Figure 2).

What evidence might exist for the presence of high rates of mass loss for stars in the Magellanic Clouds? The predicted rates of mass loss for main sequence stars of masses ~ 20 $M_{\odot}$ are not high enough to effect any significant change in the envelope mass (Kudritzki 1987). Stars which remain in the blue supergiant regime of the H-R diagram and never evolve to red supergiants, as in the calculations of Brunish and Truran (1982b), may therefore not experience significant levels of mass loss. There are, however, several indications that stars in the LMC and SMC do suffer significant mass loss. Bessel (1987) has noted that the pulsation periods for red supergiants in the SMC imply masses which are less than those one would normally infer from the luminosities, suggesting that mass loss has indeed occurred. Kudritzki (1987) has determined high helium to hydrogen ratios for two BO stars in the LMC, again indicating that substantial envelope depletion is realized at least for stars of somewhat larger masses ~ 30 $M_{\odot}$.

These observations suggest that significant mass loss may accompany phases of red supergiant evolution. Other features of such evolution are also of interest. Models that evolve to red supergiants prior to carbon ignition develop convective envelopes which act to bring processed matter from the CNO-hydrogen-burning shell to the surface. Abundance signatures of such mixing will include an increased He/H ratio and high N/C and N/O ratios, such as have been reported for other blue supergiants in the LMC by Kudritzki (1987). High mass loss rates can further emphasize these signatures. The detection of such abundance trends in the spectrum of Supernova 1987A during the phase when the spectra are dominated by hydrogen would serve to confirm the convective character of the envelope at an earlier epoch and thus support the argument that Sanduleak -69 202 was once a red supergiant. Alternatively, the facts that no such peculiar abundance ratios have been detected to date and that the spectrum remains hydrogen rich appear to support the view that significant mass loss did not occur for the supernova progenitor.

4 CONCLUDING REMARKS

We conclude from the results of our numerical studies that the evolution of a star of mass in the range ~ 15-20 $M_{\odot}$ and initial metal composition $Z = 0.25\text{-}0.5\ Z_{\odot}$, compatible with the metallicity of the LMC, is consistent with the occurrence of a blue supergiant progenitor for Supernova 1987A. Indeed, the predicted properties of such stars are quite in agreement with those observed for the presumed stellar progenitor Sanduleak -69 202 and their typical envelope structures, characterized by steeper density gradients and smaller photospheric radii, have been found to be consistent with the spectral and light curve development of Supernova 1987A (Arnett 1987; Grassberg *et al*. 1987; Hillebrandt *et al*. 1987; Shaeffer *et al*. 1987; Shigeyama *et al*. 1987; Woosley *et al*. 1987).

Models calculated with the choice of the Ledoux criterion can in principle (Woosley, Pinto and Ensman 1967) provide blue progenitors which have passed through the red, but this requires a careful tuning of mixing timescale.

A critical question which remains to be answered is that concerning the relative timescales for evolution in the domains of red and blue supergiants and the consistency of stellar evolution models with the observed distribution of red and blue supergiants in the LMC (Humphreys and Davidson 1979; Humphreys 1983). We find that a metallicity spread of a factor of a few for the young stellar component of the LMC is sufficient to allow an understanding of both (i) the presence of red supergiants at luminosities compatible with those of ~ 15-20 $M_\odot$ stars and (ii) a blue supergiant progenitor of mass ~ 15-20 $M_\odot$ for Supernova 1987A *in the absence of blue loops*. The presence of a significant population of red supergiants in the critical luminosity range in the SMC (Humphreys 1983), in a population of somewhat lower average metallicity, suggests however that the problem may be more complicated. The mass loss rates assumed in stellar evolution studies can indeed be tuned to reproduce the observed distributions of red and blue supergiants in the Galaxy and the Magellanic Clouds and to account for the presence of a blue supernova progenitor (Maeder 1987). We believe the resolution of this issue can best be provided by observations of Supernova 1987A itself. If Sanduleak -69 202 was indeed a red supergiant at one stage of its history, significant mass loss and envelope mixing by convection should have occurred and both the mass of envelope (hydrogen rich) matter and particularly its composition should provide evidence for these behaviors.

ACKNOWLEDGEMENTS

This research was supported in part by the United States National Science Foundation under grant AST 86-11500 at the University of Illinois.

REFERENCES

Arnett, W.D. (1987). Astrophys. J., 319, 136.
Baraffe, I. and El Eid, M. (1987). Private communication.
Bessel, M. (1987). Private communication.
Brunish, W.M. and Truran, J.W. (1982a). Astrophys. J., 256, 247.
Brunish, W.M. and Truran, J.W. (1982b). Astrophys. J. Suppl., 49, 447.
Brunish, W.M., Gallagher, J.S., and Truran, J.W. (1986). Astron. J., 91, 598.
Chiosi, C. (1978). In The HR Diagram, ed. A. G. D. Philip and D. S. Hayes, p. 357. Dordrecht: Reidel.
Chiosi, C. and Maeder, A. (1986). Ann. Rev. Astron. Astrophys., 91, 598.
Dufour, R. J. (1984). In Structure and Evolution of the Magellanic Clouds, ed. S. van den Bergh and K. S. de Boer, p. 353. Dordrecht: Reidel.

Grassberg, E.K., Imshennik, V.S., Nadyozhin, D.K., and Utrobin, V.P. (1987). Preprint.
Hillebrandt, W., Höflich, P., Truran, J.W., and Weiss, A. (1987). Nature, 327, 597.
Hübner, W.F., Merts, H.L., Magee, N.H. Jr., and Argo, M.F. (1977). Los Alamos Sci. Lab. Rept. LA 6760 M.
Humphreys, R.M. (1983). Astrophys. J., 265, 176.
Humphreys, R.M. and Davidson, K. (1979). Astrophys. J., 232, 409.
Isserstedt, J. (1975). Astron. Astrophys. Suppl., 19, 259.
Kudritzki, R. (1987). Private communication.
Lamb, S. A., Iben, I., Jr., and Howard, W. M. (1976). Astrophys. J., 207, 209.
Langer, N., El Eid, M., and Fricke, K. J. (1985). Astron. Astrophys., 45, 179.
Maeder, A. (1981). Astron. Astrophys., 102, 401.
Maeder, A. (1987). Preprint.
Panagia, N., Gilmozzi, R., Clavel, J., Barylak, M., Gonzalez-Riesta, R., Lloyd, C., Sanz Fernandez de Cordoba, L., and Wamsteker, W. (1987). Astron. Astrophys., 177, L25.
Shaeffer, R., Casse, M., Mochkovitch, R., and Cahen, S. (1987). Astron. Astrophys., in press.
Shigeyama, T., Nomoto, K., Hashimoto, M., and Sugimoto, D. (1987). Preprint.
Stothers, R. and Chin, Ch.-W. (1975). Astrophys. J., 198, 407.
Stothers, R. and Chin, Ch.-W. (1976a). Astrophys. J., 204, 472.
Stothers, R. and Chin, Ch.-W. (1976b). Astrophys. J., 203, 800.
Truran, J.W. and Weiss, A. (1987). In Nuclear Astrophysics, ed. W. Hillebrandt, E. Müller, R. Kuhfuss, and J.W. Truran, p. 81, Berlin: Springer.
Weaver, T.A., Zimmermann, G.B., and Woosley, S.E. (1978). Astrophys. J., 225, 1021.
West, R.M., Lauberts, A., Jørgensen, H.E., and Schuster, H.-E. (1987). Astron. Astrophys., 177, L1.
Wilson, J.R., Mayle, R., Woosley, S.E., and Weaver, T.A. (1986). Ann. N.Y. Acad. Sci., in press.
Woosley, S.E., Pinto, P.A., and Ensman, L. (1987). Astrophys. J., in press.
Woosley, S.E., Pinto, P.A., Martin, P.G., and Weaver, T.A. (1987). Astrophys. J., 318, 664.

MODELLING THE ATMOSPHERE OF SN 1987A

L.B. Lucy
European Southern Observatory, Karl-Schwarzschild-Str. 2,
D-8046 Garching bei München, FRG

Abstract. Computational investigations to complement ESO's spectroscopic and photometric monitoring of SN 1987A are reported. A Monte Carlo spectral synthesis code is used to model the supernova's spectral evolution from Feb 25 to Jun 18 and to test thereby Arnett's gas dynamical prediction of its envelope's density profile. An empirical, inverse approach for deriving this profile is also described and a preliminary solution given. In addition, last-scattering radii are computed for comparison with the CfA speckle observations and possible interpretations of the Bochum event are briefly discussed.

INTRODUCTION

SN 1987A presents us, in all probability, with a once in a lifetime opportunity to study a supernova at close quarters and thereby to advance markedly our quantitative understanding of this fundamental astrophysical phenomenon. In achieving this anticipated advance, the technique of spectral synthesis should surely figure prominently, given the high information content of the still-accumulating spectroscopic record. Specifically, in various guises, this technique should tell us about the progenitor's surface composition and prior mass loss, should test gas dynamical simulations of the explosion, and should determine the amounts and stratification of nuclear-processed ejecta.

The synthesis calculations reported here cover the supernova's early spectral development when the continuum was strong and energy deposition within the extended reversing layer was negligible. For these phases, Monte Carlo techniques developed for stellar wind problems (Abbott and Lucy 1985, 1987) are both applicable and effective. In fact, this approach has already been used to investigate the IUE spectra obtained on days 2 and 3 (Lucy 1987a), to test Arnett's (1987) model of the explosion (Lucy 1987b), and to identify features in the optical and near-infrared spectra (Fosbury et al. 1987).

BASIC MODEL

A description of the physical model together with justifications for various assumptions and approximations were presented at the ESO SN Workshop in July (Lucy 1987b). Briefly, the supernova's envelope is assumed to be undergoing homologous, spherical expansion with zero turbulence and no energy deposition above a sharply-defined photosphere (Shuster-Schwarzschild code). Line formation occurs by coherent scattering in the fluid frame and is treated in the narrow-line limit (Sobolev approximation). Departures from LTE in both excitation and ionization are approximately treated as are non-hydrogenic effects in photoionization rates (Abbott and Lucy 1985, 1987). Relativistic radiative transfer terms of O(v/c) are included, though time delays are not.

The most serious deficiency in this model is the neglect of interlocking between different energy levels of hydrogen, resulting in poor modelling of the Balmer lines. This aspect is accurately treated by Höflich (1987).

LINE LIST

The master line list used in the synthesis calculations is derived from the works of Abbott (1982) and Kurucz and Peytremann (1975). Table 1 in Lucy (1987b) gives the distribution of lines among the ions I-III for elements H-Ni in the list used for models reported at that time. Subsequently, the following additions have been made:

	I	II	III
21 Sc	369	199	37
23 V	1,673	1,202	1,099
38 Sr	137	63	0
56 Ba	90	71	0

SYNTHETIC SPECTRA

Figures 1-5 show comparisons of synthetic with observed spectra. These are reproductions of plots already published (Lucy 1987b; Fosbury et al. 1987). Additional comparisons are reported in the paper of Danziger et al. (this vol.).

For this series of synthetic spectra, the density profile is that which emerges from Arnett's (1987) simulation of the explosion. His predicted luminosities and effective temperatures are, however, ignored; instead, the adopted luminosities and photospheric temperatures are those derived from the observational record (Fosbury et al. 1987; Catchpole et al. 1987) under the assumptions that m-M = 18.5 and E(B-V) = 0.15. Finally, the chemical composition is taken to be solar (Allen 1973) but with Z reduced to $Z_{\odot}/2.75$ (Dufour 1984).

Notice that this formulation leaves no free parameters. In particular, since the time of core collapse is known, the velocity of the photosphere is not an adjustable parameter but is determined by the time of observation and the adopted distance.

One of the purposes of this series was to test Arnett's (1987) density function $\rho_1(r_1)$ and thereby his assumptions about the progenitor's structure and previous evolutionary history. Inspection of these figures shows that this function does indeed allow the successful reproduction of major features in the UV and optical spectra. Especially noteworthy are the degree of success achieved in fitting and following the rapidly changing IUE spectra on days 2 and 3 (Figs. 1 and 2) and the subsequent close matching of major absorption bands at 4000-4500 Å and 5000-5500 Å in the Mar 15 spectrum (Fig. 4). On the debit side, however, apart from the poor fit to Hα which is expected (see above), there is a striking and persistent failure in the excessive predicted width for the absorption trough of the P Cygni feature due to the CaII infrared triplet (8498-8662 Å) - see Figs. 3-5. In addition, there is a noticeable loss of detail in the computed Mar 15 spectrum in the interval 5500-6300 Å.

The poor fit for the CaII infrared triplet must be due either to the form of Arnett's function $\rho_1(r_1)$ or to some error or inadequacy in the spectral synthesis code. With respect to the latter, several remarks are pertinent: First, all steps in the CaII calculation have been checked by hand. Second, the same block of program that yields success elsewhere is also used for CaII. Third, the treatment of excitation and line formation that underlies the striking success in the far-UV is expected to be similarly valid for the CaII lines.

On the basis of such considerations, the excessive predicted CaII absorption trough has been interpreted (Lucy 1987b) as indicating that Arnett's (1987) explosion model yields too much mass with velocities $\gtrsim$ 7,000 km s^{-1}. However, before this is accepted as definitive, further improvement in the treatment of excitation would seem to be necessary: For the synthetic spectra computed thus far, excitation and ionization are determined from the internally generated Monte Carlo radiation field, but only after the latter is approximated as dilute black body radiation. In consequence, the lowering of excitation due to line blocking is averaged over the spectrum, with the inevitable loss of any strong selective effects on particular levels. But just such an effect may contribute to the observed narrowness of the CaII trough. Strong line blocking is observed in the neighbourhood of the CaII H and K lines; consequently, high in the reversing layer - i.e., at high velocities - the rate of replenishment of the 3^2D levels by radiative decays from the 4^2P^0 levels following H and K absorptions is probably significantly less than that implied by the code's current treatment of excitation.

A more accurate treatment of this feed-back between line blocking and excitation is planned. In addition, other investigators' density functions will be similarly tested.

Fig. 1: Feb 25 IUE spectrum compared with synthetic spectrum using density function of Arnett. The theoretical spectrum is displaced downwards by 1 dex. Fluxes are in units of erg cm^{-2} s^{-1} nm^{-1} and wavelengths in nm.

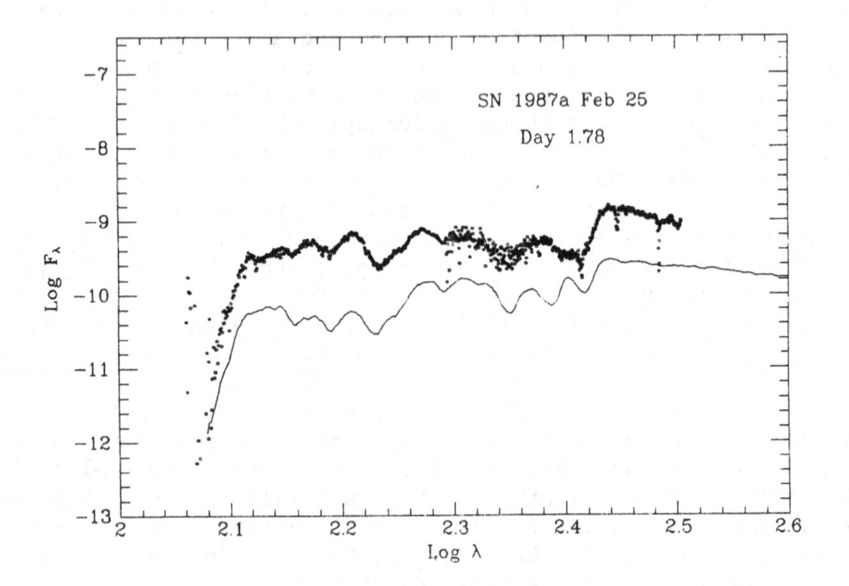

Fig. 2: Same as Fig. 1 but for Feb 26.

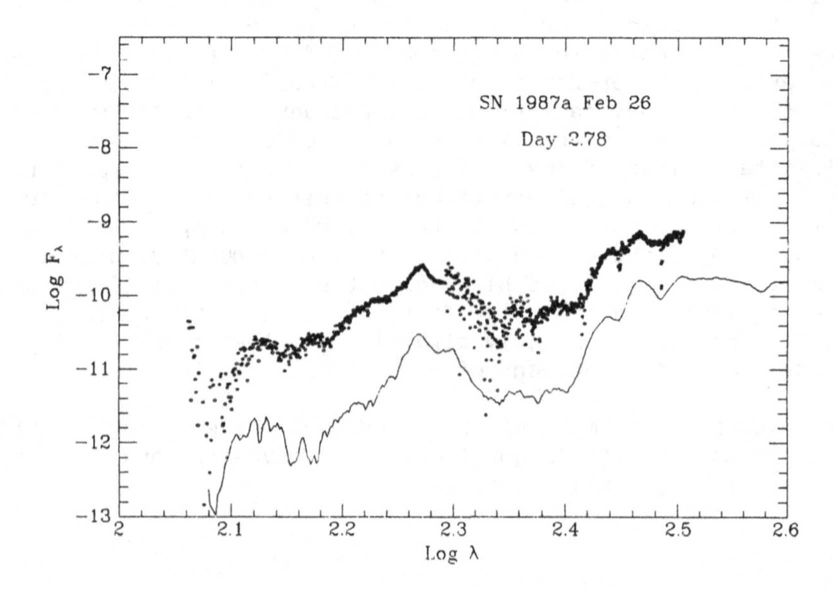

Fig. 3: Mar 3 ESO optical and near infrared spectra compared with synthetic spectrum using density function of Arnett. The theoretical spectrum is displaced dwonwards by 0.5 × the flux unit.

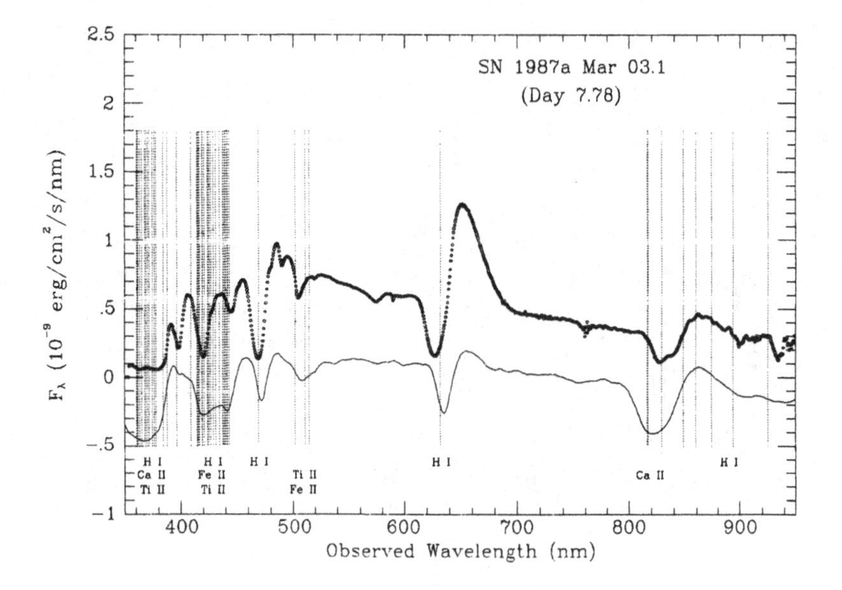

Fig. 4: Same as Fig. 3 but for Mar 15.

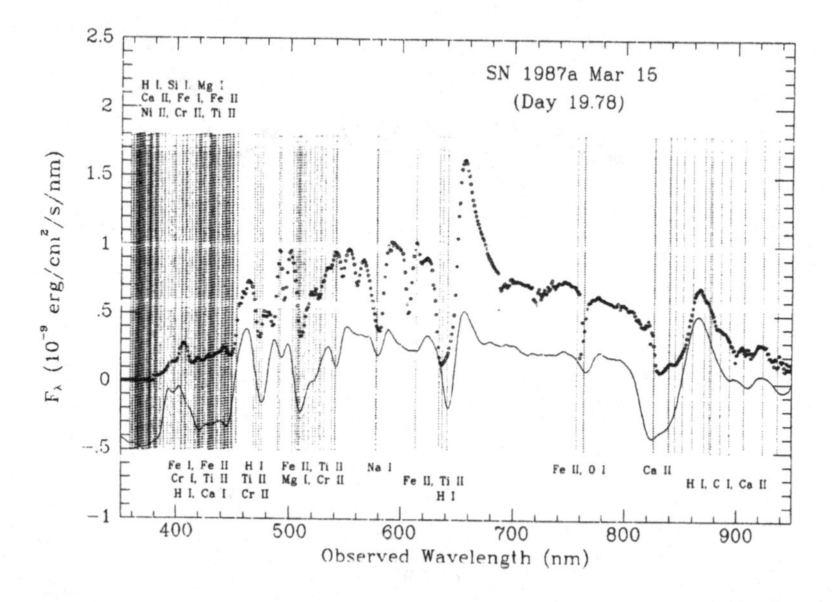

Fig. 5: Same as Fig. 3 but for Jun 18. Theoretical spectrum is displaced downwards by 1 flux unit.

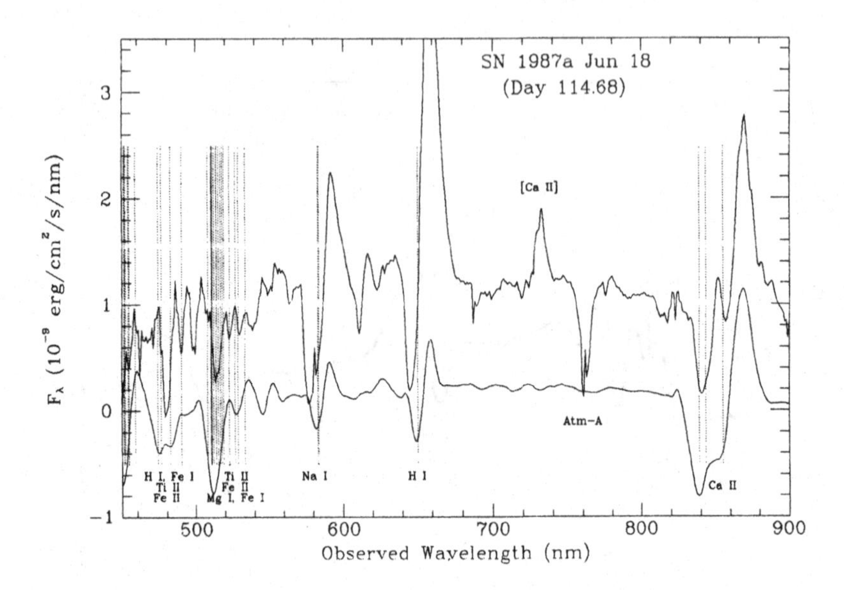

Fig. 6: Same as Fig. 1 but with E(B-V) = 0.19 instead of 0.15. Flux unit is 10^{-14} erg cm^{-2} s^{-1} Å^{-1}. The theoretical spectrum is displaced downwards by 1 dex.

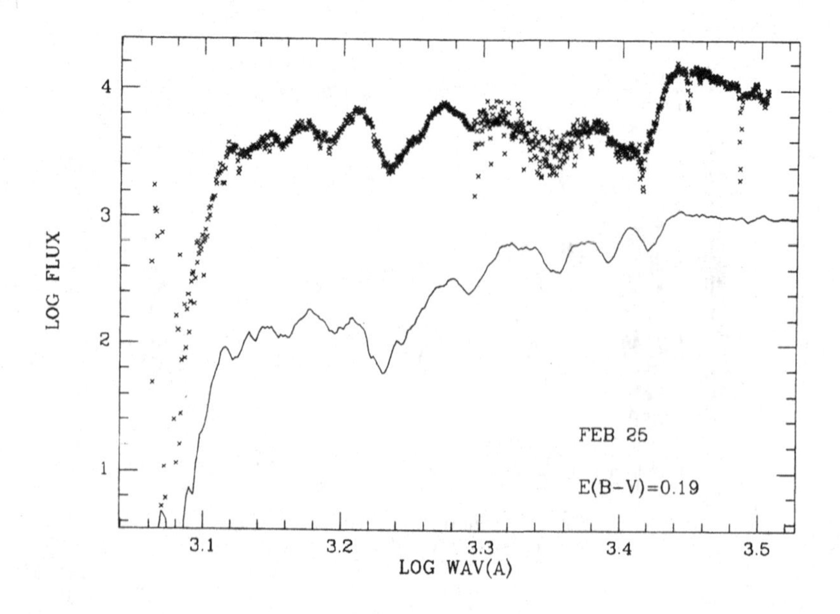

All the above calculations are for E(B-V) = 0.15. To illustrate the sensitivity to this parameter, the Feb 25 calculation has been repeated with E(B-V) = 0.19. Note that changing E(B-V) changes the luminosity and photospheric temperature derived from the data and not merely the wavelength-dependent extinction applied to the theoretical fluxes. Comparison of Figures 1 and 6 shows that the fit deteriorates and so the original choice of E(B-V) = 0.15 receives support. Evidently, an accurate theoretical determination of E(B-V) is possible.

LAST-SCATTERING RADII

Following suggestions at the GMU Workshop by J.E. Felten and M. Karovska, the spectral synthesis code has been used to compute "last-scattering" radii for models appropriate to the dates of the CfA speckle measurements (IAUC No. 4457).

Photons detected at infinity from a uniform disk of radius R have mean impact parameter $\bar{p} = 2/3$ R. Accordingly, from photons in the Monte Carlo calculation that escape to infinity with rest wavelength λ, the radius of the equivalent uniform disk can be defined as $R_\lambda = 3/2\ \bar{p}_\lambda$. From these monochromatic radii, values for comparison with the observed radii are then derived by modelling the filters as box-shaped with a full width of 100 Å. The results are as follows, where R_λ is given both as a multiple of the photospheric radius and in solar units:

	λ(Å)	R_λ	log $R_\lambda/R_\odot$	$(\log R_\lambda/R_\odot)_{obs.}$
	4500	1.90	4.47	4.81
Apr 2	5330	1.41	4.34	4.77
	6560	1.44	4.35	(4.33)
	4500	1.90	4.60	5.09
Jun 1	5330	1.41	4.48	4.99
	6560	1.45	4.49	4.63

On Apr. 2, the source was unresolved at Hα; the radius corresponding to 1σ is therefore given above in parentheses.

For the five positive detections of finite size, the resulting observed radii are from 1.5 to 3.2 larger than the predicted radii, with the average multiple being 2.6. Rather drastic changes in $\rho_1(r_1)$ would be required for line scattering to be of continued significance at such large radii, and such changes would very likely compromise the successes with the earlier spectroscopic record.

THE INVERSE PROBLEM

Given that an essentially continuous stream of high quality data is being accumulated for SN 1987A, the inverse approach of empirically inferring the envelope's density profile can be contemplated as an alternative to the direct approach of converging on $\rho_1(r_1)$ by testing numerous explosion calculations. A possible statement of the inverse problem is as follows:

Given the luminosity history L(t) of a supernova, construct the density profile $\rho_1(r_1)$ at (arbitrary) time t_1 implied by the variation of colour temperature $T_C(t)$.

This implies that the spectrophotometric record be used to obtain $\rho_1(r_1)$, so that spectroscopic record remains available as a check to be performed with a spectral synthesis code.

In considering the inverse approach, it is tempting to make sweeping approximations - e.g., constant opacity - in order to derive formulae that allow $\rho_1(r_1)$ - or at least its form - to be derived algebraically. Such efforts have, however, little merit, since there will then be no reason to believe in the significance of departures from a density function given by an explosion calculation that reproduces the light curve. Accordingly, to be a genuine contribution, an inversion must be based on a treatment of the supernova's surface layers that is substantially more detailed than that of the gas dynamic codes. Indeed, since several relevant effects cannot readily be incorporated into such codes, one can aim by means of a sophisticated inversion at encapsulating the observational record into an empirical determination of $\rho_1(r_1)$ and then challenging the owners of gas dynamic codes to ignore the observations and focus exclusively on reproducing this function.

Although this ideal is far from realization, an interim solution has been obtained that incorporates several effects absent from the gas dynamic codes. The main features of this treatment are: 1) grey atmosphere theory using the Rosseland mean opacity $\kappa(\rho,T)$ for solar composition with $Z = Z_\odot/2.75$; 2) temperature stratification determined by radiative equilibrium with sphericity and line blanketing approximately accounted for. However, an important effect <u>not</u> treated is the delayed thermalization of the radiation field when scattering dominates.

If $\rho_1(r_1)$ is an initial trial function, the envelope's structure at time t is governed by the equations

$$\rho(r,t) = \left(\frac{t_1}{t}\right)^3 \rho_1 \left(\frac{t_1}{t} r\right) , \tag{1}$$

which expresses the assumption of homologous expansion, and

$$T^4 = \frac{1}{2} T_*^4 \left(2W + A + \frac{3}{2} \tilde{\tau}\right) , \tag{2}$$

which is a modified version of the Milne-Eddington formula. Here $W = 1/2 \left[1 - \sqrt{1-(R_*/r)^2}\right]$ is the dilution factor, A(t) allows for the "bright sky" effect of line-scattering in the reversing layer and is derived from the Monte Carlo synthesis calculations, and the optical depth variable

$$\tilde{\tau} = \int_r^\infty \kappa\rho \left(\frac{R_*}{r}\right)^2 dr \ . \qquad (3)$$

The observed luminosity L(t) enters via the equation $L = 4\pi R_*^2 \times \sigma T_*^4$ and the radius R_* is an eigenvalue to be found by imposing the constraint that $\tau(R_*) = 2/3$. With R_* known, the structure is determined, and the colour temperature is then estimated to be $T(\tau=2/3)$, where τ is the conventional, unweighted optical depth.

The above treatment of radiative equilibrium in a grey, spherical atmosphere was used previously (Lucy 1976) for dusty circumstellar envelopes and has been shown with Monte Carlo techniques to give temperatures accurate to $\lesssim 5\%$.

The above formalism has been applied to the luminosities and temperatures derived by Catchpole et al. (1987) assuming m-M = 18.5 and $A_V = 0.4$. The procedure followed is to change the function $\rho_1(r_1)$

Fig. 7: Density profile obtained by empirical inversion compared with Arnett's solution. Envelope masses above indicated points given in solar units.

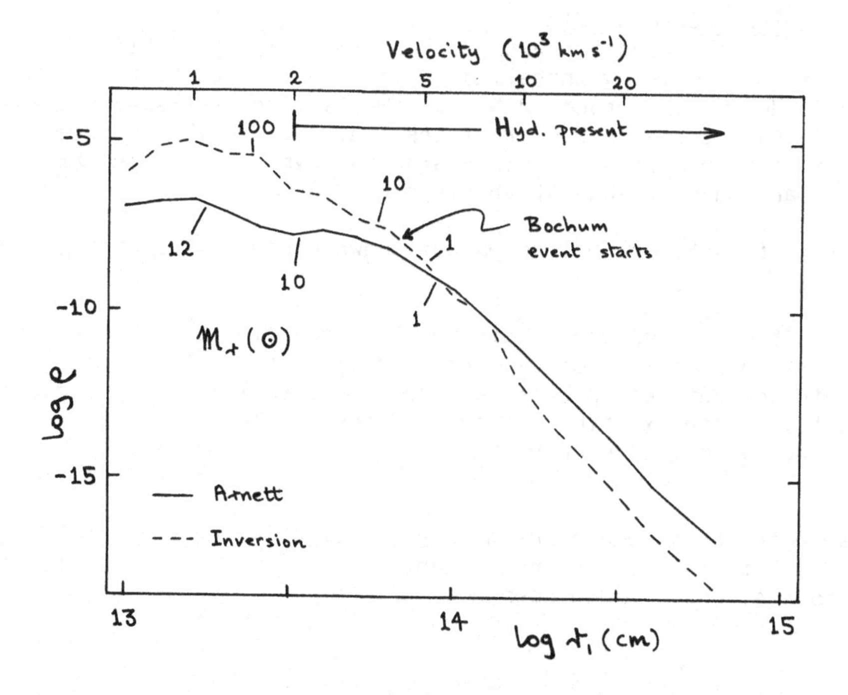

by hand until the computed run of colour temperature matches that observed. However, since the observations start at t = 1.5 days, details of the structure above the photosphere at that time cannot be determined; the solution in these layers is therefore assumed to be parallel to Arnett's on a log-log plot.

The density function thus obtained is compared with Arnett's function in Figure 7. The inversion is lower by ≈ 1.4 dex in the outer layers and this narrows the absorption trough of the CaII infrared triplet - see Figure 6 in Danziger et al. (this vol.). Note, however, that this part of the solution must be suspect because of the neglect of delayed thermalization. In deeper layers, the inversion gives markedly higher densities which translate into unbelievably high envelope masses. Since these high densities are found for layers that certainly contain hydrogen, the metallicity should not yet be enhanced, and so the fault is unlikely to be with the opacities. The tentative conclusion is that after mid-March the physics of the atmosphere is changed because an increasing fraction of the energy input from radioactive decay is being thermalized within and above the continuum-forming layers - i.e., spectrum formation develops an increasing "chromospheric" aspect.

This interim empirical determination of $\rho_1(r_1)$ is certainly not definitive. However, since it could have assigned a small mass to the hydrogen envelope but doesn't, it provides some evidence against scenarios that posit substantial prior mass loss.

Also shown in Figure 7 is the location of the photosphere when the Bochum event (see below) begins. Note that it marks the beginning of the inversion's trend to unacceptably large densities.

THE BOCHUM EVENT

In collaboration with M. Rosa (ST-ECF), possible interpretations are being investigated for the changes in the supernova's Hα profile observed between Mar 14 and 29 (Hanuschik and Dachs 1987a,b) - see also M.M. Phillips (this vol.). The changes are assumed to be intrinsic to Hα since spectral synthesis calculations do not reveal any major blends at this epoch.

Several significant conclusions can be drawn directly from the Bochum data:

1) Because the new feature within the Hα absorption trough appears at about the photospheric velocity, the event is not due to the envelope encountering circumstellar or interstellar matter (Hanuschik and Dachs 1987a) but is a pre-existing property of the envelope that is gradually revealed by the retreat (in mass coordinate) of the photosphere.

2) Because the feature appears at a velocity at which the Sobolev optical depth is large, the enhancement implies a source of additional Hα photons - a density jump does not suffice.

3) Because the violet side of the feature has constant wavelength, the source of extra Hα photons is stationary or nearly so in co-moving coordinates.

4) Because an approximately equally displaced red component subsequently appears (presumably less sharply defined in consequence of attenuation), the source has some global symmetry. However, spherical symmetry is excluded since the extra emission would then be flat-topped rather than double-peaked as observed.

Models that would produce a double-peaked feature include enhanced emission along the axis of a two-sided jet and enhanced emission within an equatorial ring. Of these, the jet is the most immediately attractive because the Bochum event shortly precedes the discovery of the speckle companion (Mar 25, Apr 2; Karovska et al. 1987) and this has been interpreted in terms of a relativistic jet (Rees 1987). On this model, the Hα emission could be attributed to the thermalization of fast particles or hard radiation or to the cooling radiation from shocks, with the radiation originating in either case at or near the walls of the channel created when the envelope was punctured by outflowing relativistic plasma.

But there are arguments against the jet model. The gradual emergence of the Bochum features is not in accord with a phenomenon whose leading edge moves at near relativistic velocity. Of course, one might argue that the punching-through occurred earlier and then contrive an explanation of why radiation from the channel is only detectable after Mar 14.

A further argument against a jet follows from the prediction that the Hα profile should be disturbed only above and below $v_* \cos i$, where i is the jet's inclination to the line of sight and $v_*(t)$ is the photosphere's velocity at Hα. This expectation is, however, contradicted by a shift of peak emission to negative velocities that develops along with the double-peaked feature (Hanuschik and Dachs 1987).

An alternative explanation assumes that hard radiation is, at the epoch of the Bochum event, finally penetrating to the atmospheric layers. If the surface within which there is a significant flux of X-rays is slightly non-spherical, then the first observable consequence could indeed be emission from an equatorial band. On this interpretation, the Bochum event marks the transition from an atmopshere in radiative equilibrium to one with increasingly significant energy deposition. In qualitative support for this conjecture is the above-discussed problem with an inversion predicated on strict radiative equilibrium as well as the growing strength of the emission line spectrum in the weeks following the Bochum event.

REFERENCES

Abbott, D.C.: 1982, Astrophys. J. **259**, 282.
Abbott, D.C., Lucy, L.B.: 1985, Astrophys. J. **288**, 679.
Abbott, D.C., Lucy, L.B.: 1987, in preparation.
Arnett, W.D.: 1987, Astrophys. J., in press.
Catchpole, R.M., Menzies, J.W., Monk, A.S., Wargau, W.F., Pollacco, D., Carter, B.S., Whitelock, P.A., Marang, F., Laney, C.D., Balona, L.A., Feast, M.W., Lloyd Evans, T.H.H., Sekiguchi, K., Laing, J.D., Kilkenny, D.M., Spencer Jones, J., Roberts, G., Cousins, A.W.J., van Vuuren, G., Winkler, H.: 1987, Mon. Not. R. astr. Soc. **229**, 15P.
Dufour, R.J.: 1984, IAU Symposium No. 108, S. van den Bergh and K.S. de Boer (eds.), Reidel.
Fosbury, R.A.E., Danziger, I.J., Lucy, L.B., Gouiffes, C., Cristiani, S.: 1987, ESO Workshop on SN 1987A, ed. I.J. Danziger, p.139.
Hanuschik, R.W., Dachs, J.: 1987a, Astron. Astrophys. **182**, L29.
Hanuschik, R.W., Dachs, J.: 1987b, ESO Workshop on SN 1987A, ed. I.J. Danziger, p.153.
Höflich, P.: 1987, ESO Workshop on SN 1987A, ed. I.J. Danziger, p.449.
Karovska, M., Nisenson, P., Noyes, R., Papaliolios, C.: 1987, IAUC No. 4382.
Kurucz, R.L., Peytremann, E.: 1975, Smithsonian Ap. Obs. Special Report 362.
Lucy, L.B.: 1976, Astrophys. J. **205**, 482.
Lucy, L.B.: 1987a, Astron. Astrophys. **182**, L31.
Lucy, L.B.: 1987b, ESO Workshop on SN 1987A, ed. I.J. Danziger, p.417.
Rees, M.: 1987, Nature **328**, 207.

SN 1987A: A STRIPPED ASYMPTOTIC-BRANCH GIANT IN A BINARY SYSTEM?

P. C. Joss, Ph. Podsiadlowski, J. J. L. Hsu & S. Rappaport
Massachusetts Institute of Technology, MA 02139, USA

Abstract. We propose that the progenitor of SN 1987A was a heretofore undetected red star in orbit with a blue supergiant (star 1 in the field of Sk −69°202). In this picture, the progenitor was the remnant of an asymptotic-branch giant that had lost most of its hydrogen-rich envelope to its blue companion via type C mass transfer (Lauterborn 1970). We have constructed a detailed evolutionary scenario which strongly supports the viability of our proposition. We find that the original mass of the supernova precursor was $\sim 10 - 15\,M_\odot$ (unless a large fraction of its mass was ejected from the binary system), and its final mass (just before the supernova event) was $\sim 3 - 6\,M_\odot$. The system remained bound, with an orbital period of $\sim 3 - 10$ yr and an eccentricity of $\sim 0.1 - 0.4$. We demonstrate below that our picture can provide plausible qualitative explanations for several anomalies in the observational properties of this supernova.

Prior to the supernova event, the field of Sk −69°202 contained three known stars, denoted as stars 1, 2, and 3 in the nomenclature of West *et al.* (1987). Star 1 (by far the brightest star in the visual) has been classified as a B supergiant (Isserstedt 1975a; Rousseau *et al.* 1978), but the available photometric data indicate that the light from this field is too red relative to the intrinsic colors of a star of this spectral type and luminosity class (Johnson 1966). We have placed limits on the brightness of a heretofore unseen red star (star 4 in the above nomenclature) by means of the following procedure (see also White & Malin 1987). We constructed composite spectra of stars 1 and 2, using the photometric data obtained by Isserstedt (1975a) and Rousseau *et al.* (1978) and the empirical color relations given by Johnson (1966). We allowed a range of B2-B5 in the spectral type of star 1, and we assumed that the spectral type of star 2 is no later than B3 (see Blanco *et al.* 1987). To further allow for the uncertainties in the colors, we considered the possibility that the luminosity class of each star is either I or V. We did not explicitly include star 3 in our fits. However, since star 3 is also of early spectral type (Walborn *et al.* 1987), it almost certainly cannot be the source of excess red light from the field.

We find that the dominant source of uncertainty in our fits is the amount of reddening in the direction of Sk −69°202. If the reddening is assumed to be sufficiently high $[E(B-V) \geq 0.16]$, the presence of a red star brighter than $V = 16$ is not needed to obtain a satisfactory fit to the photometric data (see also West *et al.* 1987), though a $V = 15$ red star is consistent with the observations. For lower levels of reddening, a red star brighter than $V = 16$ is *required* to obtain consistency with the data, unless the color of star 1 is very peculiar for its spectral type (see also Blanco *et al.* 1987). For $E(B-V) \approx 0.12$ a $V = 14.5$ red star is consistent with the data; for $E(B-V) \approx 0.08$ (a value typical of the LMC as a whole; see Isserstedt 1975b), even a $V = 14$ red star cannot be excluded. If the spectral type and class of the red star is that of an early to middle M supergiant, as we expect in the context of the evolutionary scenario described below, then its absolute bolometric magnitude could be as low as ~ -6 to -7 (for an assumed bolometric correction of $\sim 1.5 - 2$ mag (Flower 1975), an assumed distance modulus of ~ 18.6, and an assumed

V absorption of ~ 0.4 mag). We have also examined the implications of R and I band photometric data and objective-prism spectroscopic data (Humphreys *et al.* 1987; Blanco *et al.* 1987) and find that these data do not significantly alter the above conclusions. We thus find that a suggestively luminous red star in the field of Sk $-69°202$ is consistent with the available photometric data, in view of the uncertainties in the amount of reddening in the direction of this field. Further, we hypothesize that such a star was, indeed, present as a binary companion to star 1 and was the actual presupernova.

The essential feature of our model is the presence of a nearby companion (star 1), in an orbit with an initial separation of ~ 4 A.U. and an initial period of ~ 2 yr. The companion strongly influences the evolution of the supernova progenitor by accreting much of its hydrogen-rich envelope via type C mass transfer (Lauterborn 1970). (Prior to the occurrence of SN 1987A, an asymptotic-branch giant was, in fact, the generally preferred theoretical model for a type II presupernova; see, for example, Falk & Arnett 1977; Woosley & Weaver 1985). We invoke type C transfer because this scenario explicitly allows star 4 to reach the asymptotic giant branch before mass transfer commences; however, we cannot exclude the viability of alternative binary scenarios wherein mass transfer commences at an earlier evolutionary stage (see also Fabian *et al.* 1987).

We now briefly describe how this type of binary model can explain, at least qualitatively, many of the anomalous features of SN 1987A. (1) A reduction of the envelope mass to $\sim 1-3\ M_\odot$ prior to the supernova event would have lowered the total mass of ejected material, but would have preserved the presence of Balmer lines in the supernova spectrum. (2) A lower envelope mass would have increased the initial velocities imparted to the ejecta, and higher expansion velocities would, in turn, have produced a more rapid evolution of the ultraviolet light curve; these inferences are in accord with the observational data (Blanco *et al.* 1987). (3) The early and low peak of the hard X-ray (10 – 30 keV) light curve (Dotani *et al.* 1987) may indicate a relatively low progenitor mass and a low-mass hydrogen-rich envelope. (4) The occurrence of the supernova in a close binary would provide a plausible explanation for the substantial asymmetries in the expansion of the supernova shell that have been inferred from spectroscopic observations (see, for example, Hearnshaw & Haar 1987). (5) Although the rapid rise time of the optical light curve and the low plateau luminosity have been advanced (Woosley *et al.* 1987) as an argument that the radius of the presupernova was at least a factor of ~ 4 smaller than in our scenario ($\sim 2 \times 10^{13}$ cm), published models (Woosley *et al.* 1987) suggest that a large initial radius can be counterbalanced by a reduction of the mass of the hydrogen-rich envelope and a decrease in the kinetic energy imparted to the ejecta. Our models therefore appear to be consistent with the constraints imposed by the early light curve of the supernova. (6) Jet models (see, for example, Rees 1987) to explain the mysterious companion object of the supernova (Nisenson *et al.* 1987) may require a close binary companion. (7) Barium excesses similar to that reported by Williams (1987) for the supernova progenitor have previously been observed only for stars which are members of close binary systems (McClure 1983).

Another advantage of our model is that we do not have to (1) assume a metallicity for the presupernova (cf. Arnett 1987; Hillebrandt *et al.* 1987) that is substantially lower than that of most early-type stars in the LMC (Becker *et al.* 1984; Harris 1983; Smith 1980), or (2) introduce a non-standard treatment of convection within the presupernova or a high *ad hoc* mass-loss rate from the presupernova (cf. Woosley *et al.* 1987). We also note that irrespective of any specific observational evidence, the fact that a large fraction of all stars occur in binaries (see, for example, Abt & Levy 1978) suggests the advisability of determining whether a viable binary model for SN 1987A can be constructed.

For specificity, we have restricted most of our model calculations to a binary wherein the initial primary (presupernova) star has a mass, M_4, of 12 $M_\odot$ and the initial secondary star (now star 1) has a mass, M_1, of 10 $M_\odot$. We have also carried out some additional calculations with the initial value of M_4 as low as 10 $M_\odot$ and as high as 15 $M_\odot$ and with the initial value of M_1 in the range $4\,M_\odot \leq M_1 \leq M_4$. If the initial value of M_4 had been less than $\sim 8\,M_\odot$, star 4 would probably have failed to become a type II supernova (Iben & Renzini 1983); on the other hand, our calculations indicate that if the initial value of M_4 had exceeded $\sim 15\,M_\odot$, then the presupernova would have been too bright in the visual for consistency with the observations. The range of initial masses for star 1 is constrained by the requirements that this star would not have evolved more rapidly than star 4 but could, nevertheless, have become sufficiently massive to be a B supergiant prior to the supernova event.

Our calculations begin with an approximate numerical model that we constructed for a $12\,M_\odot$ asymptotic giant-branch star (star 4) with a hydrogen-exhausted core of mass $2.7\,M_\odot$, a metallicity of $Z = 0.02$ in the stellar envelope, and a convective mixing length equal to the pressure scale height. Star 4 is assumed to be in orbit with a $10\,M_\odot$ star (star 1). We do not calculate the detailed evolution of star 1, nor do we attempt to determine the amount of mass loss from either component that may have occurred at earlier evolutionary stages. We choose an initially circular orbit with a separation, a, such that the system would come into contact during the asymptotic giant-branch evolution of star 4, resulting in type C mass transfer (Lauterborn 1970) as described above. We use a Henyey-type code (Kippenhahn *et al.* 1967), which we have modified for our present purposes, to follow the subsequent evolution of star 4. The concomitant evolution of the orbital parameters is followed by the methods developed by Rappaport, Joss, and Webbink (1982) and Rappaport, Verbunt, and Joss (1983) and is specified by two free parameters, the fraction, β, of the mass lost by star 4 that is accreted by star 1 and the specific angular momentum, α, of any matter lost from the system in units of $2\pi a^2/P$, P being the orbital period. We assumed that the values of α and β were constant throughout each evolutionary calculation. Further details of our theoretical assumptions and computational procedure will be published in a subsequent paper (Joss *et al.* 1987). It is noteworthy that the mass-transfer binary system AZ Cas has measured properties (Cowley *et al.* 1977) that resemble, in several respects, those of the model presupernova system that we propose.

Since M_4 is initially greater than M_1, the mass transfer, when it commences, is generally unstable and proceeds on a dynamical timescale (Paczyński 1970). Our code is unable to treat the hydrodynamics of the unstable mass-transfer phase; we therefore use the following approximate technique to follow the binary evolution during this phase (see also Iben & Tutukov 1987). During each time step of the evolutionary calculation, we checked to determine whether the radius, R_4, of star 4 exceeded that of its Roche lobe, R_L (whose value we determined by application of a formula due to Eggleton (1983)). Whenever $R_4 > R_L$, we removed the outermost $\sim 0.2\%$ of the mass of star 4; the appropriate portions of the removed matter were taken to be accreted by star 1 or ejected from the system, as specified by the prescribed value of β. The remainder of star 4 was then allowed to relax to a state of hydrostatic equilibrium on a timescale that was assumed to be short compared to all relevant timescales other than the dynamical timescale for this star; in particular, we assumed that the distribution of specific entropy throughout star 4 was fixed. This procedure neglects the possible hydrodynamic effects of mass motions on the mass-transfer process, but we believe it to be an adequate approximation for our purposes.

We discontinued the unstable mass-transfer process when R_4 again became less than R_L. For a specified initial structure for star 4 and a specified evolutionary epoch at which the system comes into contact, the value of R_4 is uniquely determined by the amount of mass that the star has lost. However, the total amount of mass lost from star 4 is a function of α, β, and the initial value of M_1 (see Figs. 1 and 2). Immediately following the unstable mass-transfer stage, star 4 undergoes an epoch of stable Roche-lobe overflow which proceeds initially on the thermal timescale of the residual stellar envelope and subsequently on the timescale of nuclear evolution of the stellar core. We are able to follow these evolutionary phases directly using our modified Henyey-type code. We find that for initial values of M_4 less than or of the order of 12 $M_\odot$, this epoch terminates and R_4 begins to decrease prior to the supernova event; for initial values of M_4 higher than $\sim 15\,M_\odot$, the system remains in contact and R_4 continues to increase until star 4 becomes a supernova. We terminate our evolutionary calculations at or shortly before the final nuclear burning stages that immediately precede the supernova event itself.

The results of our evolutionary calculations are summarized in Figs. 1-2. Figure 1 shows the variation of R_4 with residual mass of the hydrogen-rich envelope, for an initial primary mass of 12 $M_\odot$, a range of initial secondary masses, and chosen illustrative values of α and β ($\alpha = 1$, $\beta = 0.9$). Figure 2 displays the residual envelope mass, $M^f_{\rm env}$, of star 4 shortly before the supernova event in our standard model as a function of α and β; it is evident that there is a substantial swath of plausible values of α and β for which $M_{\rm env} \simeq 1 - 3\,M_\odot$, as required in our scenario.

Our evolutionary calculations lend credence to the binary scenario that we have proposed for SN 1987A. We note, in particular, that the spectral type and luminosity class of star 1 would be consistent with those observed if this star were currently on the main sequence (see, for example, Meylan & Maeder 1982) or in any post-main-sequence evolutionary phase up to the epoch of core helium burning. Any such evolutionary state would, in turn, be consistent with theoretical expectations if, as expected, star 1 were originally the less massive of the binary components. (However, the thermal readjustment timescale for the envelope of star 1 after the mass-transfer phase is of the same order as the remaining lifetime of the primary; hence, the luminosity and spectral class of star 1 may have been significantly altered by the preceding mass-accretion episode at the time of the supernova event.) We conclude that a substantial range of values of α and β lead not only to acceptable values for $M_{\rm env}$, but also to final masses for star 1 in the range of $18 \pm 4\,M_\odot$, as suggested by our fits to the photometric data for the presupernova. Furthermore, in our standard model, star 4 just prior to the supernova event is an underluminous early to middle M supergiant with a V magnitude of $\sim 14 \pm 0.5$; these theoretical results are consistent with our analysis, as described above, of the constraints imposed by the available photometric and spectroscopic data for the presupernova field.

In our models, the final mass of star 4 (just prior to the supernova event) was $\sim 3 - 6\,M_\odot$. On the basis of simple dynamical considerations, we find that a spherically symmetric supernova event should have left the system in a bound orbit, with a new orbital period of $\sim 3 - 10$ yr and an eccentricity of $\sim 0.1 - 0.4$.

We predict that if our model is correct, star 1 will reappear inside the thinning supernova shell within the next few years. Finally, we note that if the supernova has left a pulsar remnant, it will not only become straightforward to confirm or refute the duplicity of SN 1987A, but also to verify the orbital parameter values that we have suggested and to distinguish between our model and alternative binary models (such as that proposed by Fabian *et al.* 1987).

We are very grateful to L. Nelson for providing us with a copy of his stellar evolution code and for many helpful discussions concerning the modification and implementation of the code. An earlier version of this paper has been submitted for publication in *Nature*. This work was supported in part by the US National Science Foundation under grant AST-8419834 and by the US National Aeronautics and Space Administration under grants NSG-7643 and NGL-22-009-638.

Reference list
Abt, H. A. & Levy, S. G. (1978). Astrophys. J. Suppl., 36, 241.
Arnett, W. D. (1987). Astrophys. J., 319, 136.
Becker, S. A., Mathews, G. J. & Brunish, W. M. (1984). In IAU Symp. No. 105, Observational Tests of Stellar Evolution Theory, ed. A. Maeder & A. Renzini, 83. Dordrecht: Reidel.
Blanco, V. M., Gregory, B., Hamuy, M., Heathcote, S. R., Phillips, M. M., Suntzeff, N. B., Terndrup, D. M., Walker, A. R., Williams, R. E. Pastoriza, M. G., Storchi-Bergmann, T. & Matthews, J. (1987). Astrophys. J., 320, 589.
Cowley, A. P., Hutchings, J. B. & Popper, D. M. (1977). Pub. A.S.P., 89, 882.
Dotani, T., *et al.* (1987). Preprint.
Eggleton, P. P. (1983). Astrophys. J., 268, 368.
Fabian, A. C., Rees, M. J., van den Heuvel, E. P. J. & van Paradijs, J. (1987). Nature, 328, 323.
Falk, S. W. & Arnett, W. D. (1977). Astrophys. J. Suppl., 33, 515.
Flower, P. J. (1975). Astron. Astrophys., 41, 391.
Harris, H. C. (1983). Astron. J., 88, 507.
Hillebrandt, W., Höflich, P., Truran, J. W. & Weiss, A. (1987). Nature, 327, 597.
Hearnshaw, J. B. & Haar, J. (1987). IAU Circ. No. 4352.
Humphreys, R. M., Jones, T. J., Davidson, K., Ghigo, F. & Zumach, W. (1987). IAU Circ. No. 4325.
Iben, I., Jr. & Renzini, A. (1983). Ann. Rev. Astron. Astrophys., 21, 271.
Iben, I., Jr. & Tutukov, A. V. (1987). Astrophys. J., 313, 727.
Isserstedt, J. (1975a). Astron. Astrophys. Suppl., 19, 259.
Isserstedt, J. (1975b). Astron. Astrophys., 41, 175.
Johnson, H. L. (1966). Ann. Rev. Astron. Astrophys., 4, 193.
Joss, P. C., Podsiadlowski, Ph., Hsu, J. J. L. & Rappaport, S. in preparation.
Kippenhahn, R., Weigert, A. & Hofmeister, E. (1967). In Methods in Computational Physics, 7, ed. B. Alder, S. Fernbach & M. Rothenberg, 129. New York: Academic Press.
Lauterborn, D. (1970). Astron. Astrophys., 7, 150.
Meylan, G. and Maeder, A. (1982). Astron. Astrophys., 108, 148.
McClure, R. D. (1983). Astrophys. J., 268, 264.
Nisenson, P., Papaliolios, C., Karovska, M. & Noyes R. (1987). Astrophys. J. Lett, 320, L15.
Paczyński, B. (1970). In IAU Colloq. No. 6, Mass Loss and Evolution in Close Binaries, ed. K. Gyldenkerne & R. M. West, 139. Copenhagen: Copenhagen University Publications.
Rappaport, S., Joss, P. C. & Webbink, R. F. (1982). Astrophys. J., 254, 616.
Rappaport, S., Verbunt, F. & Joss, P. C. (1983). Astrophys. J., 275, 713.
Rees, M. J. (1987). Nature, 328, 207.
Rousseau, J., Martin, N., Prévot, L., Rebeirot, E., Robin, A. & Brunet, J. P. (1978). Astron. Astrophys. Suppl., 31, 243.
Smith, H. A. (1980). Astron. J., 85, 848.
Walborn, N. R., Lasker, B. M., Laidler, V. G. & Chu, Y.-H. (1987). Astrophys. J. Lett., 321, L41.
West, R. M., Lauberts, A., Jørgenson, H. E. & Schuster, H.-E. (1987). Astron. Astrophys. Lett., 177, L1.
White, G. L. & Malin, D. F. (1987). Nature, 327, 36.
Williams, R. E. (1987). Astrophys. J. Lett., 320, L117.
Woosley, S. E., Pinto, P. A. & Ensman, L. Astrophys. J. (submitted).
Woosley S. E. & Weaver, T. A. (1985). Proc. 5th Moriand Astrophys. Conf. on Nucleosynthesis and its Implications on Nuclear and Particle Physics, ed. J. Audouze & T. van Thuan, 145. Dordrecht:Reidel.

Figure 1. *Solid curve:* Calculated radius of star 4 as a function of residual mass, $M_{\rm env}$, for the hydrogen-rich envelope (increasing toward the left), during the dynamical mass-loss phase, for an initial total mass of 12 $M_\odot$ and an initial core mass of 2.7 $M_\odot$. *Dashed curves:* Radius of the Roche lobe of star 4 as a function of $M_{\rm env}$, for various values of the initial mass, M_1^i, of star 1 and for chosen illustrative values of α and β ($\alpha = 1$, $\beta = 0.9$). We assume that the evolution of star 4 follows the solid curve toward the right during the dynamical mass-loss phase, and then follows the appropriate dashed curve toward the right during the subsequent phases of dynamically stable mass transfer. The dot-dashed curve corresponds to $M_1^i = 10\,M_\odot$, and the arrows indicate the evolutionary track followed by our standard model (see text). *Dotted curve:* Maximum radius achieved by star 4 for various values of M_1^i, as indicated, and for other parameter values as stated above.

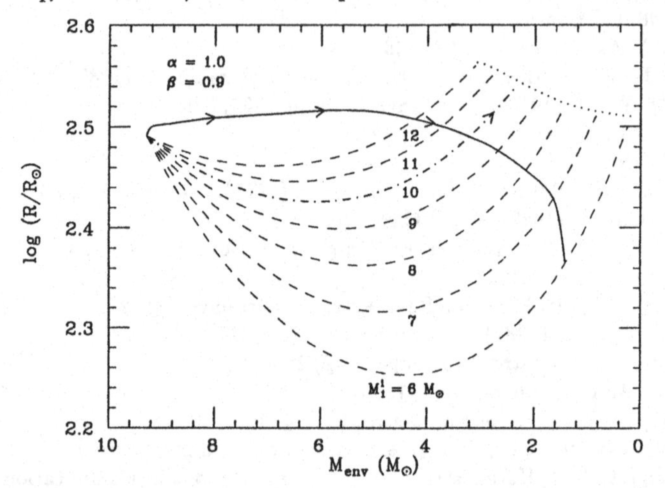

Figure 2. *Solid curves:* Loci of values of α and β which result in specified values for the final envelope mass, $M_{\rm env}^f$, of star 4 just prior to the supernova event in our standard model. Each curve is labeled with the value of $M_{\rm env}^f$ in units of solar masses. *Dashed curve:* Locus of values of α and β, in our standard model, for which the final mass of star 1 is 14 $M_\odot$, which is the lower limit allowed by our fit to the photometric observations. (The maximum possible value for the final mass of star 1 in our standard model is 19.3 $M_\odot$, which is substantially lower than the observational upper limit of $\sim 22\,M_\odot$.) *Shaded region:* Domain in α-β parameter space which simultaneously obeys the constraints that $1\,M_\odot \leq M_{\rm env}^f \leq 3\,M_\odot$ and that the final mass of star 1 exceeds 14 $M_\odot$.

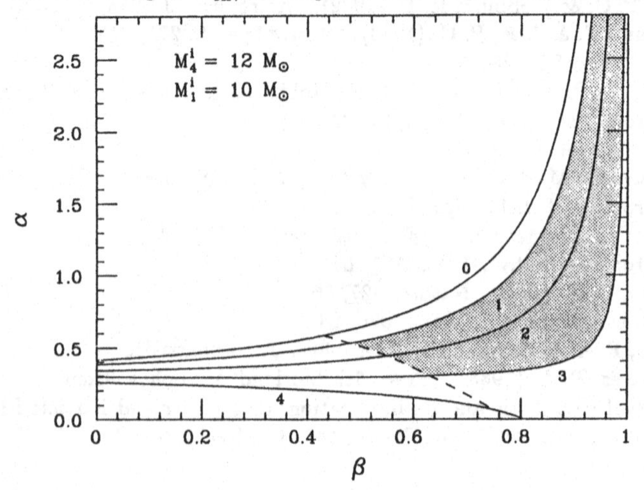

PULSAR FORMATION AND THE FALL BACK MASS FRACTION

Stirling A. Colgate
Theoretical Astrophysics, Los Alamos National Laboratory
Los Alamos, New Mexico 87545

Abstract

The satisfying picture of the explosion of 1987a following collapse to a neutron star and neutrino emission is difficult to reconcile with the subsequent behavior of the ejected mass. It is shown that the inner solar mass of ejected matter should progressively fall back onto the neutron star after it collides with the outer 10-15 $M_\odot$ and after an initial phase of explosion driven by a hot bubble for $t \gtrsim 2 \times 10^3$ s. This fall back is augmented due to heating by the radioactive decay of ^{56}Ni. The matter accreted onto the neutron star rapidly cools due to neutrino emission and merges with the neutron star, thus resulting in zero back pressure to the free falling matter. The predicted fall back mass $\gtrsim 1$ $M_\odot$ is far larger than than is consistent with observations. The most likely explanation is that the hot radiation dominated bubble is continuously heated by neutrino emission from continuing accretion. This accretion continues until the presumed pulsar magnetic field exceeds the bubble pressure.

Introduction

The extraordinary confirmation of the general theory of supernova by the sequential developments of 1987a has been a joyous story for all of us involved. The experimental observation of the earlier pulses from the Mount Blanc collaboration is one exception. There are several more boots to drop: The gamma rays have yet to be seen, and the emergence of the pulsar may be the greatest finale. (The positron escape or transparency will be an unseen lesson starting at a time ≅ 17 times the gamma ray 1 mean free path time (≅ 1.2 y) or ≅ 21 y from now with its probable 56-day signature.) The expected emergence of the neutron star as either a pulsar or x-ray source raises several fundamental questions concerning the hydrodynamics of the explosion mechanism and behavior of a presumed initial magnetic flux.

The Pulsar Formation Problem

The formation of a pulsar following the initial collapse to a neutron star involves several stages that may not have been modeled adequately. These stages are the fall back onto the neutron star of matter previously ejected and the reconnection of magnetic flux that presumably initially threads the ejected and collapsed matter.

One assumes the collapse to a neutron star is the standard 15 $M_\odot$ iron core blue giant (Woosley 1987, Arnett 1987, Hillebrandt et al. 1987) with the associated neutrino emission so dramatically confirmed by the signals from 1987a.

The Supernova Explosion Mechanism

The mechanism of the explosion is of importance to the following discussion because in order to limit initially the fall back mass fraction, the region of matter separating the neutron star from the ejected matter must be a hot, low density expanding "bubble." Otherwise, as we shall see, the reimplosion mass flux will be unacceptably large. There are two mechanisms for the explosion: the bounce on the neutron star core initially and found inadequate by Colgate and White (1966, hereafter C&W) and resurrected by Bethe et al. (1979) and the accretion-derived neutrino heating C&W and Wilson (1971).

Since the bounce shock mechanism marginally causes a viable explosion, it is hard to see how the mass fraction at the separating boundary can be maintained on a high adiabat some 10 to 100 times the shock entropy, (the amount needed to make a hot bubble) without major continuing heating during expansion by the neutrino flux. Such a heating is then sufficient to cause the explosion regardless of the bounce shock. In this picture the bounce shock then becomes a short hiatus in the collapse allowing the time for a partial deleptonization of the core and an increase in binding energy before the major energy release in neutrinos by subsequent accretion commences.

Bounce Good Fortune vs. Reversal Terminated Accretion

In this picture the explosion is not the great good fortune of nature that somehow found 1/2% positive energy difference between the collapsed neutron star binding energy and the neutrino emitted energy. This small energy difference is then the positive energy necessary to eject matter with $\cong 1 \times 10^{51}$ ergs leaving behind 100% in binding and 99.5% lost in neutrinos. Instead I believe continuing collapse and neutrino emission from the accretion shock on the deleptonized tightly bound, $c^2/10$, neutron star continues to heat imploding matter to a high enough entropy that the pressure reverses the implosion, turning off the accretion heating (neutrino) mechanism. This reversal is the explosion. In other words the velocity reversal necessary to turn off accretion need only represent a small velocity change from negative to positive and have a small kinetic energy, but it is sufficient to turn off the energy release mechanism. This the small 1/2% kinetic energy of the ejecta is a natural consequence of an energy release mechanism, the implosion, that is turned off by a small reversal in velocity, rather than the unlikely serendipitious difference between two large numbers, the binding and neutrino emission. The explosion conditions at times longer than this have not yet been calculated by Mayle and Wilson (1987). It will now be considered.

Explosion, Early Conditions

The adjacent external solar mass surrounding the neutron star, hereafter called the driving solar mass, has a sonic time constant of roughly 1/10 s, which is also the characteristic time scale of the bounce and accretion. Hence, one expects approximately a solar mass to be ejected with enough energy to create the supernova. This solar mass must have enough interval and kinetic energy to (1) overcome its gravitational binding to the combined mass of itself plus the neutron star, and (2) enough additional energy to both overcome the remaining $\cong 13\ M_\odot$ binding energy and eject it with the canonical 1×10^{51} ergs. The internal energy of the driving solar mass is chosen sufficient to overcome its own binding. With these criteria and choosing a radius of the driving solar mass of $r = 10^8$ cm, one has the following approximate conditions for the start of the explosion:

$$r_o = 10^8 \text{ cm}$$

$$\rho_o = 5 \times 10^8 \text{ g cm}^{-3}$$

$$\varepsilon_o = 2 \times 10^{18} \text{ ergs g}^{-1}$$

$$v_o = 1.4 \times 10^9 \text{ cm s}^{-1}$$

The Hot Explosion Bubble

The extreme optimistic view of the explosion is that a hot bubble of radiation gas of near zero nucleon density initially expands between the neutron star and the driving solar mass so that initially the reimplosion nucleon mass can be neglected. The condition for such a hot bubble is that the radiation pressure exceed the pressure inside the driving solar mass, or $T \gtrsim 1.7 \times 10^{10}$ K. In addition one notes that at $T_{10} \gtrsim 1.7$ the pair density exceeds the electron density of the driving solar mass. At low neucleon density in the bubble the pair rest mass becomes the accreted mass and the optical depth to the emitted neutrinos is roughly 1%. The cooling by neutrino pair emission will be discussed later, but the optimistic assumption is that the hot radiation bubble initially pressure supports and ejects the driving solar mass.

However, after some time the driving solar mass collides with the remaining 13 $M_\odot$. This is a progressive collision with a polytropic density gradient so that the reflected shock in the driving solar mass will be weak and possibly nonexistent. If it were strong, corresponding to the full velocity change of $v_o = 1.4 \times 10^9$ cm s^{-1}, it would only make the following arguments stronger since the stronger shock means higher specific internal energy and a larger sound speed which determines the blow-back mass flux.

Post Collision Conditions

The driving solar mass will have collided with most of the remaining mass, $\cong 10\ M_\odot$ by $r_1 \cong 10^{12}$ cm, with a mean velocity $v_1 = 3 \times 10^8$ cm s^{-1} in $\cong 3\times 10^3$ seconds. The approximate conditions after some expansion are

$$r_1 = 10^{12} \text{ cm}$$

$$v_1 = 3 \times 10^8 \text{ cm s}^{-1}$$

$$\rho_1 = 2 \times 10^{-2} \text{ g cm}^{-3}$$

$$\varepsilon_1 = 2 \times 10^{17} \text{ ergs g}^{-1}$$

$$M_1 = 10\ M_\odot$$

Inside this is the driving solar mass with a near uniform pressure since it has now had time to expand backwards, filling what was the radiation bubble after cooling by neutrino emission. This expansion backwards occurs because of the back shock, which is weak because of the polytropic density gradient. The weak back shock will heat the driving solar mass above its expansion adiabat by at least a factor of times 2. This adiabat corresponds to the collided mass pressure of $\rho_1 \varepsilon_1/3 = 1.3 \times 10^{15}$ dynes cm^{-2}. The corresponding density $\rho_2 = 1/2 \rho_o (\rho_1 \varepsilon_1/\rho_o \varepsilon_o)^{3/4}$; radius $r_2 = (M_\odot/4\rho_2)^{1/3}$ and $v_2 = v_1 r_2/r_1$ becomes

$$r_2 = 10^{11} \text{ cm}$$
$$v_2 = 3 \times 10^7 \text{ cm s}^{-1}$$
$$\rho_2 = 0.7 \text{ g cm}^{-3}$$
$$\varepsilon_2 = 6 \times 10^{15} \text{ ergs g}^{-1}.$$

Heating by ^{56}Ni

The inner mass fracion of this driving solar mass is presumably ^{56}Ni, although the ^{56}Ni may be mixed throughout the driving solar mass because of instabilities. An initial shock corresponding to the conditions $\varepsilon_o = 2 \times 10^{18}$ erg g^{-1} and $\rho_o = 5 \times 10^8$ g cm^{-3} corresponds to a temperature of $T \cong (2\varepsilon_o \rho_o/11a)^{1/4} \cong 1$ MeV. At this temperature and density all α particle nuclei, i.e. carbon and heavier will burn to ^{56}Ni in the expansion time of $r_o/v_o \cong 1/10$ second. Hence, we expect this material to be heated by the radioactive decay of ^{56}Ni of $\dot{\varepsilon}_{56} = 4.0 \times 10^{10}$ ergs g^{-1} s^{-1}. This heating will exceed the adiabatic cooling of the energy density following isobaric conditions at $t_1 = t_2 = r_1/v_1 = r_2/v_2 = 3 \times 10^3$ s after which one expects ε to follow the radiation dominated adiabat, $\varepsilon = \varepsilon_2 t_2/t$. Hence, ^{56}Ni heating dominates at t_3, when

$$\dot{\varepsilon} t_3 = \varepsilon_2 t_2/t \qquad (1)$$

or when $t_3 = 2 \times 10^4$ s. Thereafter the competition between expansion and heating leads to an internal energy of roughly $\varepsilon = \varepsilon_{56} t/2$.

Blow Back Condition

In Colgate (1971) the blow back condition is given assuming that the neutron star surface acts as a perfect sink for any mass flux, i.e. as a black hole. We will justify this assumption shortly. Assuming this, then the second condition is that the velocity of a backwards rarefaction mass flow exceeds the outward isobaric velocity trajectory minus escape velocity. The final condition is that a rarefaction wave (sound characteristic) must reach the mass point in question.

Free-Fall Zero Entropy Change

The assumption here is that once matter is injected onto a free fall, zero angular momentum trajectory towards a gravation sink, it will continue to free fall regardless of its internal energy. This is equivalent to saying that no entropy is produced by shocks or transport and that ε/r is a decreasing function of r. The former is assured because in free fall $\rho \propto \rho_o r^{-1/2}$ (Colgate and Petschek 1981) and $\varepsilon \propto (\rho/\rho_o)^{\gamma-1} \propto r^{-(\gamma-1)/2} = r^{-1/6}$. The latter is so because the radiation diffusion is small and there is no back pressure. This free fall condition or zero entropy change, regardless of q, the artificial viscosity, was checked in C&W, but not usually in other hydrocodes.

Rarefaction Expansion Velocity Distribution

A rarefaction wave converts the internal energy of a fluid into a "rarefaction fan" (Courant and Fredricks 1963) whose limiting velocity is

$$v_{max} = -(2/(\gamma-1)\ c_s = 6\ c_s \quad . \tag{2}$$

where c_s is the sound speed and γ the usual ratio of specific heats, or 4/3 in the case of a radiation dominated gas. The mass fraction corresponding to this limiting velocity is a decreasing function of the mass overtaken by the rarefaction wave, which progresses backwards into the fluid at velocity c_s. Here it is assumed that the expansion in the backwards direction occurs in a time short compared to that of the unperturbed expanding fluid. This may be a poor approximation. Hence, for the reimplosion mass flux condition it is simpler to choose the point in the rarefaction fan where the velocity corresponds to a kinetic energy equal to the initial internal energy of the fluid. Then all the mass overtaken by the wave can expand at this velocity. This corresponds to a velocity v_{ex} of

$$v_{ex} = c_s(2/[\gamma(\gamma-1)])^{\frac{1}{2}} = 2.1\ c_s \quad . \tag{3}$$

In order for the velocity of the blow back matter to exceed the explosion trajectory velocity, minus the escape velocity, we have the condition for the inner driving solar mass,

$$2.1\ c_s \geq v_2\ F^{1/3} - v_{escape} \quad . \tag{4}$$

Here F is the mass inside the rarefaction wave satisfying this condition. However, $c_s^2 = \gamma(\gamma-1)\varepsilon$ and so the blow back condition becomes

$$(2\varepsilon)^{\frac{1}{2}} = v_2\ F^{1/3} - v_{escape} \quad . \tag{5}$$

In the time interval t_2 to t_3 before ^{56}Ni heating dominates, this gives

$$F = (2\ \varepsilon_2 t_2/t)^{3/2}\ v_2^{-3}$$

$$= 8 \times 10^6\ t^{-3/2} \qquad t_2 < t < t_3$$

and

$$F = 3 \times 10^{-7}\ t^{3/2} \qquad t > t_3 \quad . \tag{6}$$

Here we have neglected the correction due to escape velocity for algebraic simplicity. Including it only makes more mass blow back. Both before and after $t = t_3 = 2 \times 10^4$ s, the limiting mass fraction is greater than one, and so all the driving mass can potentially fall back if uncovered by a rarefaction wave, i.e. the sound speed is greater than the expansion velocity.

Hence, to obtain the actual fall back mass, the sound characteristic must be integrated from time t_2 where the presumed isobaric condition commences. In Colgate (1971) the integration of the sound characteristic is given and shown in Fig. 1, starting at zero time.

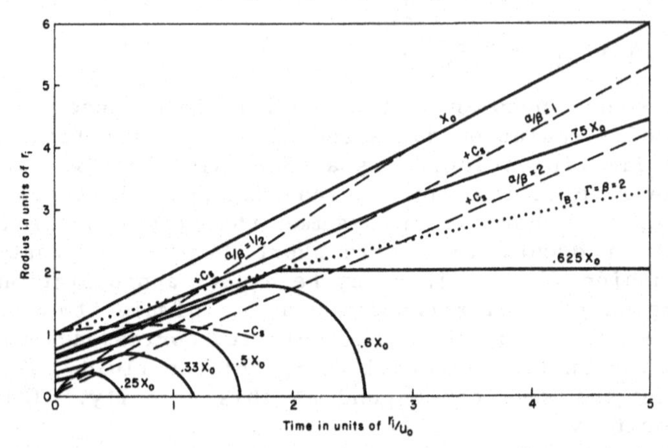

FIG. 1.—Radius versus time of the explosion history in linear coordinates and using the reduced variables $X = r/r_i$ and $\tau = tU_0/r_i$. *Heavy lines*, Lagrange coordinates of various mass fractions denoted by the initial radius fraction of the outer boundary X_0. The inner mass fractions reimplode when overtaken by the outgoing rarefaction wave, denoted by $+C_s$. Three such waves (*dashed curves*) are shown for various ratios of α/β, where α/β is the ratio of internal to kinetic energy. The escape-velocity boundary r_B is shown as a dotted curve for the condition $\Gamma = \beta = 2$. The reimplosion terminates when the rarefaction wave passes the escape-velocity boundary.

Here the start of the rarefaction wave takes place later at $t_2 = 3 \times 10^3$ s. Several approximations can now be made to give a simpler estimate of the reimplosion mass flux.

The mass fraction uncovered by a rarefaction wave is

$$dF/dt = 4\pi\, r_s^2\, \rho\, c_s/M_\odot \tag{7}$$

where r_s is the radius of the characteristic and ρ the instantaneous density. As in Colgate 1971, the velocity of the characteristic becomes

$$\dot{r}_s = v + c_s = v_2\, F^{1/3} + (5/9\ \varepsilon)^{1/2} \tag{8}$$

Integrating gives

$$r_s = v_2 F^{1/3}(t - t_2) + (\varepsilon_2 t_2 (t - t_2))^{1/2} + 1/3(\dot{\varepsilon}_{56})^{1/2}\, t^{3/2}$$

for

$$t_2 < t < t_3$$

or

$$r_s = 3 \times 10^7\ F^{1/3}(t-t_2) + 4 \times 10^9\ (t^{\frac{1}{2}}-t_2^{\frac{1}{2}}) + 7 \times 10^4\ t^{3/2}\ \text{cm}. \quad (9)$$

This can be substituted into equation (7) and integrated to find F, but a simpler result is to neglect the expansion velocity, v, relative to the sound speed in the time of interest. This further underestimates the reimplosion mass, but even so if $r_s = v_2\ t$, F = 1, and all the driving solar mass will have collapsed onto the neutron star. The neutron star would then be close to collapsing to a black hole. This occurs in Eq. (9) with the first term on the right hand side neglected at $t = 5 \times 10^4$ s, only slightly greater than the time of transition to radioactive heating. The time to reimplode the remaining mass of the star is only five times longer and hence, both scenarios are completely inconsistent with observations. Hence, something must prevent the mass accretion onto the neutron star.

Mass Accommodation at the Neutron Star

As pointed out in Colgate 1971, the mass accretion onto the neutron star gives rise to a hot accretion shock at the surface of a hot, accreting, subsonic atmosphere, Fig. 2.

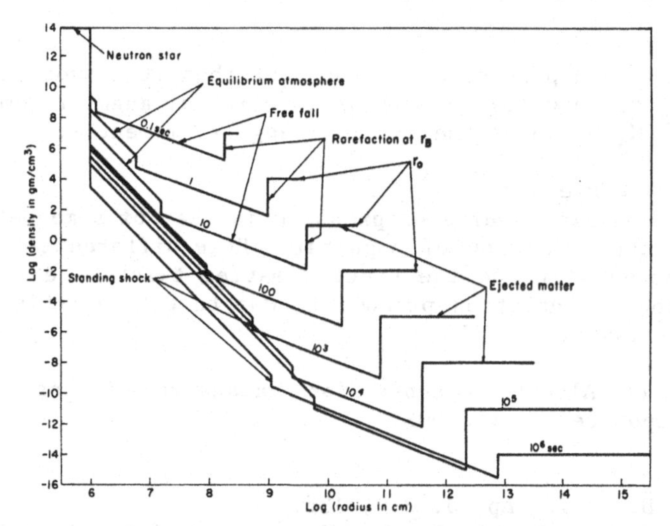

FIG. 2.—Density distribution at later times as a function of radius. Outer boundary is $r_0 = U_0 t$. The rarefaction wave is approximated as a discontinuous decrease in density at the escape-velocity boundary r_B. Free fall of the matter takes place to some radius where a standing shock matches onto an equilibrium neutron-star atmosphere $\rho \propto r^{-4}$. The rate of mass accumulation is determined by this atmosphere at the surface of the neutron star and in turn is equated to the flux of matter in free fall.

This atmosphere increases in temperature adiabatically as a function of depth to the point where the internal energy is emitted as neutrinos by electron pair annihilation. The mass flux of a solar mass in $t = 5 \times 10^4$ s corresponds to a shock density of

$$\rho_s = 7\ M_\odot/(4\pi\ R_{Ns}^2(c/3)\ t) = 2 \times 10^6\ \text{g cm}^{-3} \quad (10)$$

with a specific internal energy of $c^2/10$ and $T_s = 10^{10}$ K. The temperature at which cooling by pair neutrino emission in a time of accretion, 5×10^4 s is (Chiu and Stabler 1961)

$$t_{pair} = \varepsilon(d\varepsilon/dt)^{-1} = (11/4)\ aT^4/(3kT\ n_e\ \sigma_\nu\ c) = 440\ T_{10}^{-5}\ \text{s}$$

$$\text{for } kT \gtrsim mc^2/2 \quad . \tag{11}$$

Thus when $t_{pair} = 5 \times 10^4$ s, the temperature at which neutrino cooling keeps up with the accretion rate will be $T_{10} \cong 0.5$, i.e. less than T_s. Hence, the atmosphere will be near isothermal because of the high power of the temperature in the cooling rate and the pressure will be matter dominated close to the shock surface at $\cong 2.5\ \rho_s$. The scale height then is $h = kT/g = 3 \times 10^3$ cm. Hence, the $\cong 18$ scale heights, $(\ell n(\rho_{NS}/\rho_s))$, necessary to accommodate the solar mass is small compared to the radius, and the accreting mass will merge with the neutron star. Of course at some density other neutrino processes will become dominant and the temperature will decrease further. Thus free fall of matter onto the neutron star will not result in a back pressure due to internal energy of the accretion shock. Thus there is too much late time accreted mass despite several simplifying assumptions each of which would reduce the blow back mass.

Angular momentum would prevent accretion, but then it is not likely that the first 1.4 $M_\odot$ could collapse well below critical angular momentum and the next outer 1 $M_\odot$ be above the critical angular momentum.

Magnetic Pressure

One remaining source of pressure is that of a magnetic field associated with the formation of a pulsar. Here differential rotation between the neutron star and the ejected matter twists the connected magnetic flux and reconnection presumably due to the resistive tearing mode becomes the issue.

I am indebted to Albert Petschek for extensive discussions. This work was supported by the DOE.

References

1. Arnett, W. D., 1978, Ap. J. 319, 136.
2. Bethe, H., Brown, G., Applegate, J., and Lattimer, J. 1979, Nucl. Phys. A, 324, 487.
3. Colgate 1971, Ap. J. 163, 221.
4. Colgate, S. A. and Petschek, A. G. 1981, Ap. J. 248, 771.
5. Chiu, H. Y., and Stabler, R. C. 1981, Phys. Rev. 122, 1317.
6. Courant, R. and Fredricks, K. O. 1963, "Supersonic Flow and Shock Waves," Interscience, New York.
7. Hillebrandt, W., Höflich, P., Truran, J. W., Weiss, A., Nat. 327, 597.
8. Mayle, R. and Wilson, J. R., Ap. J. 381, 288.
9. Wilson, J. R. 1971, Ap. J. 163, 209.
10. Woosley, S. 1987, Ap. J. in press.

An Unusual Hard X-Ray Source in the Region of SN1987A

Y. Tanaka

Institute of Space and Astronautical Science
4-6-1 Komaba, Meguro-ku, Tokyo 153, Japan

Abstract. An unusual hard X-ray source was discovered in the region of SN1987A from the X-ray astronomy satellite Ginga. The error box of 0.2° x 0.3° includes the supernova. The energy spectrum is very hard above 10 keV and quite unusual for any of the known classes of X-ray source. The source intensity showed a steady increase from July through August till early September, while no further increase was seen in the observations in late September and mid October. The positional agreement, a steady brightening and an unusual hard spectrum argue for this source being SN1987A. The flux in the range 10 - 30 keV on September 2 - 3 when the source was brightest is approximately 5×10^{-11} ergs/cm²s.

We have been regularly searching for X-rays from the supernova 1987A in the Large Magellanic Cloud (LMC) from the X-ray astronomy satellite Ginga (Makino et al. 1987) since February 25, 1987, right after the optical discovery (Shelton 1987; Nelson and Jones 1987). Ginga carries a set of proportional counters of a total 4000 cm² effective area and a field of view of 2° x 4° full width. SN1987A (hereafter abbreviated as SN) is about 0.6° away from LMC X-1. This source is one of the brightest X-ray sources in the LMC with an intensity of approximately 20 mCrab in the 1 - 10 keV range, and is also time variable. In order to distinguish the SN from LMC X-1, two different modes of observation have been employed; (1) slow scans along a path through the SN and LMC X-1, and (2) pointing observations at a position about 1° offset from LMC X-1 on the side of the SN as shown in Fig. 1, which gives an exposure to the SN with approximately half the maximum sensitivity and little contribution from LMC X-1.

In Fig. 2, we show the result obtained from the scanning observations on September 2 and 3 in the form of histograms of the count rate with time in different energy bands. In the energy range below 10 keV, the counts from LMC X-1 dominate. On the other hand, as the energy increases the peak position shifts towards the line position near the SN. The contributions of LMC X-1 and SNR0540-69.3 are minor in the range above 10 keV and quickly diminish with increasing energy. The histograms clearly indicate the presence of a new source having a very hard spectrum separated from LMC X-1. We are certain that this source did not exist in the March - April period (see Fig. 3). By fitting the collimator response function to the observed count-rate histograms, we determined

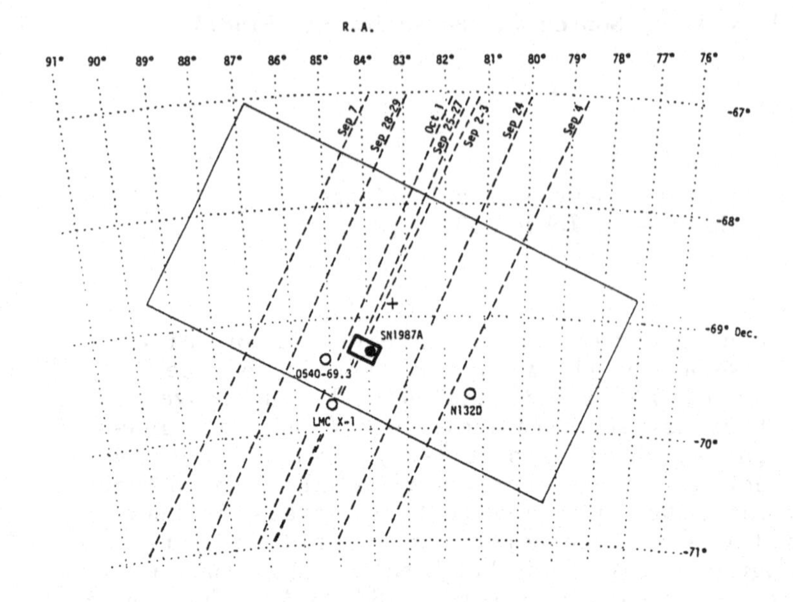

Fig. 1.
The sky map of the LMC region. The scan paths (dashed lines), the target position (+) for the offset pointing mode and the full field of view (2° x 4°) are indicated. The thick solid rectangle shows the 90% confidence error box of the hard source.

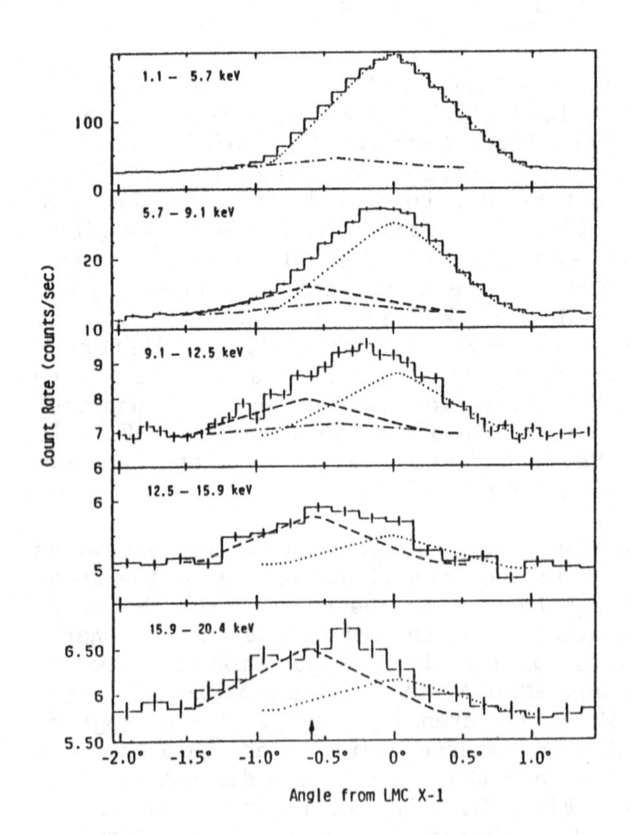

Fig. 2.
Count rate histograms obtained from the scans on September 2 - 3 are shown in five different energy bands. The dotted lines (LMC X-1), dash-dotted lines (SNR0540-69.3) and dashed lines (the hard source) indicate the fits to the collimator response function. Position angle of SN1987A is marked by an arrow.

the contribution of the variable source LMC X-1 and the intensity as well as the line position of the hard X-ray source separately. The contribution of the nearby supernova remnant 0540-69.3 is also taken into account using the result of a separate Ginga observation of this remnant. The result of the fit is shown by the dashed lines in Fig. 2. The line position of the hard source thus determined is in agreement with that of the SN within ± 0.1° (90% confidence limit). The discovery of this hard source was announced on an IAU circular (Makino 1987).

In order to determine the error box of the hard source, we conducted separate scans on September 4 - 7 and September 24 - October 1 along several different paths which were parallel to each other with separations of 0.3° to 05°. These scan paths are shown in Fig. 1. From the comparison of the peak count rate observed on each scan path, we determined the 90% confidence error box of the hard source of the size 0.2° x 0.3° as shown in Fig. 1. This error box includes the SN.

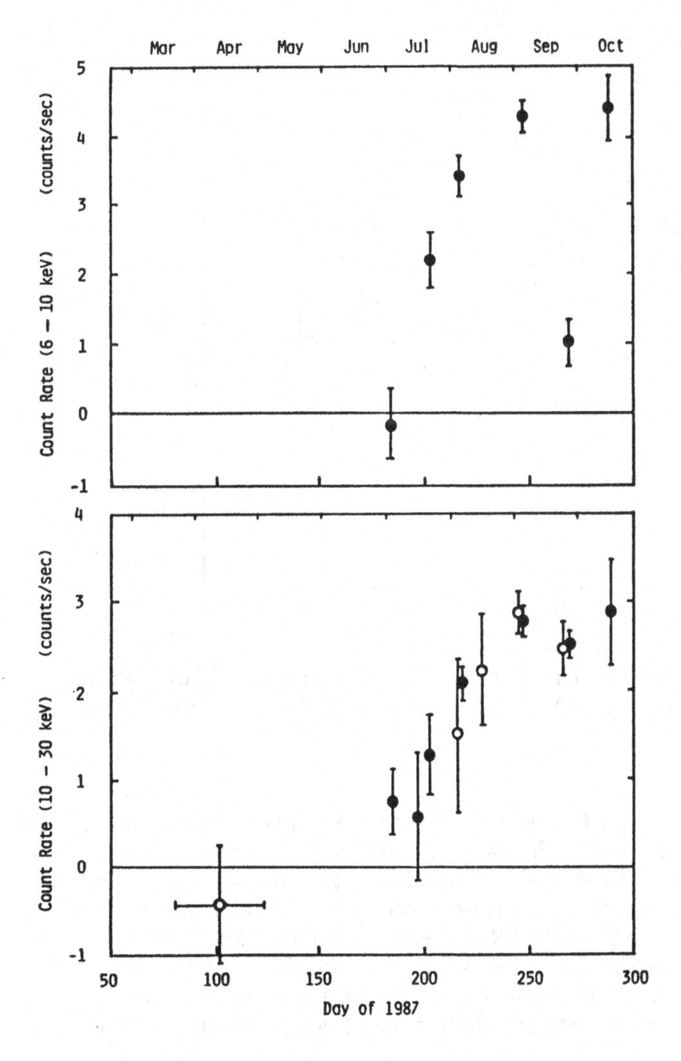

Fig. 3.
The light curves of the hard source in terms of the count rate in two energy ranges, 6 - 10 keV (upper panel) and 10 - 30 keV (lower panel) obtained from pointing (filled circle) and scanning (open circle) observations.

Figure 3 shows the light curves of the hard source in terms of the count rate (counts/cm^{2}s) in two energy ranges, 6 - 10 keV and 10 - 30 keV. The result is consistent with a steady increase of the source intensity since some time in June through early September. However, the observations on September 24 -26, three weeks after the preceeding ones on September 2 - 3, showed a marked intensity decrease in the lower energy range. The intensity came back up again in the observaton on October 15. There is a hint that the intensity in the higher energy range on September 24 - 26 also decreased slightly, but the decrease is statistically insignificant. At present, it is unclear whether or not the intensity is near a peak. In any event, the result appears to indicate that the light curve is not smooth. Observations in search for shorter time variabilities is under way.

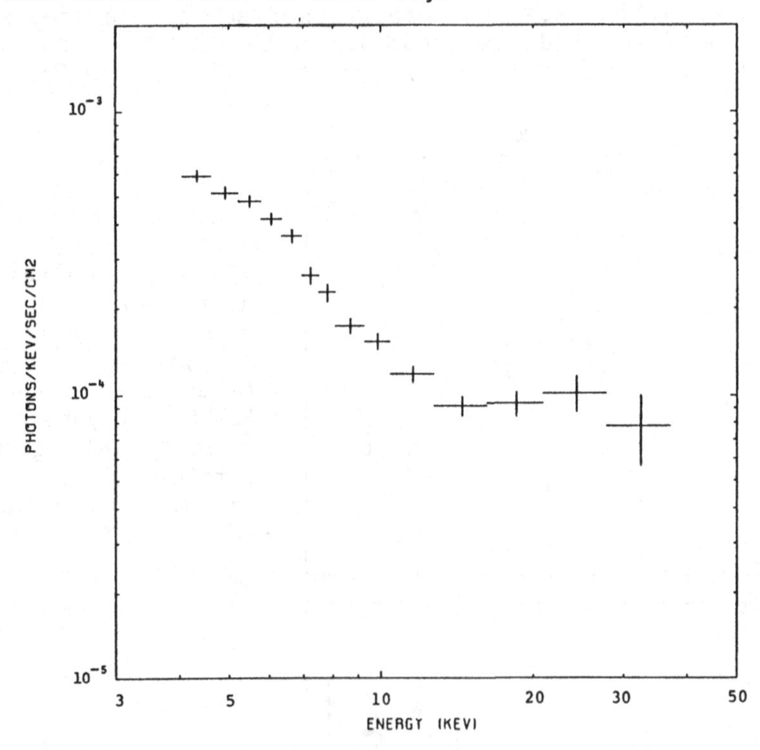

Fig. 4. Photon number spectrum of the hard source obtained from pointing observations on September 3, corrected for the detection efficiency.

The energy spectrum of the new source is quite unusual. Fig. 4 shows the energy spectrum of the source obtained from a pointing observation on September 3. Note that this is a photon number spectrum corrected for the energy-dependent detection efficiency. Two nearby supernova remnants, N132D and 0540-69.3, were in the field of view in this observation. The result of the LMC survey from the Einstein Observatory (Long et al. 1981) show that the contribution from the sources within our field of view other than these two supernova remnants is negligible. The contributions of N132D and 0540-69.3 are subtracted using the results of the separate Ginga observations of these supernova remnants. However, the spectrum below 4 keV is still subject to an uncertainty, because it is difficult to eliminate unambiguously a slight contamination from the brighter and variable source LMC X-1 through reflection on the collimator wall.

The observed spectrum is of an extraordinary shape, exhibiting a break near 10 keV. It turns very hard towards high energies and is essentially flat above 12 keV. This is unlike any of the known classes of X-ray source. The observed spectrum can be fitted to a composite of a thermal bremsstrahlung spectrum with kT of approximately 4 keV and a flat spectrum; a constant independent of energy. In this fit, the same amount of neutral absorbing column was assumed for both hard and soft components. If one allows different absorption columns for the two components, the kT value can be different from the above value. There is no doubt that these two components come from the same source, because of the positional agreement and qualitatively similar light curves observed for two energy ranges, 6 - 10 keV and 10 - 30 keV. We do not find a significant low-energy cut off due to photoelectric absorption down to 4 keV. This would imply that the column density of cool matter is not much larger than 10^{23} H atoms/cm^2 for cosmic abundances of elements. In addition, a signature of an iron emission line is present. The energy flux in the range 10 - 30 keV is approximately 5 x 10^{-11} ergs/cm^2s.

The entire LMC was previously surveyed from Einstein Observatory by Long, Helfand and Grabelsky (1981). The present error box still includes two faint, unidentified Einstein sources. However, the positional agreement, a steady brightening, and an unusual hard spectrum of the source support the identification of the hard source with SN1987A or an object directly related to the SN. If it is the case, the measured intensity as of September 2 - 3 corresponds to a luminosity of approximately 1.5 x 10^{37} ergs/s in the range 10 - 30 keV for a distance of 50 kpc.

Most of the above results are published elsewhere (Dotani et al. 1987). The origin of the x-ray emission from the SN is not obvious yet. The distinct difference between the two spectral components suggests that they derive from separate origins. The lack of a significant absorption would imply that the region responsible for the low-energy component is not surrounded by an appreciable amount of cool matter. A plausible model may be that the hard component is the X-rays which are produced in the interior and diffuse out of the expanding shell, while the soft component is the thermal emission from the outer shell heated by the reverse shock (Nomoto et al. 1987). It is very important to find out any intensity variation on a time scale much shorter than 10 - 20 days (typical separation of the Ginga observations), which might have to do with a quick change in the structure of the shell. An interesting question is whether or not the apparent plateau observed for the component above 10 keV from early September through mid October is the real maximum. If it were so, it is an order of magnitude or two lower than theoretically predicted (e.g., Xu et al. 1987; Itoh et al. 1987).

References

Dotani, T. et al. (1987), to appear in Nature, November 1987.

Itoh, M., Kumagai, S., Shigeyama, T., Nomoto, K., and Nishimura, J. (1987), to appear in Nature, November 1987.

Long, K.S., Helfand, D.J., and Grabelsky, D.A. (1987), Ap.J. **248**, 925-944.

Makino, F. and the Astro-C Team (1987), Astrophys. Letters, **25**, 223-233.

Makino, F. (1987), IAU Circular No. 4447.

Nomoto, K., Shigeyama, T., Hayakawa, S., Itoh, H., and Masai, K. (1987), this Workshop.

Shelton, I. (reported by W. Kunkel and B. Madore); Jones, A., Nelson (reported by F.M. Bateson) (1987), IAU Circular No. 4316.

Xu, Y., Sutherland, F., and McCray, R. (1987), Ap. J. in press.

THE HARD X-RAY SPECTRUM OF SN 1987A

J. Trümper, C. Reppin, W. Pietsch J Englhauser, W. Voges
Max-Planck-Institut für Physik und Astrophysik,
Institut für extraterrestrische Physik, 8046 Garching, FRG

E. Kendziorra, M. Bezler, R. Staubert
Astronomisches Institut der Universität, 7400 Tübingen, FRG

R. Sunyaev, A. Kaniovskiy, V. Efremov, M. Grebenev, A. Kuznetsov, A. Melioranskiy, D Stepanov, I. Chulkov,
Space Research Institute, Moscow USSR

Abstract: We report the discovery of hard X-rays in the energy range from 20 to 350 keV by the HEXE and PULSAR X-1 instruments on the MIR-KVANT Röntgen observatory. The hard X-rays were first observed on August 10, 1987 and thereafter SN 1987A became the main target of the observatory. The measured spectrum is extremely hard. At high energies the photon spectrum has a power law index of ~ 1.4. At lower energies the spectrum becomes flatter and there is indication of a cut-off below 25 keV The luminosity in the above energy band is ~ $2x10^{38}$ erg/s. The flux shows little variation between August 10 and beginning of October.

Introduction

A set of X-ray detectors covering a wide energy band (2 to 800 keV) has been launched aboard the KVANT scientific module. This module was docked onto the MIR station on April 12, 1987. The overall payload which is called KVANT Röntgen observatory has been described by Sagdeev et al. (1986).

The measurements reported here have been made with the HEXE and PULSAR X-1 instruments at energies $\geq$ 20 keV. The results of the TTM coded mask camera are discussed separately by Skinner et al. (1987) in these proceedings. A more detailed comparison between the data obtained for SN 1987A and theoretical models is given by Sunyaev et al. (1987).

Observations

HEXE (High Energy X-ray Experiment) is a detector system provided by the Max-Planck-Institut für Extraterrestrische Physik (MPE) and the Astronomisches Institut Tübingen (AIT). It consists of four phoswich detectors (3.2 mm NaJ (TL), 50 mm CsJ (TL)) having an effective area of 200 cm^2 each. Their energy resolution is about 20 % FWHM at 60 keV. The highest time resolution is 0.3 m sec. Background reduction is performed by rise time discrimination, passive graded shielding and a five-sided plastic scintillator anticoincidence shield. The four detectors form two groups which have separate tungsten collimators defining a field of view of 1.6° x 1.6° FWHM. The two collimators can separately be tilted by 2.3° which allows simultaneous source and background measurements with half of the detector on and half the detector area off the sources at a time. This swaping is done in 2 minute intervals. We call this observation mode "collimator rocking". More detailed descriptions of this instrument which is the successor of the MPE/AIT balloon-HEXE have been given by Reppin et al. (1983) and Reppin et al. (1987).

PULSAR X-1 is a phoswich detector system as well using a thicker NaJ crystal (30 mm NaJ, 30 mm CsJ) in order to give a good high energy response. It has been developed by the Space Research Institute in Moscow and consists of 4 detector units with 250 cm^2 area each. The energy resolution is 25 % at 122 keV, and the nominal energy range is 50 800 keV. The field of view is defined by collimators of 3° x 3° FWHM. Since these collimators are fixed, background measurements are performed by tilting the MIR station by 12°. This can be done at 4 minute intervals.

Due to its large mass and moments of inertia the MIR station has a good pointing stability. The attitude drift is typically smaller than one arc minute per orbit. The absolute pointing accuracy is a few arc minutes. Attitude measurement is done with the star tracker of the TTM instrument package.

Astronomical observations started in the beginning of June, 1987. During one orbit the source can be seen for typically 20 minutes. Up to September 15, a total of 115 orbits has been spent on SN 1987A. The first measurements of SN 1987A were made on June 8, July 16, and August 1. Unfortunately, these observations were too short for a sensitive detection.

Figure 1. Count rates obtained with HEXE in the collimator rocking mode. The data points are alternatively for on source and off source pointings.

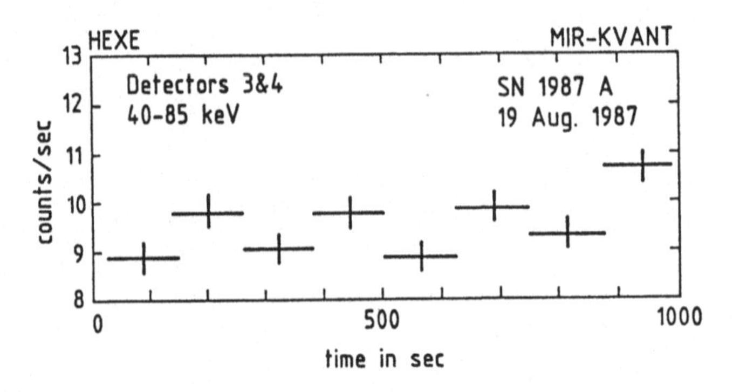

The hard X-ray source at the position of SN 1987A was first detected by the HEXE instrument on August 10. As illustrated by figure 1 the source is easily detected during one orbit using the collimator rocking mode. The spectrum obtained in the August 10-21 period by HEXE in this mode is shown in figure 2. It is based on 9600 sec of data. The conversion from pulse height spectra has been performed using a response matrix based on laboratory calibrations. Crab nebula observations performed with HEXE in September were analysed in the same way and gave a satisfactory fit. Although the SN 1987A spectrum given in fig. 2 is still preliminary we estimate that its systematic errors are less than 30 %.

Figure 2 The preliminary energy spectrum of the hard X-rays. Squares: 3σ upper limits from TTM. Crosses: detections with HEXE (1σ error bars). Diamonds: detections and 3σ upper limits obtained with PULSAR X-1. The histograms show the results of Monte Carlo simulations by Grebenev and Sunyaev (1987) - brief details: 0.1 $M_\odot$ of ^{56}Co, expanding envelope 16 $M_\odot$ with metallicity 1.3 solar, mean expansion velocity 4150 km s^{-1}. 180 days (1) and 210 days (2) after the explosion. The histogram bins in the region of the Fe fluorescence line are 0.5 keV wide.

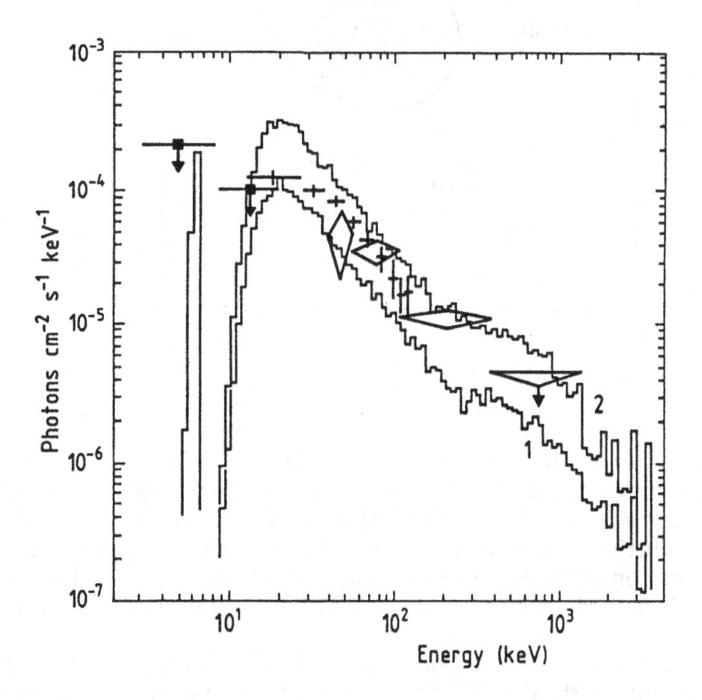

As the TTM observations yielded only upper limits (Skinner et al. 1987) we localised the position of the hard X-ray source by a number of HEXE pointings with offsets of up to 0.8° from the position of SN 1987A. Figure 3 shows the result of an analysis of 54 pointings in the form of probability contours for the source position. The supernova lies within the 1 σ contour, whereas LMC X-1 and the 50 msec pulsar are well outside the 3 σ contour.

Figure 3. The position of the source of high-energy flux obtained by analysing 54 HEXE observations at the offset positions shown (crosses). Contours are at 1σ, 2σ and 3σ confidence. The positions of LMC X-1 and 0540-693 are excluded at the 3 sigma level.

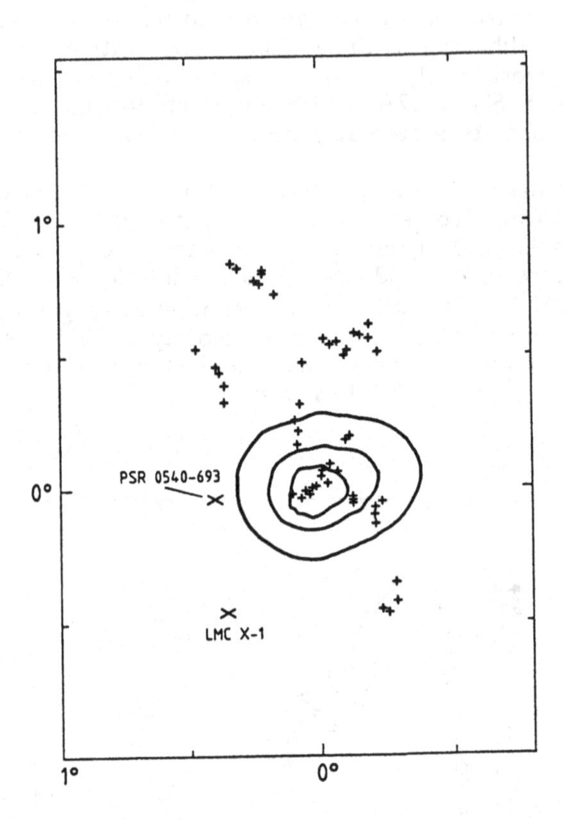

To extend the spectral information to higher energies the PULSAR X-1 instrument was used. Background observations were made by rocking the whole MIR station. Since this mode is inconvenient for TTM and HEXE only relatively few observations have been made in this mode (a total of 5000 sec on-source) between August 16 and September 15. This resulted in two 5 σ detections in the 50-100 keV and 100-350 keV energy bands.

As can be seen from figure 2 the PULSAR X 1 spectral points are consistent with the HEXE data and extend the measured spec trum up to 350 keV. We also note that there is a good agreement in absolute intensity with the GINGA data in the 20-30 keV interval as presented by Dotani et al. (1987). For a comparison between the GINGA and TTM data see Skinner et al. (1987).

An analysis of the HEXE data obtained in the August 10 September 15 period shows that the intensity remained approximately constant. Any flux increase during this period cannot have been more than ~ 30 %.

Discussion

The hard X-ray luminosity integrated over the 20-350 keV energy band of SN 1987A is ~ $2x10^{38}$ erg/s which is a small fraction of the bolometric luminosity (~ 10^{42} erg/s).

The hard X ray flux is unlikely to result from thermal emission of a hot shock heated plasma (Sunyaev et al. 1987, see also Aschenbach et al. 1987). In principle it could originate from the activity of a young pulsar (Pacini 1987), if the expanding shell were already transparent enough. Another possibility is that the hard X-rays are produced by downwards comptonization of hard photons which perculate out through the expanding shell. The hard photon source could be either ^{56}Co or a pulsar and/or its synchrotron nebula.

Figure 4. As fig. 2, except that the simulations assume that the central energy source is a pulsar with a 1-1000 keV luminosity of $8x10^{41}$ erg s^{-1} and a spectrum similar to that of the Vela pulsar (photon index 1.5). The predictions are for 180, 240, 300, 360, 420, 720 days after the explosion (labelled 1-6 respectively)

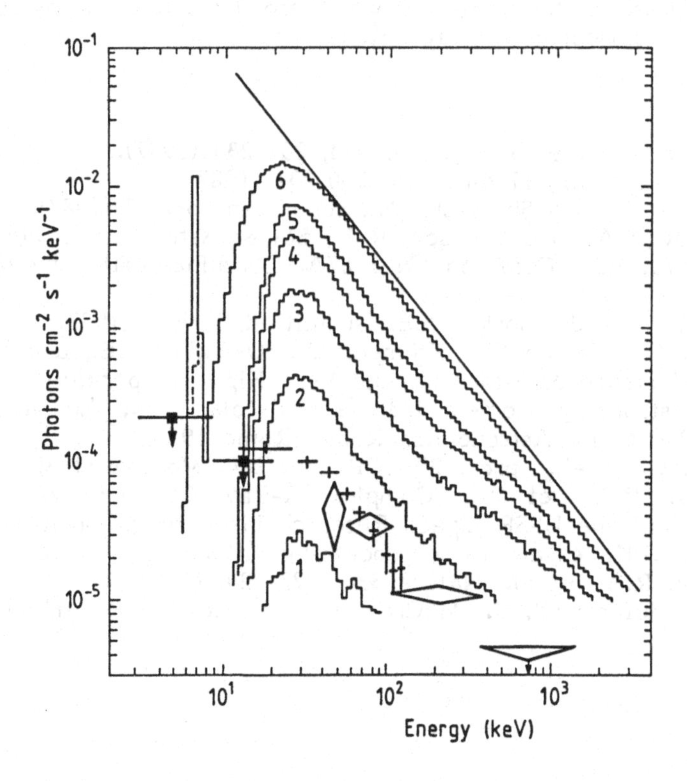

Monte Carlo calculations of the expected time dependent spectra have been performed by a number of authors (Grebenev & Sunyaev 1987; Xu et al. 1987, Ebisuzaki & Shibazaki 1987, Pinto & Woosley 1987). In the case of ^{56}Co these models predict a maximum flux to occur after ~ 1 year. In the case of a pulsar/synchrotron nebula origin of the injected hard radiation they predict a steady increase up to the point where the shock becomes Thomson optically thin (see fig. 4).

In figure 2 we show the expected spectra for 6 and 7 months after explosion obtained by Grebenev & Sunyaev (1987) for a ^{56}Co source model. They assumed a mass of 0.1 $M_\odot$ ^{56}Co and a spherically symmetric shell of 16 $M_\odot$ expanding with an average velocity of 4150 km/s. As can be seen from figure 2 the agreement in spectral shape and absolute fluxes is quite good. On the other hand the observed stability of the X ray flux over the August 10 - September 15 period is not reproduced by this model. We note that this is likewise true for most of the current models which predict a quite strong hard X-ray flux increase during the fall of 1987.

However, the current models contain a number of uncertainties concerning expansion velocities, the amount of mixing (between inner and outer layers), clumping or nonspherical distribution of the ejected matter (Grebenev & Sunyaev 1987, Pinto & Woosley 1987). We note that it should be possible to narrow down these uncertainties by simultaneous fitting to the optical and X-ray data.

References

1) Aschenbach, B., et al., Nature **330**, 232-233 (1987).
2) Dotani, T., et al., Nature **330**, 230-231 (1987).
3) Ebisuzaki, T., and Shibazaki, N., submitted to ApJ (1987).
4) Grebenev, S.A. and Sunyaev, R.A., Soviet Astr. Lett.**13**,945-963 (1987).
5) Pacini, F., ESO Conf. on SN 1987A, Garching, edited by I.J. Danziger (1987).
6) Pinto, P.A., and Woosley, S.E., submitted to ApJ (1987).
7) Reppin, C., Pietsch, W., Trümper, J., Kendziorra, E., and Staubert, R., Proc. Workshop Non-thermal and Very High Temperature Phenomena in X-ray Astronomy. Rome (1983), (eds. Perola, G.C., and Salvati, M.) 279-282 (Istituto Astronomica Roma, Rome 1985).
8) Reppin, C. et al., proc. **20,** ICRC, p. 284, Moscow (1987).
9) Sagdeev, R.Z., et al., Complex X-Ray Observatory; Academy of Sciences of the USSR. Space Research Institute, preprint **1177** (1986).
10) Skinner, G.K. et al., these proceedings (1987).
11) Sunyaev, R.A., et al., Nature **330**, 227-229, (1987).
12) Xu, Y., Sutherland, P. McCray, R., and Ross, R.R., submitted to ApJ (1987).

CODED MASK X-RAY OBSERVATIONS OF SN1987A FROM MIR

G.K.Skinner
University of Birmingham, Birmingham B15 2TT, U.K.

O.Al-Emam
University of Birmingham, Birmingham B15 2TT, U.K.

A.C.Brinkman
Space Research Laboratory, Utrecht, The Netherlands

E.Churazov
Space Research Institute, USSR Academy of Sciences, Moscow

M.Gilfanov
Space Research Institute, USSR Academy of Sciences, Moscow

J.Heise
Space Research Laboratory, Utrecht, The Netherlands

R.Jager
Space Research Laboratory, Utrecht, The Netherlands

W.A.Mels
Space Research Laboratory, Utrecht, The Netherlands

N.Pappe
Space Research Institute, USSR Academy of Sciences, Moscow

T.G.Patterson
University of Birmingham, Birmingham B15 2TT, U.K.

R. Sunyaev
Space Research Institute, USSR Academy of Sciences, Moscow

A.P.Willmore
University of Birmingham, Birmingham B15 2TT, U.K.

N.Yamburenko
Space Research Institute, USSR Academy of Sciences, Moscow

Abstract. We report observations made of the supernova SN1987a with the TTM coded mask imaging X-ray telescope on the Mir-Kvant observatory 'Röntgen'. Upper limits for the flux below 20 keV from SN1987A are below the extrapolation of the higher energy emission observed from the region by other instruments, implying a cutoff or change of slope. Useful limits can be placed on the density of the region into which the supernova is expanding.

The Kvant module, launched 1987 March 31 and subsequently attached to the Mir space station, carries a package of coaligned X- and gamma-ray instruments comprising the Mir-Kvant observatory 'Röntgen'. Observations are made by orienting the entire MIR complex to point the instruments in the desired directions. In view of the high inclination of the orbit, this is usually done for a period of 1000-1500 s centred on a passage over the equator, when the particle background is lowest. Up to ~10 such observations can be made in a day.

One of the instruments is a coded mask telescope, TTM, (Brinkman et al., 1985) which provides the only means currently available for clearly resolving the X-ray sources in the region of SN1987a from each other. TTM provides imaging with a resolution of 1.8 arc minutes full width at half maximum (FWHM) over a 7.8° FWHM square field of view. It uses a Xenon filled multi-wire proportional counter with a geometrical area of 650 cm^2 and a sensitive depth of 5.5 cm atmospheres, giving a useful response over the range 2-32 keV, although the sensitivity is highest in a band approximately 3-12 keV. The mask contains 32767 square holes on a 255×257 element array with a pitch of 1 mm. The hole pattern is based on a Singer cyclic difference set, even though the mask is the same size as the detector and so the optimum coding properties of a cyclic system using such a pattern are not available.

Successful observations of the region of SN1987A with TTM were made on 1987 June 8, June 9, June 10 and July 16. Following the detection of hard X-rays by the HEXE experiment, SN1987A became the main target of the Mir-Kvant observatory and more TTM data was obtained on 1987 August 10, 11, 12, 13, 16, 17, 19, 20, 21, 22 and 23.

The modest size and wide field nature of the instrument means that high sensitivity can only be achieved by combining relatively large amounts of data together. For the present work images were first obtained from individual observations, correcting individual photon positions for small changes in pointing (~1-2 arc minute) during the observation. The absolute pointing directions for the separate observations were then checked using the known positions of the bright sources LMC X-1, X-2, X-3 and X-4 and were taken into account in forming a weighted mean image. Figure 1 shows part of such an image obtained using data from observations during the period August 10-23. The weighted mean epoch for the 55 observations used in forming the image is approximately 1987 August 19.0 and the total observing time about 40000 s.

The only sources which are detected in the vicinity of SN1987A are LMC X-1 and the supernova remanent 0540-69.3 which contains the 50 ms pulsar discovered by Seward _et al._ (1984). No significant emission is detected from SN1987A. The image in Figure 1 was obtained using photons from the entire energy range of the instrument, but similar ones have been derived in narrower energy bands. This process is time consuming but will in due course allow spectra of individual sources

Figure 1. A 2.5° by 8.5° section from an image in the energy range 2-32 keV obtained with the TTM instrument during the HEXE observations. There is no peak at the position on SN1987a. Vertical height indicates significance of detection, not absolute intensity.

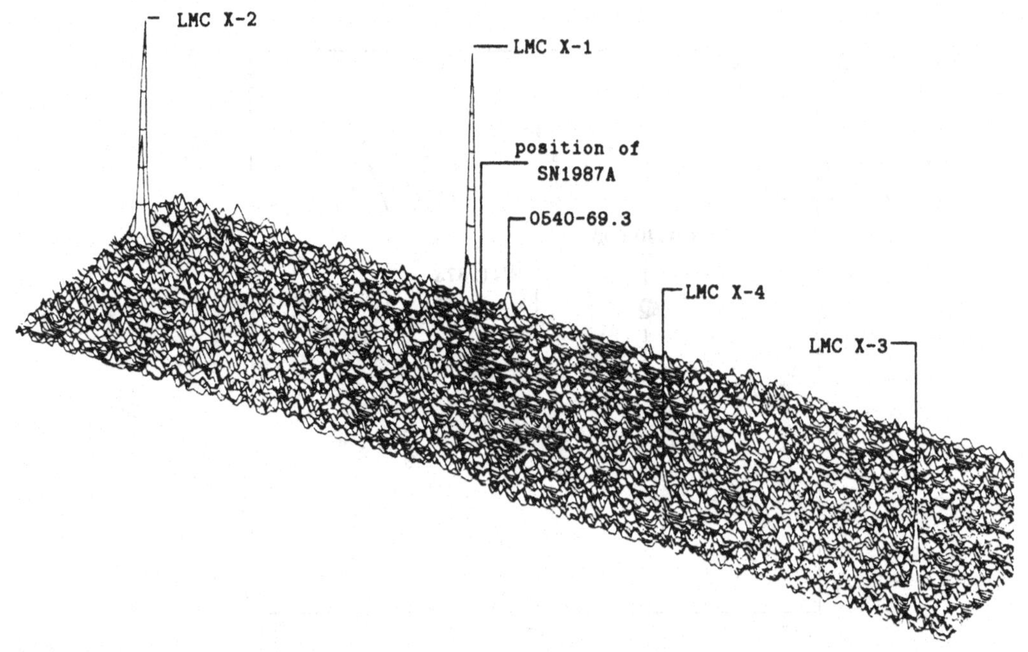

to be examined and upper limits to be established as a function of energy.

It has not yet been possible to make observations of the Crab Nebula with TTM to confirm the instrument response, so 0540-69.3 provides the best calibration of the instrument sensitivity. This system is a close analogue of the Crab nebula, containing an X-ray and optical pulsar (Seward et al.,1984; Middleditch et al., 1986), and an optical synchrotron nebula (Chanon et al., 1984). As in the case of the Crab, the optical luminosity of the nebula matches an extrapolation of the E^{-2} X-ray photon spectrum. Again like the Crab, the pulsar has a somewhat flatter spectrum, judged both from X-ray measurements and from optical/X-ray comparisons. Einstein IPC X-ray observations of this object showed no variability, whilst HRI measurements showed an extended component (Long et al., 1981). Thus we feel secure in using intensity measurements made with the Einstein SSS by Clark et al. (1982), despite the different epoch and the fact that some extrapolation is needed from the 0.5-4 keV band in which the observations were made to 7 keV where the TTM response is centred.

Taking the spectrum of 0540-69.3 to be 0.006 $E^{-1.8}$ photons $cm^{-2}\ s^{-1}\ keV^{-1}$ (Clark et al., 1982; Seward et al.,1984) we find a

Figure 2. Contour map of the 2° square region of the image in Fig. 1. surrounding SN1987a. Cooordinates are offsets from the position of the supernova. Apart from LMC X-1 and 0540-693, the levels seen are consistent with the expected noise level. Contours are at 0.3, 0.4, 0.5, 0.6, 0.7, 0.8, 1.0, 2.0, 4.0 & 6.0 uncorrected counts per second.

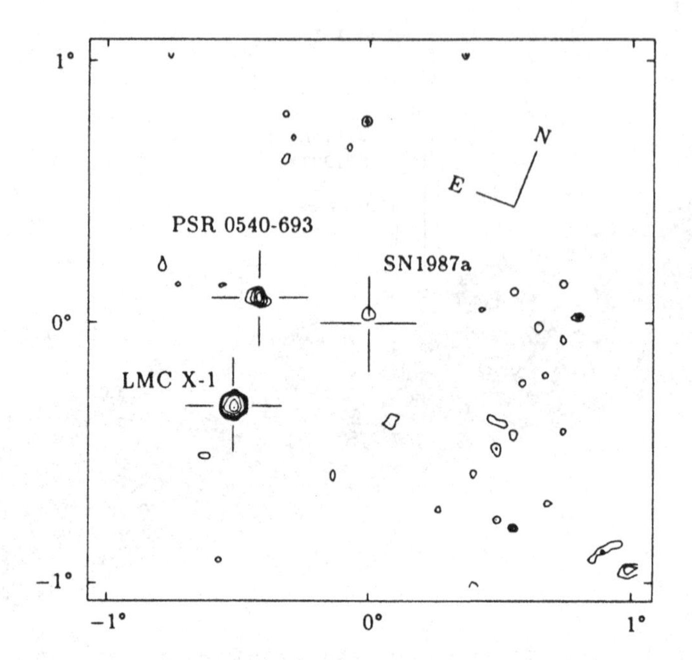

3σ upper limit of 5×10^{-4} times the intensity of the Crab nebula for any emission from SN1987a with a similar spectrum. For a source with a flat spectrum we derive upper limits in the 2-10 and 10-20 keV bands of 2×10^{-4} and 1×10^{-4} photons cm^{-2} s^{-1} keV^{-1} respectively. These limits are consistent with the hard X-ray detection by the HEXE experiment on the same mission (Trümper et al., this volume) and with the 10-20 keV flux reported by Makino et al. (1987) using GINGA, but not with the low energy component of the more recent GINGA spectrum (Tanaka et al., this volume). A source with the latter spectrum should appear in the TTM images with a significance of greater than 6 sigma and it is difficult to understand how it might be missed.

The TTM observations exclude the presence in the region of SN1987a of any source apart from LMC X-1 with a normal X-ray spectrum and strong enough to explain the hard X-ray flux observed by HEXE and GINGA. Provided that it can be shown that the hard X-rays cannot originate in LMC X-1 then it is highly probable that they come from SN1987A. The upper limits at low energies then suggest that the spectrum is heavily cut-off below 10-20 keV by photoelectric absorbtion.

A shock wave will be propogating outwards into the medium

surrounding the supernova and the TTM upper limits may be used to estimate the density of the medium. The electron temperature is expected to be given by

$$kT_e = 10\left(\frac{N}{10^5\ cm^{-3}}\right)^{0.4}\left(\frac{V}{10^4\ km\ s^{-1}}\right)^{0.8} keV,$$

where N is the post-shock electron density V is the expansion velocity. We find that for temperatures anywhere in the range $3<kT_e<300$ keV the emission measure of the region behind the shock must be $N_e^2V<3\times10^{59}$ cm^{-3}. Estimating the volume assuming a shock velocity of greater than 10^4 km s^{-1}, we conclude $N_e<2.7\times10^5$ cm^{-3}. As there is a fourfold compression in the shock, this implies that the density of the surrounding region is less than 3×10^5 cm^{-3}.

In conclusion, we find no evidence for X-ray flux from SN1987A at energies less than 20 keV. This allows useful limits to be placed on the density of the region into which the supernova ejecta are expanding and favours models for the hard X-ray flux in which the X-rays are down-Comptonised photons of even higher energies which originate deep within the supernova but which escape before they are photoelectrically absorbed.

References

Brinkman,A.C., Dam,J., Mels,W.A., Skinner,G.K. & Willmore,A.P. (1985). Coded mask imaging spectrometer for the Saljut mission. In Proceedings of Workshop on Non-thermal and Very High Temperature Phenomena in X-ray Astronomy, Rome, 1983, eds. G.C.Perola & M. Salvati, pp.263-270, Rome: Instituto Astronomica Roma.

Chanon, G.A., Helfand, D.J. & Reynolds, S.P. (1984). An optical synchrotron nebula around the X-ray pulsar 0540-693 in the Large Magellanic Cloud. Astrophys. J., 287, L23-L26.

Clark, D.H., Tuohy, I.R., Long, K.S., Szymkowiak, A.E., Dopita, M.A., Mathewson, D.S., & Culhane, J.L. (1982). X-ray spectral classification of supernova remnants in the Large Magellanic Cloud. Astrophys. J., 255, 440-446.

Long, K.S., Helfand, D.J., & Grabrelski,D.A. (1981). A soft X-ray study of the Large Magellanic Cloud. Astrophys. J., 248, 925-944.

Makino, F. (1987). I.A.U. Circular 4447.

Middleditch, J., Pennypacker, C.R., & Burns, M.S. (1986). Optical color, polarimetric, and timing measurements of the 50 ms Large Magellanic Cloud pulsar, PSR0540-69. Astrophys. J., 315, 142-148.

Seward, F.D., Harnden, F.R., & Helfand, D.J. (1984). Discovery of a 50 millisecond pulsar in the Large Magellanic Cloud. Astrophys. J., 287, L19-L22.

GAMMA-RAY OBSERVATIONS OF SN 1987a WITH AN ARRAY OF HIGH-PURITY GERMANIUM DETECTORS

W. G. Sandie, G. H. Nakano, & L. F. Chase, Jr.
Lockheed Palo Alto Research Laboratory
3251 Hanover Street
Palo Alto, California 94304 U.S.A.

G. J. Fishman, C. A. Meegan, R. B. Wilson, & W. Paciesas
High Energy Astrophysics Laboratory
NASA Marshall Space Flight Center
Huntsville, Alabama 35812 U.S.A.

G. P. Lasche
DARPA
Arlington, Virginia 22209 U.S.A.

Abstract. A balloonborne gamma-ray spectrometer comprising an array of high-purity n-type germanium (HPGe) detectors having geometric area 119 cm^2, resolution 2.5 keV at 1.0 MeV, surrounded by an active NaI(Tl) collimator and Compton suppressing anticoincidence shield nominally 10 cm thick, was flown from Alice Springs, Northern Territory, Australia, on May 29-30, 1987, 96 days after the observed neutrino pulse. The average column depth of residual atmosphere in the direction of SN 1987a at float altitude was 6.3 g cm^{-2} during the observation. SN 1987a was within the 22-deg full-width-half-maximum (FWHM) field of view for about 3300 s during May 29.9-30.3 UT. No excess gamma rays were observed at energies appropriate to the Ni(56)-Co(56) decay chain or from other lines in the energy region from 0.1 to 3.0 MeV. With 80% of the data analyzed, the 3-sigma upper limit obtained for the 1238-keV line from Co(56) at the instrument resolution (about 3 keV) is 1.3×10^{-3} photons cm^{-2} s^{-1}. The corresponding limit for the 847-keV Co(56) line is 1.7×10^{-3} photons cm^{-2} s^{-1}, owing to higher background at this energy. The upper limits scale approximately as the square root of the line width assumed in the search, and kinematic broadening at the source may be of the order of 6 keV. These data imply that there was less than 2.5×10^{-4} solar masses of Co(56) exposed to the Earth at the time of the observation. Additional balloon-borne observations are planned.

INTRODUCTION

Gamma-ray line emissions provide the most valuable characterization of nuclear synthesis and inelastic processes occuring in supernova development. The Supernova (SN 1987a) is Type II, exhibiting an extensive hydrogen envelope. The Supernova is expected to produce a copious gamma-ray display as the rapidly expanding hydrogen envelope dissipates to reveal the underlying mantle (Woosley et al., 1987 a, b, c, d). The observation of neutrinos detected at Kamiokande (Hirata et al., 1987) and IMB (Svoboda 1987) supports supernova formation via core collapse; furthermore the time interval between the neutrino detection that signaled core collapse and the optical flare-up impose restrictions on the size of the pre-Supernova star and on the energetics of the explosion. Theoreticians (Woosley et al., 1987 a, b, c, d) and others have extensively modeled the Supernova, concluding that the progenitor star is most likely the blue super giant Sanduleak -69 202 as originally supposed.

For a progenitor of mass 15 solar masses, the models synthesize a "neutronized" core of about 1.4 solar masses. Only about 7.5×10^{-2} solar masses of the total amount of Ni(56) synthesized forms in a thin shell outside the core and is potentially visible. This thin shell is ejected with a velocity of about 1000 km s^{-1} and underlies the residual mass of ejecta moving radially outward at much greater velocity (about 10,000 km s^{-1}), which forms the visible nebula. Initially the mass of ejecta concentrated about the core is opaque to gamma rays produced by the decay of short-lived Ni(56), half-life 6.1 days. These gamma rays contribute to the nebula luminosity. As the mass of ejecta moves radially outward, becoming more tenuous, the opacity decreases, ultimately revealing gamma-ray line emission from longer-lived Co(56) in the underlying shell. The rise time of the line flux depends upon the amount of Ni(56) synthesized and the opacity of the ejecta. The fall time is expected to follow the e-folding decay time of Co(56), which is about 114 days. To predict the time-dependent intensity of line emissions requires detailed modeling of the evolution of the Supernova; this has been carried out (Woosley, 1987 a. b, c, d; Pinto & Woosley, 1987; Arnett 1987; Chan & Lingenfelter, 1987; Xu et al., 1987; Gehrels et al., 1987; Ebisuzaki & Shibazaki, 1987) predicting peak intensities for the 847-keV line of 0.0002 to 0.008 photons cm^{-2} s^{-1} occuring 200 to 400 days after core collapse. The predicted peak flux is within the sensitivity of the instrument described here.

INSTRUMENT DESCRIPTION

The gamma-ray spectrometer is designed to detect gamma rays with useful efficiency over the energy interval 25 to 8000 keV. The instrument comprises a three-by-three array of nine 27% N-type high-purity germanium (HPGe) detectors cooled to cryogenic temperature by the utilization of a cooler containing 15 L of liquid nitrogen. The HPGe detectors have a total geometric area of 214 cm^2 and a field of view (FOV) of 22 deg, full-angle, corresponding to a geometric aperture-area product of 28.8 cm^2 sr.

For the observations described here, five of the nine detectors were functional, reducing the geometric area to 119 cm^2. The cooled HPGe detectors have a typical resolution of 2.5 keV FWHM at 1.33 MeV; the cryogen is initially frozen nitrogen with a thermal hold time longer than 24 h through triple point and about 96 h through the liquid phase. The calculated area efficiency products (expressed as percent of geometric efficiency) at energies of interest are 34% (511 keV), 23% (847 keV), 18% (1236 keV), 21% (1038 keV), 14% (1771 keV), and 9% (2599 keV). To reduce background, the HPGe array is surrounded by a live NaI(Tl) "crystal ball" antishield with a nominal thickness of 10 cm. The live antishield serves to reject both external, out-of-apterture gamma rays, and, in-aperture, Compton-scattered events. The 22-deg FOV of the HPGe detectors is defined by collimator holes in the upper NaI(Tl) shield. Charged particles, in-aperture, are rejected by a 4-mm-thick plastic anticoincidence shield. The NaI(Tl) antishield is viewed by 16 3-in. phototubes and the plastic antishield is viewed by 4 1-in. phototubes. Data are recorded in list mode, 13 bits for each HPGe detector pulse amplitude accompanied by two 8-bit antishield amplitudes, formatted into a 128-kbs Miller-DMM data stream, and transmitted to the ground.

OBSERVATIONS

A balloon payload, containing the spectrometer, was launched from Alice Springs, Northern Territory, Australia, at 19:19 (UT) May 29, 1987 by the field crew of the National Scientific Balloon Facility (NSBF). The balloon maintained nominal float altitude, approximately 120,000 ft, for about 9 h. Instrument pointing accuracy was verified by observing the position of the sun with a precision sun sensor.

The background spectrum obtained for approximately 7 h at float altitude for the five detectors combined is shown in Fig. 1. The spectrum in Fig. 1 demonstrates the high-resolution performance of the instrument throughout the flight.

The Supernova was within the 22-deg FOV of the spectrometer for a period of about 7000 s, taking into account the angular response of the instrument, the effective time on the Supernova at full geometric aperture, 119 cm^2, was approximately 3300 s. Measurements were made by alternately pointing the spectrometer at the Supernova (source plus background) and then swinging away from the Supernova in azimuth while maintaining elevation (background). Difference spectra obtained by subtracting the background count rates from the source-plus-background count rates are given in Figs. 2 and 3. These data represent approximately 80% of the useful flight data thus far analyzed. No excess counts from the Supernova were observed. Three-sigma upper bounds on the line flux, F(keV), from the Supernova for Co(56) of interest at the instrument resolution are:

F(847) 1.7×10^{-3} photons cm^{-2} s^{-1}
F(1238) 1.3×10^{-3} photons cm^{-2} s^{-1}
F(2599) 1.1×10^{-3} photons cm^{-2} s^{-1}
F(1038) 1.5×10^{-3} photons cm^{-2} s^{-1}
F(1771) 1.7×10^{-3} photons cm^{-2} s^{-1}

Fig. 1 Typical Gamma-Ray Spectrum Obtained at Altitude Over the Australian Desert. The lines in the spectrum, observed with high resolution, are due to background.

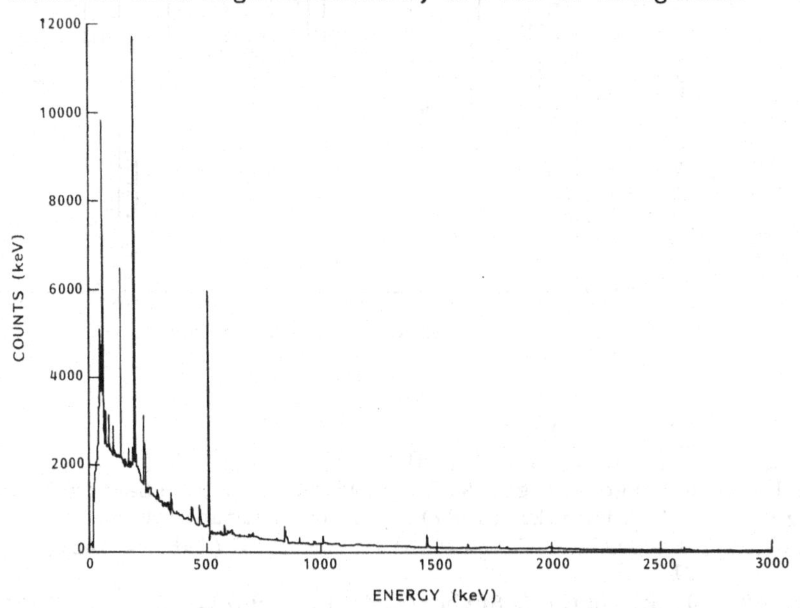

Fig. 2 Net Source Spectrum in 3-keV Bins in the Vicinity of the 847-keV Co(56) Line.

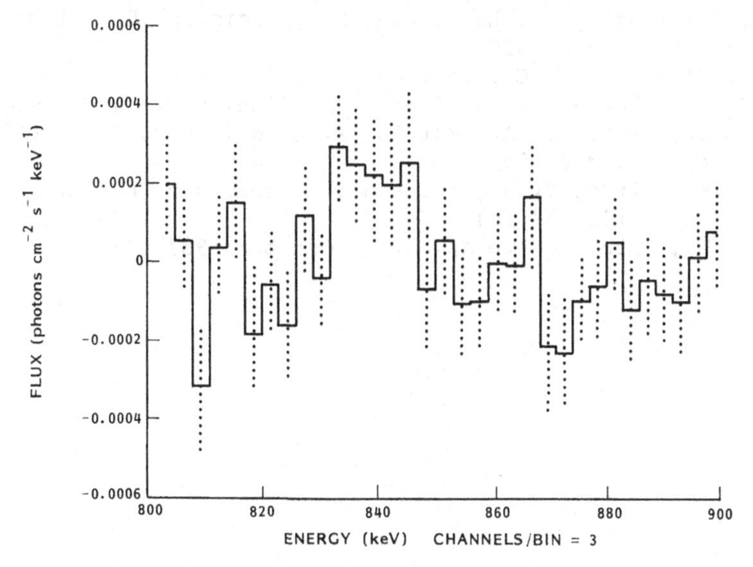

Fig. 3 Net Source Spectrum in 3-keV Bins in the Vicinity of the 1238-keV Co(56) Line.

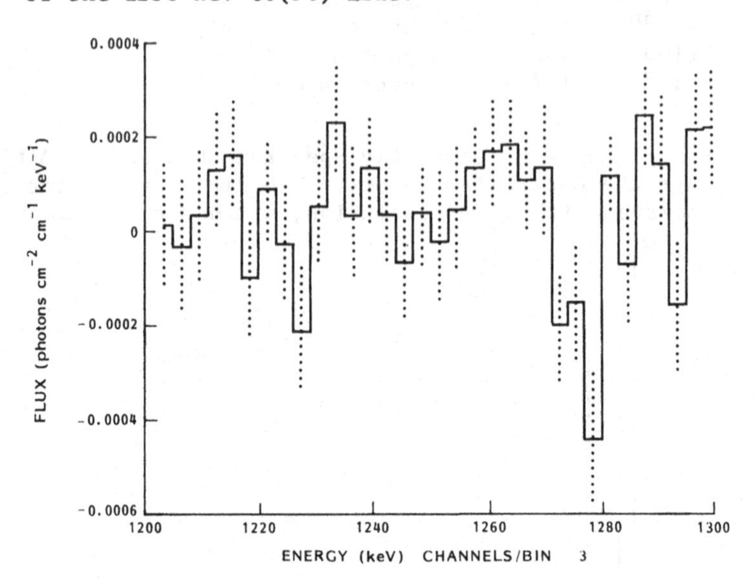

REFERENCE LIST

Arnett, W. D. (1987). Ap. J., 319, 136.

Chan, K. W. & Lingenfelter, R. E. (1987). Ap. J. (Letters), 318, L51.

Ebisuzaki, T. & Shibazaki (1987), private communication.

Gehrels, N., MacCallum, C. J. & Leventhal, M. (1987). Ap. J., 322, 215.

McCray, R., J. M. Shull & Sutherland, P. (1987). Ap. J. (Letters), 317, L73

Pinto, P. A. & Woosley, S. E., submitted to Ap. J., 8 October 1987.

Shigeyama, T., Nomoto, K., Hashimoto, M. & Sugimoto, D., (1987). Nature, 328, 320.

Svoboda, R. (1987). IAU Circular 4320

Woosley, S. E. & Weaver, T. A. (1980, a). Ann. Rev. Astr. Ap., 24, 205

Woosley, S. E., Pinto, P. A., Martin, P. G. & Weaver, T. A. (1987,b). Ap. J., 318, 664

Woosley, S. E., Pinto, P. A., & Ensman, L. submitted to Ap. J., 22 April 1987 (c)

Woosley, S. E., Submitted to Ap. J., 5 October 1987 (d)

THE EMERGENCE OF X-RAYS AND GAMMA-RAYS FROM SUPERNOVA 1987A

J. Michael Shull and Yueming Xu
Center for Astrophysics and Space Astronomy, and
Joint Institute for Laboratory Astrophysics
University of Colorado and National Bureau of Standards
Boulder, CO 80309 (USA)

Abstract We discuss attempts to arrive at a unified scenario for the optical, X-ray, and γ-ray emission of SN 1987A. We discuss the X-ray and γ-ray spectra that should emerge from SN 1987A, as the expanding envelope expands to reveal the inner debris of the explosion. It is now almost certain that the supernova light curve is dominated by the radioactive decay of ^{56}Ni to ^{56}Co to ^{56}Fe, although an X-ray emitting pulsar is possible at some level. Gamma rays from ^{56}Co degrade in energy due to Compton scattering in the mantle and envelope, only emerging as lines when the Thomson optical depth at 1 MeV drops to $\tau_T \approx 1$. Hard X-rays emerge between 20 - 400 keV owing to Comptonization in the envelope. Theoretical spectra are in reasonable agreement with recent observations of hard X-rays by the *Ginga* and *Mir* satellites, but the early turn-on of X-rays suggests that the mantle and envelope may be "leaky", perhaps as a result of Rayleigh-Taylor instabilities and clumping. The soft X-ray spectrum should be dominated by a 6.4 keV Fe K_α fluorescence line. The reported *Ginga* detection of 4-10 keV X-ray emission is also discussed.

1 INTRODUCTION

In the first months following the explosion of SN 1987A, numerous groups made theoretical predictions of the emergence of X-rays and γ-rays from the opaque envelope. McCray, Shull, and Sutherland (1987) discussed two likely scenarios: (1) that the core produces ^{56}Co γ-ray lines which Comptonize and emerge as 20-400 keV X-rays from the envelope; and (2) that an X-ray emitting pulsar exists at the center. Similar calculations have been reported by Xu *et al.* 1987, Gehrels *et al.* (1987), Pinto and Woosley (1987), and Ebisuzaki and Shibazaki (1987). The emergent X-ray spectra in the two scenarios are not dissimilar, but the X-ray light curves can distinguish between these scenarios (Xu *et al.* 1987).

Among the exciting observations reported at this conference are three which affect these interpretations. First, we learned that the optical light curve is now tracking an exponential decay, with $\tau_o = 113.6 \pm 0.6$ day (Feast 1987), suggesting that ^{56}Co is a dominant energy source. Second, we learned that hard X-rays have been detected (Tanaka

1987; Trümper 1987; Dotani *et al.* 1987), with spectra similar to theory but with a turn-on somewhat earlier than predicted by Comptonization models. Finally, Tanaka announced that *Ginga* also detected a "soft X-ray" (4-10 keV) excess, fitted with a 4 keV thermal spectrum and containing a hint of an Fe line between 6 and 7 keV. The origin of the soft X-rays is unknown, although possible explanations include a leaky envelope, shocks, or reprocessing of hard X-rays by circumstellar matter.

In this review, we will describe Comptonization theories for the emergence of the hard X-rays and γ-rays. Taking advantage of the new observations, we will then present new results for theoretical X-ray and γ-ray spectra and "light curves" based on revised interior models kindly provided by Stan Woosley and Phil Pinto. The predicted spectra are in reasonable agreement with the data, but the early turn-on suggests that the mantle or envelope may be leaky due to clumping.

2 BASIC IDEAS

Theoretical models for SN 1987A require an internal energy source with luminosity $L \approx 10^{41-42}$ erg s^{-1} to explain the late-time light curve. Two possibilities have been proposed (McCray, Shull, and Sutherland 1987; Arnett 1987; Shigeyama *et al.* 1987; Woosley, Pinto and Ensman 1987): (1) radioactive decays of ^{56}Ni and ^{56}Co; or (2) a luminous X-ray emitting pulsar. For the first model, a mass $M_{Ni} \approx 0.075 M_\odot$ of ^{56}Ni is required. Each ^{56}Co decay results in a characteristic spectrum of γ-ray lines (Gehrels, MacCallum, and Leventhal 1987), with energies ranging from 0.511 MeV to 3.452 MeV and a mean lifetime $\tau_o \approx 114$ days. A typical decay yields $N_l \approx 3$ lines having net energy $\sim$3 MeV. The γ-ray photon luminosity is then,

$$S_\gamma = \left(\frac{M_{Ni}}{m_{Ni}}\right)\left(\frac{N_l}{\tau_o}\right)\exp(-t/\tau_o) = (6 \times 10^{46}\ \mathrm{s}^{-1})\left(\frac{M_{Ni}}{0.1 M_\odot}\right)\exp(-t/\tau_o). \tag{1}$$

In the pulsar scenario, a straightforward estimate for a Crab-like magnetic field and spin, gives a total luminosity $L_x \approx (2 \times 10^{41}$ erg s$^{-1})$ with a photon spectrum $dJ/d\epsilon \propto \epsilon^{-2}$ for photon energies 1 eV $\leq \epsilon \leq$ 1 MeV. Of course, there is no guarantee that a pulsar is formed with a sufficient magnetic field or rotation period to produce a hard spectrum. In fact, current fits to the optical light curve appear to rule out a pulsar with a very short spin period.

In each scenario, γ-rays and X-rays from the core scatter repeatedly in the mantle and envelope, and at early times are destroyed by photoelectric absorption. For purposes of illustration, we assume a simple homogeneous model, in which an envelope of mass M_e (solar units) expands homologously at velocity $V_{exp} = (10^4$ km s$^{-1})V_4$. At time $t =$ (100 days)t_{100}, the outer radius, mean hydrogen density, and Thomson optical depth are,

$$R = V_{exp}t = (8.64 \times 10^{15}\ \mathrm{cm})V_4 t_{100} \tag{2}$$

$$n_H = (3M_e/4\pi R^3 \mu) = (3.17 \times 10^8\ \mathrm{cm}^{-3})M_e(V_4 t_{100})^{-3} \tag{3}$$

$$\tau_T = (1.2 n_H \sigma_T R) = (2.19)M_e(V_4 t_{100})^{-2}. \tag{4}$$

The envelope thins to $\tau_T = 1$ at a time,

$$t_{thin} \approx (148 \text{ days}) M_e^{1/2} V_4^{-1}, \tag{5}$$

marking the emergence of γ-ray lines. The Klein-Nishina optical depth at 1 MeV is $\tau_\gamma \approx 0.3\tau_T$, so the photon luminosity of unscattered lines should be $S_\gamma(t) \approx S_0 \exp(-0.3\tau_T)$. The Compton scattered γ-ray lines form a hard X-ray continuum which fades and hardens as the envelope thins due to expansion.

A simple analytic model shows how the X-ray emergence proceeds. In Compton scattering of a photon of energy E, the change in wavelength $\Delta\lambda = (h/m_e c)(1 - \cos\theta)$ implies a change in energy $\Delta E \approx -(E/m_e c^2)E$. Thus, the total number of scatterings, from initial energy ϵ_o down to a final energy ϵ is,

$$N_{sc} = \int_{\epsilon_o}^{\epsilon} \frac{dE}{(-E^2/m_e c^2)} \approx (m_e c^2)/\epsilon, \tag{6}$$

for $\epsilon \ll \epsilon_o$. For a random walk, we have $N_{sc} \approx \tau_T^2$, so that the 20 keV X-rays emerge after $511/20 \approx 25$ scatterings, or a "straight-line" Thomson depth $\tau_T \approx 5$.

The emergent X-ray flux will reach a maximum when τ_T is just large enough to Comptonize the γ-rays to X-ray energies that are just hard enough to escape the envelope without photoelectric absorption. For a random walk, the Thomson scattering optical depth along the total path length is $\tau_p \approx \tau_T^2$. An approximate fit to the photoelectric opacity (Morrison and McCammon 1983) relative to τ_p is,

$$\tau_a(\epsilon) \approx (1500\tau_p)\zeta\epsilon^{-3} \; ; \epsilon > 8 \text{ keV} \tag{7}$$

$$\tau_a(\epsilon) \approx (320\tau_p)\zeta\epsilon^{-2.5} \; ; 0.5 < \epsilon < 8 \text{ keV}, \tag{8}$$

where ϵ is measured in keV and where ζ is the metallicity factor ($\zeta = 1$ for solar abundances, $\zeta \approx 10^2$ for the mixture of heavy elements in the mantle of SN 1987A, but $\zeta \approx 0.25$ for the envelope, formed with LMC abundances). Substituting the emergent X-ray energy, $\epsilon_{em} \approx (m_e c^2)/N_{sc} \approx (m_e c^2/\tau_T^2)$, into eq. (7), we find

$$\tau_a(\epsilon_{em}) = \frac{1500\zeta}{(m_e c^2)^3} \tau_T^8 \approx (1.12 \times 10^{-5}) \, \zeta \, \tau_T^8. \tag{9}$$

The hard X-rays emerge when $\tau_a(\epsilon_{em}) \approx 1$, and therefore when $\tau_T \approx 4.2\zeta^{-1/8}$. The characteristic photon energy is $\epsilon_{em} \approx (30 \text{ keV})\zeta^{1/4}$. Since most of the Comptonization occurs in the H/He envelope, $\zeta \approx 0.25$ and the X-ray light curve is controlled primarily by the rate at which γ-rays escape the mantle.

Monte-Carlo simulations of the Compton scattering and absorption of X-rays in an expanding envelope (McCray, Shull, and Sutherland 1987) confirm the validity of this formula. The hard X-rays emerge at $\tau_T \approx 6$, corresponding to a time,

$$t \approx t_{thin}/6^{1/2} \approx (60 \text{ days}) M_e^{1/2} V_4^{-1}. \tag{10}$$

Thus, for $V_4 \approx 1$ and envelope masses M_e between 5 and 10 $M_\odot$ (Woosley, Pinto, and Ensman 1987; Nomoto 1987), we would expect hard X-rays and γ-rays to begin leaking out of the envelope after 130 to 200 days. To make further comparisons with the observations, we turn now to more detailed models of the interior mass and velocity distribution.

3 RESULTS OF MODELS

Theoretical modeling of the X-ray emission from SN 1987A has proceeded in three phases. The first phase was a study of homogeneous expansion, as described above. The second phase used detailed models for the interior density, velocity, and elemental stratification to compute the scattering and absorption optical depths. Xu *et al.* (1987) and Pinto and Woosley (1987) have performed Monte-Carlo simulations of the transfer of X-rays and γ-rays through various interior models of varying envelope mass. The early emergent X-ray spectrum is due entirely to down-Comptonization of the γ-rays or hard X-rays that can penetrate the envelope. The models with $M_e \approx 10 M_\odot$ predicted X-ray emergence some 2 to 3 months later than is now observed.

Attempts to reconcile this early emergence and the subsequent X-ray light curves with interior models have led to a third phase of modelling. These models have generally followed three approaches: (1) lowering the mass of the envelope; (2) "mixing" the radioactive debris in the mantle and partially into the envelope; and (3) examining the effects of "clumping" the mantle in order to produce a range of optical depths. These hybrid models preserve some aspects of the homogeneous models; for example, once the optical depths are determined at a given epoch (say 100 days), the constant expansion velocities assumed for each shell imply that $\tau \propto t^{-2}$ thereafter.

There are some constraints on these new approaches. Lowering the envelope mass M_e will allow an earlier turn-on, but the lower optical depths mean that the emerging X-rays undergo insufficient Compton scatterings to downgrade to 10 - 20 keV. Mixing the mantle allows the γ-rays to penetrate the envelope more easily, and thus directly influences the turn-on. However, there may be insufficient energy from the radioactive core to mix the heavy elements into the H-He envelope as far as required. Clumps in the mantle may arise from Rayleigh-Taylor instabilities associated with the ^{56}Ni "bubble", but modeling them in a convincing fashion is probably the most difficult task awaiting us.

In this section, we present results from new Monte Carlo models employing the first two approaches, and speculate on the effects of the third. Pinto and Woosley (1987) discuss two models (5L and 5LM) in which the explosion energy was reduced (6×10^{50} ergs) and an envelope of mass $M_e = 5 M_\odot$, was added to a $6 M_\odot$ helium core with $M_{Ni} = 0.075 M_\odot$. Model 5L gave a good fit to the optical light curve, but had its slowest hydrogen moving at 1800 km s^{-1} (compared to the observed 2100 km s^{-1}) and the X-rays still emerged too late. We have performed Monte-Carlo simulations of the X-ray transfer for model 5LM, in which the heavy elements are mixed throughout the mantle. The X-ray and γ-ray light curves are presented in Figs. 1 and 2, and the spectra at selected epochs are shown in Figs. 3 and 4. According to our predictions and those of Pinto and Woosley (1987), the γ-ray lines peak at a flux near 10^{-3} cm^{-2} s^{-1} between 400 and 450 days following the explosion. Since this flux is at the threshold of sensitivity, the γ-ray detections may be difficult.

The prospects for late detection of X-rays from the inner mantle (or a pulsar) are illustrated in Fig. 5, which gives the photoelectric opacity at 1, 3, and 10 yrs. However, the envelope may be leaky, and soft X-rays could emerge sooner than we predict.

The X-ray spectra (Fig. 3) show the expected hardening of spectral index between

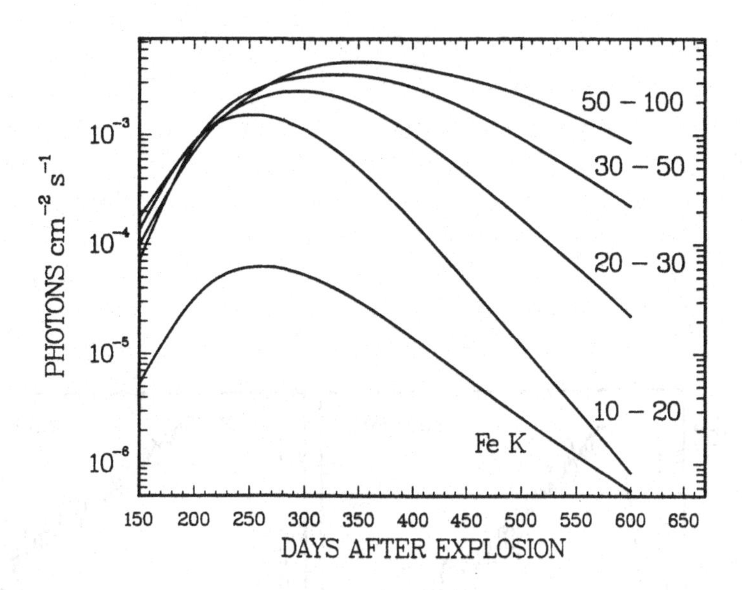

Figure 1: X-ray fluxes in various energy bins (labelled in keV) versus time for Woosley-Pinto model 5LM. Also shown is the 6.4 keV K_α fluorescence line of Fe.

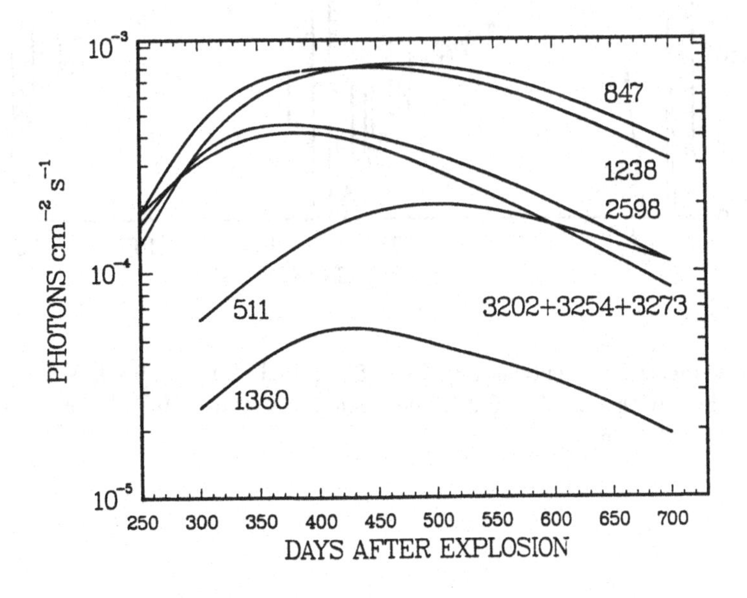

Figure 2: Light curves for selected γ-ray lines of ^{56}Co (847 and 1238 keV are the strongest), plus the 511 keV e^+ annihilation line.

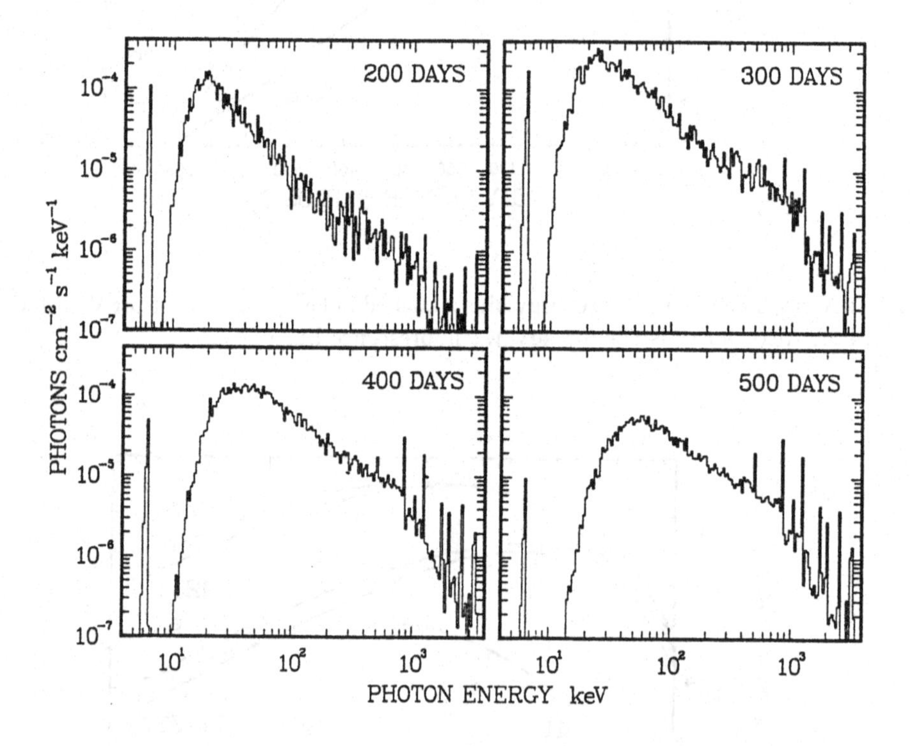

Figure 3: X-ray spectra at 4 epochs for Model 5LM, including the 6.4 keV Fe-line, for 200 equally spaced bins with $\Delta E/E = 0.0342$ between 4 keV and 4 MeV. Note the appearance of γ-ray lines after 300 days.

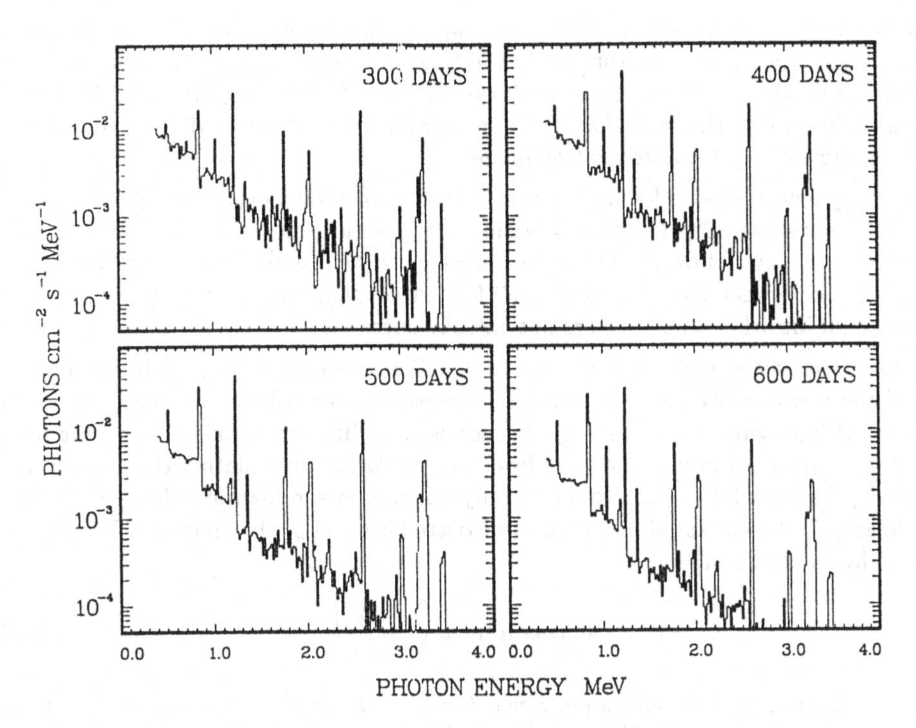

Figure 4: Gamma-ray spectra at 4 epochs for Model 5LM. The energy bins are linearly spaced, with $\Delta E = 0.018$ MeV. Note the appearance of lines at 511, 847, 1238 keV, as well as weaker lines up to 3.452 MeV.

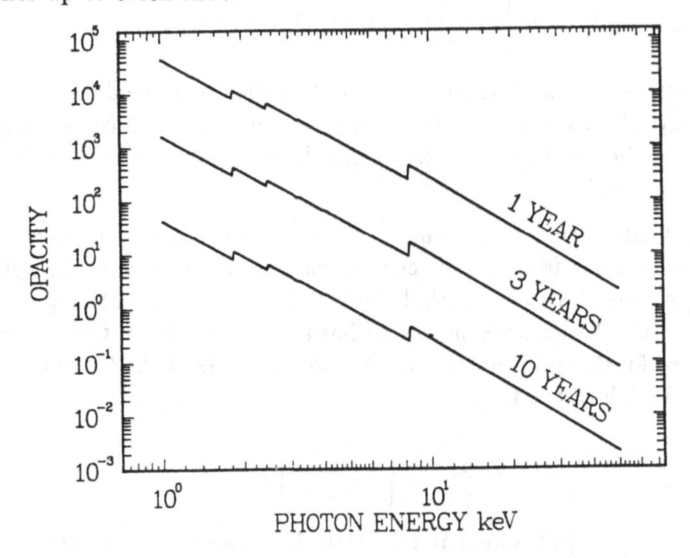

Figure 5: Photoelectric absorption optical depths of the envelope in Model 5LM after 1, 3, and 10 yrs. Note the K-shell edges of Si (1.84 keV), S (2.47 keV), and Fe (7.1 keV).

200 and 500 days as the envelope thins. Chi-square fits to the Monte-Carlo spectra of hard X-rays yield power-law photon spectra $J(\epsilon) \propto \epsilon^{-\alpha}$, where α declines steadily, from 1.53 ± 0.07 (200 days), 1.42 ± 0.04 (300 days), 1.20 ± 0.05 (400 days), and 1.04 ± 0.04 (500 days). Monitoring the behavior of the spectral index with time will provide a good observational test of the Comptonization model.

The early emergence of the hard X-rays is still problematic, even in Model 5LM with its lower-mass envelope. The *Ginga* observations suggest turn-on at about 130 days with a peak at about 200 days following the explosion, whereas the Monte-Carlo theory predicts a 10 - 30 keV peak between 250 and 300 days. Efforts to explain the early turn-on by mixing mantle material into the envelope were not entirely successful.

Instead, we may be seeing the effects of a distribution of optical depths in the mantle. If some of the γ-rays can escape the mantle more easily than others, we would witness a distribution of light curves such as Fig. 1, each with a different peaking time. We have investigated an idealized model of escape from the mantle, approximating the distribution of optical depths from fully transparent to fully opaque. If the mantle is filled with fully opaque "clumps" of constant size and total projected area A_{tot}, the emergent X-rays will be "gated" by the fraction,

$$f_\gamma = \left[1 - (A_{tot}/4\pi R^2)\right] = \left[1 - \frac{\alpha^2}{t^2}\right], \tag{11}$$

of γ-rays which penetrate the envelope. Since $R = V_o t$, where V_o is the velocity of mantle material, we have $\alpha^2 = (A_{tot}/4\pi V_o^2)$. As the envelope expands, the clumps intercept fewer of the γ-rays and the X-ray flux will have the time dependence,

$$J_x(t) = J_o \left[1 - \frac{\alpha^2}{t^2}\right] \exp(-t/113.6 \text{ days}). \tag{12}$$

We have fitted the *Ginga* data (Dotani *et al.* 1987) to this functional form (Fig. 6), and find $\alpha = 130 \pm 5$ days (95% confidence limits) with a reduced $\chi^2 = 1.04$. Clearly, the next step is to determine whether Rayleigh-Taylor instabilities at the core - mantle boundary can produce such clumps.

There is one final enigma remaining in the 4- 10 keV *Ginga* detection reported by Tanaka (1987). As is clear from the discussion earlier, photoelectric absorption in the expanding envelope puts a strong cutoff below 10 keV (see Fig. 2). As the envelope expansion proceeds, the X-ray peak moves to harder energies since the escaping X-rays have had less chance for down-Comptonization. Empirical fits of the lower energy spectra ($8 \text{ keV} \le \epsilon \le 40$ keV) follow the form,

$$J(\epsilon) = J_o \exp\left[-\left(\frac{\epsilon_o}{\epsilon}\right)^\beta\right], \tag{13}$$

where β ranges between 2.9 (300 days) and 2.1 (500 days), and ϵ_o ranges between 17.4 keV (300 days) and 32.7 keV (500 days).

Although it is yet too soon to tell, these "soft" X-rays may arise from the same mechanism that explains the early turn-on of harder X-rays. If there are leaks in the

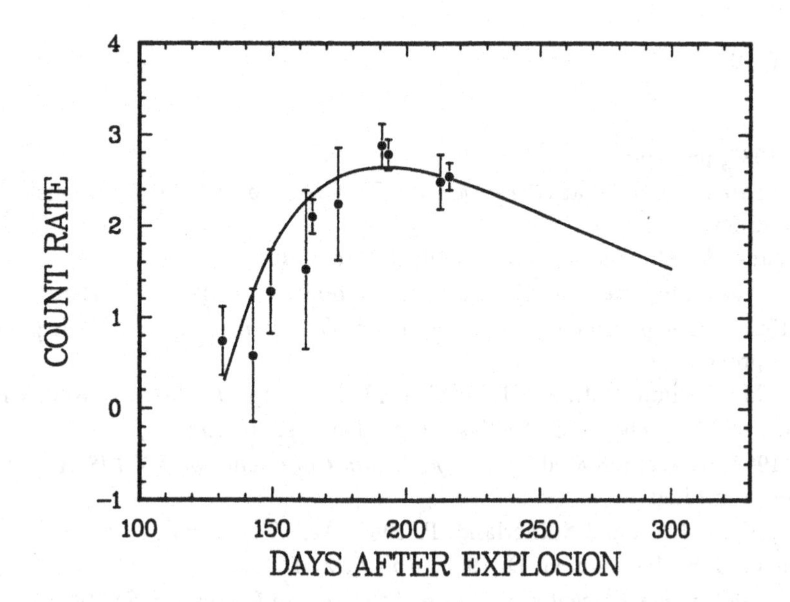

Figure 6: Fit of *Ginga* data for 10 - 30 keV counts (Dotani *et al.* 1987) to the functional form of eq. 12, with $\alpha = 130 \pm 5$ days.

core and envelope, soft X-rays might escape near the light curve peak. Alternative, and perhaps more reasonable explanations include reprocessing of the 10 - 30 keV X-rays by gas in the outer envelope or shock waves in circumstellar gas. The reports of CO emission lines between 3 - 5 μm (Larson 1987; Danziger 1987) and the *IUE* detection of N-rich ultraviolet emission lines (Kirshner and Sonneborn 1987) could be interpreted as evidence for interactions with a wind containing CNO-processed material. Certainly the energetics of an outflowing shock wave are consistent with the soft X-rays. For a shock wave of velocity (1000 km s^{-1})V_{1000} striking circumstellar gas of baryon density (10^3 cm^{-3})n_{1000} in a shell of radius (10^{16} cm)R_{16}, the radiated luminosity, $(\rho v_s^3/2)(4\pi R^2)\xi$, is approximately,

$$L_s \approx (10^{36} \text{ erg s}^{-1}) R_{16}^2 n_{1000} V_{1000}^3 \xi, \tag{14}$$

where ξ is an efficiency factor for converting the inflowing bulk energy into radiation. At 150 days following the explosion, the expanding envelope should be at $R \approx 10^{16}$ cm; a red giant wind of 10^{-6} $M_\odot$ yr^{-1} and velocity 10 km s^{-1} would have a density $\sim 3 \times 10^4$ cm^{-3} at this distance, and the shock velocities could easily exceed 1000 km s^{-1}. The B-star wind would alter this, however. Further analysis of the *Ginga* data, particularly the wavelength and strength of an Fe line or absorption edge, could help distinguish among a leaky mantle, thermal reprocessing, and X-ray fluorescence.

This work was supported by NASA Astrophysical Theory grants (NSG-7128 and NAGW-766) at the University of Colorado. We thank Stan Woosley and Phil Pinto for sharing the results of their numerical models.

REFERENCES

Arnett, D. 1987, preprint.

Danziger, J. paper presented at *George Mason Conference on SN 1987A*, published in these procedings.

Dotani, T., and the *Ginga* team 1987, submitted to *Nature.*

Ebisuzaki, T., and Shibazaki, N. 1987, preprint submitted to *Ap. J. (Letters).*

Feast, M. 1987, paper presented at *George Mason Conference on SN 1987A*, published in these procedings.

Gehrels, N., MacCallum, C.J., and Leventhal, M. 1987, *Ap. J. (Letters)*, **320**, L19.

Kirshner, R., and Sonneborn, G. 1987, submitted to *Ap. J. (Letters).*

Larson, H. 1987, paper presented at *George Mason Conference on SN 1987A*, published in these procedings.

McCray, R., Shull, J.M., and Sutherland, P. 1987, *Ap. J. (Letters)*, **317**, L73.

Morrison, R.L., and McCammon, D. 1983, *Ap. J.*, **270**. 119.

Nomoto, K. 1987, paper presented at *George Mason Conference on SN 1987A*, published in these procedings.

Pinto, P., and Woosley, S.E. 1987, *X- and γ-ray Emission from Supernova 1987A*, preprint.

Shigeyama, T., Nomoto, K., Hashimoto, M., and Sugimoto, D. 1987, *Nature*, **328**, 320.

Tanaka, Y. 1987, paper presented at *George Mason Conference on SN 1987A*, published in these procedings.

Trümper, J. 1987, paper presented at *George Mason Conference on SN 1987A*, published in these procedings.

Woosley, S.E., Pinto, P., and Ensman, L. 1987, preprint, to appear in *Ap. J.*, **324**.

Xu, Y., Sutherland, P., McCray, R., and Ross, R.R. 1987, preprint, to appear in *Ap. J.*, **327**.

X-RAY EMISSION FROM A PULSAR NEBULA

Franco Pacini, Arcetri Astrophysical Observatory and University of Florence (Italy)

Our theoretical understanding of Type 2 Supernovae suggests the presence of a collapsed object, possibily a neutron star, inside SN 1987a. Additional arguments for the formation of such an object stem from the detection of a prompt flux of neutrinos shortly before the optical event and the general argument that an internal energy source is required in order to maintain the optical luminosity of the Supernova against expansion losses.

A possible source of internal energy is the radioactive decay of Ni^{56} and Co^{56}. As several authors have stressed in the course of this Conference, an appropriate amount of radioactive elements can account for the observed optical and UV light curve. One expects then a copious flux of X-rays, resulting from the degradation of the γ-rays emitted by the radioactive material when the photons diffuse in the expanding SN material.

It is not clear whether the observed X-properties of SN 1987a, as reported by J. Trumper, G. Skinner and Y. Tanaka during this Conference, conform to the predictions of the simple model based on radioactivity. Among the various difficulties (rise time of the X-ray emission, spectral shape, total X-ray power), we stress in particular that the expected low energy cut off, below roughly 10 KeV, is in serious conflict with the data reported by the Ginga's group who has observed a steep rise in the low energy X-ray spectrum. This additional flux below 10 Kev has been attributed to an additional component, namely circumstellar gas heated up to ~ 4KeV by the SN shock. If so, the near coincidence of the rise epoch for the X-ray flux below 10 KeV (due to heated material) and above 10 KeV (due to radioactivity) becomes a puzzling coincidence. Even more puzzling (and probably fatal for the two components model) is the fact that the Ginga satellite has observed during September 1987 a marginal decrease in the X-ray flux above 10 KeV in coincidence with a clear large decrease in the flux below 10 KeV (see the paper by Tanaka). Finally, there is no evidence at present of the strong radioemission expected when the shock reaches the circumstellar material.

Because of these difficulties and because of the fact that, sooner or later, an hypothetical neutron star should manifest itself, it is worthwhile to investigate the expected properties of the non-thermal nebula produced by a central pulsar when its rotational energy is converted into magnetic fields and relativistic particles. A strong synchrotron radiation can then be expected from the Supernova, arising either from its internal parts or from jets reaching the outside of the envelope. The evolution of this non-thermal nebula has been considered in the past (F. Pacini and M. Salvati, 1973), with special emphasis on the nature of Radiosupernovae (Bandiera et al 1984). We defer the reader to the original papers quoted above for details. Similar considerations can be made for the X-ray emission and one expects that the high energy flux from the pulsar nebula should arise sooner than the radioemission, largely because of opacity effects (Pacini 1987). While more details will be published elsewhere (Bandiera, Pacini and Salvati, in preparation), we report here that a simple homogeneous expanding pulsar nebula would have a spectrum $S_\nu \propto \nu^{-\alpha}$ with $\alpha \geqslant 0.5$ ($\alpha = 0.5$ corresponds to a perfectly flat injected particle's spectrum). The spectral shape has been computed assuming that the lifetime of the radiating electrons is less than the age of the nebula i.e. that the break frequency lies below the X-ray band. The X-ray flux should initially increase $S_\nu \propto t^{\frac{1}{2}}$ and then become nearly constant. The time evolution has been computed assuming a naked nebula, i.e. neglecting the possible effects due to opacity (it is possible that in real life the rise of the X-ray emission is faster than the law quoted above if the influence of opacity is still important).

The evolution of the X-rays from a pulsar nebula should therefore be far different from that expected in the radioactivity model which predicts a rapid decrease of the X-ray flux following the initial rise. Furthermore, in the latter model, one expects the emergency of the γ-rays produced by the radioactivity as soon as the envelope is sufficiently thin. A non-thermal nebula should, instead, in the coming months, give rise to radioemission.

It has been repeatedly stressed that the non-thermal flux from the central pulsar nebula should remain hidden for many years or even decades because of the opacity in the surrounding SN envelope. If so, our calculations can only be relevant to future observers (in the years 2000?) However, as witnessed by the appearance of the Crab Nebula, it is also possible that the expanding envelope breaks into pieces and that open lines of sight develop very soon in the evolution of the shell (see, e.g. Bandiera et al 1983). If this occurs, today's astronomers are possibly already involved in some sort of cosmic peeping-tomism which enables them to look through the envelope, towards the internal volume occupied by the particles and fields generated by the pulsar. It is tempting to explain the luminosity decrease observed by Ginga in September 1987 as a temporary eclipse caused by clouds, filaments and changing opacity effects (which necessarily influence more the low energy, rather than the high energy flux) and predict that the X-ray flux should soon go up again.

In any case, the expectations for the near-future evolution of SN 1987a are widely different in the pulsar-nebula and in the radioactivity model. By 1988, we should be able to find out which one better represents the present reality, although of course a combination of the two also appears possible.

References

Bandiera, R., Pacini, F., Salvati M., Astron. Astrophyus. 126, 7 (1983).
Bandiera, R., Pacini, F. Salvati, M. Ap. J. 285, 134 (1984).
Pacini, F., Salvati, M., Ap. J. 86, 249 (1973).
Pacini, F., in Proc. of ESO Workshop on SN 1987a, Garching, July 1987 (in press).

Signatures of Particle Acceleration at SN1987a

T.K. Gaisser* Alice Harding+ Todor Stanev*

*Bartol Research Institute, University of Delaware
Newark, DE 19716
+NASA/GSFC, Greenbelt, MD 20771, U.S.A.

1 Introduction

It has been recognized for some time that young supernova remnants may be bright sources of energetic photons and neutrinos produced by collisions of particles accelerated inside the remnant.[1,2] For example, if the central engine is a rapidly rotating neutron star with a strong, nonaligned magnetic field, then the power available is

$$L_p = \epsilon\, L \approx \epsilon\, 4 \cdot 10^{43}\, \frac{\text{ergs}}{\text{sec}}\, \frac{B_{12}^2}{P_{\text{ms}}^4}, \tag{1}$$

where B_{12} is the magnetic field at the surface of the neutron star in units of 10^{12} Gauss and P_{ms} is its rotation period in milliseconds. Here L_p is the luminosity in accelerated ions and ϵ is the efficiency for particle acceleration. Because of the strong dependence of L on P_{ms} any pulsar-driven mechanism will be strongest in the first few years before the pulsar begins to slow down appreciably. One would also expect the highest energies to be accessible initially when the system is compact and the magnetic fields high. On the other hand, if the initial rotation of the neutron star is too slow ($P > \simeq 10\,\text{ms}$ for SN1987a at ≈ 50 kpc) then L_p will be too low to produce a detectable signal.

The accelerated particles themselves will not be directly observable for two reasons. First, they are likely to be contained in the expanding envelope by diffusion in turbulent magnetic fields. Second, even the particles that escape would be deflected by large-scale galactic and inter-galactic magnetic fields and become part of the general pool of cosmic rays. We must depend on neutral secondaries for a signal. Photons will be produced from decay of neutral pions produced in collisions of protons in the material of the envelope and by radiation of accelerated electrons and positrons. Neutrinos can come only from decay of mesons, so their observation would *require* acceleration of protons (or heavier nuclei). The upward muon signal that would be produced by interactions of neutrinos in the Earth can be searched for with large underground detectors in the Northern hemisphere[3]. The evaluation of the expected spectrum of energetic photons is somewhat more model-dependent.[4,5] than that of neutrinos. Unlike neutrinos, photons can come from radiation by electrons, and they can be attenuated on the way out

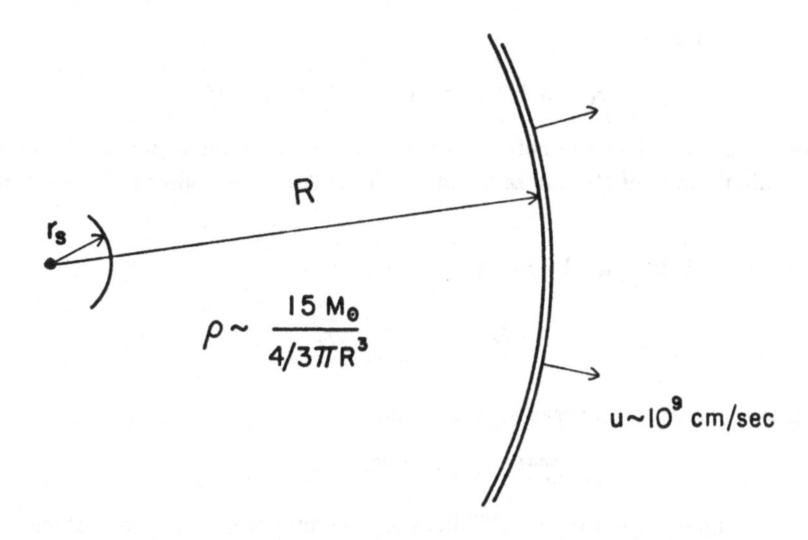

Figure 1: Schematic illustration of young supernova remnant.

of the production region. It is nevertheless likely that the photon signal will be more readily detectable than the neutrino signal because of the small cross section of neutrino interactions at the detector.

2 Model for acceleration

To calculate the expected photon flux we need to know the spectrum of parent protons and electrons, the nature of the region in which they propagate after acceleration, and the magnetic field and radiation environment which determines the subsequent fate of produced photons. The signal depends primarily on the power in accelerated protons and electrons, and various models of acceleration may give observable signals. We have evaluated[4] photon production in a model originally due to Rees and Gunn[6] that has been applied in detail to the Crab Nebula as a mechanism for transferring some of the rotational energy of the pulsar into relativistic electrons in the nebula.[7] The model is illustrated schematically in Fig. 1. In vacuum the central pulsar would emit magnetic dipole radiation with power given by Eq. 1. Because the frequency of the dipole radiation is lower than the ambient plasma frequency, this power goes instead into driving a relativistic wind of electrons and positrons. A shock forms at r_s where the ram pressure of the wind is balanced by the pressure in the nebula due to the accumulated energy outflow from the pulsar. We postulate that acceleration of charged particles (both ions and electrons) occurs by a first order Fermi mechanism at the shock. The maximum energy for protons is obtained by equating the acceleration time to the time for protons to diffuse away from the shock. For electrons the maximum energy occurs where acceleration time equals synchrotron loss time.

The results of Ref. 4 (see also Ref. 8) are summarized as follows:

- The shock radius is

$$r_s \approx 3.3 \times 10^{15}\text{cm}\, \tau_{\text{yr}}\, u_9^{3/2} \approx 0.1 \times R, \tag{2}$$

where τ_{yr} is the age of the nebula in years, u_9 is the outer expansion velocity of the nebula in units of 10^9 cm/sec, and $R = ut$ is the outer radius of the expanding nebula.

- The magnetic field near the shock is estimated as

$$B_s \approx \frac{10\, Gauss\, B_{12}}{P_{ms}^2 \tau_{\text{yr}} u_9^{3/2}}. \tag{3}$$

- The maximum proton energy accessible is

$$E_p^{max} \approx 1.9 \cdot 10^6\, \text{TeV}\, B_{12}\, P_{\text{ms}}^{-2}. \tag{4}$$

This is an upper estimate for E_p^{max} because the minimum diffusion coefficient has been used to calculate the acceleration time. The acceleration time is inversely proportional to magnetic field at the shock, which is large (Eq. 3). This is what allows such large potential values of E_p^{max}.

- The maximum energy for electrons is estimated as

$$E_e^{max} \approx 5\, TeV\, P_{ms} B_{12}^{-1/2}\, \tau_{\text{yr}}^{1/2}\, u_9^{3/4}. \tag{5}$$

- As shown in Ref. 8, particles with energy less than

$$E_{\text{esc}} \simeq 10^{10}\, \text{GeV} \times \frac{B_{12} u_9^{1/2}}{\xi P_{\text{ms}}^2} \tag{6}$$

will be contained in the expanding envelope for times less than the age of the supernova. Here $\xi > 1$ is a parameter that characterizes the degree to which the diffusion coefficient of the energetic particles exceeds its miminum value, which is one-third the Larmor radius times the particle velocity.

3 Estimate of secondary signals

Because of the large value of the escape energy in Eq. 6 we can assume that virtually all accelerated particles are contained in the expanding envelope. The analysis of Ref. 1 then applies: After the first few days the density is low enough so that all produced mesons decay. Only neutral particles escape from the envelope. Protons are contained and any remnant protons re-interact. Thus until the time at which the interaction rate, $c\sigma M/(M_H\, \frac{4}{3}\pi u^3 t^3)$, falls below the expansion rate, we can assume that all accelerated protons interact at least once. (Here M is the total mass of the shell.) After this time, only a fraction of the accelerated protons interact to produce photons and neutrinos. Thus

$$f = min[0.28 \frac{M}{M_\odot} u_9^{-3} \tau_{\text{yr}}^{-2},\ 1] \tag{7}$$

Table 1: Photon fluxes at Earth ($cm^{-2}s^{-1}$) for $L_p = 10^{40}$ ergs/sec

	$\gamma = 2$	2.2	2.4	2.6	2.8	3.0
>100 MeV	2.1(-6)	5.0(-6)	6.7(-6)	7.3(-6)	7.4(-6)	7.3(-6)
>1 TeV	3.5(-10)	1.3(-10)	3.0(-11)	5.9(-12)	1.1(-12)	2.1(-13)
>200 TeV	5(-13)	8.9(-14)	7.5(-15)	5.6(-16)	3.9(-17)	2.7(-18)

is the fraction of L_p that emerges in photons and neutrinos. For $t > 0.5$ years $u_9^{-3/2}\sqrt{\frac{M}{M_\odot}}$ Eq. 7 implies that the power in proton-induced secondary particles will begin to decrease quadratically with time. The suppression factor in Eq. (7) must then be combined with the decrease in power due to pulsar spin-down. For t such that $f = 1$, up to the full power of the wind could go into photons (and neutrinos) through accelerated particles. Eqn. (1) therefore sets the scale for estimating expected signals.

As an aside, we note that neutrons can also escape directly from the envelope provided

$$\gamma c \tau_n > ut, \tag{8}$$

where τ_n is the neutron lifetime and γ its Lorentz factor. Neutrons with $E < \cong 1\,\mathrm{TeV}\,\tau_{yr}$ decay before escape and give protons that can re-interact.

The photon and neutrino spectra are simply proportional to the power L_p in accelerated particles. As an example we summarize the results for photon signals in Table 1 for a normalization of $L_p \approx 10^{40}$ ergs/sec for protons with energies between 1 and 10^8 GeV injected into the nebula 50 kpc away. This would be achieved, for example, with $\epsilon \simeq 0.1$ and $P \simeq 4$ ms. The numbers in Table 1 scale linearly with power. In calculating these numbers we have included only photons from decay of neutral pions produced by collisions of accelerated protons that have been injected into the nebula. The parameter γ is the assumed differential spectral index of the proton beam.

Figure 2 shows the expected photon flux from proton interactions in SN1987a for a particular normalization, as in Table 1. Shaded regions show respectively the energy regions accessible to atmospheric Cherenkov experiments and to air shower detectors. The dip is due to pair production by the high energy photons on the intervening microwave background.

Accelerated electrons will also produce photons from the synchrotron radiation that limits their acceleration to E_e^{max} as given in Eq. 5. The electron spectrum will exhibit a pile-up peak[9] just below the electron cutoff energy. We estimate the flux of synchrotron photons from the pile-up, assuming that all energy loss comes in photons of the synchrotron peak frequency of $6.4\,MeV (= \frac{1}{3}\nu_c)$, which in this model is independent of all parameters and constant in time. The flux of such photons at Earth is proportional to the total dipole energy in accelerated electrons. It is

$$F(6.4MeV) = 12\, B_{12}^2\, P_{ms}^{-4}\, \epsilon_e \,\text{photons}\, cm^{-2}s^{-1}, \tag{9}$$

where ϵ_e is the fraction of the dipole energy that goes into pile-up electrons. The photon production spectrum is the single particle emissivity, which is proportional to $\nu^{\frac{1}{3}}$ below ν_c. Such fluxes will be easily detectable with rocket experiments in the Southern hemisphere when the optical depth of the shell becomes smaller than unity and may even exceed the expected line fluxes from radioactive decays.[10]

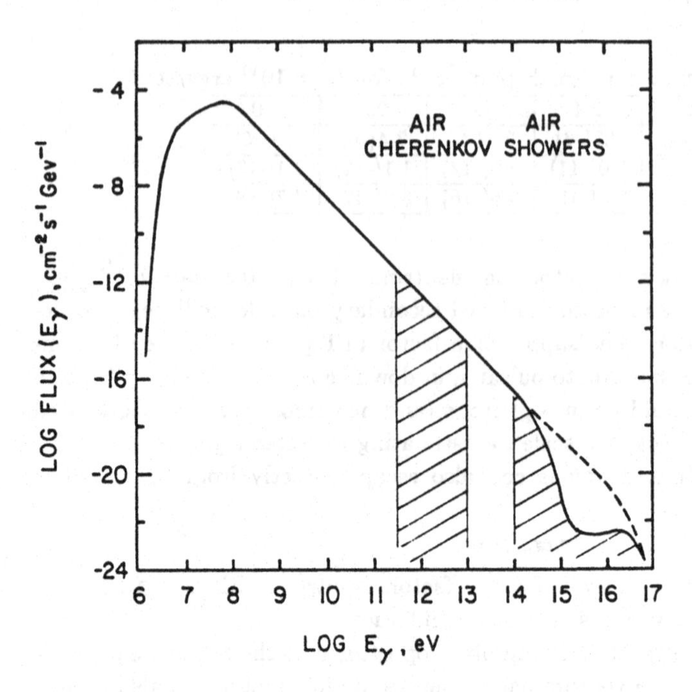

Figure 2: Energy spectrum of photons from proton interactions (no photon cascading) for $L_p = 10^{40}$ ergs/sec at 50 kpc.

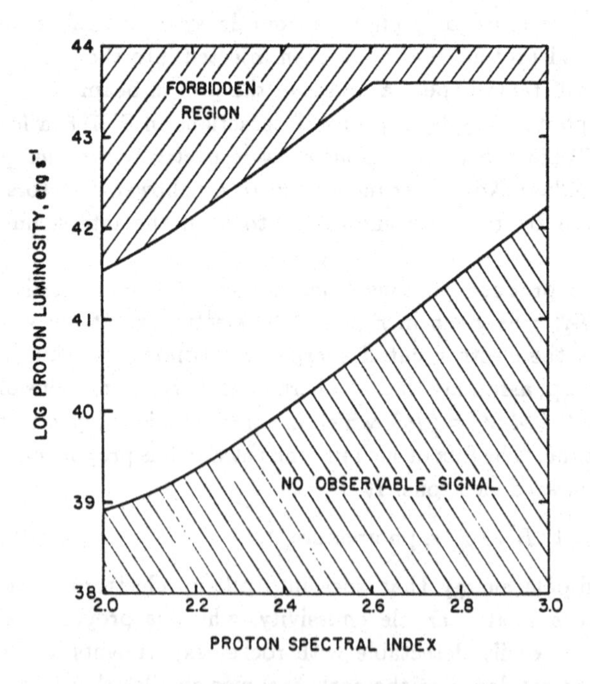

Figure 3: Limits on particle luminosity from SN1987a for various spectral indices of the accelerated particles.

4 Discussion

Because of the large opacity of the shell at early times, a photon signal would be suppressed at first even if the accelerator turned on immediately. At later times, because of adiabatic losses of the magnetically trapped particles in the expanding envelope,[1] both photon and neutrino signals will begin to decrease after some characteristic time. There is thus a window during which the signal of secondary photons will be at its maximum. For the uniform density model this window of opportunity is between a few months and about a year.[1] For a more realistic model of the envelope,[5,10,11] the density is higher in the more slowly expanding inner regions, and the window therefore occurs later, roughly between one and ten years. The details of how the signal declines after maximum are quite sensitive to the model of the shell. For any given model, however, we believe that the duration of maximum signal will be longer than shown in Ref. 5. This conclusion holds if the accelerated particles are confined in the shell, as we have argued, rather than propagating straight through it, as assumed in Ref. 5. In any case, the best time to look for a signal of particle acceleration in SN1987a is as soon as possible.

Neutrinos would be a sure sign of acceleration of protons (or nuclei), but they have small interaction cross sections. Present limits from Kamioka,[12] IMB[13] and Baksan[14] based on the first six months of observation do, however, put interesting upper limits on L_p. The experimental limits are at the level of 10^{-13} neutrino-induced, upward muons per cm^2 per sec. We use the results of Ref. 3 to convert this to a limit on neutrino flux and hence to a limit on L_p. These limits are shown in Fig. 3.

The lower band in Fig. 3 is the level below which air Cherenkov signals would be too low for detection. The allowed possibilities include an interesting range of luminosities and spectral indices. If the observed decline of the optical light curve of SN1987a is interpreted to be due to the ^{56}Ni, ^{56}Co decay chain, this also can be used to put upper limits on other sources of light such as a central pulsar. These limits would be more restrictive than the neutrino limits, but depend on the efficiency for conversion of pulsar wind energy into optical radiation.

Three air Cherenkov experiments in the Southern Hemisphere have the capability of searching for $\cong$TeV signals from SN1987a, the University of Durham (UK) detector at Narrabri[15], the Potchefstroom detector in South Africa[16] and the Adelaide experiment[17]. We have assumed a threshold for detection of 3×10^{-11} photons per cm^2 per second above 1 TeV to draw the limit of sensitivity in Fig. 3. We note that photon signals that have an origin in collisions in the envelope are unlikely to show any periodicity, and this makes the signal more difficult to detect than if it were pulsed with a known period. We are grateful to K.E. Turver for a discussion on this point.

Air shower experiments such as proposed by the JANZOS collaboration[18] for New Zealand or the one already operating at Buckland Park in Australia can search in the PeV (10^{15} eV) range. The Buckland Park group have already reported[19] a limit on the flux of air showers coincident with the collapse in SN1987a. Because the air shower detectors have thresholds approaching the region where interactions on the intervening background become important, it is more difficult to estimate the sensitivity of these detectors to the power at the supernova. A particularly attractive air shower experiment is that proposed for the South Pole by a collaboration of University of Leeds and Bartol. Advantages are low threshold and constant, small local zenith angle (21°) for observing SN1987a.

We are indebted to S.E. Woosley, P.A. Pinto and L. Ensman for providing us with the detailed tables describing some of the models of the shell of SN1987a. We gratefully acknowledge discussions with W.D. Arnett, G. Auriemma, A.E. Chudakov, E. Dwek, F. C. Jones, J. Matthews, J.Ostriker, R.J. Protheroe, K. E. Turver and J. van der Velde. This research was supported in part by the U.S. Department of Energy and by NSF.

5 References

1. Berezinsky, V. S. and Prilutsky, O. F.; Astron. Astrophys. 66, 325-334 (1978).
2. Shapiro, M. M. and Silberberg, R. in Relativity, Quanta and Cosmology, edited by F. DeFinis (Johnson Reprint Corporation, New York, 1979) v. 2, pp. 745-782.
3. Gaisser, T. K. and Stanev, T.; Phys. Rev. Lett. 58, 1695-7 (1987).
4. Gaisser, T.K., Harding, A.K. and Stanev, T., Nature, 329, 314 (1987).
5. Nakamura, T., Yamada, Y. and Sato, H., preprint KUNS 877, July 1987.
6. Rees, M. J. and Gunn, J. E., Mon. Not. Roy. Astr. Soc., 167, 1-12 (1974).
7. Kennel, C. F. and Coroniti, F. V., Ap. J., 283, 694-709 (1984).
8. Auriemma,G., Gaisser, T.K. and Lipari, P. preprint, October, 1987.
9. Schlickeiser, R., Astron. Astrophys. 136, 227-236 (1984).
10. Woosley, S.E., Pinto, P.A. and Ensman, L., Submitted to Ap. J. (1987).
11. Berezinsky, V.S.and Ginzburg, V.L., submitted to Nature (1987).
12. KAMIOKANDE-II Collaboration, UT-ICEPP-87-05 and UPR-0145E (1987).
13. IMB Collaboration, private communication.
14. Baksan Experiment, A.E. Chudakov (private communication).
15. Carstairs, I. et al., in *Very High Energy Gamma-ray Astronomy*, (ed. Turver, K. E.) pp. 221-223 (Reidel, Dordrecht, 1987).
16. DeJager, H. I., DeJager, O. C., North, A. R., Ranbenheimer, B. C., Van der Walt, D. J. and Van Urk, G.; S. African Journal of Physics, 9, 107-117 (1986).
17. Clay, R. W., Elton, S. D., Gregory, A. G., Paterson, J. R. and Protheroe, R. J.; Proc. Astron. Soc. Australia, 6, 338-343 (1986).
18. JANZOS Collaboration Proposal, April 23, 1987.
19. Bird, D.J., et al., submitted to Proc. Astronomical Soc. of Australia, June, 1987.

GAMMA-RAYS FROM THE PULSAR IN SN 1987A

K. Brecher
Boston University, Boston MA 02215, U.S.A.

Abstract. Theoretical arguments, as well as direct observations, suggest that SN 1987A has left behind a neutron star. Here we will examine what gamma-ray flux may be expected to arise from a hot, magnetized, rapidly spinning neutron star/pulsar during the first few years after the initial supernova event. If the pulsar itself significantly contributed to powering the optical light curve of SN 1987A during its first few months, it must be strongly magnetized ($B \geq 10^{12}$ gauss) and rapidly rotating (with spin period $t \approx 2 - 6$ ms). Pair production by pulsar accelerated electrons scattering on the soft x-ray photons emitted from the neutron star surface will give rise both to a copious flux of electron-positron pairs, and to a strong flux of Compton scattered gamma-rays. If the supernova shell has fragmented early, as we argue here that it has, within about a year of the initial supernova event, the supernova remnant should become optically thin to MeV gamma-rays, leading to the possible detection of a strong, few millisecond gamma-ray pulsar with continuum flux $F(\geq 1 \text{ MeV}) \approx 10^{-3}$ $cm^{-2}s^{-1}$.

Arguments for a Neutron Star/Pulsar

Supernova explosions, even of Type II, need not lead to a remnant neutron star. However, two major observations are certainly consistent with the formation of a neutron star in SN 1987A. First, the observed neutrino flux is in good agreement with theoretical calculations of stellar core collapse leading to the formation of a neutron star (Burrows & Lattimer, 1987). Second, while the optical light curve can be fit by models involving heating of the ejecta in the initial explosion, followed by subsequent heating due to nucleosynthesis decay products (Arnett, 1987), it may also be fit assuming energetic input from a central pulsar. Whether any remnant neutron star should be rapidly rotating and/or strongly magnetic is not known. If the core of the progenitor supergiant contained sufficiently strong magnetic fields and/or angular momentum, it is quite plausible to leave behind a rapidly rotating strongly magnetic neutron star (Brecher and Chanmugam, 1978).

Gamma-Ray Emission from a Hot Young Neutron Star

Assuming that a rapidly rotating, strongly magnetic neutron star has formed in the SN 1987A event (Brecher 1987), how should it appear during its first few years? Unlike older pulsars which have already cooled down, or which are not radiating significantly as non-thermal sources, a young neutron star should still be "hot", at least locally near its magnetic poles. According to the best estimates for interior neutrino and surface electromagnetic neutron star cooling, the mean surface temperature should be less than a few million degrees within a year after formation. However, if the source is accelerating electrons which subsequently pair form while leaving the source, positrons can heat the polar cap regions to much higher temperatures, perhaps as high as several times 10^7 °K. In such circumstances, electrons accelerated by the pulsar can pair form on the radiation emitted from the pulsar surface. This will lead to a positron-electron cascade. The resulting pair-formed electrons and positrons will also Compton scatter on the background soft x-ray photons as well as on photons radiated by other electrons. The net result is a strong electron-positron wind, as well as a strong flux of gamma-ray photons.

These properties of a hot young pulsar were first discussed by Brecher and Mastichiadis (1983), and subsequently examined in detail in Mastichiadis, Marscher and Brecher (1986) and Mastichiadis, Brecher and Marscher (1987). The conclusion of these models was that a hot young pulsar will be a copious source of electron-positron pairs, and that approximately 99% of the energy produced by such an object will be radiated in the form of gamma-rays with energies greated than about an MeV. How do these ideas apply to the possible pulsar in SN 1987A? It may be that such an object has a weak field ($B \leq 10^{12}$ gauss) and /or slow rotation ($t \geq 10$ ms), in which case the pulsar would run by "normal" pulsar action (whatever that is). However, for strong fields and rapid rotation, the self heating effect should make pair formation on the photons the dominant physical interaction for pulsar accelerated electrons, in which case the picture presented above should be applicable. SN 1987A had a peak luminosity, some 100 days after the initial event, of approximately 10^{42} ergs s^{-1}. If all, or a significant fraction, of this flux is due to the pulsar, then the energy loss rate of the pulsar must satisfy $dE/dt \approx B^2R^6\Omega^4c^{-3} \approx 10^{42}$ erg s^{-1}. On the other hand, the heating must continue for some months, so that the pulsar spin down time must be greater than a few months. Therefore, $t_{spin\text{-}down} \approx Mc^3/B^2R^4\Omega^2 \geq 10^6$ sec (say). Assuming a typical neutron star of mass M and radius R, and taking the neutron star magnetic field B to lie in the range 10^{12} - 10^{13} gauss, one finds the pulsar spin period $t \approx 2 - 6$ ms. A pulsar with 2 ms spin period and magnetic field of 10^{12} gauss would in fact have a luminosity decrease e-folding time of several years, considerably longer than the 111 day decay-time for the ^{56}Ni decay products which have been suggested as the source of continuing optical luminosity of the SN at late times. How and when would such a pulsar manifest itself directly?

When Will the Gamma-ray Pulsar Appear?

According to several estimates (Xu et. al. 1987; Chernoff et. al.1987), the supernova remnant should remain optically thick to gamma-rays from its central region for several years after the original explosion. In the interim, any gamma-rays emanating from the central region should be degraded by Compton scattering as they leave the nebula, and be visible initially as a flux of hard x-rays. In this picture, only after about five years should the gamma-rays themselves (either from a cental pulsar or from nucelosynthesis decay products) become detectable directly. Several observations argue against this point of view, and suggest that the nebula will become optically thin to gamma-rays within a few months, if it is not transparent already. The first observations concern the apparent size of the object. A month or so after the explosion, the effective photometric radius (as determined by combining the flux and temperature) was found to be about 10 AU. However, the size as determined by speckle interferometric observations (Karovska et. al 1987) was about 500 AU. Some 100 days after the explosion, the photometric and speckle radii were roughly 20 AU and 1000 AU, respectively. While the source was becoming optically thin to optical radiation after 100 days, so that the discrepancy might only be an optical deapth effect, at 40 days after the event the source was still optically thick. The early difference between the speckle and photometric radii can still be reconciled if, within the first month after the explosion, the source had already broken into filamanets, knots or wisps, similar to what is seen in the Crab nebula. Providing the filling facter is of order 10^{-2} or 10^{-3} (as in the Crab), the discrepancy between the photometric and speckle radii can be accomodated as resulting from the emission from separate blobs or filaments which fill a radius comparable to the physical supernova front (as determined by the expansion velocity), while filling only a small fraction of the volume. What else argues in favor of early fragmentation of the supernova? As reported at this meeting, the Japanese Ginga observations of x-rays appeared far earlier than predicted assuming a smooth expanding nebula. They have also been reported to be disappearing already in September 1987. Assuming that the nebula has already broken up, the x-ray observations suggest that any centrally produced gamma-rays are leaking out of the nebula volume through open paths. Finally, the late light curve after about 250 days may have begun to deviate from the standard exponential decay. While this could result from many causes including the lack of significant ^{56}Ni decay gamma-ray input into the light curve), it might well result from the fragmentation of the expanding nebula.

How can we understand fragmentation of the nebula so early in its history? To date, the generally accepted suggestion concerning why the Crab nebula or Cas A appear so fragmented concerns the sweeping-up of interstellar material, resulting in a Rayleigh-Taylor instability at the outside edge of the expanding nebula several hundred years after the original event (Gull, 1975). This is clearly irrelevant in the present context. Here we suggest that it is the presence of the pulsar itself which has caused the supernova ejecta to fragment very early. As the ejecta expands, the gas cools to the observed temperature of 5,000 ^{o}K. However, surrounding the pulsar (but interior to the expanding supernova ejected shell) relativisitic particles (electrons and positrons, but also the gamma-rays, pulsar produced magnetic fields and, perhaps, accelerated protons or ions) fill the central cavity. When the pressure of the light (but relativistic fluid) exceeds the gas pressure on the inside of the expanding ejecta, the expanding shell must begin to fragment as the relativistic fluid tries to push its way in. Assuming the ejecta has a mass $M \approx 10\ M_{\odot}$, internal temperature

$T \approx 5000^{o}K$, and that the pulsar has a luminosity $L \approx 10^{41} - 10^{42}$ erg s^{-1}, then this fragmentation should occur after the explosion at time $t_f \approx MkT/Lm_H \approx 10^5 - 10^6$ s, that is, within days or weeks after the initial explosion.

Conclusions

In conclusion, we propose that the early light curve of SN 1987A was largely powered by the presence of a central pulsar with spin period $t \approx 2 - 6$ ms. This pulsar should be a strong source of continuum gamma-rays. Assuming that the pulsar does indeed power the light curve, then we find $F(\geq 1 \text{ MeV}) \approx 10^{-3}$ $cm^{-2}s^{-1}$. Such a pulsar would cause the break-up of the supernova ejecta into a fragmented shell, allowing the gamma-rays to leak out of the source earlier than previously predicted. If we are lucky enough to have the pulsar beam pointing in our direction, we would expect to see a pulsed source of gamma-rays above about 1 MeV appear within about a year of the initial event. (A more detailed prediction would require detailed knowledge of the actual shape of the ejecta.) Finally, such an energetic pulsar might offer an explanation for the enigmatic "mystery" spot (Nisenson et. al. 1987). If, during the first few weeks after the explosion, radiation and particles were trapped in the central supernova region their pressure might have built up significantly. Then, owing to the Rayleigh-Taylor instability developing on the inside of the ejected shell, they could have "punched" their way out in one direction after a month or so. This might fluoresce surrounding gas in one direction with an intensity greater than what might be expected given the (then) external luminosity of the ejecta as a whole.

References

Arnett, W. D. (1987). Ap. J., 319, 136.
Brecher, K. (1987). B.A.A.S., 19, No. 2, 735.
Brecher, K. & Chanmugam, G. (1978). Ap. J., 221, 969.
Brecher, K. & Mastichiadis, A. (1983). In Positron Electron Pairs in Astrophysics, ed. M. L. Burns et. al., New York, American Institute of Physics, 287.
Burrows, A. & Lattimer, J. M. (1987). Ap. J. Lett., 318, L63.
Chernoff, D. F., Shapiro, S. L., Wasserman, I. (1987). Preprint.
Gull, S. F. (1975). M.N.R.A.S., 171, 263.
Karovska, M., Nisenson, P., Papaliolios, C., & Standley, C. (1987). I.A.U. Circ. No. 4450.
Mastichiadis, A., Brecher, K. & Marscher, A. (1987). Ap. J., 314, 88.
Mastichiadis, A., Marscher, A. & Brecher, K. (1986). Ap. J., 300, 178.
Nisenson, P., Papaliolios, C., Karovska, M, and Noyes, R. (1987). Ap. J. Lett., 320, L15.
Xu, Y., Sutherland, P., McCray, R. & Ross, R.R. (1987). Preprint.

SUPERNOVA 1987a AND THE EMERGENCE OF THE HOT NEUTRON STAR: A SEMI-ANALYTIC APPROACH

D.F. Chernoff
S.L. Shapiro
I. Wasserman
Center for Radiophysics and Space Research, Cornell University, Ithaca, New York, 14853, USA

Abstract. Supernova 1987a is modeled as a Sedov blast wave in a star with a power law density distribution. The luminosity is determined numerically once the shock breaks out of the star. Analytic scaling relations are derived based on simple physical arguments. These agree with our detailed numerical calculations and permit exploration of a wide range of model parameters. The neutrino and early V-band observations are used to deduce the range of possible values for the mass and radius of the pre-supernova star and for the shock velocity. From the range of satisfactory fits alone, we find that the shock velocity lies in the range 3 - 5 x 10^8 cm/s and the pre-supernova radius lies between 10^{10} and 5 x 10^{12} cm. The luminosity in the V-band is almost independent of the pre-explosion structure of the envelope or of the mass of the envelope. These calculations support the conventional core collapse scenario in which the supernova shock is propelled into an overlying stellar envelope with initial velocity near the escape speed from a hot neutron star.

The time-dependent opacity over an energy range 100 eV to 1 MeV is modeled for the expanding envelope, assuming no energy input from radioactive nucleotides or an underlying pulsar. The 1 - 3 KeV thermal X-ray emission expected from a cooling neutron star will be exponentially attenuated for about 100 years for the parameters we have deduced. Should higher energy emissions from the star be present (greater than 10 KeV), then these should start to become visible within a period of a few years.

DISCUSSION

Supernova 1987a will provide our most detailed look at the explosive end-point of stellar evolution. We consider the simplest possible models of the explosion to demonstrate how the basic physical parameters of the exploding star, the stellar mass and radius and the energy of explosion, are constrained by the visual observations of the first few days. The role of the neutrino observations (the Kamioka detector, Hirata 1987, and the IMB detector, Bionta 1987) is crucial in our analysis because it provides a zero-point time for the catastrophic disturbance within the star. The earliest visual observations (Jones 1987) and McNaught 1987 have already been used to fit highly specific models of the supernova explosion (Arnett 1987, Woolsely 1987a,b). We have not aimed at the same degree of precision, but instead derived scaling results, which we use with numerical calculations, to study an extensive range of initial conditions which are consistent with the early observations.

We demonstrate that the simplest possible description, diffusing radiation in an expanding spherical envelope, is sufficient to recover the essential characteristics of the luminosity curve over the first few days. We assume that the preshock stellar envelope obeys an power-law density profile ($\rho \propto r^{-N}$, $N \approx 3$) and that the postshocked material satisfies the Sedov blast wave similarity solution. Using only that information, we show that the range of consistent initial conditions is large. After about ten days, the evolution becomes far more complex and we do not attempt to treat it. However, once we have deduced the range of plausible initial conditions, it is then straightforward to consider the long-time behavior of the expanding atmosphere. By this we mean features that are dependent only on the bulk envelope mass and its rate of expansion many dynamical times after the explosion, but before the ejecta begins to be affected by the interstellar medium. We treat the evolution of the optical depth for absorption and scattering as a function of frequency. We address the general question of when a source of frequency ν and luminosity L_ν first becomes visible at the center of the remnant. We assess the possibilities of detecting thermal x-rays from the hot neutron star surface and gamma rays from a pulsar.

Our key conclusions are: (1) The observed value of m_V at one day essentially fixes the shock velocity to be $v_s \approx 3 \times 10^8 \mathrm{cm/s}(r_s/2 \times 10^{12}\mathrm{cm})^{-2/15}$, where r_s is the pre-supernova radius, (2) variations in the stellar mass in the range 5-25 $M_\odot$ have little effect on the previous results, and (3) following core collapse at the stellar center, the shock was launched with a low initial velocity compared to the escape velocity at the surface of a cold neutron star.

Our key results concerning the long-time behavior of the envelope opacity over a wide range of frequencies are summarized in Figure 1. We conclude that the "soft" X-rays (1-3 KeV) from a hot central neutron star will not be visible for about 100 years. This does not preclude the possibility (now confirmed by Ginga) of seeing X-rays from the envelope, such as degraded gamma-ray lines given off by decaying radioactive nucleotides. If harder energy emissions are present ($\gtrsim$ 10 KeV) from the central source, then these should begin to be visible any time now.

Figure 1. The time dependent behavior of the effective opacity at energies between 160 eV and 1 MeV are shown for a model with $M_s = 15\ M_o$, $r_s = 2 \times 10^{12}$ cm, $E = 10^{51}$ ergs and N = 2.5. The scaling regime for the opacity is indicated by the straight parallel lines after about three months.

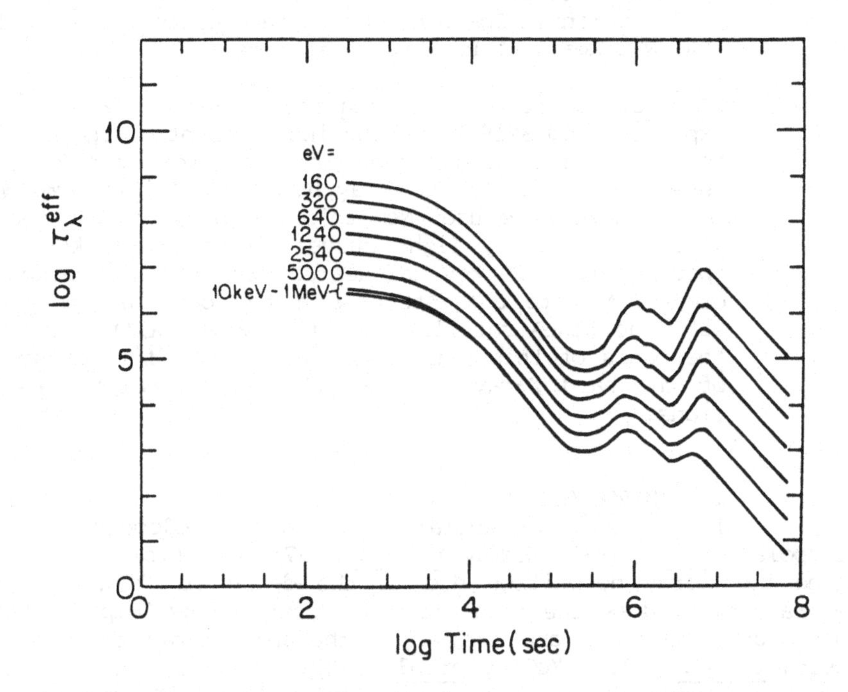

ACKNOWLEDGMENTS

We acknowledge useful conversations with Drs. L. Bildsten, E.E. Salpeter and J.C.L. Wang. We thank Dr. Wang for sharing his complications of SN 1987a observational data with us. We are grateful to Drs. F. Makino of the Ginga team and G. Skinner of the Roentgen mission team for communicating instrumental specifications of their X-ray observatories to us. This research was supported in part by NSF Grant No. AST-84-15162 at Cornell University.

REFERENCES

Arnett, W.D. 1987, preprint.

Bionta, R. et al. 1987, Phys. Rev. Lett., 58, 1494.

Hirata, K. et al. 1987, Phys. Rev. Lett., 58, 1490.

Jones, A. 1987, IAUC, 4340.

McNaught, 1987, IAUC, 4316.

Woolsley, S.E., Pinto, P.A., Martin, P.G. and Weaver, T.A. 1987a, Ap. J. Lett. 318, 664.

Woolsley, S.E., Pinto, P.A., Ensman, L. 1988, Ap. J. Lett, 324, (in press).

THE EFFECTS OF MIXING OF THE EJECTA ON THE HARD X-RAY EMISSIONS FROM SN1987A

T. Ebisuzaki[1]
ES-65, Marshall Space Flight Center, Huntsville, AL 35812

N. Shibazaki[1]
ES-65, Marshall Space Flight Center, Huntsville, AL 35812
[1]NAS/NRC Resident Research Associate

Abstract. We have calculated the X- and gamma-ray emissions expected from SN1987A taking into account mixing of material in the ejecta. Nuclear gamma-rays emitted by ^{56}Co are scattered down to the hard X-ray band by multiple Compton scatterings. We have used Nomoto's 11E1Y6 model for the ejecta of SN1987A. X-ray light curves in the 10-30 keV band and spectra above 20 keV calculated with an inner mixed region of 5 ± 1 $M_{\odot}$ are consistent with the observations performed with the GINGA satellite and the KVANT/ROENTGEN mission. On the basis of this comparison, we discuss further evolutions of the hard X-ray, gamma-ray, and optical/infrared emissions.

1 INTRODUCTION

The recent exponential decay in the bolometric luminosity (Catchpole et al. 1987; Hamuy et al. 1987) suggests that SN1987A is powered by energy deposition due to the decay of ^{56}Co. The nuclear gamma-rays emitted by the ^{56}Co are down scattered by Compton scattering in the outer envelope and emerge in the hard X-ray band 10-200 keV (Woosley et al. 1987a, McCray et al. 1987). Several authors calculated X-ray light curves for the models provided by Dr. S. E. Woosley (Gehrels et al. 1987; Ebisuzaki and Shibazaki 1987; Xu et al. 1987). According to them, the flux in the 10-30 keV band reaches a peak around one year after the explosion. However, the GINGA satellite (Makino 1987b) and the KVANT-ROENTGEN mission (Truemper 1987) detected an excess X-ray flux from the direction of SN1987A in the mid August, 1987. According to Dotani et al. (1987) and Makino (1987c), the flux in the 10-30 keV band observed by GINGA increased monotonically through July and August, 1987, while observations in late September, 1987, do not show any further increase. According to Truemper (1987), the hard X-ray flux is almost constant within 30 %, from August 21 to September 15, 1987. These observations suggest that the peak in the 10-30 keV band took place around 0.5 year after the explosion. This is inconsistent with the results obtained for the Woosley's models.

This discrepancy may be resolved if radioactive ^{56}Co is mixed into the outer region of the ejecta. Mixing is also suggested by hydrodynamical simulations of the ejecta and the optical light curve (Woosley et al. 1987b; Nomoto et al.1987). The purpose of this paper is to describe how the X-ray light curve and spectrum depend on mixing. Similar works are performed by Itoh, Kumagai, Shigeyama, Nomoto, and Nishimura (1987) and Pinto and Woosley (1987).

2 ASSUMPTIONS

Our method of calculations is found in Ebisuzaki and Shibazaki (1987a,b). We use the model 11E1Y6 model provided Dr. K. Nomoto (Nomoto, Shigeyama, and Hashimoto 1987) in order to describe density distribution and chemical composition. According to him, the 11E1Y6 model (0.07 $M_{\odot}$ of ^{56}Ni, a metal rich core of 2.4 $M_{\odot}$, a total mass of 11.3 $M_{\odot}$, and an expansion energy of 1.0×10^{44} J) provides the best fit to the optical light curve of SN1987A among the models he has constructed so far. The ^{56}Ni decays to ^{56}Co with a half life of 6.1 days. In the hydrogen-rich envelope of the 11E1Y6 model, X = 0.39, and Y = 0.6, and ζ = 0.25, where X and Y are the mass fraction of hydrogen and helium, respectively, and ζ is the metallicity normalized by the value of solar abundance. This suggests that a moderately hydrogen-depleted layer is exposed due to a certain amount of mass loss. Although hydrodynamical calculations suggest mixing of ^{56}Co and the optical light curve also favors it, the extent of mixing is very uncertain. Therefore, we introduce a parameter M_{mix} which specifies the mass of the mixed region. We assume that the inner region with mass M_{mix} is well mixed but that the velocity structure is not changed by this mixing. In order to estimate the X-ray flux at the earth, we assume a distance of 52 kpc.

3 COMPARISON WITH THE OBSERVATIONS

3.1 X-Ray Light Curves

Figure 1 shows the X-ray light curves in the 10-30 keV band for various masses of the mixed core, M_{mix}. The X-ray flux reaches a peak earlier for models with higher values of M_{mix}. The time of the peak is mainly determined by the outer unmixed layer. The soft photons below 30 keV are negligible in the inner mixed core because of the enhanced photoelectric absorption. The outer unmixed envelope, therefore, has to be optically thick enough to scatter harder photons down to the 10-30 keV band. According to our calculations, the optical depth of the unmixed layer is 5 ± 1, when the flux in the 10-30 keV band reaches a peak. Itoh, Kumagai, Shigeyama, Nomoto and Nishimura (1987) performed similar calculations for Nomoto's 11E1Y6 using a Monte Carlo method. Their light curves for the 10-30 keV band are consistent with ours within 20 %.

The filled circles with error bars in Figure 1 show the X-ray fluxes in the 10-30 keV band observed by Large Area Proportional Counters (LAC) on board the GINGA satellite. According to Makino (1987a), the total effective area of LAC is 0.38 m^2. If we take into account a low efficiency of the LAC in the 10-30 keV band, the effective area in 10-30 keV reduces to 0.13 m^2, where we assumed a flat spectrum as observed. The X-ray fluxes plotted in Figure 1 are calculated applying this effective area to the counting rates given by Dotani et al. (1987). The observational points are between the light curves for M_{mix} = 4 $M_{\odot}$ and 6 $M_{\odot}$. This suggests M_{mix} is around 5 $M_{\odot}$.

3.2 X-Ray Spectrum

In Figure 2 is shown the energy spectrum observed by the GINGA satellite (Dotani et al. 1987) on September 2, 1987 and by the KVANT-ROENTGEN mission (Sunyaev et al. 1987) in late August and early September, 1987. The flux obtained by GINGA is consistent with those obtained by HEXE/KVANT between 10 and 30 keV. The LAC/GINGA also detected a soft X-ray flux down to 4 keV. This detection, however, is inconsistent with the upper limit of the TTM/KVANT observation (Skinner 1987).

Solid curves in Figure 2 show the energy spectra calculated

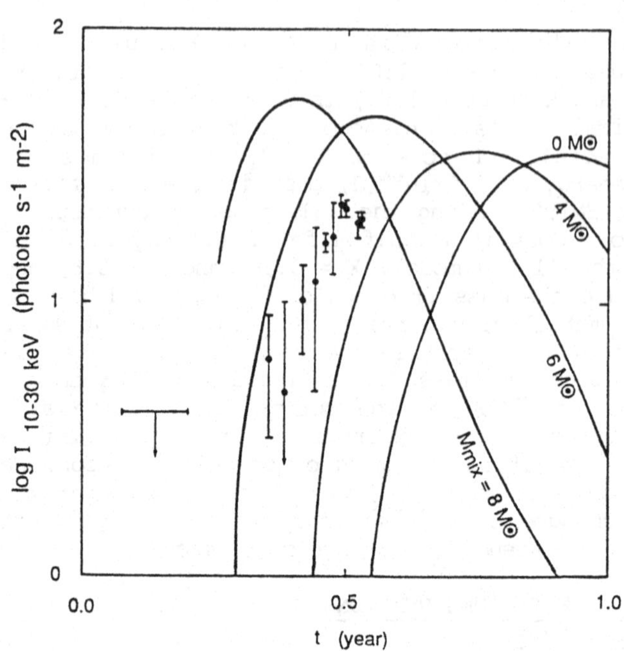

Figure 1: Calculated X-ray light curves in the 10-30 keV band for various masses, M_{mix}, of the inner mixed region. The filled circle with error bars (90 % confidence level) show the X-ray flux in the 10-30 keV band observed by GINGA.

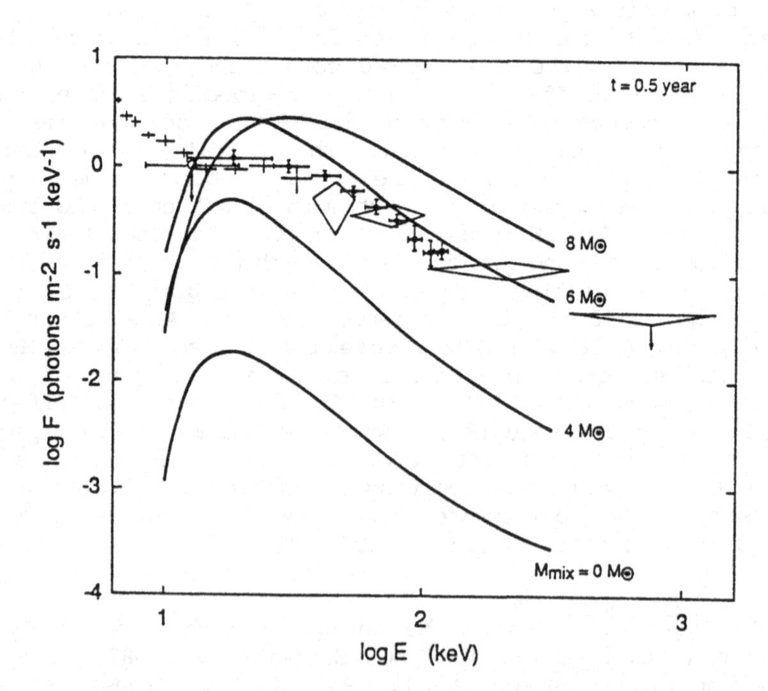

Figure 2: X-ray spectrum observed by GINGA satellite and KVANT/ROENTGEN mission; cross: LAC/GINGA, open circle: TTM/KVANT, filled circle: HEXE/KVANT, and PULSAR X-1/KVANT. Solid line show the calculated spectra for various masses, M_{mix} of the mixed region.

for the various M_{mix} at 0.5 year after the explosion, i.e., when the above observations were performed. Spectrum calculated for $M_{mix} = 6\ M_\odot$ reproduces the observed one especially above 40 keV. Above 20 keV, the observational points are between the curves for $M_{mix} = 4\ M_\odot$ and 6 $M_\odot$. This also suggests that M_{mix} is around 5 $M_\odot$. Below 20 keV, all of the calculated spectra rapidly decrease with decreasing energy because of photoelectric absorption. The soft X-ray flux down to 4 keV can not be produced by down scattered nuclear gamma-rays (McCray et al. 1987; Xu et al. 1987; Ebisuzaki and Shibazaki 1987). The soft component detected by GINGA might be due to the shock heating proposed by Itoh, Hayakawa, Masai, and Nomoto (1987).

4 DISCUSSIONS

As described in the previous section, Nomoto's 11E1Y6 model with an inner mixed region of 5 $M_\odot$ provides X-ray light curves and spectra which are consistent with the observations. On the basis of this comparison, we predict the evolution of X-ray, gamma-ray, and optical/infrared emissions as follows. It is very important to observe them continuously in order to determine unknown parameters of the ejecta.

4.1 Hard X-ray Spectrum and Light Curve

As the outer unmixed layer become transparent, the emergent spectrum becomes hard and peaks around 50 keV. This is because the spectrum formed in the inner mixed region with strong photoelectric absorption emerges. The X-ray flux in the 30 - 200 keV band will reach the detection limit (14 photons $m^{-2}\ s^{-1}$) for the planned balloon experiments by Fishman (1987) between 0.4 and 0.6 year after the explosion. The flux in this band can be seen until 1.5 year after the explosion.

The X-ray flux in the 10-30 keV band will decrease significantly at 0.8 year after the explosion. However, this may not occur if the outer unmixed region is asymmetric as suggested by Shull (1987). The shape of the light curve in the 10-30 keV band is independent of the inner mixed region because Compton scatterings in the outer layer completely wash out the information about the inner region.

4.2 Gamma-Ray Lines

The intensity of the 845 keV line will reach the detection limit (2 photons $m^{-2}\ s^{-1}$) of planned balloon experiments (e.g. Sandie 1987) between 0.5 and 0.7 year after the explosion. This line can be seen until 2 years after the explosion. The time evolution of the gamma-ray line contains information about the distribution of ^{56}Co in the ejecta (Pinto and Woosley 1987). Detailed discussions will be given elsewhere (Shibazaki and Ebisuzaki 1987).

4.3 Optical and Infrared Emissions

According to Woosley (1987), the decline of the bolometric luminosity derived from the optical/infrared observations is good agreement with energy deposition due to 0.07 $M_\odot$ of ^{56}Co. According to our calculations, the deviation of this bolometric luminosity from the predicted exponential decay will reach 10 % between 0.7 and 0.9 year after the explosion as seen in Figure 3.

References

Catchpole, R. et al. 1987, submitted to M. N. R. A. S.

Dotani, T. et al. (GINGA team) 1987, submitted to Nature.
Ebisuzaki, T. and Shibazaki, N. 1987a, Ap. J. in press.
Ebisuzaki, T. and Shibazaki, N. 1987b, submitted to Ap. J. (Letters).
Fishman, G. 1987, private communication.
Gehrels, N., MacCallum, C. J., and Leventhal, M. 1987, preprint.
Hamuy, M., Suntzeff, N. B., Gonzalez, R., and Martin, G. 1987, submitted to Ap. J.
Itoh, H., Hayakawa, S., Masai, K., and Nomoto, K. 1987, Publ. Astron. Soc. Japan, **39**, 529.
Itoh, M., Kumagai, S., Shigeyama, T., Nomoto, K., and Nishimura, J. 1987, submitted to Nature.
Makino, F. 1987a, talk presented at Taos conference on Multi- Wavelength Astronomy, Aug. 1987, Taos, New Mexico.
Makino, F. 1987b, IAU Circ., **4447**.
Makino, F. 1987c, IAU Circ., **4466**.
McCray R., Shull, J. M., and Sutherland, P. 1987, Ap. J. (Letters), **317**, L73.
Nomoto, K., Shigeyama, T., and Hashimoto, M. 1987, Proceedings of ESO Workshop on SN1987A, Garching, 6-8 July, 1987, ed. J. Danziger (ESO, Garching).
Pinto, P. A. and Woosley, S. E. 1987, submitted to Ap. J.
Sandie, W. G., 1987, private communication.
Shibazaki, N. and Ebisuzaki, T. 1987, in preparation.
Shull, J. M. 1987, this volume.
Skinner, G. K. 1987, this volume.
Sunyaev, R. et al. in preparation.
Truemper, 1987, this volume.
Woosley, S. E. 1987, submitted to Ap. J.
Woosley, S. E., Pinto P. A., Martin, P. G., and Weaver, T. A. 1987a, Ap. J. **318**, 664.
Woosley, S. E., Pinto, P. A., and Ensman, L, 1987b, Ap. J., **324**, 000.
Xu Y., Sutherland, P, McCray, R., Sutherland, P., and Ross., R. R., 1987, Ap. J. (Letters) in press.

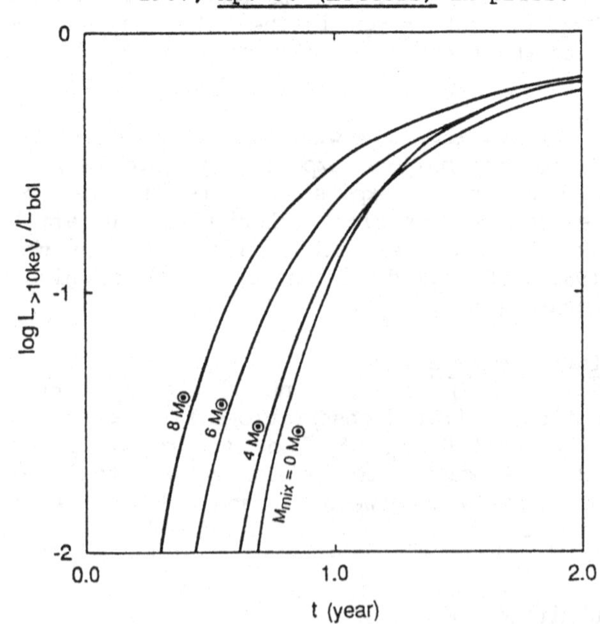

Figure 3: Luminosity of hard photons above 10 keV normalized by luminosity emitted by ^{56}Co plotted for various values of M_{mix} against the time elapsed from the explosion. If we assume a steady state for the radiation field in the ejecta, this ratio represents the deviation of the bolometric luminosity obtained by optical and infrared observations from the predicted exponential decay.

A BROAD BAND X-RAY IMAGING SPECTROPHOTOMETER TO OBSERVE SN1987A

Gotthelf, E.V., Lum, K.S.K., McMahon, P.M., Martin, C.,
Novick, R., Shafer, N., Szentgyorgyi, A.H.
Columbia University

Fenimore, E.E., Roussel-Dupre, D.
Los Alamos National Laboratory

Abstract. A Broad Band X-Ray Imaging Spectrophotometer (BBXRIS), to be flown as a sounding rocket payload in late 1987, has been built to make high spatial and spectral resolution observations of SN1987a. The focal plane instrument, an imaging gas scintillation proportional counter, offers energy resolution of 8% (FWHM) at 6 keV, and is sensitive to x-rays from 0.1 to 20.0 keV. Soft x-rays (0.1 - 1.8 keV) are imaged by a nested pair of Wolter Type I grazing incidence mirrors with a resolution of 8 arcmin, a field of view of 2.6^{o}, and an effective area of 70 cm^2 at 0.2 keV. A specially designed collimator is able to resolve SN1987a from LMC X-1 and provide approximately 70 cm^2 of effective area from 0.1 - 20.0 keV.

Introduction

For several years the Columbia Astrophysics Laboratory has been developing a Broad Band x-Ray Imaging Spectrophotometer (BBXRIS), the purpose of which is to perform high resolution measurements of cosmic sources over a broad energy range. An earlier version of this instrument was used to make the first spectrally resolved image of SN1006 (Vartanian, et al., 1985). An improved version of this instrument has been built to observe the supernova SN1987a in the United States' first SN1987a rocket campaign from Woomera, Australia. The soft x-ray channel (0.1 - 1.8 keV) is imaged by a pair of nested Wolter Type I grazing incidence mirrors (Fig.1). To increase both throughput in the low energy channel and bandwidth of the BBXRIS, a collimator has been mounted coaxially inside the mirrors with the intent to spatially resolve SN1987a from nearby sources. The focal plane instrument is an Imaging Gas Scintillation Proportional Counter (IGSPC), which has been scaled up from the version used for the observation of SN1006. The BBXRIS is capable of approximately 10 arcmin spatial resolution and the energy resolution at 6 keV is 8% FWHM. A high precision clock has been incorporated into the payload which is sufficiently accurate to permit a search for millisecond periodicity over the duration of a typical rocket flight.

In this paper, we discuss the payload, its calibration, and the scientific goals of the observation.

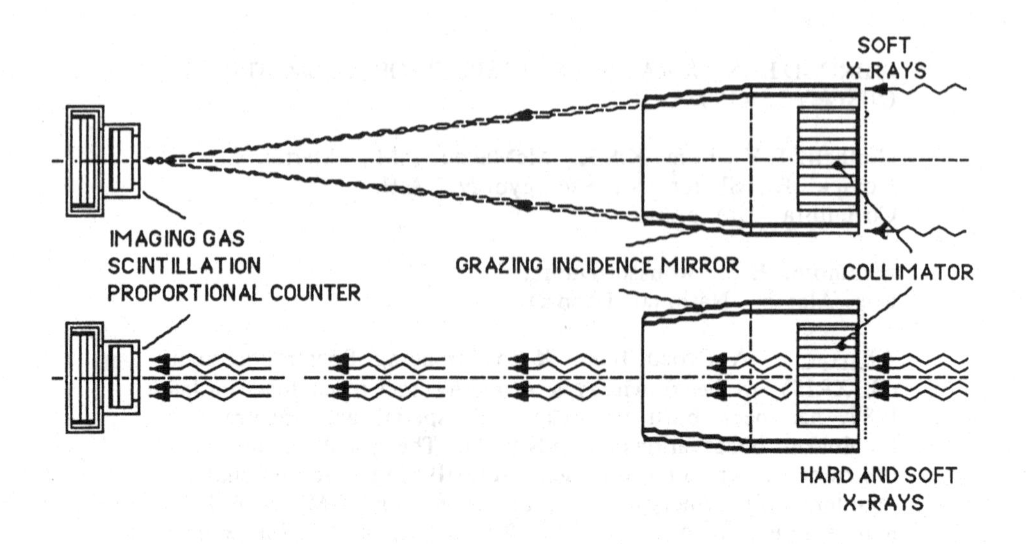

Figure 1 : A schematic of the concept behind the BBXRIS which combines low-energy grazing-incidence imaging with high-energy coded-aperture imaging in one instrument.

Payload description

The soft x-ray channel is imaged with a nested pair of Wolter Type I grazing incidence mirrors. These mirrors have been flown previously to observe SN1006 (Vartanian, et al., 1985) and have recently been recoated and repolished by Applied Optics Corporation for the SN1987a campaign. The mirrors were fabricated from aluminum with an optical surface consisting of an undercoat of nickel and an overcoat of gold. x-rays in the energy band 0.1 - 1.8 keV are imaged by these mirrors with an angular resolution of 8 arcmin (FWHM). The field of view of the mirrors is 2.6 ^{o}. The geometric area of the mirrors is 300 cm^2, and folded with the response of the IGSPC, an effective area of 70 cm^2 is obtained at 0.2 keV. At this energy a 5σ line sensitivity is achieved at a flux of 1×10^{-3} photons/cm^2-sec within the duration of a typical rocket flight.

To achieve greater flux sensitivity at the low energy channel and to increase the bandwidth of the instrument, a collimator has been incorporated into the design, and is mounted coaxially within the x-ray mirrors. This collimator has been specially designed to spatially resolve SN1987a from the bright x-ray source LMC X-1 which is only 35 arcmin away. To achieve this end, a slat collimator has been built which spatially resolves x-rays from the direction of the SN1987a from the x-rays arriving from the direction LMC X-1 at the focal plane of the instrument (Fig. 2). The open area of the slats of the collimator has been further subdivided into square cells 0.25 inches on a side, which limit the field of view of the collimator to 0.75^{o} FWHM. The geometric area of the collimator is 96 cm^2, and the effective area over the 0.8 to 12 keV band is

70 cm^2. The line sensitivity for the BBXRIS is typically 3 x 10^{-4} photons/cm^2-sec over its spectral range.

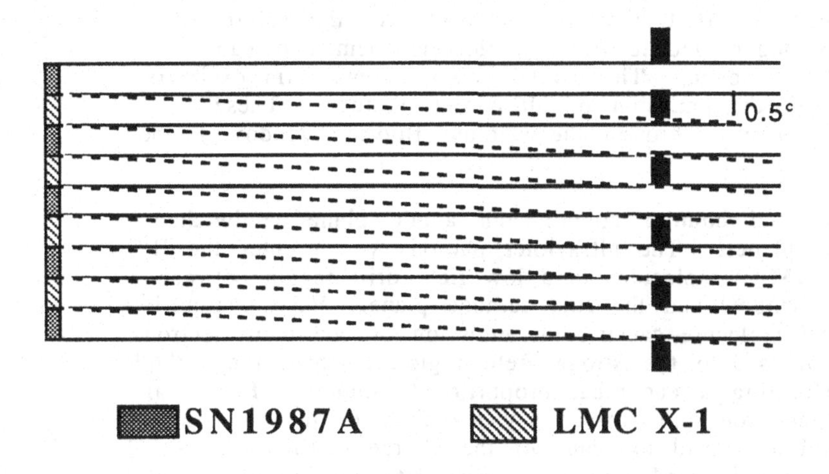

Figure 2: Schematic of slat collimator for resolving SN1987a and LMC X-1.

The IGSPC (Fig.3), the focal plane instrument of the BBXRIS, is a scaled up version of the IGSPC described by Hailey et al. (1983). This instrument has an entrance window area of 250 cm^2, making it suitable for use with a collimator. The gas scintillation section of the detector (GSPC) is filled with high purity xenon, while the imaging proportional counter (IPC) section is filled with a gas mixture composed of 80% argon, 20% methane and a trace of tetrakis dimethylamino ethylene (TMAE). The gas scintillation and imaging proportional counter sections are separated by a 12 mm thick calcium fluoride window, which is transparent to the ultraviolet scintillation photons.

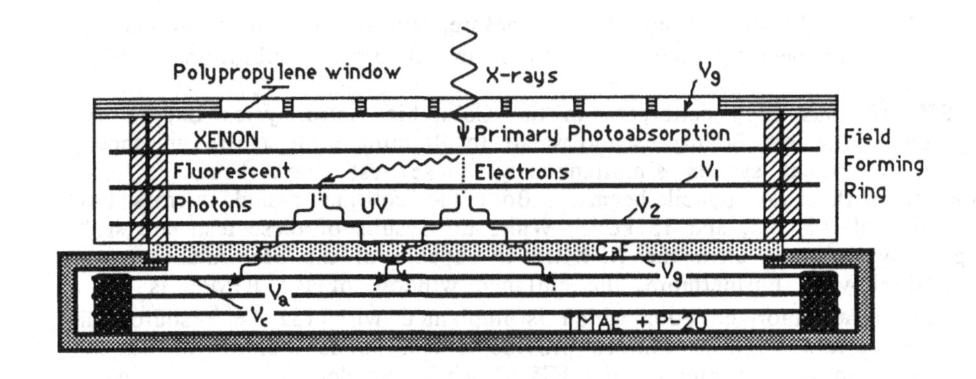

Figure 3: Schematic of the IGSPC Detector.

The GSPC contains two high voltage grids. X-rays enter the gas scintillator through a 1.5 micron thick polypropylene entrance window and are photoelectrically absorbed in a weak field drift region. The resulting electrons drift through the first high voltage grid into the scintillation region, where the stronger electric field accelerates them and causes further excitation of the xenon. The excited xenon atoms form excimers which deexcite through the emission of ultraviolet photons. These ultraviolet photons then pass through the calcium fluoride window to be detected by the IPC.

The imaging proportional counter contains an anode plane sandwiched between two cathode planes. The ultraviolet photons which enter the IPC are absorbed by the TMAE molecules in a low field drift region between the calcium fluoride window and the first cathode plane. Each photon is absorbed by a TMAE molecule causing the emission of one photoelectron. The electrons then drifts into the strong field region between the cathode and anode planes, forming a canonical proportional counter. Positional information is obtained from the crossed x-y geometry of the cathode planes. First, second and third moments of the charge distribution on each cathode plane are calculated by the on line data acquisition system, which is a flight component of the payload. This information is used to determine the location of the incident x-ray on the focal plane. The energy of the incident x-ray is calculated by summing the charge collected on the cathode plane. Background discrimination is performed by analyzing the shape of the observed pulse.

The energy resolution of the IGSPC is 8% at 6 keV and the position of the incident x-ray is determined to 1.0 mm at this energy. As the plate scale is 3.0 mm per degree, this spatial sensitivity is consistent with the resolving power of the optical system.

To provide the capability to search for millisecond periodicity in the x-ray emission from SN1987a, a high resolution clock has been incorporated in the BBXRIS. This clock, which consists of an oven stabilized crystal oscillator and a frequency multiplier, has been designed and built at the Columbia Astrophysics Laboratory and is stable to better than one part in 10^8. This stability has been checked before, during, and after thermal testing, and has been checked to one part in 10^7 during vibration testing.

The BBXRIS has been calibrated in the Columbia Astrophysics Laboratory x-ray tube. As this facility is too small to illuminate the entire entrance aperture of optical system simultaneously, these tests have been performed with x-ray pencil beams. Both the collimator and mirrors have been tested at 0.8, 1.5, and 15 keV. While the results of these tests are still being analyzed, it is clear the BBXRIS is capable of the performance described above. Furthermore, the entrance window of the IGSPC is protected by a motorized door which is implanted with 132 Fe^{55} sources in a circular pattern. These sources provide a continuous monitor of the energy and spatial resolution of the IGSPC while the door is closed. The mirrors have been extensively tested at both optical and x-ray wavelengths for an earlier rocket flight (Vartanian, 1985) in addition to the aforementioned calibration.

Scientific Goals

Data from the MIR and GINGA satellites are in rough agreement with current theoretical pictures above 10 keV (Dotani et al., 1987, Skinner et al., 1987, Pinto and Woosley, 1987). However, below 10 keV the results are contradictory and surprising. While GINGA would indicate that SN1987a is a relatively bright source of x-rays in the 4.0 - 10.0 keV band, data from MIR has produced upper limits well below the GINGA flux. Furthermore, neither of these instruments is capable of exploring the energy region below 4.0 keV. The BBXRIS is uniquely capable of simultaneously exploring the energy region from 0.2 to 20.0 keV, hopefully resolving the current conflict between MIR and GINGA data, as well as providing new information at lower x-ray energies. Furthermore, the energy resolution of the BBXRIS should make it possible to detect the presence of Iron fluorescence emission suggested by the GINGA data, as well as to detect continuum emission The resolving power of the optics of the BBXRIS make it possible to unambiguously reject nearby sources in the SN1987a field, a problem which has proved difficult for the satellite instruments. Should a pulsar with a period of a millisecond or greater be observable in the BBXRIS energy band, the high resolution clock incorporated in the payload will allow us to search for it.

Conclusions

The soft x-ray luminosity revealed by the GINGA satellite has proved both surprising and enigmatic. The superior energy resolution of the BBXRIS should help resolve the questions that these data now pose. In particular, the question of the existence of an Fe fluorescence line will be answered. Furthermore, the BBXRIS will extend the spectral range of current x-ray observations. If pulsar emission is present in the data, the BBXRIS has sufficient temporal resolution to detect its presence to the highest plausible frequencies for a young pulsar.

The BBXRIS was fully integrated during October 1987 and is expected to be launched from Woomera, Australia near the end of November 1987 on a Terrier Black Brandt rocket in the first U.S. Supernova Rocket Campaign. Reflights are planned for March 1988 and January 1989.

REFERENCES

Dotani, T., et. al., 1987, ISAS RN 371.

Hailey, C., et. al., 1983, NIM **213**, 397.

Pinto, P. and Woosley, S., 1987, Ap.J., Submitted.

Skinner, G., et. al., 1987, These proceedings.

Vartanian, M.H., 1985, Ph.D. Dissertation (Columbia University).

Vartanian, M.H. et al. 1985 Ap.J., Lett., 288, L5.

This paper is contribution number 343 of the Columbia Astrophysics Laboratory.

THE X-RAY ENVIRONMENT OF THE PRECURSOR OF SN1987A

F. R. Harnden, Jr. and F. D. Seward
Harvard-Smithsonian Center for Astrophysics

Figure 1 presents a 1979 November Einstein Observatory X-ray image of the Sanduleak -69 202 region in the Large Magellanic Cloud, and Figure 2 provides a key identifying X-ray objects in the field. These imaging proportional counter (IPC) data yield an 0.2-4 keV X-ray luminosity upper limit for the pre-supernova star of $< 9 \times 10^{34}$ ergs s^{-1}. The image demonstrates that soft X-ray observations of SN 1987A will require ~10 arcmin spatial resolution in order to distinguish the supernova from other sources of diffuse and point-like emission in the vicinity.

REFERENCES

[d] Cowley, A. P., Crampton, D., Hutchings, J. B., Helfand, D. J., Hamilton, T. T., Thorstensen, J. R., and Charles, P. A. (1984). Ap. J., 286, 196.

[b] Long, Knox S. and Helfand, David J. 1979, Ap. J. Letters, 234, L77.

[c] Long, Knox S., Helfand, David J., and Grabelsky, David A. 1981, Ap. J., 248, 925.

[a] Seward, F. D., Harnden, F. R., Jr., and Helfand, David J. 1984, Ap. J. Letters, 287, L19.

[e] Shelton, Ian, *et al.* 1987, I.A.U. Circ. No. 4316.

Figure 1. A mosaic of several Einstein Observatory images. Portions of the combined IPC image are obscured to varying degrees by the window support structure, but the immediate vicinity of the pre-supernova star (cf. Figure 2) is free of such effects. Many X-ray objects populate this region of the LMC, making soft X-ray observations of the region difficult with non-imaging detectors.

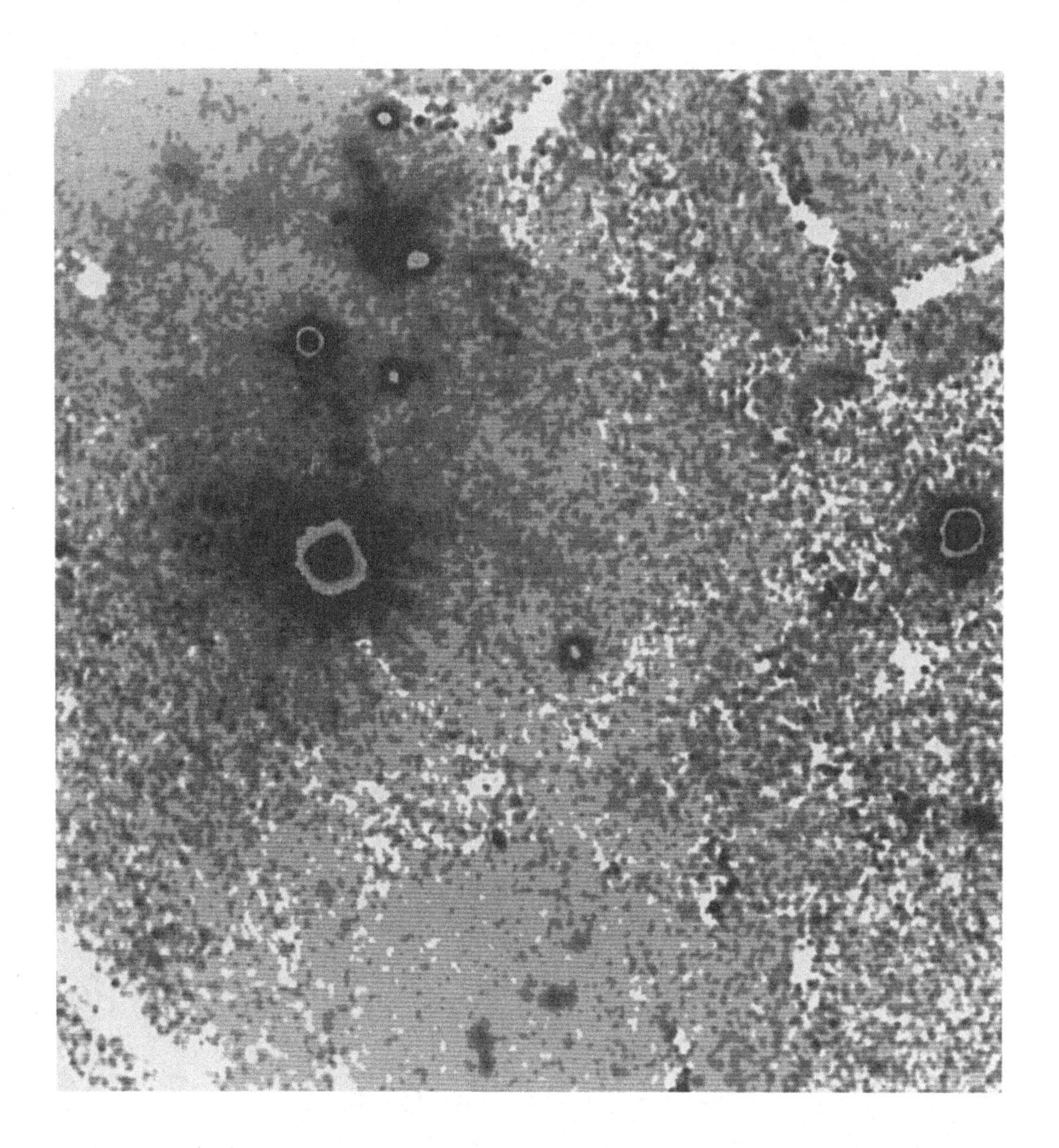

Figure 2. X-ray sources in Figure 1 (letters a-e denote references).

1	SNR; 50 msec pulsar	a			
2	SNR N157B	b	9	foreground star (G2 V)	d
3	SNR 0534-69.9	c	10	foreground star (G2 III)	d
4	SNR N132D	b	11	foreground star (dK7e)	d
5	SNR DEM 238	c	12	background Seyfert galaxy	d
6	SNR DEM 249	c	13	LMC X-1	c
7	SNR DEM 299	c	14	30 Dor complex	c
8	SNR N135	b	15	site of SN 1987A	e

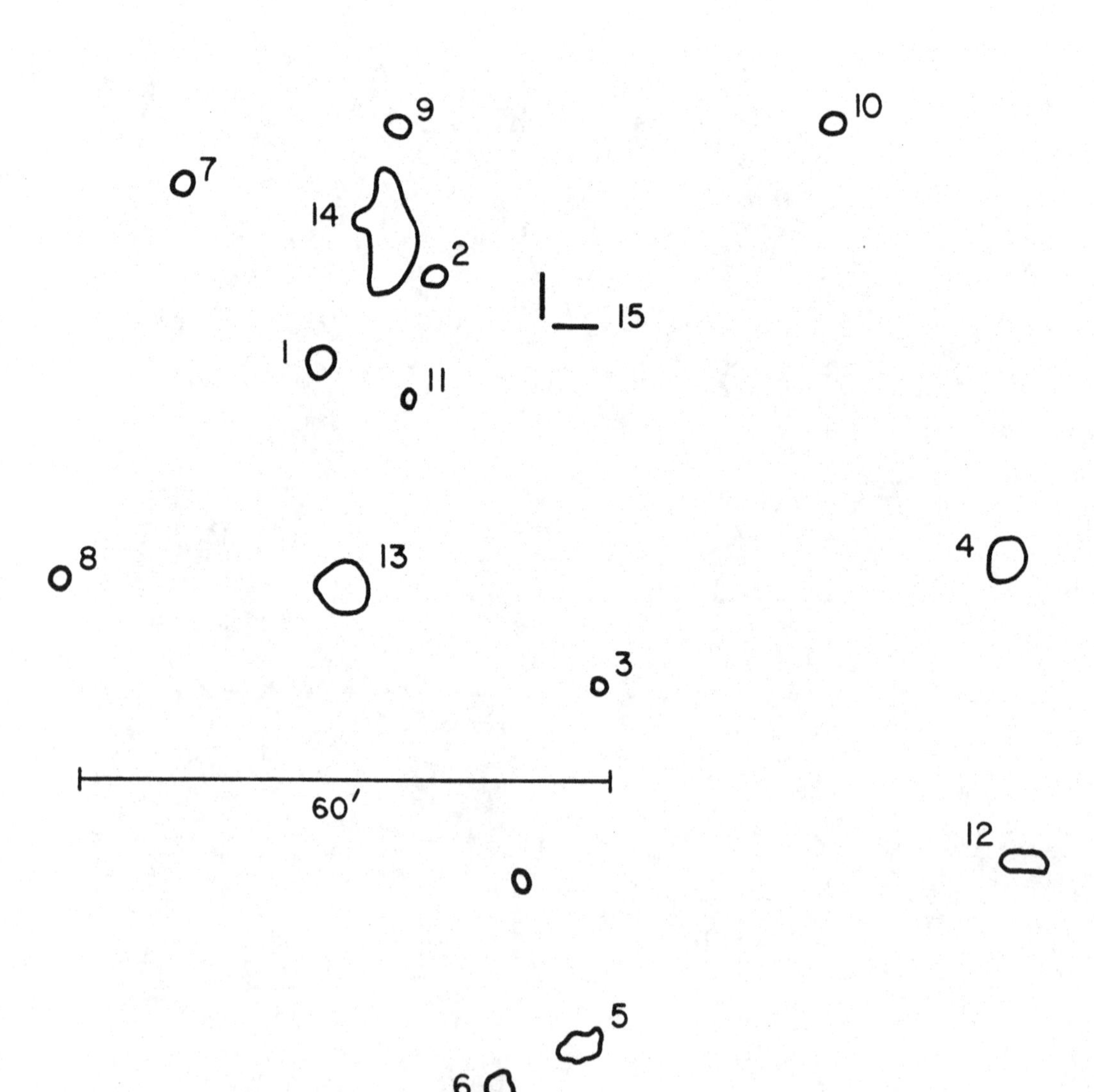

POSSIBLE s-PROCESS GAMMA RAY LINES IN SUPERNOVAE

M. J. Harris
S M Systems and Research Corporation
8401 Corporate Drive, Suite 510
Landover, Maryland 20785

ABSTRACT

We calculate the abundances in the helium burning shells of presupernova stellar models by Woosley and Weaver (1986 a,b). For stellar masses above $\simeq 20\ M_\odot$ the $^{22}Ne(\alpha,n)^{25}Mg$ reaction produces enough neutrons on a sufficiently short time-scale for the s-process to produce ^{59}Fe ($\tau_{1/2}$ = 45 d) and ^{60}Co ($\tau_{1/2}$ = 5.3 yr). These isotopes are expected to survive the passage of the shock, and gamma rays from their decays should be detectable from most Galactic Type II supernovae with the Gamma Ray Observatory (GRO) Compton telescope.

The calculated abundances of these nuclei in the helium zone are shown in Fig. 1 as functions of helium-burning temperature. The ^{60}Co is expected to be produced only by stars in a narrow mass range around $\simeq 20\ M_\odot$, and its lines at 1.173 and 1.332 MeV will therefore be an important mass indicator. The precise strength of these lines will also, if measured, yield information about the presupernova structure of the helium zone. The 1.099 and 1.292 MeV lines from ^{59}Fe will fall off very rapidly in strength after the explosion, but should remain visible from supernovae of $\gtrsim 20\ M_\odot$ from the epoch ($\simeq$200 d) at which the envelope becomes transparent to gamma rays up to $\simeq$200 d thereafter. Because of its great distance and low metallicity, these lines are not expected to be observable from SN 1987A.

References

Woosley, S. E. & Weaver, T. A. (1986a). In Radiation Hydrodynamics in Stars and Compact Objects, ed. D. Mihalas and K.-H. Winkler, p. 91. Berlin: Springer-Verlag.

Woosley, S. E. & Weaver, T. A. (1986b). Ann. Rev. Astron. Astrophys., 24, 205.

Figure Caption

Fig. 1. Abundances by mass fraction of ^{59}Fe and ^{60}Co produced as a function of helium burning temperature. Arrows at top denote the helium burning temperatures of the 15 and 25 $M_\odot$ presupernova models of Woosley and Weaver. Arrow at the side represents the abundances of ^{59}Fe and ^{60}Co which are marginally detectable by GRO (flux level of 5×10^{-5} cm^{-2} s^{-1}) assuming that they exist throughout a convective helium zone of mass 2 $M_\odot$. Decay of ^{59}Fe for 200 d prior to detection is assumed. N.B. It is coincidental that this minimum detectable abundance is the same for both species.

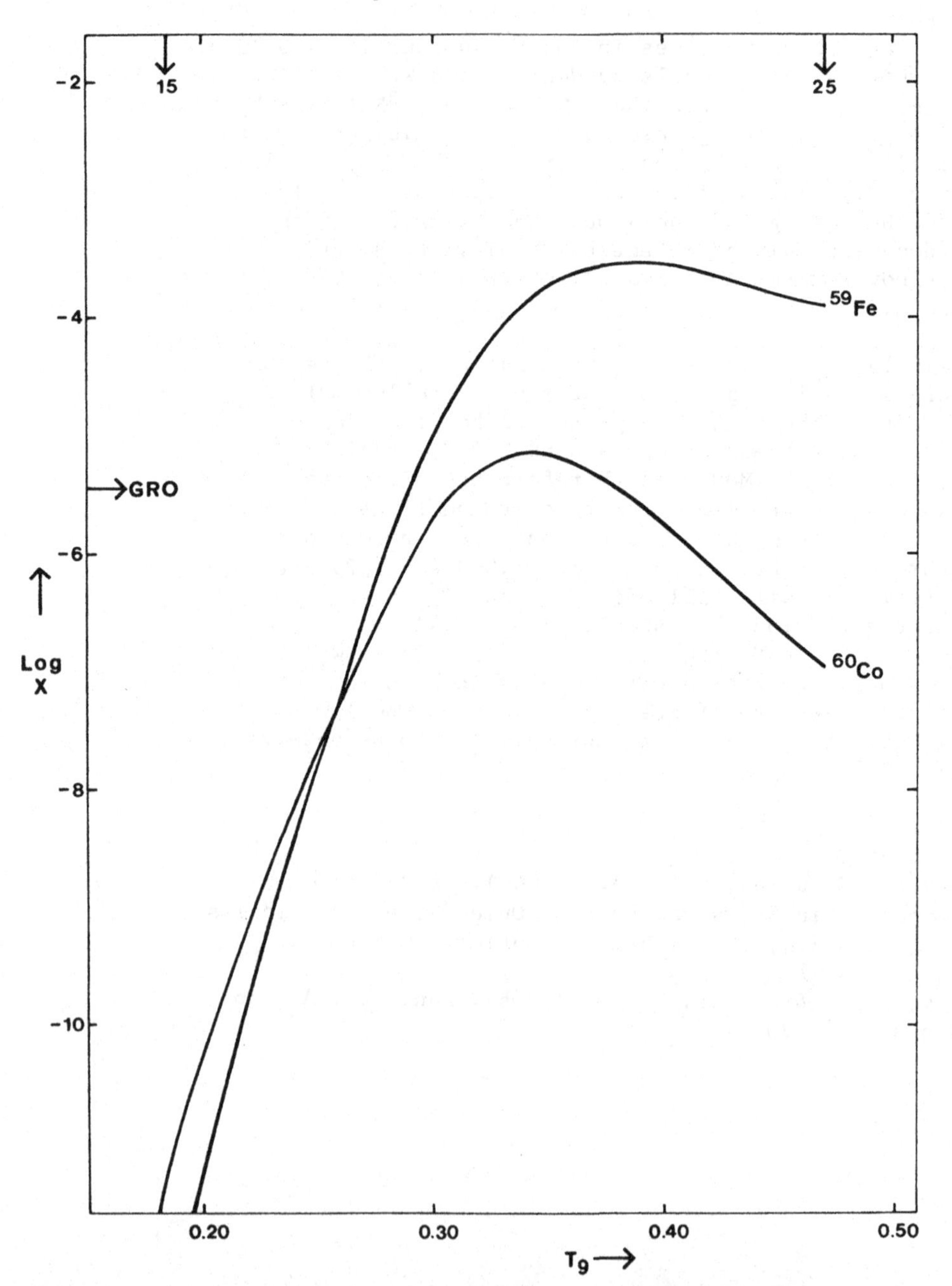

HARD X- AND GAMMA-RAYS FROM SUPERNOVA 1987A

S. Kumagai[1], M. Itoh[2], T. Shigeyama[1], K. Nomoto[1], and J. Nishimura[2]

[1]Department of Earth Science and Astronomy, University of Tokyo
[2]Institute of Space and Astronautical Science

Abstract. The X-ray light curve and spectrum from SN 1987A due to Compton degradation of γ-rays from the ^{56}Co decay are calculated and compared with the Ginga and Kvant observations. If mixing of ^{56}Co into outer layers has taken place, the X-rays emerge much earlier than in the case without mixing and the resulting hard X-rays are in reasonable agreement with observations.

MODELS

The X-rays observed from SN 1987A show several notable features (Dotani et al. 1987; Sunyaev et al. 1987). The spectrum could consist of separate soft and hard components (Dotani et al. 1987). The soft component might be thermal radiation from the collision between the supernova ejecta and the circumstellar materials (H. Itoh et al. 1987; Masai et al. 1987). The hard component could be due to Compton degradation of γ-rays from the ^{56}Co decay, but its emergence is much earlier than the recent predictions (Xu et al. 1988; Ebisuzaki and Shibazaki 1988a). However, the X-ray light curve due to radioactive decays is sensitive to explosion energy E, mass of the hydrogen-rich envelope M_{env}, and the distribution of elements. Itoh, Kumagai, et al. (1987) has found that, if ^{56}Co is mixed into outer layers, resulting hard X-rays are in reasonable agreements with observations. Here more detailed spectral evolution of X- and γ-rays will be shown.

Based on the comparison between the optical light curve model and observations, we selected a hydrodynamical model 11E1 (Nomoto et al. 1987; Shigeyama et al. 1987, 1988) where $E = 1 \times 10^{51}$ erg, the mass of ^{56}Co is 0.07 $M_\odot$ and the progenitor had a 6 $M_\odot$ helium core with a hydrogen-rich envelope of 6.7 $M_\odot$. We assumed that some ^{56}Co is mixed into the outer layers. During the shock propagation through the star, the expanding core materials are decelerated by the low density envelope. Then a reverse shock and a density inversion form. The materials of core and helium layer possibly undergo Rayleigh-Taylor instability (Woosley et al. 1988; Nomoto et al. 1987). If the mixing has taken place, the resulting light curve is in better agreement with the observations (Nomoto et al. 1987; Shigeyama et al. 1988). The extent of mixing is very uncertain and thus an unknown parameter in our calculation.

X-RAY LIGHT CURVE

We calculated X-ray light curves and spectra with Monte Carlo method. Figure 1 shows the intensity changes in the 10 - 30 keV range and the 847 keV γ-ray line for three cases of mixing, i.e., Case OX (dash-dotted) where ^{56}Co is mixed up to the outer edge of the oxygen-rich core (M_{mix} = 2.4 $M_\odot$), Case HE (dashed) with mixing up to the outer edge of the helium layer (M_{mix} = 4.6 $M_\odot$), and Case HY with mixing into part of the hydrogen-rich envelope (M_{mix} = 6.0 $M_\odot$). Figure 2 shows the light curves of several energy ranges for Case HY. The scattering optical depths at the outer edge of the cobalt layer are 5.3 (Case OX), 2.4 (HE), and 1.3 (HY) at t = 200 d, which are much smaller than without mixing.

Figure 1 shows that emergence of the X and γ-rays is earlier with more extensive mixing because of smaller optical depth of the cobalt layer (see also Pinto & Woosley 1988; Ebisuzaki & Shibazaki 1988b). Then, the intensity peak of the X-ray comes in the order t = 200 d, 250 d, and 320 d for cases HY, HE, and OX, respectively. The X-ray emergence observed around t = 130 d is consistent with case HY, while the observed flux at t = 190 - 220 d is close to that of case HE. Case OX is inconsistent with the observation. This implies that $M_{mix} = 4.6 - 6\ M_{\odot}$ is required to account for the X-ray observation with the our hydrodynamic model.

X-RAY SPECTRA

Figure 3 shows the calculated evolutionary spectra at t = 200 - 500 d for case HY and compares the model at t = 200 d with the observations. The observed spectrum below 12 keV (Dotani et al. 1987) cannot be accounted for by the radioactive model because of the photoelectric absorption of low energy X-rays. The power law spectrum, $E^{-1.4}$, at 30 - 200 keV shows a good fit to the observed spectrum by Kvant. On the other hand, the peak flux around 20 keV for case HY is about a factor of 2 - 3 too high. For case HE, the calculated spectrum is consistent with the observed one at 10 - 30 keV, while the high energy portion of the spectrum is a little too low. Again $M_{mix} = 4.6 - 6\ M_{\odot}$ is necessary to explain the observed spectrum.

HARD X-RAYS AND GAMMA-RAYS

For $M_{mix} = 4.6 - 6\ M_{\odot}$, Figs. 2 and 3 predict that the 847 keV γ-ray line flux will reach 4×10^{-4} photons $cm^{-2}\ s^{-1}$ around t = 240 - 330 d and thus will be detected by the planned balloon observations. Detection of Hard X-rays is also expected. With increasing high energy flux, the deviation of the optical luminosity from exponential decline will become significant unless the pulsar heating becomes appreciable.

It should be emphasized that the mixing process would involve unknown effects due to the changes in density and abundance distribution, formation of the clumpy medium, etc. These effects might be sources of some quantitative discrepancies between simple models and the observations. Further observations will provide some information about these effects.

REFERENCES

Dotani, T., & the Ginga team (1987). Nature, **330**, 230.

Ebisuzaki, T., & Shibazaki, N. (1988a,b). Ap. J. Lett., submitted.

Gehrels, N., MacCallum, C.J., & Leventhal, M. (1987). Ap.J. Let. **320**, L19

Itoh, H., Hayakawa, S., Masai, K., & Nomoto, K. (1987). Publ. Astron. Soc. Japan, **39**, 529.

Itoh, M., Kumagai, S., Shigeyama, T., Nomoto, K., & Nishimura, J. (1987). Nature, **330**, 233.

Masai, K., Hayakawa, S., Itoh, H., & Nomoto, K. (1987). Nature, **330**, 235.

McCray, R., Shull, J.M., Sutherland, P. (1987). Ap. J. Lett., **317**, L73.

Nomoto, K., Shigeyama, T., & Hashimoto, M. (1987). in Proc. ESO Workshop on SN 1987A, ed. I.J. Danziger (ESO), p. 325.

Pinto, P., & Woosley, S.E. (1988). Ap. J., submitted.

Shigeyama, T., Nomoto, K., & Hashimoto, M. (1988). Astron. Ap. submitted.
Shigeyama, T., Nomoto, K., Hashimoto, M., & Sugimoto, D. (1987). Nature, **328**, 320.
Sunyaev, R. et al. (1987). Nature, **330**, 227.
Woosley, S.E., Pinto, P., & Ensman, L. (1988). Ap. J., **324**, in press.
Xu, Y., Sutherland, P., & McCray, R. (1988). Ap. J. Lett., in press.

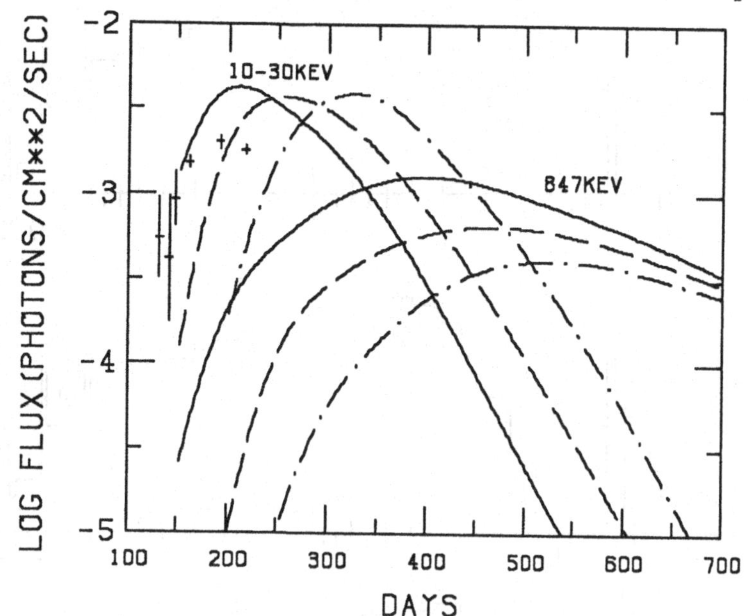

Figure 1: Calculated X- and -ray light curves for E = 1 x 10^{51} erg and M_{env} = 6.7 $M_{\odot}$. Three cases of mixing are: Case OX (M_{mix} = 2.4 $M_{\odot}$; dash-dotted), Case HE (4.6 $M_{\odot}$; dashed), and Case HY (6.0 $M_{\odot}$; solid). The crosses are the observed X-ray light curve (Dotani et al. 1987).

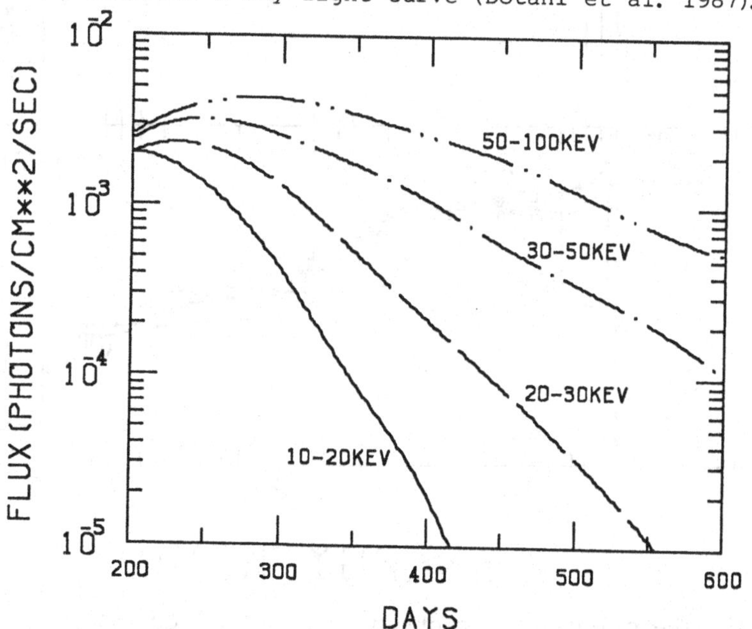

Figure 2: The light curve for several energy bands for case HY.

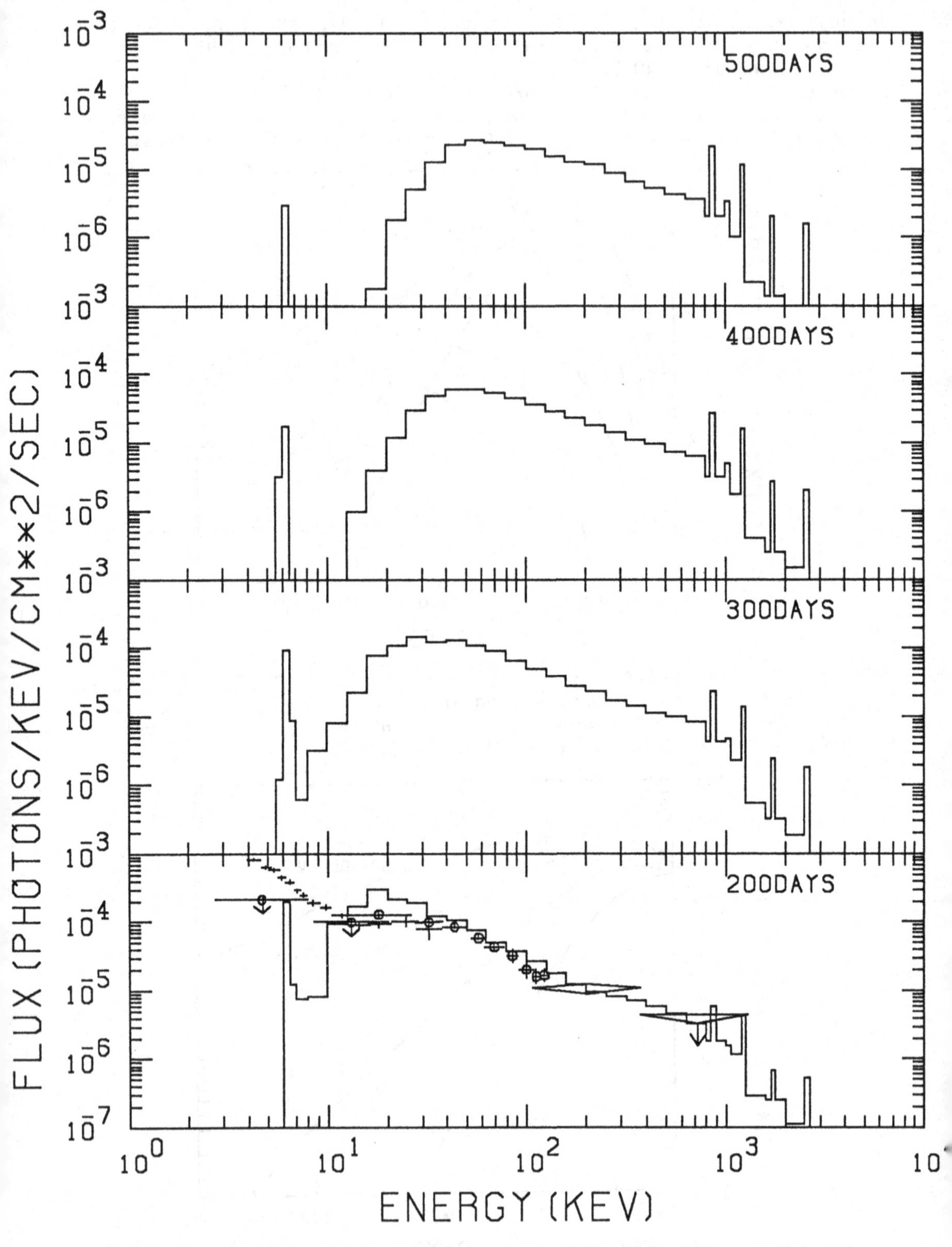

Figure 3: The calculated spectrum at t = 200, 300, 400, and 500 d for case HY. The crosses indicate the spectrum observed by Ginga and the open circles and the diamonds are obtained by Kvant.

THERMAL X-RAYS FROM SN1987A?

K. Nomoto and T. Shigeyama
Dept. of Earth Sci. and Astron., Univ. of Tokyo, 153 Japan

S. Hayakawa
Department of Astrophysics, Nagoya University, 464 Japan

H. Itoh
Department of Astronomy, University of Kyoto, 606 Japan

K. Masai
Institute of Plasma Physics, Nagoya University, 464 Japan

Abstract. The X-ray spectrum of SN1987A observed with the Ginga satellite may be explained by the ejecta-circumstellar matter collision model at photon energies below 15 keV. The harder X-rays may be ascribed to Compton degradation of the gamma-rays from ^{56}Co.

INTRODUCTION

Recent detection of X-rays in the region of SN1987A with Ginga (Dotani et al. 1987) gives impetus to the investigation of the high-energy phenomena in, and around, SN1987A. Several scenarios have been proposed to give rise to intense X-ray emission from SN1987A. They include Compton degradation of the gamma-rays from decaying ^{56}Co (McCray et al. 1987), the nonthermal radiation from the central pulsar (Ostriker 1987), and the collision of the supernova ejecta with the circumstellar medium (CSM) (H. Itoh et al. 1987a). The so-called mystery spot (IAU Circ. No. 4382, 1987) might also emit X-rays. If the X-rays originate in the colliding ejecta and a dusty CSM, they may be accompanied by strong infrared emission from shock-heated dust of the CSM (Itoh 1987). We here compare the predictions of the ejecta-CSM collision model with the results of recent X-ray spectral observations of SN1987A.

MODEL

The observed spectrum can be explained by a composite of a thermal bremsstrahlung spectrum and a flat spectrum (Dotani et al.1987). We show that the thermal component can be reproduced by the ejecta-CSM collision model approximately. The physical and numerical aspects of the model have been discussed elsewhere (H. Itoh et al. 1987a,b; Nomoto et al. 1987; Masai et al. 1987a,b). An illustrative numerical model is presented below. We here adopt a hydrodynamic model for SN1987A, in which the mass and explosion energy of the ejecta are 11 $M_\odot$ and 1.8 x 10^{51} erg, respectively. We approximate the density profile in the outer region by a power law with index -9.75 and extend it to an initial expansion speed of 2 x 10^4 km s^{-1}. The CSM is assumed to have an inner boundary at a radius of $10^{-2.5}$ pc. This value is chosen so that the calculated X-ray luminosity is insensitive to the time, t, since the supernova explosion at t ~ 1/2 yr as observed with Ginga. If the expansion speed of the wind which has formed the CSM was 10 km s^{-1}, the supernova explosion is inferred to have taken place $10^{2.5}$ yr after the cessation of the wind. The mass-loss rate divided by the wind speed is set equal to 3 x 10^{-6} $M_⊙$ yr^{-1}/10 km s^{-1} so that the observed 10-20 keV luminosity, $L_{10-20\ keV}$, at t ~ 1/2 yr [= 7 x 10^{36} erg s^{-1} at a distance of 55 kpc (IAU Circ. No. 4447, 1987)] is reproduced approximately. In the present calculation, we assume that electron and ion temperatures are equilibrated everywhere.

RESULTS

The calculated time evolution of $L_{10-20\ keV}$ is shown in figure 1 for the whole remnant, the reverse-shocked ejecta, and the blast-shocked CSM. The ejecta dominate the X-ray emission from the remnant because of the relatively high density. The rate of the decrease in $L_{10-20\ keV}$ after t = 1/2 yr would be smaller (larger) if the density in the unshocked CSM varied with radius r less (more) steeply than r^{-2}. Such a density distribution would result if the mass-loss rate of the progenitor star decreased (increased) with time. Figure 2 shows the temperature averaged over the shocked ejecta and CSM, respectively, under the weight of the bremsstrahlung. Corresponding to the blast-shock speed of about 10^4 km s^{-1}, the temperature in the shocked

Figure 1. Time evolution of the 10-20 keV luminosities of the whole remnant (solid line), the reverse-shocked ejecta (dashed line), and the blast-shocked CSM (dotted line).

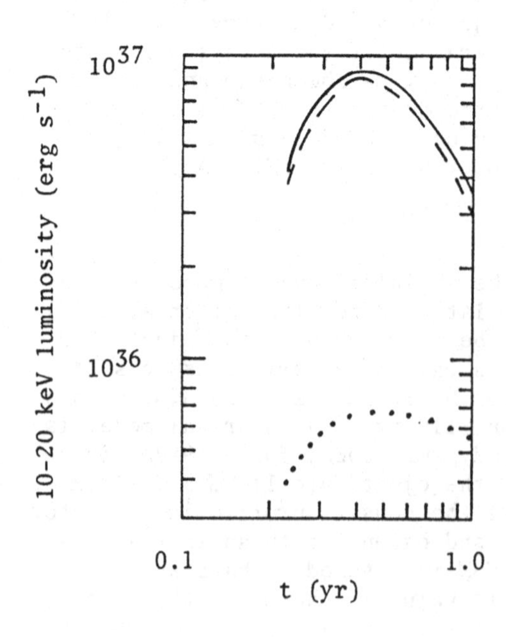

CSM is of the order of 10^2 keV. If electrons were assumed to be heated only through Coulomb collisions with ions, as in previous calculations, it would be lower by about one order of magnitude. The electron temperature in the shocked ejecta is affected much less. Figure 3 shows the calculated X-ray spectrum at t = 0.53 yr. It is similar to the observed spectrum (Dotani et al. 1987) at photon energies below 15 keV. The agreement will be improved by tuning model parameters. At photon energies above 15 keV, the calculated spectrum, even in the single-temperature model, is much steeper than the observed spectrum. If a thermal bremsstrahlung spectrum corresponding to a temperature of about 10 keV is subtracted from the observed spectrum, the residual resembles the spectrum to be expected for the X-rays due to Compton degradation of the gamma-rays from ^{56}Co (Masai et al. 1987a). Such X-rays may be detectable with Ginga at t $\sim$ 1/2 yr if ^{56}Co has been mixed into outer layers of SN1987A (M. Itoh et al. 1987).

The above interpretation of the observed spectrum has two problems at present. First, an upper limit to the photon flux at 1 keV obtained with a rocket experiment (IAU Circ. No. 4452, 1987) is much lower than the extrapolation of the spectrum observed with Ginga. The inferred cutoff of the spectrum between photon energies of 1 and 4 keV is not predicted by the ejecta-CSM collision model. Secondly, the light curves observed for two energy ranges above and below 10 keV are similar

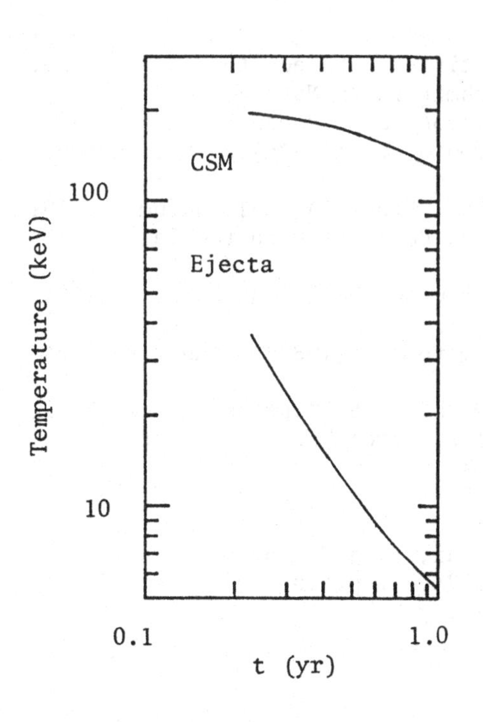

Figure 2. Time evolution of the temperature averaged over the reverse-shocked ejecta and over the blast-shocked CSM, respectively, with the weight of the bremsstrahlung. The electron and ion temperatures are assumed to be equilibrated everywhere, in the present model.

to each other (Dotani et al. 1987). The synchronism of the two X-ray emission processes is not explained either. Further observations of SN1987A are to be desired.

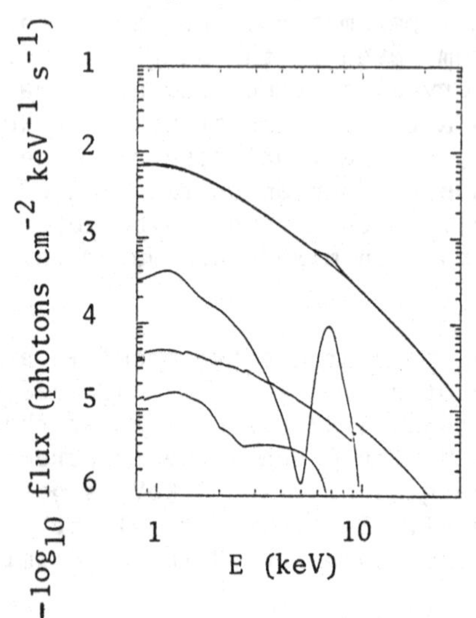

Figure 3. The X-ray spectrum at t = 0.53 yr. The total and line spectra are convolved with a Gaussian to a resolution of $0.49(E/1\ \text{keV})^{0.5}$ keV (FWHM), where E is the photon energy. Also plotted are the continuum spectra due to bremsstrahlung, radiative recombination, and two-photon decay. The metallicity is assumed to be one-third of the cosmic value (Allen 1973). The distance and hydrogen column density to SN1987A are assumed to be 55 kpc and 2×10^{21} cm^{-2}, respectively.

REFERENCES

Allen, C.W. (1973). Astrophysical Quantities, 3rd ed. London: Athlone.

Dotani, T. & the Ginga team (1987). submitted to Nature.

Itoh, H. (1987). submitted to Publ. Astron. Soc. Japan.

Itoh, H., Hayakawa, S., Masai, K. & Nomoto, K. (1987a). Publ. Astron. Soc. Japan, 39, 529.

Itoh, H., Masai, K. & Nomoto, K. (1987b). Proc. IAU Colloquium No. 101, The Interaction of Supernova Remnants with the Interstellar Medium, in press.

Itoh, M., Kumagai, S., Shigeyama, T., Nomoto, K. & Nishimura, J. (1987). submitted to Nature.

Masai, K., Hayakawa, S., Itoh, H. & Nomoto, K. (1987a). submitted to Nature.

Masai, K., Hayakawa, S., Itoh, H., Nomoto, K. & Shigeyama, T. (1987b). Proc. IAU Colloquium No. 108, Atmospheric Diagnostics of Stellar Evolution, in press.

McCray, R., Shull, J.M. & Sutherland, P. (1987). Astrophys. J. Letters, 317, L73.

Nomoto, K., Hayakawa, S., Itoh, H., Masai, K. & Shigeyama, T. (1987). Proc. ESO Workshop on SN1987A, in press.

Ostriker, J.P. (1987). Nature, 327, 287.

DETECTABILITY OF EARLY THERMAL RADIATION FROM A NEUTRON STAR IN SN 1987A

Ken'ichi Nomoto
Department of Earth Science and Astronomy, University of Tokyo

Sachiko Tsuruta
Department of Physics, Montana State University

Abstract. Cooling of a young neutron star right after its birth is examined. Our theoretical calculations show that within less than a month after the explosion the surface temperature falls significantly below the detection limit of Ginga due to the plasmon neutrino emission near the surface. However, it will remain high enough to be detected easily by ROSAT, AXAF, and other future X-ray satellites within the next 100 years.

NEUTRON STAR FORMATION IN SN 1987A

The supernova 1987A in the Large Magellanic Cloud has provided a new opportunity to study the evolution of a young neutron star right after its birth. The observed neutrinos (Hirata et al. 1987; Bionta et al. 1987) are likely to be emitted from a hot proto neutron star (e.g., Burrows and Lattimer 1987; Suzuki and Sato 1987). From the total neutrino energy, the mass of the remnant compact star in SN 1987A is estimated to be 1.2 - 1.7 $M_\odot$ (Burrows and Lattimer 1987; Sato and Suzuki 1987). This suggests the formation of a neutron star, though a black hole formation is not completely ruled out.

NEUTRON STAR COOLING AND THERMAL X-RAY EMISSION

A proto neutron star first cools down by emitting neutrinos that diffuse out of the interior within a minutes (Burrows and Lattimer 1986). After the neutron star becomes transparent to neutrinos, the neutron star core with > 10^{14} g cm^{-3} cools predominantly by Urca neutrino emission (Tsuruta 1986). However, the surface layers remain hot because it takes at least 100 years before the cooling waves from the central core reach the surface layers. In other words, during these early stages the surface layers are thermally decoupled from the core and, hence, the change in the surface temperature T_s is practically independent of the core cooling (Nomoto and Tsuruta 1981, 1986, 1987).

Thermal X-rays are emitted from the surface. The detection limit for X-rays from SN 1987A by the Ginga satellite is 3 x 10^{36} erg s^{-1} (Makino 1987; Tanaka 1987). If the thermal X-rays are to be observed by Ginga, the surface temperature should continue to be as high as T_s > 8 x 10^6 $(R/10\ km)^{-1/2}$ K until the ejecta becomes transparent. The exact value of the initial surface temperature depends on various factors during the violent stages of explosion, cooling stages of the proto neutron star through diffusive neutrinos, and possible re-infalling of the ejected material. Therefore, until the surface layers become thermally relaxed T_s may satisfy the above condition.

Figure 1 shows the cooling behavior of neutron stars during the first one year after the explosion. The total luminosity of the surface photon radiation L_γ^∞ (left) and the surface temperature T_s^∞ (right), both to be observed at infinity, are plotted as a function of time, for three nuclear models PS (stiff), FP (intermediate), and BPS (soft). The temperature scale refers to the FP model.

From Figure 1 it is clear that the surface radiation falls significantly below the Ginga detection limit within a few to $\sim$20 days for all the models considered. Such a decrease is caused by the plasmon neutrino emission from the outer layers of $\rho = 10^9 - 10^{10}$ g cm^{-3}. We note that during these early stages the surface temperature is independent of the complicated thermal behavior in the central core.

The surface layers at the lower densities ($\rho < 10^{11}$ g cm^{-3}) are so thin that the time scale of thermal conduction is shorter than the time scale of the plasmon neutrino cooling. Consequently, the surface cooling quickly follows the plasmon neutrino cooling in the layers just beneath the surface. This mechanism of surface cooling by the plasmon neutrino process clearly wipes out the initial conditions.

We emphasize that the surface radiation and temperature in the above early cooling era are determined rather accurately. This is due to the thermal decoupling between the core and the outer layers where the plasmon neutrino process dominates and the accurately determined plasmon neutrino emission rate (Beaudet et al. 1967). The effects of magnetic fields are not included in our current theoretical models, but these effects on cooling are predicted to be negligible.

Consequently, it is unlikely that Ginga would detect the thermal X-ray emission directly from the surface of a neutron star in SN 1987A even if the ejecta should become transparent right now.

DETECTABILITY WITH FUTURE X-RAY SATELLITES

Looking beyond Ginga, it should be important to search by other future X-ray satellites for a thermal soft X-ray point source in SN 1987A. This is because due to the finite time scale of thermal conduction, it will take at least another 100 years before the efficient cooling of the central core by the Urca process will be registered at the surface. Until then the observed surface temperature will remain at the level of at least two to four million degrees (see Figure 9 in Nomoto and Tsuruta 1987). These temperatures correspond to the observed luminosity of $> 10^{35}$ erg s^{-1} (approximately the asymptotic level reached in Figure 1 after the initial plasmon cooling from the outer layers). Note that the expected detection limit for ROSAT is 10^{34} erg s^{-1} (Trumper 1981), while for AXAF it is less than 10^{32} erg s^{-1} (NASA 1980).

We conclude that future satellite observations of SN 1987A with high sensitivity soft X-ray detectors (< 1 keV) should be critical in order to test the evolution theories of young neutron stars. This goal may very well be realized within the next 10 years, by the programs such as ROSAT, SXO, GRANAT, SPECTRA, and AXAF. SN 1987A is giving us this unique and invaluable opportunity too costly to be missed.

This research was supported in part by the Japanese Ministry of Education, Science and Culture through Research Grant No. 62540183, by the National Science Foundation through Grant No. AST-8602087, and by MONTS through Grant No. 16404216.

REFERENCES

Beaudet, G., Petrosian, V., and Salpeter, E.E. (1967). Ap. J., **150**, 979.
Bionta, R.M. et al. (1987). Phys. Rev. Lett., **58**, 1494.
Burrows, A., and Lattimer, J. (1986). Astrophys. J., **307**, 178.
Burrows, A., and Lattimer, J. (1987). Astrophys. J. Lett., **318**, L63.
Hirata, K. et al. (1987). Phys. Rev. Lett., **58**, 1490.
NASA (1980). TM-78285 (NASA, MSFC)
Nomoto, K., and Tsuruta, S. (1981). Astrophys. J. Lett., **250**, L19.
Nomoto, K., and Tsuruta, S. (1986). Astrophys. J. Lett., **305**, L19.
Nomoto, K., and Tsuruta, S. (1987). Astrophys. J., **312**, 711.
Makino, F., and the Ginga team (1986). Astrophys. Lett., **25**, 223.
Sato, K., and Suzuki, H. (1987). Phys. Lett., submitted.
Suzuki, H., and Sato, K. (1987). Publ. Astron. Soc. Japan, **39**, 538.
Tanaka, Y., and the Ginga team (1987). Paper presented at the meeting of the Astronomical Society of Japan.
Trumper, J. (1981). in Proc. Uhuru Memorial Symposium (NASA, GSFC).
Tsuruta, S. (1986). Comments on Astrophys., **11**, 151.

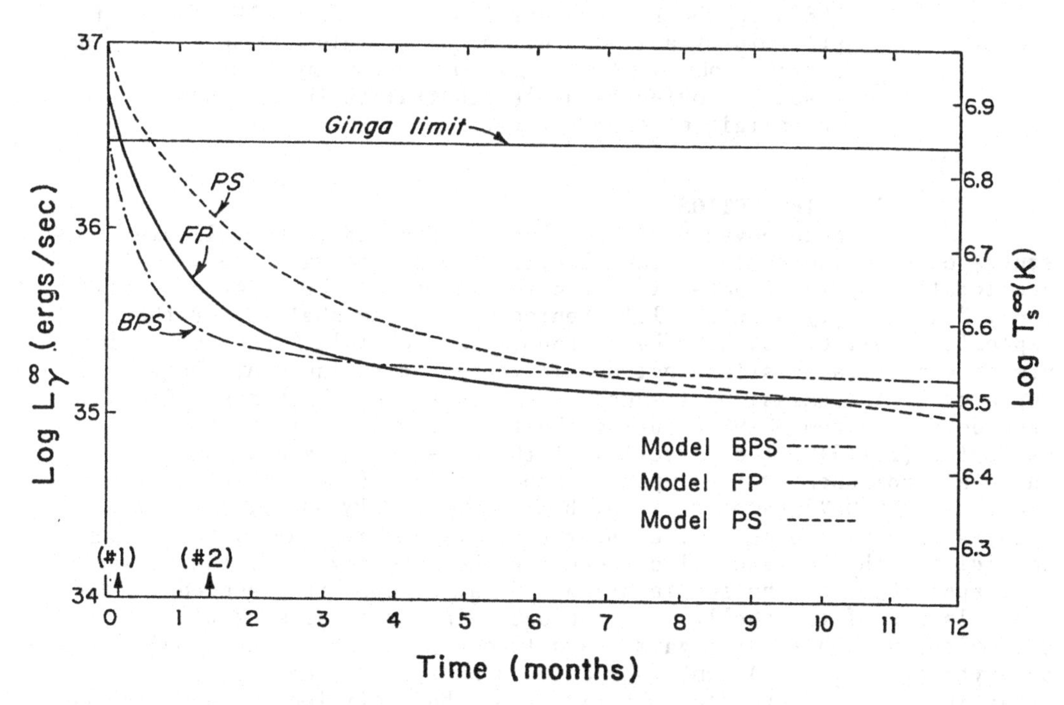

Figure 1: Observed photon luminosity vs. time during a year after the supernova explosion predicted from standard cooling theory of neutron stars. Shown are three representative nuclear models PS (dashed), FP (solid), and BPS (dot-dashed). The detection limit from Ginga is shown as a horizontal line. The stage numbers (#1, #2) refer to the stages with t = 3.7 days and 1.4 months, respectively.

SEARCH FOR HIGH ENERGY GAMMA RAY EMISSION FROM SN1987A

R.K. Sood, J. Thomas, L. Waldron, B.V. Denehy and
R.K. Manachanda
University College, ADFA, Univ. of NSW, Canberra,
Australia.

R. Staubert and E. Kendziorra
Astronomisches Institut der Universitat Tubingen,
Tubingen, W. Germany.

P. Ubertini and A. Bazzano
Istituto di Astrofisica Spaziale, CNR, Frascati, Italy.

G.K. Rochester and T.J. Sumner
Imperial College, London, U.K.

G.M. Frye, T.L. Jenkins, R. Koga and P. Albats
Case Western Reserve University, Cleveland, OH 44106,
USA.

Abstract. An experiment for the detection of high energy gamma rays (50-500 MeV) from the supernova event 1987a in the LMC was carried out on April 19, 1987. The preliminary analysis of the data yield an upper limit of 3×10^{-4} photons/cm^2 s for the gamma ray flux. The results provide valuable constraints for SN models for the origin of cosmic rays.

1 INTRODUCTION

Supernovae have long been considered to be a likely source of both the nucleons and electrons in cosmic rays (ref. 1,2). Acceleration by shock waves in the outer layers of the supernova shell (ref. 3,4,5) or by a pulsar left behind within the shell after the explosion (ref. 6,7,8) have been proposed as possible mechanisms. The presence of relativistic electrons in young supernovae would be revealed by the emission of synchrotron radiation, which has indeed been detected from SN1987A during the first few days after the explosion (ref 9). The detection of the expected more abundant nucleonic component relies upon the observation of the emission of high energy (> 50 MeV) gamma rays which are produced by way of π^0-meson decay as a consequence of the interaction of the relativistic charged particles with the material surrounding the supernova. If the pulsar is responsible for the acceleration and gamma rays are produced in the inner layers of the shell, then the overlying material must be sufficiently thin for the gamma rays to escape if they are to be observed (ref. 10). A number of authors have predicted that the expanding shell of expelled material would be sufficiently thin and the emission of high energy gamma rays would commence within a few weeks of the occurrence of a Type II supernova. It was therefore believed to be important that an early observation of SN1987A be made in high energy gamma rays.

2 INSTRUMENTATION

The gamma ray detector which was used to make the observations had been flown several times between 1967 and 1975 (ref. 11, 12). Two scintillators and a directional Cerenkov counter which responded only to downward moving electrons formed a coincidence system and also defined the geometric opening angle, which extended to 50°. The FWHM opening angle was 60° with a fairly flat response to 11°. The energy range covered extended from 50 MeV to 500 MeV. A 30-gap optical spark chamber of sensitive area 400 cm^2 was placed above the scintillators. The converter plates were made of stainless steel, total thickness 0.0187 m, except for the bottom four, which were aluminum. The 4mm gap between the plates was filled with a helium-neon gas mixture at 1.2 bar. The plates converted the gamma rays to electron-positron pairs, either one of which triggered the application of a high voltage to the chamber when it passed through the coincidence system, producing sparks along any ionized tracks. For each event two orthogonal views of the spark patterns were recorded on 16 mm film along with the time according to an on-board crystal clock and the output from a 3-axis fluxgate magnetometer which determined the orientation of the payload with respect to the Earth's magnetic field. The drift of the on-board clock was calibrated against a rubidium clock which was itself checked against a time standard available locally. Anticoincidence scintillators placed above and on all four sides of the spark chamber prevented it from being triggered by incoming charged particles. The detector elevation was fixed at 45° for the flight, while SN1987A was tracked in azimuth.

3 OBSERVATIONS

The detector was launched on a 21 Mcf balloon from Alice Springs, Australia, on 19th April. An almost constant float altitude of 2.2±0.1 mb was maintained over the period 0057 UT to 0710 UT. The wide field of view allowed SN1987A to be observed for the entire flight duration. Since the spark chamber data were stored on board it was deemed prudent to alter the azimuthal bearing of the detector away from the supernova occasionally so that the telemetered gamma ray event rate could be used to establish some kind of a flux level from this source in the event of a camera failure, or loss of the instrument package.

A preliminary analysis of the event rate revealed no indication of a gamma ray flux from SN1987A (ref. 13). All the observed counting rate variations could be attributed to changes in cut-off rigidity during the flight (correlation coefficient of 0.82). Out of ~ 48,000 recorded events on the spark chamber film, 7900 were visually identified as resolved gamma ray electron positron pairs. A 3-sigma upper limit of 3×10^{-4} photons/cm^2 s has been established from the counting rate profile. Initial analysis of the spark chamber data is almost complete; this should improve our sensitivity by a factor of ~ 10. Pending the finalization of these results, present discussion will focus on the upper limit quoted above.

4 DISCUSSION

In order to compare theoretical predictions with our observations it is necessary to make some assumptions about the conditions pertaining to SN1987A. It is generally accepted that the progenitor was the blue B3 I supergiant Sk-69 202, rather than the more usual M supergiant, and that this star may have been metal deficient (ref. 14). In the following discussion, we assume an ejected mass of 7 $M_\odot$ though there are strong indications that this value could be as low as $4M_\odot$ (ref 15). An average velocity of 10^4 km/s up to the day of the balloon flight (day 55) is assumed (ref. 16). The distance to SN1987A is taken to be 48 kpc (ref. 17).

We first consider the model developed by Cavallo and Pacini (ref. 8) for gamma ray emission from young supernovae wherein the cosmic ray acceleration is provided by a central pulsar. Gamma ray production of interest to us takes place via interactions of these particles with the surrounding supernova shell. It is assumed that the cosmic ray luminosity L_{cr} stays constant for a long time, and the value of L_{cr} of 1.3 x 10^{45} ergs/s is derived from arguments of equipartition with the ratio of the shell kinetic energy at maximum luminosity to the age of the supernova. The parameters assumed above lead to a line of sight of ~ 150 gm and a photon flux of 2.1 photons/cm^2 s on day 55, several orders of magnitude above our upper limit. Taking a more realistic value of L_{cr} of 10^{43} ergs/s (ref. 18) reduces the calculated flux by only two orders of magnitude.

In the model of Gunn and Ostriker (ref. 6), it is again postulated that cosmic rays are accelerated by a central pulsar. The cosmic ray luminosity is derived from the spin down of the pulsar. The model explaining the observed optical light curve suggests the presence of a central energy source of ~10^{41} ergs/s. Assuming the source to be in the form of a pulsar one can make a reasonable estimate of the resultant gamma ray flux. The calculated value of 5 x 10^{-3} photons/cm^2 s is again above our established limit.

Taking a different approach from Cavallo and Pacini (ref. 8) and Gunn and Ostriker (ref. 6), Chevalier (ref. 5) has investigated the production of gamma rays by cosmic rays accelerated in spherical blast waves and driven waves associated with a supernova explosion. The coupling between the cosmic rays and the shock wave is parameterized by the coefficient w. In order to calculate the gamma ray flux from blast waves for SN1987A, we have taken the total energy to be the lower limit of 2 x 10^{51} ergs from light curve modelling of optical observations (ref. 18), and a value for the thermal gas density of 2 x 10^3/cm^3. The latter quantity assumes a presupernova mass loss rate of 10^{-7} M/year at a velocity of ~100 km/s (ref. 9) consistent with the properties of a blue supergiant. The calculated flux of 1.8 x 10^{-6} photons /cm^2 s is below the sensitivity of the experiment. This figure assumes strong coupling between the cosmic rays and the shock wave (w=1). However, if the total energy in the supernova was as high as 4-5 x 10^{53} ergs as suggested by Arafune and Fukugita (ref. 19), the expected flux would have been above the ultimate sensitivity of our detector for w =1, a scenario which we hope to be able to test shortly.

Next we examine the model of Berezinsky et al. (ref. 20). These authors argue for supernova accelerated particles being the main source of galactic cosmic rays, with a value of L of 5 x 10^{41} to 10^{43} ergs/s for each source being derived from the observed cosmic ray energy density and the frequency of supernovae. The calculated gamma ray flux from SN1987A should then lie between (2.4 and 104) x 10^{-4} photons/cm^2.s for cosmic ray spectral indices between 2.1 and 2.6, well within the sensitivity of the experiment.

Finally, we turn to the work of Ginzburg and Ptuzkin (ref. 7) who have attempted to explain the unexpectedly high flux of antiprotons observed in the primary cosmic radiation (ref. 21 and 22). These authors have postulated that the antiproton production takes place, not in the interstellar medium, but as secondaries in compact sources with a high concentration of relativistic protons. Their model specifically involves antiproton generation in young supernova remnants, and predicts that p-p interactions would lead to copious production of gamma rays in the very early stages. In translating their calculations to the LMC galaxy, assumptions have to be made about the frequency of supernovae (η_{SN}) in that galaxy and also the mean amount of matter traversed by cosmic rays in its interstellar medium (X). Taking their assumed values of η_{SN} = 1/220 years and X = ~5 gm/cm^2 for the LMC, the flux expected from SN1987A turns out to be 10^{-2} photons/cm^2 s.

5 CONCLUSIONS

The transparency of the shell of SN1987A is dependent on the star parameters assumed. However, observational evidence for these is sufficiently strong to ensure that the gamma ray flux limit quoted above is low enough to test some models for the origin of cosmic rays. Complete analysis of our spark chamber data will enable us to further constrain these models. Our calculations of the time dependent density distributions of the expanding shell material, to be presented later, also indicate that SN1987A will remain bright in gamma rays above our sensitivity threshold, up to the time of our next planned flight in March 1988.

6 ACKNOWLEDGEMENTS

The observations would not have been possible without the expertise and dedication of J. Panetierri and A. Meilak of University College, ADFA, and M. Mastropietro of IAS. We also thank Prof. V.D. Hopper for his many words of wisdom.

7 REFERENCES

1. Baade, W. and Zwicky F. (1934). Proc. Nat. Acad. Sci. Amer. 20, 259.
2. Ginzburg V.L. and Syrovatskii S.I. (1964). Origin of Cosmic Rays, Pergamon Press.
3. Colgate S.A. (1975). Origin of Cosmic Rays, Reidel Pub. Co.
4. Colgate S.A. and Johnson M.H. (1960). Phys. Rev. Lett. 5, 235.
5. Chevalier R.A. (1983). Ap. J. 272, 765.
6. Gunn J.E. and Ostriker J.P. (1969). Phys. Rev. Lett. 22, 728.

7. Ginzburg V.L. and Ptuskin V.S. (1987). J. Astrophys. Astr. 5, 99.
8. Cavallo G. and Pacini F. (1980). Astron. Astrophys. 88, 367.
9. Storey M.C. and Manchester R.N. (1987). Nature, in the press.
10. Berezinsky V.S. and Prilutsky O.F. (1978). Astron. Astrophys. 66, 325.
11. Albats P. et al. (1972). Nature 240, 221.
12. Frye G.M. and Wang C.P. (1969). Astrophys. J. 158, 925.
13. Sood R.K. et al. (1987). I.A.U. Circular 4405.
14. Hillebrandt W. et al. (1987). Nature Lett. 327, 597.
15. Wood P.R. and Faulkner D.J. (1987). Proc. Astr. Soc. Australia, in the press.
16. Blanco V.M. et al. (1987). Cerro Tololo Observatory preprint.
17. Visvanathan, Private communication.
18. Shigeyama T. et al. (1987). Nature 328, 321.
19. Arafune J. and Fukugita M. (1987). Phys. Rev. Lett. 59, 367.
20. Berezinsky V.S. et al. (1985). 19th ICRC 1, 305.
21. Buffington A. et al. (1981). Ap. J 248, 1179.
22. Golden R.L. et al. (1979). Phys. Rev. Lett. 43, 1196.

HARD X-RAY OBSERVATION OF THE SUPERNOVA 1987A

P. Ubertini, A. Bazzano, C. La Padula
Istituto di Astrofisica Spaziale, CNR, Frascati, Italia

R. Sood, J. Thomas, L. Waldron
Physics Department, A.D.F.A., Univ. of N.S.W., Canberra, Australia

R. Staubert, E. Kendziorra
Astronomisches Institut, University of Tubingen, Germany

G. Rochester, T.J. Sumner
Physics Department, Imperial College, London, UK

G. Frye, R. Koga, P. Albats
Case Western Reserve University, Cleveland, USA

Abstract. On April 19th, 1987, the Supernova 1987A was observed between 19.04 and 19.30 UT with a hard X-ray and a high energy (>50 MeV) gamma ray detectors flown together on a stratospheric balloon launched from the Balloon Launching Station, Alice Springs, Australia. The balloon floated at an altitude corresponding to 2.0 mbar for seven hours. The hard X-ray detector consisted of a 2.0 bar Xenon-filled MWPC with a sensitive area of 500 cm² in the range 15-180 KeV and spectral resolution of 8% at 130 KeV. The results of the final analysis of the X-ray data are presented, including the timing analysis, together with an upper limit for spectral line intensities at 122 and 136 from the decay of Co57 expected to be present inside the shell of the supernova.

Introduction

The observation of the initial stage of a nearby supernova offers a unique opportunity to study the final stage in the evolution of a massive star. Of particular interest is what happens as the expanding envelope, still surrounding the collapsed nucleus, becomes optically thin to the outgoing hard X-ray photons. Different phenomena are predicted to occur depending upon which mechanism is responsible for the electromagnetic emission. The emission may be due either to the decay of radioactive material leading to the production of the 847 and 1238 KeV Co60 and the 122 and 136 KeV Co57 emission lines, as suggested by the recent optical behaviour of the SN light curve, or else to newly formed rapidly rotating pulsar emitting syncrotron radiation over a wide range of energies (Cavallo & Pacini 1980, McCray et al. 1987, Woosley et al. 1987, Xu et al. 1987). In both cases, in the initial stage, the optically thick envelope will scatter the outgoing radiation thereby Comptonizing the original X-ray lines and/or continuum. In order to study the early behaviour of SN87A a hard X-ray and high energy gamma ray detectors were combined in the same balloon payload. This first high energy observation of the supernova was made only 55 days after the Supergiant SK-69 202 exploded. The two upper limits so far reported (Sood et al. 1987, Ubertini et al. 1987) have clearly demonstrated that at the time of the balloon flight the residual amount of material covering the remnant in the line of sight exceeded the expected 70 g/cm^2 (Xu et al. 1987). Also, if there were any inhomogenity or mixing of the mantle due either to the presence of a companion star moving accross the rapidly expanding envelope or to any other form of dinamic instability then it was not sufficient to bring enough Co57 close to the expanding surface for it to be observed by our experiment at the level of 2.7×10^{-3} ph/(cm^2 s) (3 sigma upper limit), which was the sensitivity of the channels corresponding to the 122-136 KeV lines. In the case of an optically thin envelope this upper limit would have meant that less than 1.3×10^{-3} M⊙ of Co57 had been synthesised.

Figure 1:
The balloon experiment during test and calibration

Figure 2:
The MWPC counting rate and pressure vs time

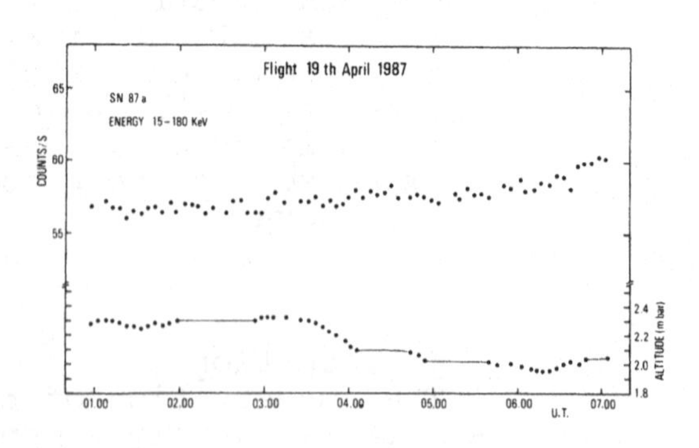

The experimental configuration

The balloon borne payload used to perform these measurements is shown in figure 1 undergoing tests and calibration at the Balloon Launching Station, Alice Springs (Australia). The instrument comprised four major subsystems:
- the spark chamber
- the hard X-ray detector
- the orientation system
- the telemetry, telecommand and flight systems.

The two detectors, consisting of scintillator telescope and multiplate spark chamber for the gamma ray and a Multiwire Proportional Chamber (MWPC) filled to 2.0 bar with Xenon-Argon-Isobutane for the hard X-ray, were coaligned. The gamma ray detector, sensitive in the range 50-500 MeV, had an effective area of 400 cm^2on axis and a field of view of 60 deg FWHM. The X-ray detector had a geometric area of 500 cm^2 in the operative range 15-180 KeV. The time resolution was 10 microseconds. The crystal clock was calibrated against a UT rubidium clock generator.

The hard X-ray detector

The detector used is one of two identical modules built at IAS in 1980 which have already been flown several times succesfully on balloons and which have been widely described in the literature (Ubertini et al. 1981). The sealed MWPC are made from a stainless steel vessels which have an entrance window composed of an aluminium honeycomb sandwich of total thickness 400 microns, thin enough to be 50% transparent to 15 KeV photons. The internal sophisticated anticoincidence event rejection system kept the residual background down to 7×10^{-4} ct/(cm^2 s KeV) during the flight at 2.2 mbar residual atmospheric pressure. The spectral resolution as a function of energy is $\Delta E(\%)=93/E^{1/2}$ and corresponds to 8% at the Co57 lines energies (122 and 136 KeV). The field of view of the detector was limited by a hexagonal copper collimator with an aperture of 10 deg FWHM, coaligned with the optical axis of the spark chamber. The two instruments were tilted at a fixed angle of 45 deg to the vertical in order to maximise the exposure during the SN87A observation. The azimuth of the gondola was oriented to the supernova azimuth except when rotated off-source for the X-ray detector background determination.

Data analysis

The payload reached a ceiling altitude corresponding to a residual atmospheric pressure of 2.2 mbar at 00.55 UT on the 19th April. The float altitude was controlled to mantain the residual pressure within $\pm.2$ mbar for the whole flight duration. In figure 2 is shown the integral counting rate for the MWPC in the range 15-180 KeV together with the pressure as a function of time. A general

increase of the counting rate of about 2% was observed between beginning and the end of the observation due to the variation in atmospheric background caused by the change in geomagnetic cut-off with longitude. The background data obtained from the MWPC, binned in four different energy channels, were normalised to a constant atmospheric pressure of 2.2 mbar and then corrected for the effect of the variation in geomagnetic cut-off. Then the (BGD + S) -(BGD) excesses were computed for the whole observation of the source. Finally the upper limits in the four energy channels were computed, taking into account the atmospheric absorption, the exposed area and the detector matrix response. These upper limits are shown in figure 3.
A Fourier analysis was performed to search for pulsed emission at periods between 4 and 100 milliseconds giving the white distribution as shown in figure 4.

Figure 3:
The spectral upper limits as detected by the MWPC

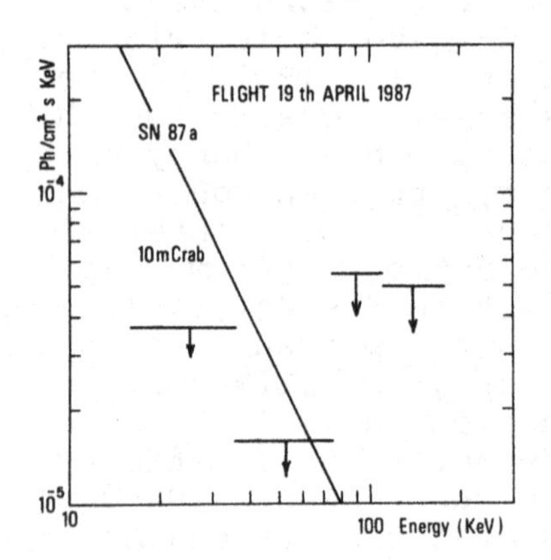

Figure 4:
The harmonics amplitude versus frequency

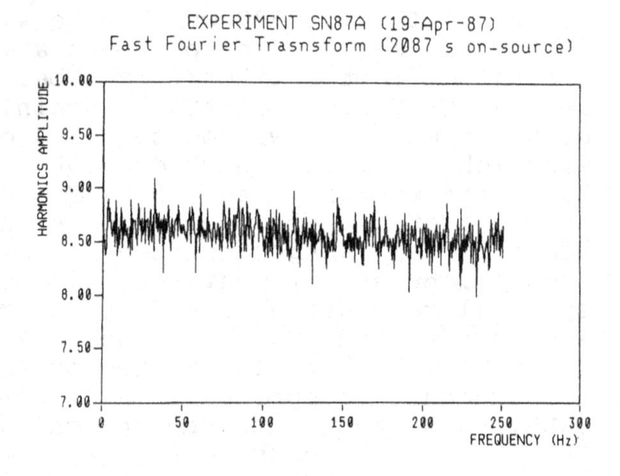

Conclusions

Historically theoretical predictions were very uncertain about the expected X-ray flux a few days after a SN explosion and about the time scale of the envelope fragmentation and mixing.
Our data show that:

- The optical depth of the SN87A was still very high 55 days after the explosion (> 70 g/cm^2) becouse of the lack of any direct or comptonized continous hard X-ray flux
- There is no line emission at 122 or 136 KeV at the level of $2.7x10^{-3}$ ph/(cm^2 s)
- The total luminosity of the SN in the hard X-ray range was less than $3x10^{37}$ erg/s in the range 15-75 KeV and less than 10^{38} erg/s up to 180 KeV.

Acknowledgements

The authors acknowledge Miss. L. Barbanera, and Mr. L. Boccaccini, M. Mastropietro and R.Patriarca of IAS and A. Meilak and J. Panettierri of A.D.F.A for their support.

References

Cavallo, C. & Pacini, F., Astron. & Astrophys.,88,(1980),367
McCray, R. et al., Ap.J.,317,(1987),L73
Sood, R.K. et al., IAU Circular N. 4405, 12 june 1987
Ubertini, P. et al., Sp. Sci. Inst.,5,(1981),237
Ubertini, P. et al., IAU Circular N. 4387, 8 may 1987
Woosley, S.E. et al., Ap.J.,318,(1987),664
Xu, Y. et al., Ap.J. Lett.,(1987), in press

NASA PLANS FOR OBSERVATIONS OF SN1987a

Guenter R. Riegler
Astrophysics Division, National Aeronautics and Space Administration (NASA), Washington, D. C. 20546, U. S. A.

Abstract. The NASA plans for observations of Supernova SN1987a are described. The objective of the observing program is to obtain data across the full spectrum in order to develop a thorough understanding of the early post-explosion development of SN1987a. This objective is being implemented through a series of coordinated campaigns of gamma-ray, X-ray, ultraviolet and infrared observations from balloons, rockets and aircraft. Rocket-borne X-ray observations, in particular, are expected to obtain data which are not obtainable with existing orbiting instrumentation.

INTRODUCTION

NASA's Astrophysics Division responded to the discovery of SN1987a by rapidly establishing a Supernova Program to direct and coordinate all of NASA's SN1987a-related activities. By March 2, the scientific community was informed of the opportunity to launch existing gamma-ray, X-ray and ultraviolet experiments on balloons and rockets for observations of SN1987a. On March 5, the decision was made to extend the planned operations of the Kuiper Airborne Observatory (KAO) in New Zealand to include supernova observations. A Supernova Science Working Group was formed to advise NASA on the scientific aspects of the Supernova Program. By the end of March, a program for observations of SN1987a at all wavelengths from the radio to the gamma-ray bands was developed and initial steps for funding of these efforts were under way.

The physical processes in a supernova, and observable features for SN1987a have been discussed by a number of speakers during thisworkshop (see elsewhere in these Proceedings). This paper summarizes the plans for observations under the NASA Supernova Program. The reader should bear in mind that, with the exception of radio astronomy observations which use the Deep Space Network, NASA does not support ground-based astronomical observations.

PLANS FOR OBSERVATIONS

The overall goal of the observations is to make a balanced set of measurements at all wavelengths in order to study the immediate post-explosion phase of the supernova. These measurements and their interpretation will form the basis for a long-term study of SN1987a, to be conducted with the "Great Observatories" series of large orbiting instruments, i. e. the Hubble Space Telescope (HST), the Gamma-Ray Observatory (GRO), the Advanced X-ray Astrophysics Facility (AXAF), and the Space Infrared Telescope Facility (SIRTF).

Radio Observations

As part of NASA's Deep Space Network (DSN), several large antennae are being linked with Australia's Parkes Antenna to operate in interferometry and very long baseline interferometry (VLBI) configurations.

One of the 34 m antennae at Tidbinbilla (the 64 m antennae are being upgraded to 70 m dishes), operated as an interferometer with Parkes, has a threshold sensitivity of 10 mJy and an angular resolution of 0.03 arc sec at 2.3 and 8.4 GHz.

Infrared Observations

Two observing campaigns for the Kuiper Airborne Observatory (KAO) are planned in each of fiscal years 1988 and 1989.

The next campaign will take place in November 1987; eight observing flights of the supernova for four infrared spectrometers will be conducted from Christchurch, New Zealand, at wavelengths from 4 to 200 microns. The Principal Investigators for the four instruments are F. Witteborn (NASA/Ames Research Center; 4.5 - 12.5 μm), H. Moseley (NASA/Goddard Space Flight Center; 20 - 70 μm), E. Erickson (NASA/Ames; 23 - 70 μm), and P. Harvey (U. Texas, 40 - 200 μm).

The objectives of the infrared observations are to measure the "infrared excess" which has been observed during the first KAO campaign in April 1987, to separate an infrared "echo" from infrared emission, to observe dust formation, and to study the interaction of the blastwave with the ambient medium.

Visual and Ultraviolet Observations

Routine monitoring will continue with the International Ultraviolet Explorer (IUE) at wavelengths as short as ≈ 1200 Å. The "ASTRO" complement of three UV instruments is currently manifested as a shuttle attached payload for flight in early 1989. ASTRO contains the Hopkins Ultraviolet Telescope (425 - 1250 Å), the Wisconsin Ultraviolet Photopolarimetry Experiment (1200 - 3200 Å), and the Ultraviolet Imaging Telescope (1200 - 3200 Å).

Rocket-borne observations with two Far Ultraviolet spectrometers for the 900 - 1250 Å band are scheduled for a series of four campaigns between November 1987 and early 1989. The purpose of these intermediate- and high-resolution measurements is to study the outer regions of the supernova shell,the interaction with the ambient medium, and the properties of the interstellar medium. The Far Ultraviolet spectrometer of W. Cash (U. Colorado; 900 - 1250 Å) is scheduled for flight on or about Nov.18, 1987 and during subsequent campaigns, while the Interstellar Medium Absorption Profile Spectrograph of E. Jenkins (Princeton U.; 950 - 1150 Å) will be launched in January 1988 or thereafter, depending on receipt of an adequate "trigger" signal.

X-Ray Observations

Three different X-ray instruments are presently being readied for rocket flight campaigns beginning in November 1987. Figure 1 shows the planned timing for the X-ray rocket flights in relation to recently published models for the X-ray continuum emission from down-Comptonized gamma-rays from the decay of ^{56}Co or a possible remnant pulsar. A number of other mechanisms may also give rise to thermal or non-thermal X-ray emission.

The first flight (G. Garmire, Pennsylvania State U.) will contain an X-ray telescope mirror with a charge-coupled device (CCD) sensor for the 0.2 - 2 keV energy range, and is scheduled for Nov. 12, 1987. Subsequent flights could take place as early as Dec. 1987, March 1988 and in early 1989, depending on receipt of an adequate "trigger" signal.

The second instrument (R. Novick, Columbia U.) contains a gas scintillation proportional counter behind an X-ray mirror and a coded-aperture mask in order to achieve sensitivity from 0.2 to 20 keV. The first flight of this instrument is scheduled for late November 1987, and two subsequent flights will take place in 1988 and 1989 when an adequate trigger signal has been recorded.

A third instrument (P. Serlemitsos, Goddard Space Flight Center) contains an X-ray telescope and Si(Li) detector for very high sensitivity spectroscopic measurements in the 0.3 - 4 keV range. This instrument is scheduled for three flights beginning as early as January 1988.

Although these X-ray observations have short observing times of ≈ 5 minutes, they are still uniquely capable of making low-energy X-ray measurements with high spatial and spectral resolution. A comparison of capabilities for these rocket-borne instruments with those on the orbiting GINGA and MIR missions is shown in Figure 2. The most important capability of the rocket-borne X-ray instruments is the high spatial resolution, combined with high spectral resolution at soft and intermediate X-ray energies.

The Broad-Band X-ray Telescope (BBXRT; P. Serlemitsos, GSFC), similar to the last-mentioned rocket experiment but with sensitivity extending to 12 keV, is currently scheduled for flight in the second Shuttle High Energy Astrophysics Laboratory (SHEAL-2) in 1992.

Gamma-Ray Observations

Routine monitoring continues with the gamma-ray instrument on the Solar Maximum Mission (SMM; E. Chupp, U. New Hampshire).

One Balloon campaign with four gamma-ray instruments was completed successfully in June 1987, and yielded upper limits for continuum and line emission from SN1987a.

Now that hard X-ray emission has been observed (Truemper, 1987; Tanaka, 1987), it is likely that the current campaign (which began in late October 1987) or subsequent campaigns in January 1988, April-May 1988, etc., will be able to detect not only gamma-ray continuum, but also gamma-ray line emission from SN1987a. As shown in Figure 3, the timing for these gamma-ray balloon flights was chosen to bracket the range of theoretical predictions for gamma-ray line and continuum emission from SN1987a. The detection of line emission is considered to be the most direct verification of the nucleosynthesis model for Type II supernovae. More than one-half of the planned instruments for upcoming balloon campaigns will contain solid-state Germanium detectors for high spectral resolution, while others will have particularly high sensitivity for gamma-ray continuum detection. Figure 3 shows the anticipated sensitivity for detection of the 847 keV line from ^{56}Co; as new generations of gamma-ray instruments become available, these sensitivities are expected to improve.

References

Xu, Y., Sutherland, P., McCray, R., & Ross, R. R. (1987). Ap. J. Letters, in press.

Pinto, P. A. & Woosley, S. E. (1987). Ap. J. Letters, in press.

Tanaka, Y. (1987). "GINGA Observations of SN1987a", submitted to the Proceedings of the Workshop on SN1987a, George Mason University, Fairfax, VA.

Truemper, J. (1987). "Observations of SN1987a from MIR", submitted to the Proceedings of the Workshop on SN1987a, George Mason University, Fairfax, VA, October 12 - 14, 1987.

Figure 1. X-ray continuum emission expected for radioactive decay (heavy solid line: (Xu et al., 1987), shaded line: (Pinto and Woosley, 1987) and a remnant pulsar (thin solid line: Xu et al., 1987). Also shown is the schedule for X-ray observations with NASA rocket flights.

Figure 2. Capabilities of rocket-borne X-ray instruments (heavy lines) and orbiting instruments.

Figure 3. Gamma-ray line emission expected for radioactive decay of ^{56}Co for a range of assumptions for the progenitor star. The rectangles marked "balloons" on top of the graph indicate the times for planned gamma-ray balloon flight campaigns. The shaded band indicates the best line detection sensitivity anticipated during a campaign.

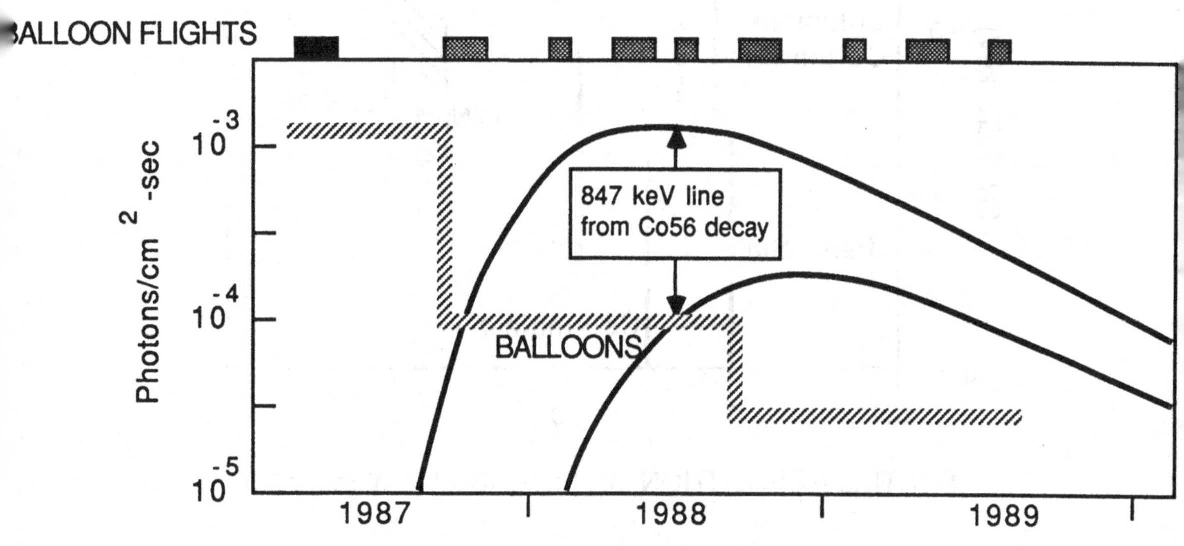

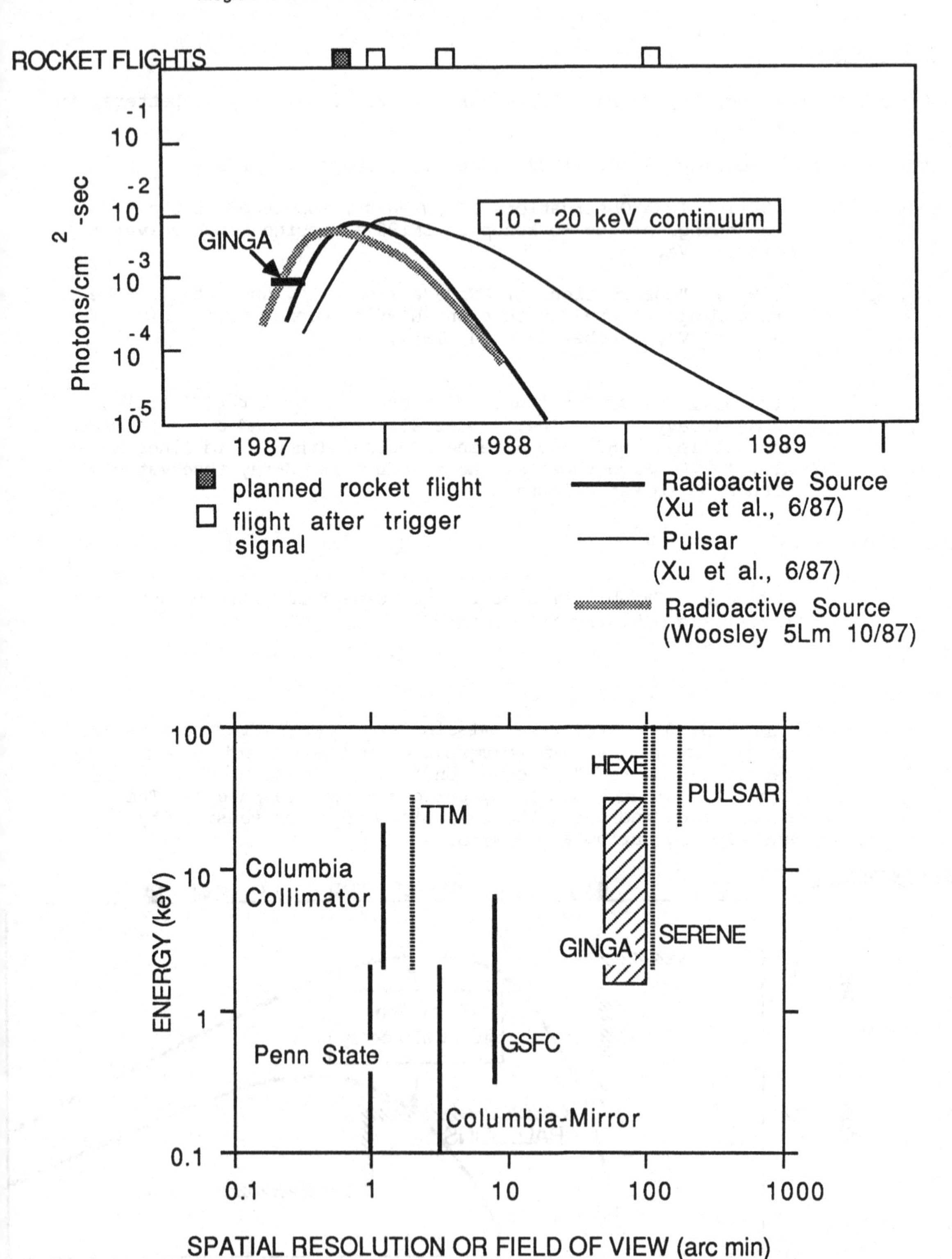
ROCKET FLIGHTS
Photons/cm^2-sec
10^{-1}
10^{-2}
10^{-3}
10^{-4}
10^{-5}
GINGA
10 - 20 keV continuum
1987
1988
1989
planned rocket flight
flight after trigger signal
Radioactive Source (Xu et al., 6/87)
Pulsar (Xu et al., 6/87)
Radioactive Source (Woosley 5Lm 10/87)
ENERGY (keV)
100
10
1
0.1
Columbia Collimator
TTM
HEXE
PULSAR
GINGA
SERENE
Penn State
GSFC
Columbia-Mirror
0.1
1
10
100
1000
SPATIAL RESOLUTION OR FIELD OF VIEW (arc min)

THE NASA SN1987A PROGRAM: GAMMA-RAY OBSERVATIONS

B. J. Teegarden
SN1987a Project Scientist
Goddard Space Flight Center
Greenbelt, MD 20771

ABSTRACT

The existing and planned experiments for making gamma-ray observations of SN1987a are reviewed. The main emphasis is on the NASA program which consists primarily of balloon-borne instruments. Some 11-13 experiments are or will be available. Four have already flown from Alice Springs, Australia with null results. Campaigns are planned on nominal six month centers with more possible if gamma-rays are detected. In addition long duration flights from Australia to South America are planned for January 88 and 89.

1. INTRODUCTION

One of the most eagerly anticipated events in the unfolding drama of SN1987a is the detection of gamma-radiation from radioactive nuclei created shortly after the onset of the explosion. Shock heating of the material just outside of the collapsing core raises the material to temperatures high enough to initiate new nucleosynthetic processes. The dominant species created is the doubly-magic nucleus ^{56}Ni. ^{56}Ni decays to ^{56}Co with a mean life 8.8 days, which subsequently decays to ^{56}Fe with a mean life of 114 days. The gamma-radiation produced in these decays is thought to play a crucial role as a sustaining source of energy driving the optical display (Woosley 1987; Woosley et al. 1987a,b). Since about day 40 after the explosion the optical light curve is believed to have been dominated by gamma-ray heating (Woosley et al. 1987a). After approximately day 120 the bolometric light curve transitioned into a exponential phase (in which it now remains) with a slope consistent with the 114 day mean life of the ^{56}Co to ^{56}Fe decay.

During the first year or so after the explosion the gamma-ray producing region lies many optical depths inside the expanding shell. The energy of these gamma-ray rays is degraded by compton interactions, and after the shell has thinned sufficiently some of them will escape as X-rays (Chan & Lingenfelter 1987; Gehrels et al. 1987; Pinto & Woosley 1987; Xu et al. 1987). X-ray detections from the Ginga and Mir satellites have recently been reported (Makino 1987; Tanaka 1987; Trumper 1987). The detailed behavior of these X-rays shows some bizarre properties that are not easily explained by the gamma-ray down-scattering model. However, the fact that they have appeared this early at such a high intensity bodes well for the eventual detection of the parent gamma-radiation.

The gamma-ray light curve (see Fig. 3) is determined by two competing processes: 1) the thinning of the expanding SN shell, and 2) the 114 day decay of the ^{56}Co. The initial rising portion of the curve is due to the thinning of the shell. When the shell begins to become optically thin the curve peaks and eventually transitions into a final exponential phase with the characteristic 114 day slope. The main uncertainties in this curve come from the amount of material in the overlying envelope and from the velocity profile of the expanding shell. Most models seem to be converging to a time-to-peak of 400-500 days and an 847-keV gamma-ray flux from a few x 10^{-4} to a few x 10^{-3} photons/cm^2-sec (Chan & Lingenfelter 1987; Gehrels et al. 1987; Pinto & Woosley 1987; Xu et al. 1987).

From the detection of gamma-rays from SN1987a we will be able to learn a number of new and important things. Their presence is required by most of the current models to explain the bolometric light curve after ~40 days. The ^{56}Co to ^{56}Fe decay produces a family of lines with known branching ratios that are independent of the conditions in the emission region. By comparing the measured line strengths with these ratios one can determine the optical depth to the emission region. With this knowledge the amount of ^{56}Ni that was produced can be directly calculated. Doppler broadening should produce gamma-ray line widths of ~5-10 keV. High-resolution gamma-ray instruments will be able to resolve these lines. It should be possible to determine the velocity of the emission region as well as perhaps learning whether or not significant mixing has taken place. Several authors (Chan & Lingenfelter 1987; Gehrels et al. 1987; Pinto & Woosley 1987) have predicted line profiles with interesting structure that could be resolved by these instruments.

2. GAMMA-RAY OBSERVATIONS ALREADY MADE

Two satellites, SMM and Mir, are now observing SN1987a in the gamma-ray region. SMM has been continuously observing the supernova since its onset whereas Mir started its observations in August. The SMM spacecraft is always pointed at the sun so that the LMC is at an angle of ~90° with respect to the instrument pointing direction. This unfavorable orientation causes significant degradation of the instrumental sensitivity to gamma-rays from SN1987a. Mir contains several instruments covering the X- and gamma-ray regions. The Pulsar X-1 instrument is a thick phoswich that can measure up to ~1200 keV. As of this writing, neither of these instruments has detected any radiation above 300 keV. The 3σ upper limits at 847 keV from SMM are typically 1.5-2.0 x 10^{-3} photons/cm^2-sec.

A number of balloon-borne gamma-ray instruments have made observations of the supernova. The first of these was a spark chamber

(Frye 1987) that was located in Australia at the time of the explosion and flew from Alice Springs in April. In May-June NASA mounted its first balloon campaign to Alice Springs, Australia. Three instruments flew, Leventhal, Prince (1987), and Sandie (1987). Again, no gamma-rays were seen. Table 1 summarizes these results:

TABLE 1. GAMMA-RAY UPPER LIMITS

	Date (1987)	Energy (keV)	Upper Limits
Continuum			
Mir (Pulsar X-1)	Aug.	400-1200	$4x10^{-6}$ ph/cm^2-s-keV
Frye	Apr.	>50,000	$3x10^{-4}$ ph/cm^2-s
Line			
SMM	Mar.-Sept.	847	$1.5-2.0x10^{-3}$ ph/cm^2-s
Prince	May	847	$1.5x10^{-3}$ ph/cm^2-s
Sandie	May	847	$1.7x10^{-3}$ ph/cm^2-s

3. NASA BALLOON PROGRAM

NASA is planning an extensive program of balloon-borne gamma-ray observations of SN1987a. Some 11-13 different instruments are (or will be) available to make these observations. A broad mixture will be flown including high-resolution Germanium spectrometers, NaI detectors, a spark chamber, and a compton telescope. Campaigns will be tied to the times of high-altitude wind turn-around (Apr. and Nov. in Alice Springs, Australia). In addition long-duration flights are planned in Jan. 87 & 88. These flights will be launched from Alice Springs, Australia and recovered in South America. Flight durations of ~2 weeks are expected. The extended exposure time afforded by this kind of flight will significantly improve the sensitivity of the observations. If gamma-rays are detected then a summer campaign will probably be added. The instruments that will fly in Nov. 87 and Jan. 88 have been selected. Instruments to fly on subsequent campaigns have been tentatively selected with final selection to take place near the end of this year. A summary of the planned campaigns is given in Table 2.

Table 3 summarizes the balloon payloads that have been provisionally selected to participate in the NASA SN1987a Program. The table has been divided into three broad categories, NaI instruments, Germanium instruments, and miscellaneous. Of the four NaI instruments, two (Frontera and Grindlay) are "thin" detectors that concentrate on the hard X-ray region (20-300 keV). The other two are

TABLE 2. BALLOON CAMPAIGNS

Date	Status	Principal Investigators
Nov., 87	Planned	Mahoney, Prince, Sandie, White
Jan., 88	Planned	Lin/Matteson (long duration)
Apr., 88	Tentative	Frontera, Johnson, Mahoney, Prince, Sandie, Teegarden/Leventhal
Aug., 88	Tentative	Teegarden/Leventhal
Nov., 88	Tentative	Aprile, Frontera, Grindlay, Johnson, Matteson, Teegarden/Leventhal
Jan., 89	Tentative	Lin/Matteson (long duration)
Apr., 89	Tentative	Matteson, Teegarden/Leventhal

"thick" and concentrate on higher energies. They have good sensitivity (typically 5×10^{-5} ph/cm^2-s @ 847 keV), but only modest resolution (~60 KeV @ 847 keV). The Germanium instruments are also subdivided into two categories. The first three are single detectors with total areas of 24-33 cm^2. Their narrow-line sensitivities are typically 1×10^{-3} ph/cm^2-s. The Lin/Matteson experiment, with its anticipated long exposure time and improved background suppression, should do substantially better than this. The second three are each arrays of from 7 to 12 detectors with total areas from 214-288 cm^2. Their narrow-line sensitivities should be within a factor of two of 1×10^{-4} ph/cm^2-s. All of these NaI and Ge experiments are actively shielded with scintillators of various types. The miscellaneous category includes three instruments. The first, a liquid Argon detector, holds great promise for the future. Its resolution is significantly better than NaI, and its volume is limited only by the lifting capacity of the balloon. Furthermore, it has position sensitivity, which is a potentially valuable tool for background suppression. The compton telescope and spark chamber are proven instruments that can extend the supernova gamma-ray spectrum to higher energies.

Examples of NaI and Ge instruments are given in Figs. 1 and 2. The first is the author's (Teegarden/Leventhal) and is called GRIS (Gamma-Ray Imaging Spectrometer). The imaging part (an active coded-aperture mask) will not be used for the supernova flights and is not shown in the Figure. The heart of the instrument is an array of seven large n-type Germanium detectors each housed in its own individual cryostat. The detector array is surrounded by a massive (364 kg) NaI anticoincidence shield having a minimum thickness of 15 cm. Seven holes in the shield define a field-of-view for the instrument of 20°

FWHM. The detectors are cooled to their operating temperature (~90K)

TABLE 3. GAMMA-RAY EXPERIMENTS SUMMARY

Experimentor	Detector Type	Detector Area (cm^2)	Energy Range (MeV)
Frontera	NaI	5400	.02-.3
Grindlay	NaI	900	.02-.3
Johnson	NaI	~800	.1-10
Prince	NaI	1320	.04-10
Mahoney	Ge	~33	.05-6.5
Leventhal	Ge	~30	.07-6.5
Lin/Matteson	Ge NaI	24 480	.02-10 .015-.18
Matteson	Ge	288	.02-10
Sandie	Ge	214	.025-8
Teegarden/ Leventhal	Ge	215-250	.02-10
Aprile	Liquid Argon	1250	.05-5
White	Compton Telescope	----	1-30
Frye	Spark Chamber	400	>50

Energy Resolution at 1 MeV: $E\Delta E$ ~500 for Ge
~15 for NaI

by a dewar filled with 50ℓ of LN_2. The experiment is pointed to 0.1 degrees accuracy by an azimuth-over-elevation pointing system having a momentum wheel for fine control of the azimuth axis. A 6° x 6° passive fine collimator (not shown in Fig. 1) may be used in subsequent flights to reduce the diffuse background contribution in the hard X-ray range. The GRIS energy range is .02 to 10 MeV, and its energy resolution at 847 keV is expected to be ~2.3 keV. Its calculated 3σ sensitivity to a narrow line at 847 keV is 1.7×10^{-4} ph/cm^2-s. GRIS will be available to observe the supernova in April, 1988.

The second instrument (shown in Fig. 2) is a prototype of one of the GRO/OSSE (Orbiting Scintillation Spectrometer Experiment) modules. It is a "thick" phoswich shielded by ~8 cm of NaI. Its energy range for spectroscopy is 0.1 - 10 MeV, and its resolution at 847 keV is ~60 keV. Broad-band measurements are also performed between 10 and 150 MeV. A tungsten collimator restricts the field-of-view to 3.8° x 11.4° FWHM. The OSSE narrow-line sensitivity between 0.1 and 10 MeV is $\sim 3 \times 10^{-4}$ ph/cm^2-s. The continuum sensitivity in the 0.1-1 MeV range is 0.06 x Crab. This instrument will be available to observe the supernova in April, 1988.

Fig. 3 is a plot of the 847 keV gamma-ray line intensity from Gehrels et al. (1987). The two most prominent lines (847 and 1238 keV) from the ^{56}Co to ^{56}Fe decay are plotted. The recent Ginga and Mir X-ray detections indicate that the gamma-ray intensity at peak may be even higher than that in Fig. 3. Pinto and Woosley (1987) calculate peak gamma-ray fluxes that are in some cases $>10^{-4}$ ph/cm^2-s. Shown for comparison are the 3σ sensitivities of SMM and the representative classes of balloon instruments. The sensitivities of the available instruments are clearly sufficient to detect gamma-rays from SN1987a if the model predictions are correct.

REFERENCES

Chan, K. W. and Lingenfelter, R. E. (1987). Calculated Gamma-Ray Line Fluxes from the Type II Supernova 1987a. Ap. J. (Letters), 318, L51.

Frye, G. A. (1987). Proceedings this Conference.

Gehrels, N., MacCallum, Crawford J., and Leventhal, Marvin (1987). Prospects for Observations of Nucleosynthetic Gamma-Ray Lines and Continuum from SN1987a. Ap. J. (Letters), 320, L19.

Makino, F. (1987). IAU Circular No. 4447.

Pinto, Philip A. and Woosley, S. E. (1987). X- and γ-Ray Emission from Supernova 1987a. Submitted to Ap. J.

Prince. T. A. (1987). Proceedings this Conference.

Sandie, W. G. (1987). Proceedings this Conference.

Tanaka, T. (1987). Proceedings this Conference.

Trumper, J. (1987). Proceedings this Conference.

Woosley, S. E. (1987). SN1987a. After the Peak. Preprint

Woosley, S. E., Pinto, P. A., Martin, P., and Weaver, T. A. (1987). Supernova 1987a in the Large Magellanic Cloud: The Explosion of a 20 $M_{\odot}$ Star Which Has Experienced Mass Loss? Ap. J., 318, 664.

Woosley, s. E., Pinto, P. A., and Ensman, L. (1988). Ap. J., 324, 000.

Xu, Y., Sutherland, P., McCray, R., and Ross, R. R. (1987). X-Rays from Supernova 1987a. Ap. J., in press.

Fig. 1. The Gamma-Ray Imaging Spectrometer (GRIS), a balloon-borne high-resolution experiment. The imaging part is not shown in the figure.

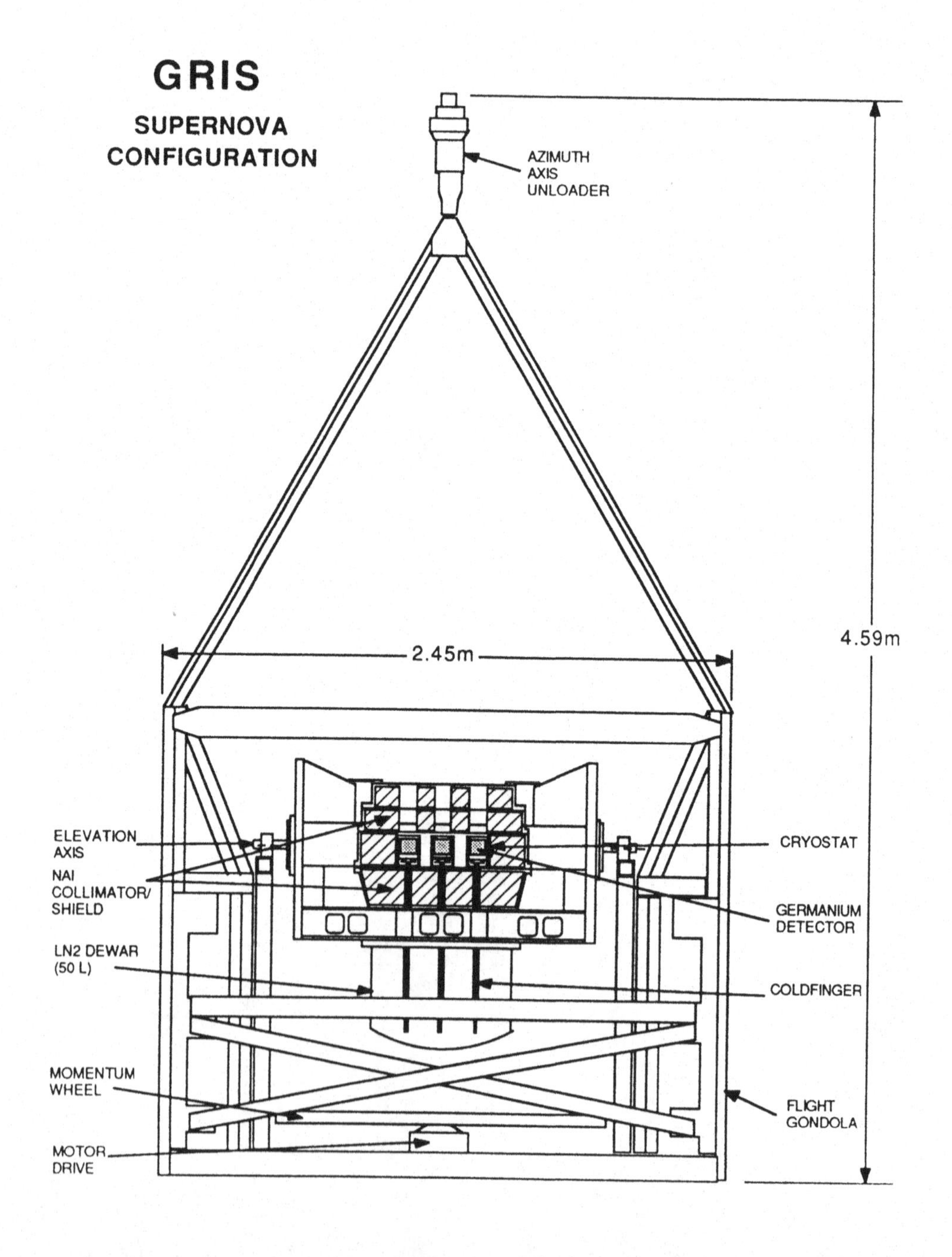

Fig. 2. The NRL gamma-ray experiment. It is a prototype module of the GRO/OSSE (Orbiting Scintillation Spectrometer Experiment).

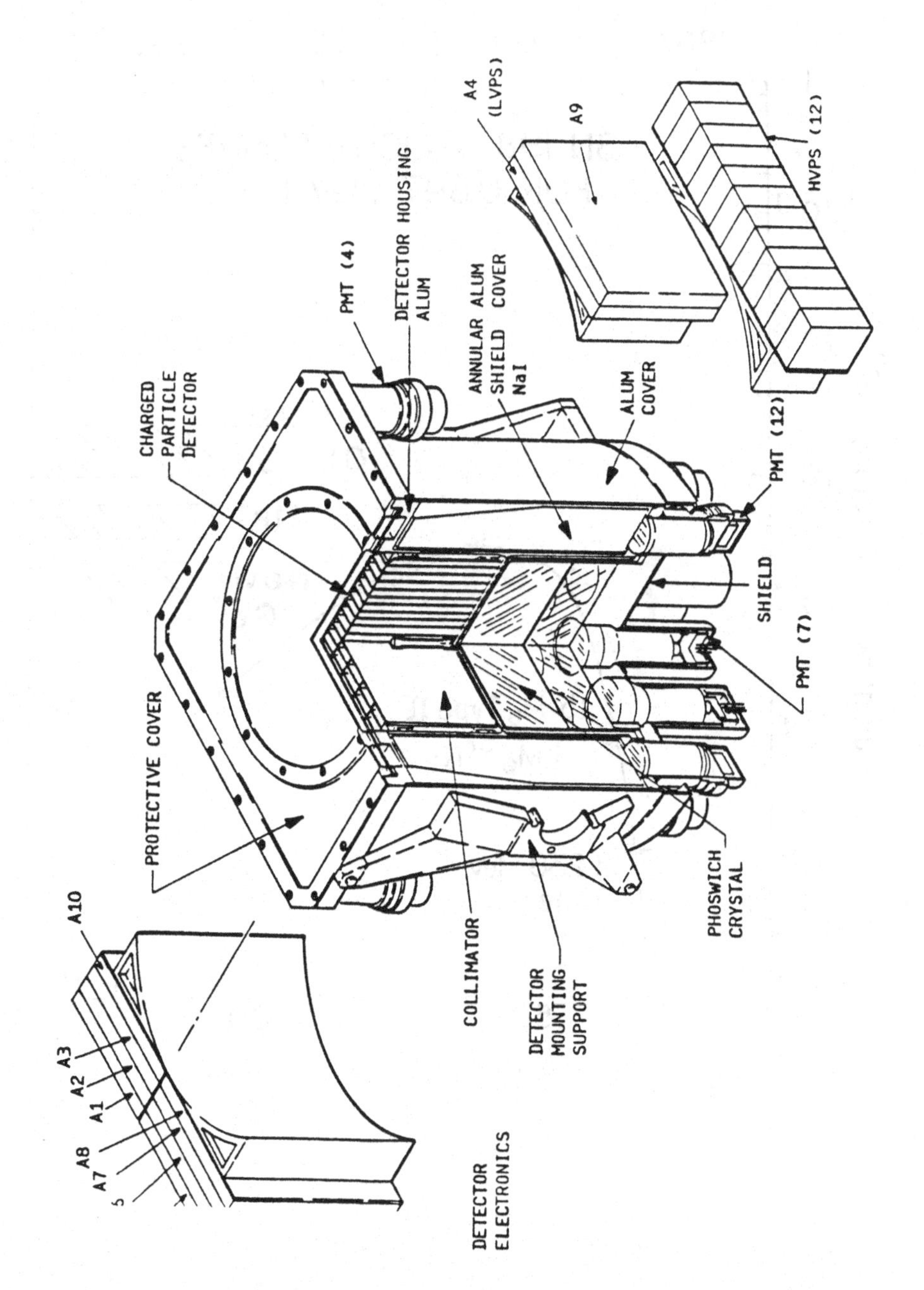

Fig. 3 Gamma-Ray intensity as a function of time from (Gehrels *et al.*). The strongest lines from the $^{56}Co-^{56}Fe$ decay are shown. Also shown are the 3σ line sensitivities of the representative gamma-ray instruments.

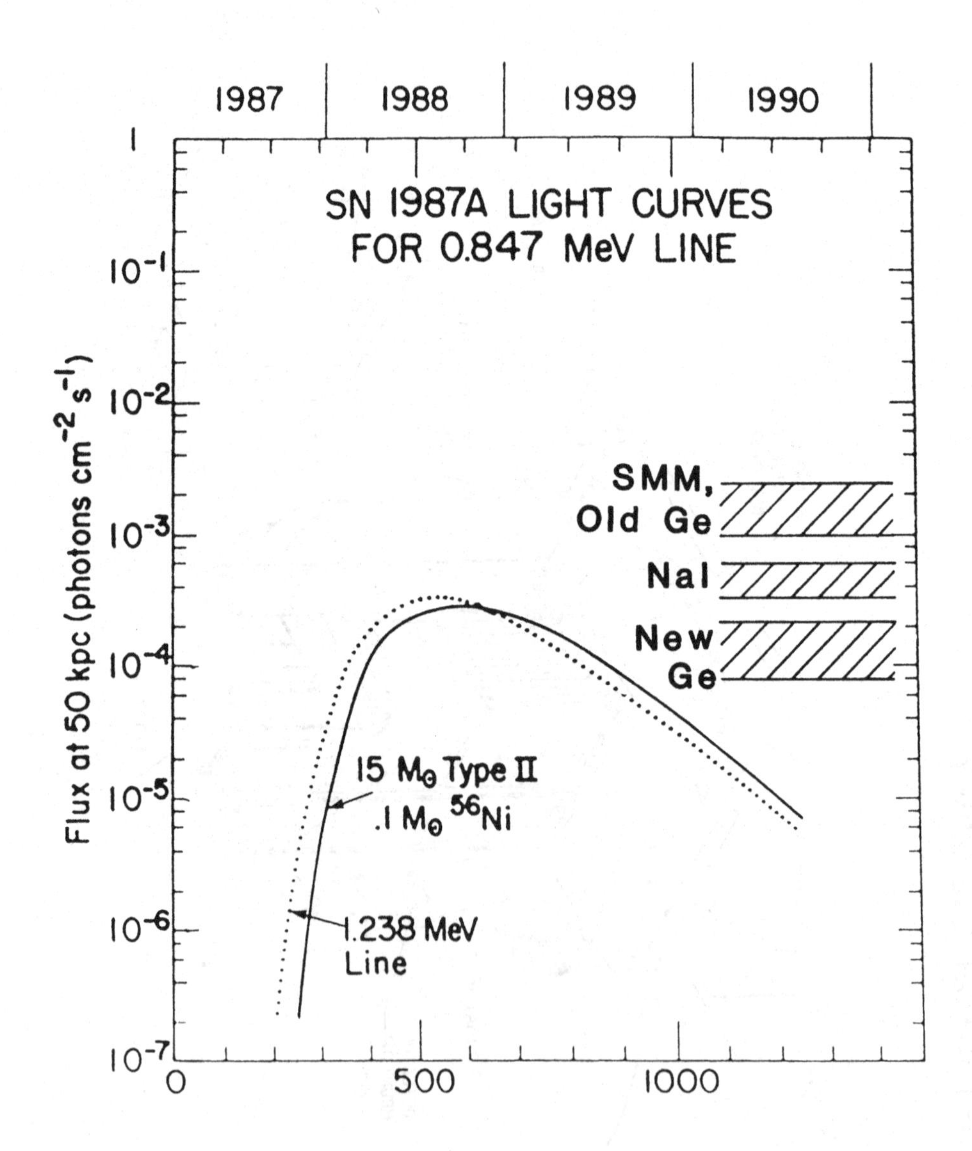

INFORMATION EXCHANGE FOR SN 1987A
A Scientifically Controlled Bulletin Board

Barry M. Lasker, Mark J. Henriksen, Nino Panagia[1], Peter M. B. Shames, Peter Shull, Jr.[2], Sarah Stevens-Rayburn, and Alan K. Uomoto[3]

Space Telescope Science Institute, 3700 San Martin Drive, Baltimore, MD 21218, USA.

Also [1]Astrophysics Division of European Space Agency and University of Catania, Italy; [2]Department of Physics, Oklahoma State University; and [3]Center for Astrophysical Sciences, The Johns Hopkins University.

Abstract. Research on the recent supernova in the Large Magellanic Cloud (SN 1987A) is facilitated by providing (1) a general-access computer based "bulletin board" for items of importance to the community, (2) a capability for communication among individual observers, and (3) access to relevant public data items available at the ST ScI, *e.g.,* directories of astronomers, bibliography for SN 1987A, and IAU Circulars relevant to SN 1987A.

These services are built upon existing communications networks and library capabilities and are managed by a science/engineering team centered at the ST ScI with additional science support being provided by participating astronomers at other institutions.

INTRODUCTION

Research programs involving SN 1987A which cover the entire spectral range, from radio to γ–ray frequencies, are being carried out by a large group of observers using ground-based telescopes, airborne facilities, balloons, rockets, and orbiting telescopes. Since these efforts are of considerable importance and interest to a number of theorists concerned with high-energy physics, stellar and galactic evolution, and the interstellar medium, the scientific community associated with SN 1987A is a large and diverse one. Additionally, one notes that (except for the orbiting telescopes supported by extensive telemetry facilities) the observations are all made from relatively isolated southern hemisphere sites and that the scientific developments related to this event normally have occurred, occur and are expected to occur in the future, on a time-scale far too rapid for the normal modes of scientific communication (letters, tapes in the mail, preprints, *etc.*) to be effective. An obvious requirement of such a geographically scattered group of investigators working on a fast-breaking topic is *communication.*

BULLETIN BOARD COMMUNICATIONS FUNCTIONS

Some of the communications requirements may be satisfied by facilitating communication among individuals, *i.e.*, a conventional electronic mail type of function, while others, related to the dissemination of general information to the larger community, are best addressed with a *computer bulletin board* function. The individual requirements are summarized below, and each is followed by a short discussion.

Communication between individual SN 1987A observers and theorists. The community of such investigators may be estimated at about 200 active scientists with a global distribution. For such a group, the effective communication method is electronic mail. (As alternatives, telephones are subject to imprecision, inconvenience, and language barriers, while ordinary written communications are far too slow.) To aid in electronic-mail communications a mail-drop facility as well as a directory containing the network addresses of SN 1987A investigators is provided.

Provision for scientific control of bulletin board (BB) functions. The policies and procedures used to operate the bulletin board are most appropriately designed and implemented by scientists familiar with the SN 1987A community and its research programs. This group (made up of scientists at ST ScI as well as other institutions) consists of a Network Scientist and a number of Deputy Network Scientists, hereafter referred to (along with their supporting staff) as the Bulletin Board (BB) Staff. The BB staff reads and responds to all incoming communications including items to be posted on the bulletin board. These items may be edited prior to posting in order to make them more consistent with the bulletin board format. However, the BB staff relies on voluntary contributions to the bulletin board by investigators and does not write reports on the work of other investigators.

Dissemination of reports from observatories, satellites, and sub-orbital missions. The periodic reports from NASA-sponsored programs, *e.g.*, the SMM reports, are available through direct channels, while the BB staff obtains similar materials from other programs, such as periodic monitoring by southern observatories, through the normal methods of scientific communication. These reports, as well as the other informational items specified later in this list, are organized and indexed appropriately by the BB staff and then posted for general access.

Forum for evolving new results. The Forum is intended to support an extended interchange among active parties, with the understanding that the materials are of the nature of a discussion, as opposed to a publication. Consistent with this is the policy that the BB *not* be cited as a reference in any other publication; rather, authors are expected to arrange with the BB contributors to cite their materials as *private communications.*

Program of Observations This function is intended to provide an outline of planned observations, a short statement of their goals, expectations, and schedules, and a list of the primary participants. In addition, follow-up reports consisting of a brief statement of the results and the names of the primary contact people involved are included.

Distribution of portions of I.A.U. Circulars pertinent to SN 1987A. This is especially important for observers at remote sites where the delivery of the *Circulars* by ordinary mail is unacceptably slow. The circular extract, prepared by the BB staff contains only text relevant to the Supernova. Circulars may be selected for reading either by number, date, or by keyword. Circulars are edited and posted on the Bulletin Board immediately upon their arrival at ST ScI.

Bibliography of SN 1987A preprints and publications. The literature of SN 1987A is currently evolving so rapidly that a carefully maintained bibliography (including preprints as well as published articles) is a very important research tool. Entries in the bibliography contain abstracts, titles, and publication information; it is possible to search references for keywords, authors, and dates.

Directory of SN 1987A investigators. A LOOKUP function for astronomers is offered as part of the Astronomy Information Service at ST ScI. In addition to the normal (address) information, the directory will list scientific interests, phone numbers, telex numbers, and network addresses. Information can be obtained on SN 1987A investigators by selecting the subset of astronomers who list SN 1987A as their scientific interest in the directory.

CONNECTIONS and USAGE

In order to provide the broadest and most convenient access for the SN 1987A community, connections are provided to the commonly used networks: SPAN, NASA Science Internet (NSI), ARPA Internet, X.25, TELEMAIL, BITNET, and UUCP. Some remarks about the individual networks follow:

- SPAN, NSI, and ARPA, which are relatively high-speed networks, are widely available in the USA and may be expected to be the most popular and convenient access method for domestic and European users.

- The X.25 connections (TELENET, NPSS [NASA Packet Switching Software], other names abroad) are essential to general international access, especially for new remote sites.

- TELEMAIL, BITNET, AND UUCP do not support remote logins and therefore message handlers that provide the capabilities to interrogate the BB are furnished.

At the time of this writing (1987 Oct 23), the facilities described herein were provisionally implemented on the ASTIS (ASTronomy Information Service) account of the RPS computer at the ST ScI. We expect that the ASTIS functions will be moved to another computer by the time these *Proceedings* are published; when that occurs, an account with the same name will be left on the

RPS computer to provide the new logon information. The ASTIS account has no password and may be reached from the ARPA Internet as STSCI-RPS.ARPA, from SPAN/HEPNET as RPS, or from the X.25 international public data nets with DTE address 31103010014012,1Y7R. (Note that the X.25 address is site and country dependent; additional network advice may be obtained from the ST ScI newsletters or from the ASTIS account itself.)

Materials to be posted on the bulletin board should be submitted by using the mail capability in the ASTIS account to send mail to SNREQ or by using one of the other mail paths which are posted on the bulletin board. Systems such as this are most useful to individual research efforts when they offer the broadest community participation; ***accordingly we most strongly invite you to send us your directory information and your scientific contributions.***

The authors are pleased to thank our colleagues at NASA and at the ST ScI for supporting the quick initiation of this project in response to the scientific opportunities afforded by SN 1987A. This project is supported by contracts NAS5–26555 and NAGW–1113 by The National Aeronautics and Space Administration to the Association of Universities for Research in Astronomy, Inc.

ANALYSIS OF THE DATA RECORDED BY THE MARYLAND AND ROME ROOM TEMPERATURE GRAVITATIONAL WAVE ANTENNAS IN THE PERIOD OF THE SN 1987 A

E.Amaldi[1,2], P.Bonifazi[2,3], S.Frasca[1,2], M.Gabellieri[2], D.Gretz[4], G.V.Pallottino[1,2], G.Pizzella[1,2], J.Weber[4], G.Wilmot[4]

1) Department of Physics, University "La Sapienza", Rome
2) Istituto Nazionale di Fisica Nucleare, Rome
3) Istituto di Fisica dello Spazio Interplanetario, CNR, Frascati
4) Department of Physics and Astronomy, University of Maryland

Abstract. The data obtained during SN 1987 A with the room temperature gravitational wave antennas of the Maryland and Rome groups have been analysed. An excess of coincidences at zero delay time above the accidentals is found in a period of 25000 s that includes the first neutrino observation by UNO. The probability that the excess is due to accidental coincidences of the background is estimated to be 3.5%.

Introduction

In presenting the results of the analysis of the data recorded in the period of SN 1987 A by the room temperature gravitational wave antennas of Maryland (M = 3100 kg, ν_R = 1660 Hz) and Rome (Geograv: M = 2300 kg, ν_R= 858 Hz) we like to stress that: (a) we are aware that today there are much more sophisticated g.w. detectors; (b) that the classical expression for the cross section of detectors of this kind gives values too small to allow the detection of the gravitational radiation expected from standard sources, but (c) these two antennas were the only g.w. detectors in operation in the period of the SN 1987 A and therefore (d) it appears to us a duty to make known what actually has been observed.

The analysis reported here covers the period from February 21, 1987, $18^h24^m33^s$ to February 23, $6^h2^m3^s$ and therefore includes the time of the UNO neutrino observation[1] (February 23, $2^h52^m36.8^s$) but unfortunately not the time of the Kamiokande[2] and IMB[3] (same day $7^h35^m41.4^s$), because electrical power problems in Maryland due to a severe snowstorm made the Maryland magnetic tape data unavailable and because in Rome at the local time 8^h35^m there were seismic disturbances.

The integration time for Geograv is equal to the sampling time Δt = 1s, while for the Maryland detector the sampling time is 0.1s with an optimum integration time of about 0.35 s. In order to simplify the data analysis we have computed from the Maryland recorded data 1 second averages. In this way, however, the SNR of the Maryland antenna is worsened by a factor close to $\sqrt{3}$.

We show in Fig.1 the data, representing energy innovations, recorded in 60 seconds starting from February 23, $2^h51^m50^s$ (UT). The Geograv data are expressed in kelvin; the Maryland data in arbitrary units, one of which corresponds to about 1/2 K. Geograv shows a peak which precedes by 1.4 ± 0.5 seconds the first neutrino observed by UNO[4]. The Maryland antenna shows a bigger peak 33 s earlier. The number written, within a circle, near the top of these two peaks, indicate the probabilities for them to accidentally occur in a preassigned second. We also notice a Maryland peak in coincidence with the above Geograv peak. The probability for this Maryland peak to be accidental is $6x10^{-2}$.

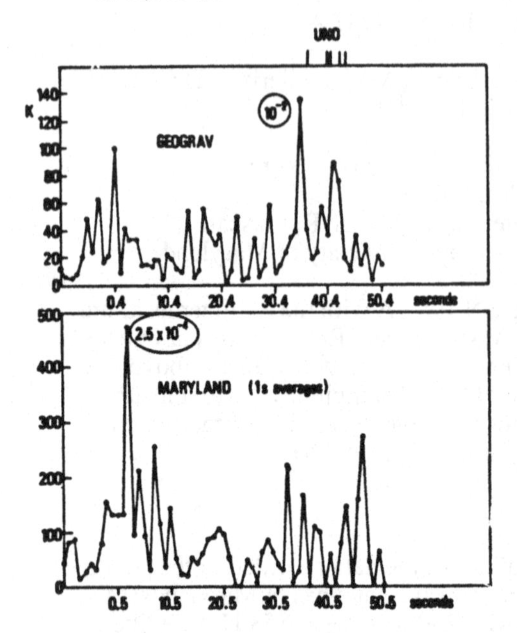

Fig.1 Energy innovations recorded by the Maryland and Rome room temperature antennas. On the top of the figure the neutrino events observed by UNO are shown.

Coincidences data analysis in the period from February 22, $23^h5^m23^s$ to Feb 23, $6^h2^m3^s$ (T = 25,000 s).

A test of an antenna performance is provided by the distribution of the energy innovation values, which, if only Brownian mechanical noise and gaussian electronic noise are present, should follow an exponential law.

Fig.2 shows the integral distribution of all the Geograv data recorded in the period indicated in the title of this section. Its very satisfactory exponential behaviour allows the determination of the effective temperature $T_{eff} = 29.5$ K.

A similar graph for the Maryland 1 s averaged data is shown in Fig.3. The corresponding effective temperature is $T_{eff} = 57.5/2 = 28.8$ K. This figure is larger than the best effective temperature of the antenna, due to the 1 s averaging. As it appears from Fig.1 the 1 s averaging of the Maryland data was made in such a way as to result only 0.1 second out of phase with respect to the Geograv data.

The coincidences between these two sets of data have been searched by counting how many times the signals of both detectors were above a given threshold within the same second.

The expected number of accidental coincidences is given by the well-known expression

$$\bar{n}\,(\Delta E) = \frac{n_1(\Delta E)\, n_2\,(\Delta E)\,\Delta t}{T} \tag{1}$$

where $\Delta t = 1$ s, $T = 25{,}000$ s and n_1 and n_2 are given by Figs.2 and 3, when their abscissae are used as variable threshold ΔE.

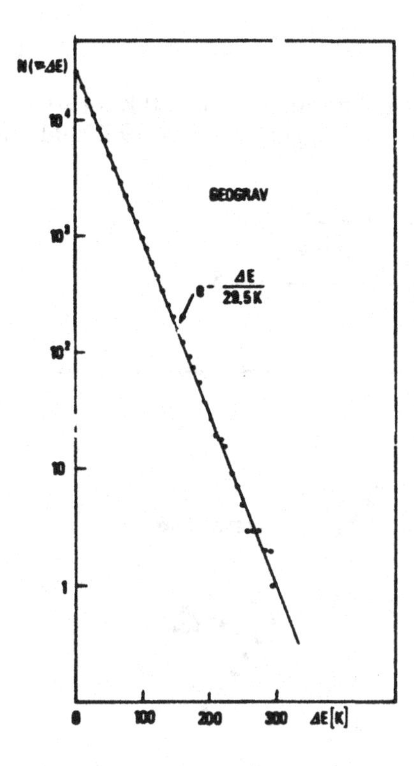

Fig.2 Integral distribution of the Geograv data during 25000 s starting from February 22, $23^h5^m23^s$ (Period 1).

The number of accidental coincidences can be determined experimentally by the well known method of shifting by a delay time δt one set of data with respect to the other and counting again the "new coincidences" $n(\Delta E, \delta t)$. Repeating this operation for a number of delays δt and averaging the corresponding values of $n(\Delta E, \delta t)$, for the same ΔE, it gives a number $n_{exp}(\Delta E)$, which is the experimental estimation of the accidental coincidences. Notice that the result of this method does not involve any assumption about the statistical distribution of the recorded data. Fig.4 shows $n_{exp}(\Delta E)$ determined by the above procedures for 100 values of δt spaced by 1 second and ranging from $\delta t = -50$ s to $+50$ s, except $\delta t = 0$.

The distribution of Fig.4 is in very good agreement with the theoretical value (1) and as expected corresponds to the exponential

$$\exp\left[-\Delta E/T_{effc}\right]$$

where

$$T^{-1}_{effc} = T^{-1}_{eff\ Geog} + T^{-1}_{eff\ Maryland} = [14.6\ K]^{-1}$$

The first attempt for coincidences was made for $\Delta E \geq 100$ K (corresponding to 100 x k joule, where k is the Boltzmann constant in joule/K). The corresponding delay histogram is shown in Fig.5. The number of coincidences for $\delta t = 0$ (zero delay) is indicated with a dot in a circle. All other dots refer to $\delta t \neq 0$. For $\delta t = 0$ one has

$$n_c\,(100\text{ K}) = 41 \tag{2}$$

which is the largest value appearing in Fig.5. From the values of $n\,(100\text{ K}, \delta t \neq 0)$ we compute their mean value, with respect to δt, $n_{exp}(100\text{ K}) = 29.2$ and the corresponding standard deviation

$$\sigma(\Delta E = 100\text{K}) = \sqrt{\sum_{\delta t \neq 0} \frac{[n(100\text{ K}, \delta t) - \bar{n}_{exp}\,(100\text{ K})]^2}{99}} = 5.0 \tag{3}$$

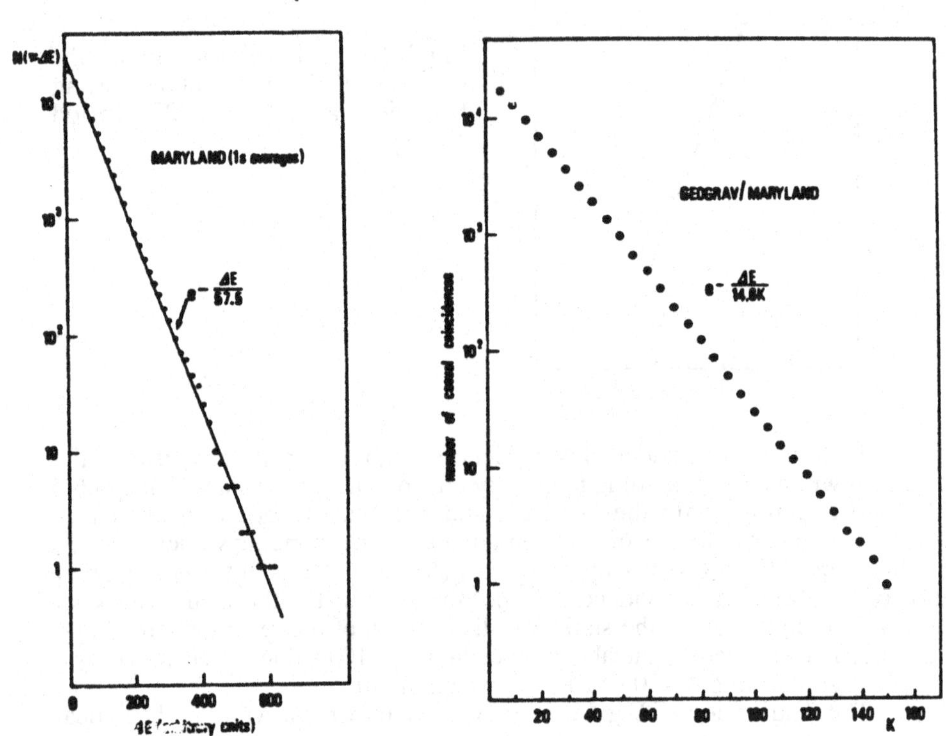

Fig.3 The same as in Fig.2 for the 1 s averaged Maryland data. One unit corresponds to about 1/2 K.

Fig.4 Integral distribution of the accidental coincidences.

Therefore, the value (3) of n_c (100 K) corresponds to

$$n_\sigma\,(100\text{ K}) = \frac{n_c\,(100\text{ K}) - \bar{n}_{exp}\,(100\text{ K})}{\sigma} = 2.4 \text{ standard deviations} \tag{4}$$

The above procedure was extended from 100 to 1000 delays, in order to improve the statistics, and repeated for various values of ΔE.

In Fig.6 we show the value of n_σ (ΔE) versus ΔE for ΔE ranging from 5 to 150K. In this figure n_σ(100 K) = 2.7 because of the different statistics (1000 delays). The numbers written under the arrows are n_{exp} (ΔE) and σ(ΔE).

A significant test of the statistical validity of these considerations is presented in Fig.7. It shows the distribution of the number of the accidental coincidences n(100 K, δt) for 999 different values of $\delta t \neq 0$, expressed in units of standard deviation (computed from eq.(3)). The experimental data represented by the dots follow very closely a gaussian function, as expected theoretically. This means that the probability of accidentally observing n_σ(100 K) = 2.7 can be estimated to be 0.35%.

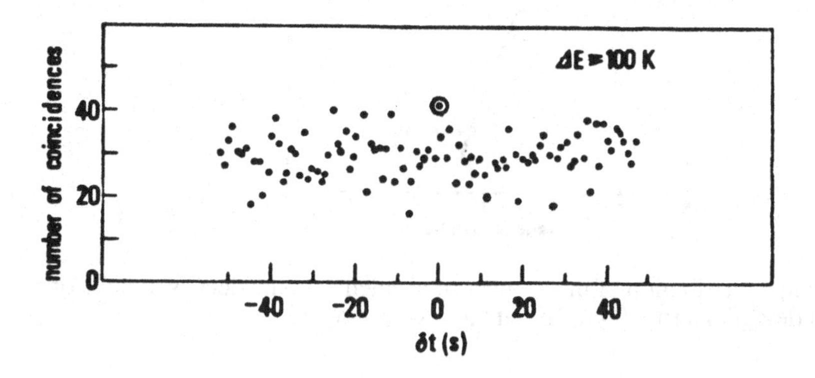

Fig.5 Delay histogram of the coincidences for the threshold $\Delta E \geq 100$ K (Period 1).

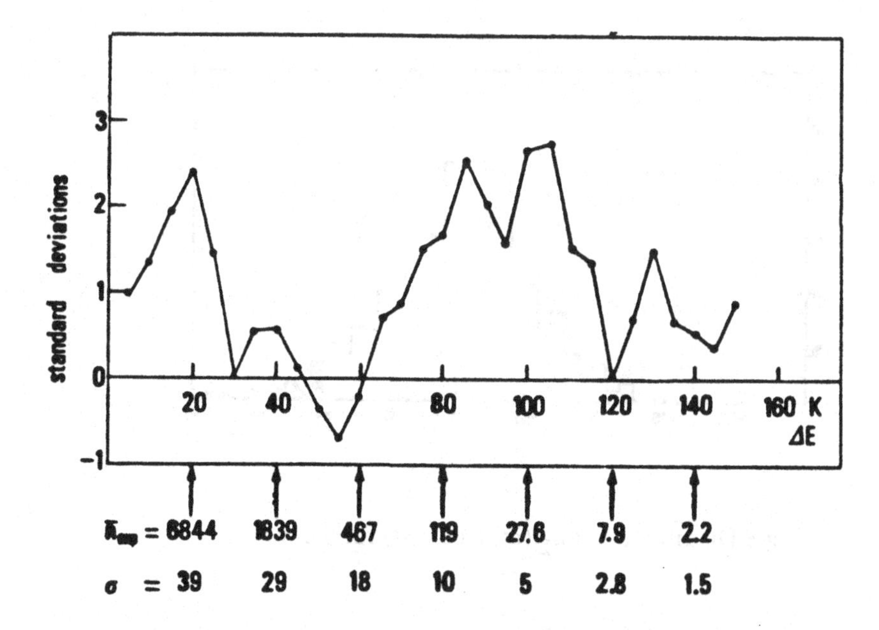

Fig.6 Number of coincidences in units of standard deviations for various energy threholds (Period 1)

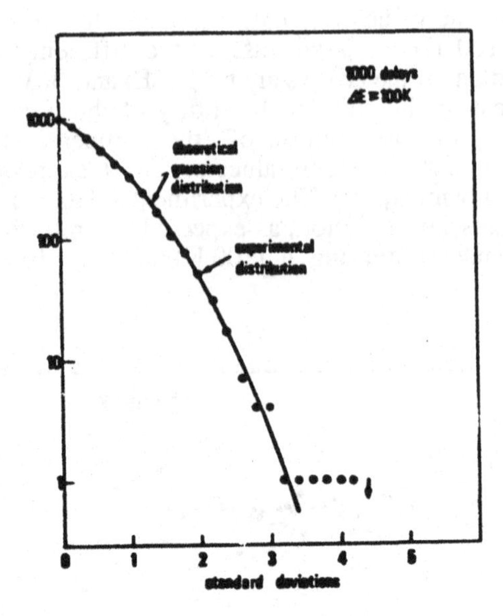

Fig.7 Experimental distribution of the number of accidental coincidences in units of standard deviations for 999 different values of delays.

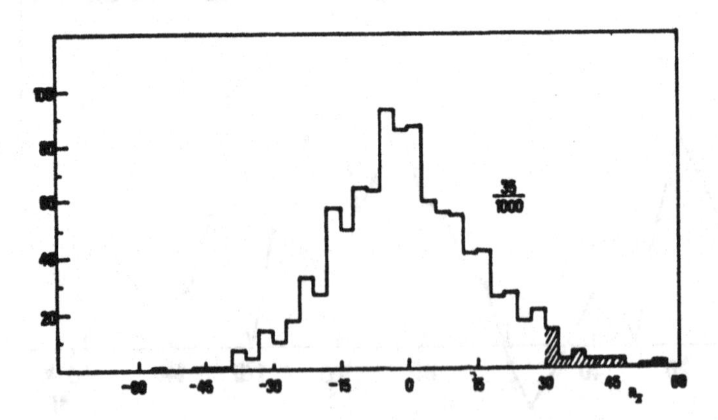

Fig.8 Distribution of n_Σ (δt) (see text) for the period 1.

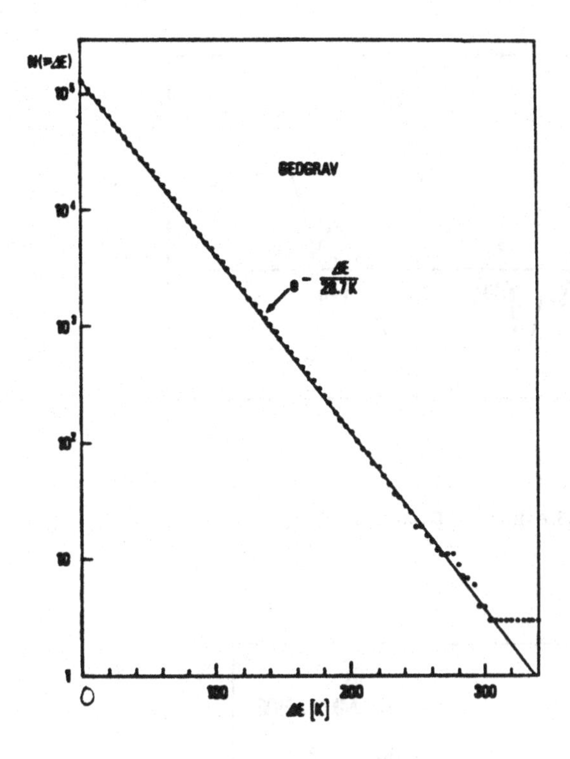

Fig.9 As Fig.2 for 100000 s starting from February 21, $18^h 24^m 33^s$ (Period 2)

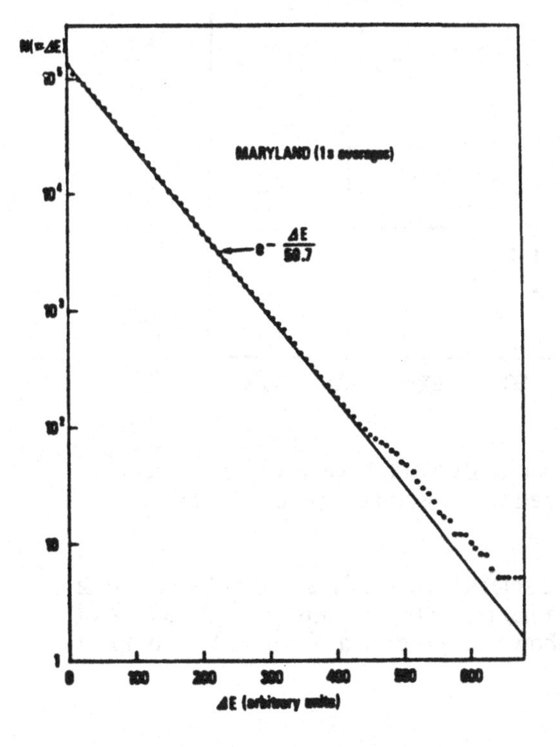

Fig.10 As Fig.3 for period 2

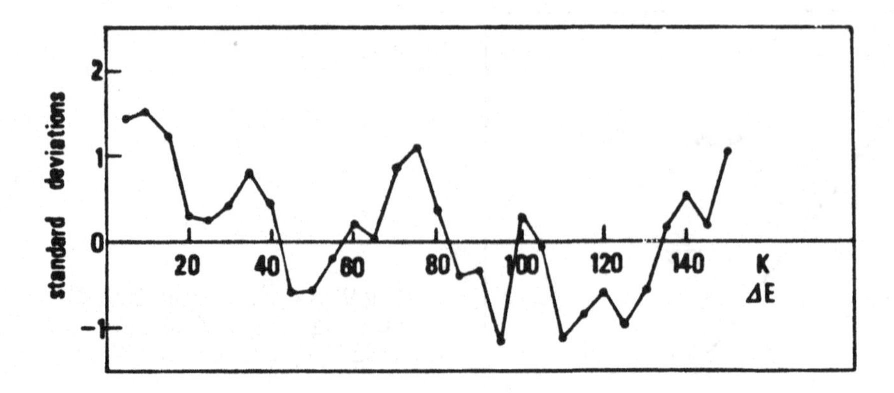

Fig.11 As Fig.6 for period 2.

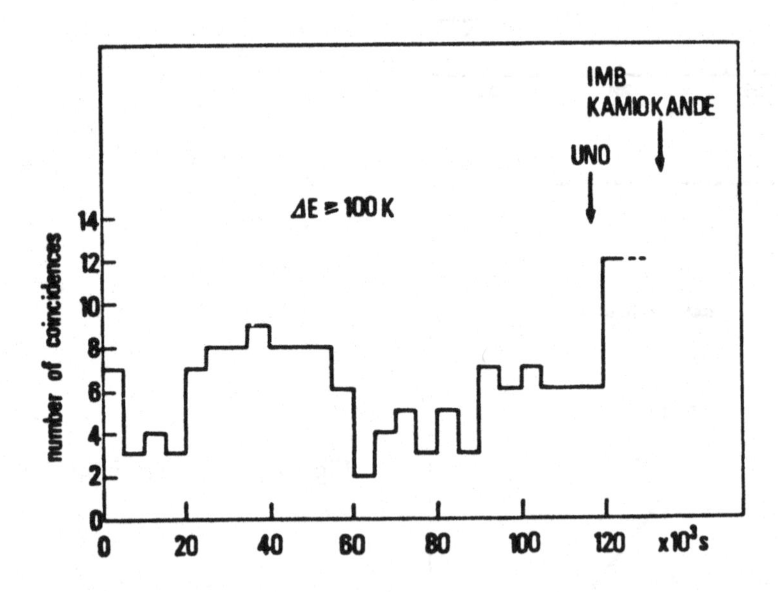

Fig.12 Number of coincidences for the threshold $\Delta E \geq 100$ K versus time for the total period of analysis. The times of the neutrino observations are indicated.

This value, however, refers to a selected value of the threshold, i.e. $\Delta E \geq 100$ K. Therefore it can not be taken as the final figure of the statistical significance of the result of the present analysis. We can, however, procede as follows. We compute the quantity

$$n_\Sigma(\delta t) = \sum_{5}^{30}{}_i \; n_\sigma(\Delta E = 5i, \delta t) \tag{5}$$

for each value of δt and plot in Fig.8 the corresponding distribution. For $\delta t = 0$, n_Σ (0) = 32, while the expected value of n_Σ (0) is 0.

From Fig.8 we see that there are 35 values of $n_\Sigma \geq 30$. Therefore we can take as an estimate of the statistical significance of the result of Fig.6

$$\frac{35}{1000} = 3.5\%, \tag{6}$$

which is ten times worse than the value obtained from Fig.7 for the predetermined threshold $\Delta E \gtrsim 100$ K.

Coincidence data analysis in the period from February 21, 1987 from $18^h24^m33^s$ to February 23, $6^h2^m3^s$ (T = 128250 s).

A similar analysis has been applied to the period indicated in the title of this section. Fig.9 and Fig.10 are similar to Fig.2 and 3. They show the satisfactory behaviour of both detectors for this more extended period. Fig.11 is similar to Fig.6 but refers to the period February 21, $18^h24^m33^s$ to February 22, $22^h7^m52^s$ (100,000 s), so that the data considered in the previous section are excluded. The amplitude of the oscillations around the line $n_\sigma(\Delta E) = 0$ are much smaller and show positive and negative values. The corresponding n_Σ ($\delta t = 0$) amounts to 2.4 to be compared with 32 found for the period analysed in the previous section. In other words in the period analysed here there is no excess at $\delta t = 0$.

In Fig.12 we have the number of coincidences for $\Delta E = 100$ K occurring, as a function of time in 5000 second-bins, in the period considered in this section. The times of observation of neutrinos by UNO, Kamiokande and IMB are indicated with arrows. The plot may suggest an increase of coincidences in the period of the SN 1987 A. Table 1 deduced from Fig.12 shows again that in the period preceding the SN there is no indication of coincidences at $\delta t = 0$.

Acknowledgement. This research has been financed by the "Ministero della Pubblica Istruzione" and by INFN in Italy and the National Science Foundation in U.S.A.

Table 1

Time interval (seconds)	n_{exp}	n_c	number of standard deviation
0 - 20000	19.3	17	-0.52
20000 - 40000	25.2	32	1.36
40000 - 60000	23.9	30	1.25
60000 - 80000	23.1	14	-1.89
80000 - 100000	19.3	21	0.38
100000 - 120000	22.1	25	0.64
120000 - 128250	11.1	20	2.67

- Bibliography -

(1) M.Aglietta, G.Badino, G.Bologna, C.Castagnoli, A.Castellina, V.L.Dadykin, W.Fulgione, P.Galeotti, F.F.Kalchukov, B.Kortchaguin, P.V.Kortchaguin, A.S.Malguin, V.G.Ryassny, O.G.Ryazhskaya, O.Saavedra, V.P.Talochkin, G.Trinchero, S.Vernetto, G.T.Zatsepin and V.F.Yakushev (1987): "On the Event Observed in the Mont Blanc Underground Neutrino Observatory During the Occurrence of Supernova 1987a", Europhys. Lett. 3 (12), 1315-1320.

(2) K.Hirata, T.Kajita, M.Koshiba, M.Nakahata, Y.Oyama, N.Sato, A.Suzuki, M.Takita, Y.Totsuka, T.Kifune, T.Suda, K.Takahashi, T.Tanimori, K.Miyano, M.Yamada, E.W.Beier, L.R.Feldscher, S.B.Kim, A.K.Mann, F.M.Newcomer, R.Van Berg, W.Zhang and B.G.Cortez (1987): "Observation of a neutrino burst from the Supernova SN 1987 A", Phys. Rev. Lett. 58, 1490.

(3) R.M.Bionta, G.Blewitt, C.B.Bratton, D.Casper, A.Ciocio, R.Claus, B.Cortez, M.Crouch, S.T.Dye, S.Errede, G.W.Foster, W.Gajewski, K.S.Ganezer, M.Goldhaber, T.J.Haines, T.W.Jones, D.Kielczewska, W.R.Kropp, J.G.Learned, J.M.Lo Secco, J.Matthews, R.Miller, M.S.Mudan, H.S.Park L.R.Price, F.Reines, J.Schultz, S.Seidel, E.Shumard, D.Sinclair, H.W.Sobel, J.L.Stone, L.R.Sulak, R.Svoboda, G.Thornton, J.C.van der Velde and C.Wuest (1987): "Observation of a neutrino burst in coincidence with Supernova 1987a in the Large Magellanic Cloud", Phys. Rev. Lett. 58,1494.

(4) E.Amaldi, P.Bonifazi, M.G.Castellano, E.Coccia, C.Cosmelli, S.Frasca, M.Gabellieri, I.Modena, G.V.Pallottino, G.Pizzella, P.Rapagnani, F.Ricci and G.Vannaroni (1987): " Data Recorded by the Rome Room Temperature Gravitational Wave Antenna, during the Supernova SN 1987 A in the Large Magellanic Cloud", Europhys. Lett. 3(12), 1325-1330.

SN 1987a AS AN EXTRAGALACTIC PROBE OF THE GALACTIC HALO AND EXTRAGALACTIC MEDIUM. PROSPECTS.

R. VIOTTI
Istituto Astrofisica Spaziale, CNR
Frascati, 00044 Italy

Abstract. Bright extragalactic sources, such as supernovae and QSOs, are ideal probes of the structure and chemical composition of the interstellar and extragalactic medium also at large Z, by means of the high resolution spectral analysis of the optical and ultraviolet interstellar absorption features. The LMC supernova has been for several months the brightest extragalactic point-like object, and has offered a unique opportunity of investigating to a greater detail the galactic halo and the ISM in the LMC. SN 1987a also represents a good example of the astrophysical inpact of future HIRES spectroscopic observations of distant SNs with the forthcoming new technology telescopes, and the Hubble Space Telescope.

SN 1987a has been (and still is in the visual at the time of this Workshop) the brightest extragalactic point-like source. This has represented a unique opportunity to study a supernova with combined very high spectral resolution and very high signal-to-noise ratio never reached in the past. It was thus possible to investigate with much more detail not only the spectral features of the SN, but also the absorption along the line of sight by gas in the galactic halo, and in the LMC. Observations in the visual and with IUE in the ultraviolet led to the identification of several atomic species with a forest of absorption components formed both in our Galaxy and in LMC (see e.g. Vidal-Madjar et al. 1987; de Boer et al. 1987). Preliminary data analysis suggests a metal deficiency in the LMC diffuse medium, as already known from other observations. Diffuse IS bands were for the first time detected in the spectrum of an extragalactic source (Vidal-Madjar et al. 1987; Vladilo et al. 1987). Vladilo et al. identified the bands at 5780, 5797, 6270 and 6258 with two components, at galactic and LMC radial velocities, with quite a small central depth, from 0.34 to 1.6 per cent of the neighbour stellar continuum. Thus only this chance of obtaining for the first time high S/N spectral observations of a high galactic latitude extragalactic object, allowed the first detection of these diffuse features in the galactic halo and in the LMC. D'Odorico et al. (1987) also found a wide and asymmetric absorption feature close to the position of the forbidden [FeX] 6374.51 A line, redshifted by the LMC radial velocity. This absorption could originate from a very hot halo of LMC, associated with the extended Tarantula Nebula (30 Dor) complex, which is the largest association of massive stars in the Local Group of galaxies.

Observation of distant QSOs has has given the opportunity of studying their very rich far-UV spectrum with ground based telescopes, and to separate the IS absorption lines into the many components formed in different absorbing media along the line of sight (e.g. the so-called Lyman forest). Supernovae frequently have a brightness comparable to that of their parent galaxy and, in addition, have the advantage of being point-like sources, thus easier to be spectroscopically observed than a galaxy. Their absolute luminosity near maximum is so high that a SN in the Virgo Cluster could be now observed with fairly good spectral resolution and S/N ratio. D'Odorico et al. (1985) observed SN 1983n in M83, and found that the IS lines were splitted in at least 11 components which can be attributed to absorption by gas in the Milky Way, in M83, and possibly by gas in the Local Group of galaxies. This example shows how fundamental is and will be the study extragalctic SN in cosmological problems.

The rapid development of focal instrumentation and especially of new technology, large collecting area ground telescopes (NTT) will soon give a better chance of observing faint objects with high spectral resolution (10,000 or more) and high S/N (300-1000). This will offer the opportunity of observing point-like sources (SNs, AGNs) at cosmological distances and of studying the nature of the ISM in external galaxies and in their haloes, and of the intergalactic medium. Several very large telescopes are in costruction or under project, with sizes of the primary mirror from 6.5 m to 4x8 m and possibly more. In addition, the Hubble Space Telescope will give the possibility to observe faint astrophysical objects in the UV at high resolution. We may thus anticipate a great progress in our knowledge of the diffuse matter in extragalactic space in the near future.

I am grateful to Paolo Molaro for his preprints, and discussions.

References.

de Boer, K.S., Grewing, M., Richtler, T., Wamsteker, Gry, C., and Panagia, N.: 1987, Astron. Astrophys., 177, L37.

D'Odorico, S., Pettini, M., and Ponz, D.: 1985, Ap. J. 299, 852.

D'Odorico, S., Molaro, P., Pettini, M., Stathakis, R., and Vladilo, G.: 1987, ESO Workshop on the SN 1987 A, in press.

Vidal-Madjar, A., Andreani, P., Cristiani, S., Ferlet, R., Lanz, T., and Vladilo, G.: 1987, Astron. Astrophys., 182, L59.

Vladilo, G., Crivellari, L., Molaro, P., and Beckman, J.E.: 1987, Astr. Astrophys., 182, L59.

SUMMARY OF GEORGE MASON UNIVERSITY SN1987a WORKSHOP*

S. van den Bergh
Dominion Astrophysical Observatory
5071 W. Saanich Road
Victoria, B.C. V8X 4M6 CANADA

1 INTRODUCTION

It is sometimes said that astronomy teaches us humility. SN1987a is certainly a good example of this. If someone had asked us in late February of 1987 to predict the maximum magnitude of this supernova we would have answered confidently (depending on our prejudices regarding the Hubble constant) that V(max) = +1.0 or V(max) = +2.5. This Conference has, however, also shown that there are some things that we can justifiably be proud of. Perhaps the most important of these are: (1) Theories of core collapse in supernovae have been brilliantly confirmed by observations of neutrinos produced by SN1987a and (2) Observations of the exponential tail of the light curve of SN1987a give strong support to the prediction that this phase of supernova light curves is powered by ^{56}Co decay.

2 THE PROGENITOR STAR

Astronometry (cf. West *et al*. 1987) now appears to show beyond reasonable doubt that the star Sk-69°202 was the progenitor of SN1987a. Walborn started off our meeting by discussing his study (Walborn *et al*. 1987) of the structure of the image of Sk-69°202 on plates obtained with the CTIO 4-m telescope. His results show that, in addition to the supernova progenitor (star 1), there is evidence for a second object with V = 15.3 located at a distance of 3 arcsec from the centre of the image of star 1 and for a third star with V = 15.7 separated by 1.5 arcsec from Sk-69°202. Furthermore Walborn's work provides tantalizing hints for the possible existence of a fourth component in this system. Van den Bergh (1987) has shown that the joint probability of having objects as bright as stars 2 and 3 located so close to Sk-69°202 is 10^{-5} i.e. it is virtually certain that the supernova progenitor was the brightest member of a physical multiple system. Since all of the stars in such a multiple star are coeval the luminosity of star 2 may be used to obtain a lower limit to the mass of the supernova. If star 2 is presently a main-sequence object then it has a mass of 10 $\mathcal{M}_\odot$ which implies that the more high-evolved supernova progenitor must have $\mathcal{M} > 10\ \mathcal{M}_\odot$. This lower limit to the mass of the supernova progenitor is, however, somewhat weakened by results presented by Sonneborn, Altner & Kirshner (1987) which show that star 2 has

*To avoid duplication I have not included material that was already covered in my summary of the ESO Workshop on SN1987a held in Garching, West Germany, July 6-8 1987.

log g = 3.0, which indicates that it is somewhat evolved.

3 OBSERVATIONS AT CTIO

On behalf of the Cerro Tololo observers Phillips showed a montage of spectra of SN1987a. These spectra provided a "film" of the spectral evolution of the supernova. A particularly interesting feature revealed by this montage was a pair of satellite lines to $H\alpha$ with velocities of ± 4000 km s^{-1}. Chalabaev speculated that this feature might due to jets moving in opposite directions or to a rotating ring, while Lucy suggested the possibility that the $H\alpha$ satellites might represent a spheroid of enhanced emission revealed by the retreating photosphere. The CTIO data also show that the lowest velocity hydrogen, which presumably marks the bottom of the hydrogen shell, has V = 2100 km s^{-1}. This low-velocity hydrogen was first seen between 25 and 40 days after outburst.

4 SPECTRAL ENERGY DISTRIBUTION

Danziger used observations obtained at ESO to discuss the evolution of the spectral energy distribution of the supernova. Recently this energy distribution has exhibited two components the first of which can be represented by a black body with a temperature of 4900°K, whereas the second may be modelled by a black body with a temperature of 700 ± 100°K. The latter component peaked approximately 100 days after the supernova outburst at a luminosity of 9×10^6 $L_\odot$. This implies that the infrared luminosity of the supernova was increasing at ~1 $L_\odot$ per second!

5 SHELL SURROUNDING SK-69°202

Both Danziger and Kirshner discussed the dense shell surrounding the supernova which was, presumably, ejected by a slow wind emitted by the red giant precursor of Sk-69°202. Material in this shell was first seen in the spectrum of the supernova in mid-May and had become quite strong by July. As a result the supernova spectrum is presently intermediate between stellar and nebular phases. The radius of the supernova shell is estimated to $\sim 1 \times 10^{18}$ cm. Excitation of this shell is believed to have taken place by the EUV flash that was emitted when the shock produced by core collapse finally emerged from the surface of the progenitor star. The total luminosity of this EUV flash is estimated to have been $\sim 10^{47}$ ergs. In this connection it is of interest to note that Peimbert & van den Bergh (1971) invoked an EUV flash with $L > 10^{50}$ ergs to account for the faint HII region observed near the supernova remnant Cas A. The fastest material detected in the supernova spectrum had V(max) ≈ 35000 km s^{-1}. This gas should hit the shell (unless it is decelerated by circumstellar material) in the 1990's resulting in strong X-ray and radio emission. Spectroscopic observations show that the supernova shell is nitrogen-rich. Such nitrogen enrichment (Walborn 1987) is observed in the atmospheres of many massive stars and does not imply that the SN progenitor was a WR star. Recent IR spectra of the supernova show strong CO emission at 1.2 μ and 4.6 μ produced in the supernova shell. Observations of CN λ3883 might be used to set an upper limit on the number of comets in any "Oort Cloud" surrounding Sk-69°202.

6 Ba AND Li ABUNDANCES

Models by Lucy show that BaII must be between 5 and 30 times overabundant relative to its abundance in other LMC stars. The exact value of this overabundance is, however, somewhat uncertain because the ionization structure in SN1987a is not well known. At the present time it is not at all clear why s-process elements should be overabundant in the supernova precursor. ESO observers have so far failed to detect interstellar Li in the supernova spectrum. It follows that Li/H $< 1.2 \times 10^{-10}$. This limit is comparable to the Li/H ratio that Spite *et al.* (1984) have recently found in stars of Population II. Clearly it would be important to push the supernova observations of lithium to fainter limits.

7 OBSERVATIONS OF THE EXPONENTIAL TAIL

On behalf of the SAAO group Feast reported on UBVRIJKL photometry of the supernova in South Africa. This photometry, which covers the range 0.36-3.5μ, encompasses almost all of the energy of the supernova. The measurements are therefore particularly well suited to determinations of the evolution of the bolometric magnitude of the supernova. Two different approaches have been used. In one of these a single black body curve was fitted to the data whereas, in a second approach, spline fits were made to the actually observed data points. Note that special care has to be exercised in the interpretation of the infrared portion of the spectral energy distribution, because some of the presently observed IR radiation may represent re-radiation of previously emitted energy. The data show that, starting about 120 days after the explosion the bolometric luminosity of SN1987a exhibits an exponential decline with an e-folding time between 104 and 120 days. This rather large uncertainty results from the difference between spline and black body fits to the data and from the remaining uncertainty in the reddening of the supernova. The observational data neatly bracket the 111.7 day e-folding time of ^{56}Co, thus providing strong confirmation of models in which the tail of the supernova light curve is powered by the γ-rays that are emitted as ^{56}Co decays into ^{56}Fe. The very last observations by Feast and by Kirshner (IUE), taken 270 days after the supernova explosion, lie below the trend line of the light curve. It will be interesting to see if future observations confirm this trend.

8 COMPARISON WITH DISTANT SNII

Panagia showed fits of the exponential tail of the light curve of SN1987a to the light curves of other supernovae that have been observed in more distant galaxies. The beauty of this fitting procedure is that it provides a direct determination of the Hubble parameter. In his poster paper Schaefer showed a montage of the light curves of many supernovae of type II. This montage served to emphasize the tremendous differences that appear to exist between the light curves of different SNII. If confirmed by future photoelectric and CCD observations these differences would tend to lower the confidence one feels in fits of the light curve of SN1987a to those of more distant supernovae.

9 FREQUENCY OF SNII

In their poster paper Schmitz & Gaskell emphasized the fact

that the frequency of SNII may have been underestimated if subluminous objects, such as SN1987a, are common. It is, however, unlikely that the presently accepted SNII frequency (van den Bergh, McClure & Evans 1987) could be in error by a factor as large as 4. The reason for this is that it would then be necessary (van den Bergh 1988) for stars with masses as small as $\sim 3\,\mathfrak{M}_{\odot}$ to become supernovae even though there is strong evidence indicating that such objects evolve non-catastrophically to the white dwarf stage.

10 CAUSE OF LOW LUMINOSITY AT MAXIMUM

It is now quite clear that the low luminosity of SN1987a at maximum light was due to the fact that the precursor of this object was a blue rather than a red supergiant. It is, however, not yet certain whether this difference was a direct result of the low metallicity of the supernova precursor. Filippenko has initiated an observing programme to search for supernovae in, presumably metal-poor, dwarf irregular galaxies to see if all supernovae of type II in such objects are subluminous.

11 OBSERVATIONS AT THE AAO

Couch reported on a variety of interesting observations of SN1987a that have been obtained at the Anglo-Australian Observatory: (1) Structure has recently been observed in the $H\alpha$ emission line profile indicating that the supernova shell is beginning to break up into clouds. (2) Polarization observations show that the supernova has an asymmetrically scattering atmosphere. If the supernova is modelled as an oblate spheroid with an axial ratio μ, then the present observations indicate that $0.77 \leq \mu \leq 0.93$. This result will affect distance determinations using the Baade-Wesselink method, which assumes emission of the supernova to come from a spherical shell. (3) High resolution ($\lambda/\Delta\lambda$ = 520000 at NaD) observations, which were obtained with a jerry-rigged spectrograph, show an unexpectedly small velocity dispersion $\sigma_V \leq 1$ km s^{-1} in interstellar clouds in the Galactic halo. (4) [Fe X] is found to be unexpectedly strong with $n \simeq 10^{17}$ atoms cm^{-2}. The most likely interpretation of this observation is that we are seeing gas in a hot bubble surrounding the 30 Doradas complex in which SN1987a is embedded.

12 X-RAY OBSERVATIONS

Skinner & Trumper reported on X-ray observations by the Kvant experiment on the MIR Space Station. Trumper et al. first observed hard X-rays on August 10. During August $\Delta I/I \leq 30\%$. The hard X-ray light curve of the supernova should, in the coming months, provide information on the amount of scattering material in the supernova shell. Observations by Skinner, which place a lower limit on the soft X-ray flux, appear to be in mild disagreement with GINGA observations reported by Tanaka. By good fortune the GINGA satellite was launched only a few weeks before the supernova exploded. Its observations appear to indicate that the X-rays from the supernova contain at least two and possibly three distinct components: (1) Soft X-rays with energies of 6-10 keV that have a thermal spectrum. These soft X-rays initially increased rapidly in intensity but started to drop noticeably 250 days after the supernova explosion. (2) A hard component with a flat energy spectrum that extends from 10 keV to at least 40 keV. Finally (3) the GINGA observations provide marginal evidence for the presence of an iron line. Presumably the soft X-rays from the supernova are formed by shocks in the ISM surrounding Sk-69°202. It is not yet clear why these shocks should not also produce some observable radio emission.

13 RADIO INTERFEROMETRY

Bartel reported on interferometric observations obtained five days after the neutrino event with an interferometer having a baseline of 5500 km between Tidbinbilla in Australia and Hartebeesthoek in South Africa. This interferometer proved too powerful for the job and fully resolved the source from which it may be concluded that the radio emitting region in the supernova had a diameter $\phi \gtrsim 2.4$ milliarcsec = 120 AU.

14 SPECKLE OBSERVATIONS

Observations of the "mystery spot" by Papaliolios were eagerly awaited by all the conference participants. This feature was seen by Papaliolios at CTIO on both March 25 and April 2. Subsequently it was observed at a low S/N ratio by Meikle on April 14. Observations by Papaliolios, Meikle and Chalabaev in May, June, July and August did not show any evidence for the existence of the "mystery spot". The reality of this object will therefore probably remain a subject of lively debate. Papaliolios' observations also yielded the following diameters for the supernova expressed in milliarcsecs:

Colour	April 2	July 1
B	12	23
V	11	18
Hα	1	8
IR	...	15

There are two puzzling features about these observations: (1) It is not clear why the Hα diameter of the supernova should be so much smaller than its diameter in the continuum and (2) The diameter of the supernova given by these observations is in order of magnitude greater than that derived from black body fits to the spectral energy distribution.

15 EXPLOSION MODELS

Woosley presented the beautiful results of his hydrodynamical calculations of the supernova explosion. The fact that these models successfully predicted the general behaviour of the supernova luminosity greatly contribute to their credibility. His "best buy" model has the following characteristics:

$\mathfrak{M}$ (main sequence) = $19 \pm 3\ \mathfrak{M}_\odot$
E (neutrinos) = 2.5×10^{53} ergs
E (explosion) = 8 x 50 ergs
$\mathfrak{M}$(Co) = $0.07\ \mathfrak{M}_\odot$
$3\ \mathfrak{M}_\odot < \mathfrak{M}$(envelope) $< 14\ \mathfrak{M}_\odot$

UV Flash L(max) = 10^{44} ergs sec^{-1}
UV Flash E = 10^{47} ergs.

16 NEUTRINO OBSERVATIONS

Contrary to previous results (Wampler *et al.* 1987) recent hydrodynamical calculations show, beyond reasonable doubt that the time at which the supernova exploded was that given by the IMB/Kamiokande experiment rather than that observed at Mont Blanc.

The neutrino observations of the supernova were discussed by Bahcall. He emphasized that theoretical interpretations of this event suffer from "neutrino starvation" because only 19 objects were observed. (It is perhaps, of interest to recall that Oort [1950] based his theory of the solar comet halo on accurate major axis determinations for only 19 long-period comets.) Bahcall emphasized the effects of the tyranny of small-number statistics by showing 100 energy versus time diagrams for 19 neutrinos produced by Monte Carlo methods that were all consistent with his "best buy" model for the supernova core collapse which was characterized by the following parameters:

$E \simeq 6 \times 10^{52}$ ergs
$\mathcal{M}^{\nu}$ (core) $\simeq 1.4 \mathcal{M}_{\odot}$
$T_o \simeq$ 4MeV
$\tau \simeq 4.5$ sec

Bahcall concluded "There is more in the supernova than in our models, but it will be hard to prove".

17 FUTURE WORK AND DESIDERATA

The present workshop clearly showed that there is an urgent need for more basic observational data on supernovae. It is not good enough to just discover supernovae in distant galaxies. We urgently need photometric and spectroscopic follow-up observations. Only four SNII presently have reasonably good UBV light curves, which shows just how serious our lack of basic information is. The fact that some of the great telescopes in the Southern Hemisphere were using 7.5 magnitude neutral-density filters for their spectral observations shows that information as detailed as that for SN1987a could easily be obtained for supernovae that are 1000 times fainter than the LMC supernova.

At the time of the workshop no γ-rays had yet been detected from the supernova. Theory predicts that the expanding shell of Sk-69°202 should become transparent to γ-rays in a few months. We wish satellite, rocket and balloon observers lots of γ-rays for Christmas!

REFERENCES

Oort, J.H. (1950). Bull. Astr. Inst. Netherlands, 11, 91.
Peimbert, M. & van den Bergh, S. (1971). Ap.J., 167, 223.
Sonneborn, G., Altner, B. & Kirshner, R.P. (1987). Preprint.
Spite, M., Maillard, J.P. & Spite, F. (1984). Astr. Ap., 141, 56.
van den Bergh, S. (1987). Nature, 328, 768.

van den Bergh, S., McClure, R.D. & Evans, R. (1987). Ap.J., 323, in press.
van den Bergh, S. (1988). Ap.J., in press.
Walborn, N.R. (1987). In Atmospheric Diagnostics of Stellar Evolution, IAU Colloquium no. 108.
Walborn, N.R., Lasker, B.M., Laidler, V.G. & Chu, Y.-H. (1987). Ap.J. (Letters), 321, L41.
Wampler, E.J., Truran, J.W., Lucy, L.B., Höflich, P. & Hillebrandt, W. (1987). Astr. Ap., 182, L51.
West, R.M., Lauberts, A., Jørgensen, H.E. & Schuster, H.E. (1987). Astr. Ap., 177, L1.
Williams, R.E. (1987). Ap.J. (Letters), 320, L117.

AN EXTRA, VERY BEAUTIFUL BUT BRIGHT STAR IN THE WRONG PLACE

Robin Bates
Producer, NOVA
WGBH-TV, Boston, MA 02134

Memory is its own master. It keeps its own rules for what is to be remembered and what is not. Its long-term component, memory proper, seems limitless, and memory traces once stored may last forever--or for a lifetime at least. First each item must pass through a short-term device whose capacity and duration are restricted to seven "chunks" and to a few seconds at most. Meaning helps. Mnemonists who memorize long strings of numbers break them into groups and give each a semantic tag: 1776 or 1492 have obvious correlates for Americans (as does 1066 for an Englishman, like me).

Communal memory collects events of global importance, preserving a million, or a billion, varying versions in molecular cold-storage. The metaphor is apt, for such incidents are usually chilling--a natural disaster, the outbreak of war, or the assassination of a president. Only on a lesser scale, in small groups, do common rememberings bring joy. To those of us who participated, the supernova of 1987 has been such an event.

The first memories come from the instant of hearing the news. Bob Kirshner, an ardent supernova observer: "It was like a dream come true. You can wait your whole lifetime for something like this to happen." Or theorist Stan Woosley: "I remember walking around for about a week and to everyone I would see, I'd say 'BOOM!'" .

For David Jauncey, co-coordinating the Australian radio astronomy effort, first--and second--reactions were more complex. "Okay, there's a supernova in the Large Magellanic Cloud. Within half an hour the penny dropped: 'There's a <u>supernova</u> in the Large Cloud!' It's something we've had four hundred years to think about, but it's not a thing that you're ready to say what you want to do."

Astronomy has a well-oiled machinery for reporting major events. You call, or telegraph, Brian Marsden at the International Astronomical Union's Central Bureau for Astronomical Telegrams at the Smithsonian Astrophysical Observatory on Garden Street in Cambridge, Massachussetts. It sounds like a great institution, but, as any one who has visited it will attest, it consists of a couple of cluttered offices, overflowing with computers, telephones, and files. If Marsden deems your discovery sufficiently important, he will let the world know by sending an IAU circular--a telegram--to several hundred subscribers. Yet, as it turns out, still there is nothing faster than word of mouth.

In New Zealand Albert Jones, a distinguished amateur astronomer with many discoveries (and now an O.B.E.) to his credit, called Frank Bateson with news of the supernova. Bateson, unable to telex the IAU, called the Anglo-Australian Observatory instead. There Tom Craggs, the chief night assistant, informed Rob McNaught, who had the supernova sitting on a photograph, unnoticed, from the night before. At the mighty Mills Cross radio telescope at Molonglo, near Canberra, Duncan Campbell heard by similar unofficial means. He "threw away" the night's observing schedule and next morning, when his director called to alert him to the supernova's existence, already he had its first fleeting radio emission recorded. As Lewis Thomas once observed, these days the telephone is the major instrument of scientific research.

For Ian Shelton, not yet a seasoned observer, the appearance on his plate of "an extra, very beautiful but bright star in the wrong place" aroused suspicion, if not downright disbelief. "The first thing that hits me is that Kodak have really done it this time! The next thing that hit me was, gee, why didn't I use <u>that</u> as a guide star? The third thing was that there <u>is</u> no guide star there. The fourth thing was that it's late and I'm probably hallucinating..."

Shelton was in good company. Here is a young Tycho Brahe, struggling to free himself from Ptolemy's crystalline spheres, glimpsing the supernova of 1572:

> As if astonished and stupefied, I stood still, gazing for a certain length of time with my eyes fixed intently upon it...When I had satisfied myself that no star of that kind had ever shone forth before, I was led into such

> perplexity by the unbelievability of the thing that I began to doubt the faith of my own eyes...

Like Tycho's, Shelton's scepticism did not subside until he had the intruder's existence and nature confirmed by others (Tycho quizzed both his servants and the occupants of a passing carriage). In the control room of Las Campanas' 40-inch telescope he found Barry Madore, Robert Jedrzejewski, and the night assistant, Oscar Duhalde. Oscar announced that he had seen the bright object a couple of hours before--thus becoming the first "official person" to see the supernova. "We stumbled out," recalls Madore, "and for a brief moment four people stood there just amazed at the sight."

My own initiation had no such drama. I read about it in The New York Times; twelve days after the discovery. The report mentioned a meeting at NASA Goddard and quoted Bob Kirshner at Harvard. Next morning, a Monday, I telephoned him. Was it as big a story as it seemed, I asked. You bet, said Bob. Mid-afternoon and several phone calls later I talked to Paula Apsell, NOVA's Executive Producer and my boss, who happened to be in London. She knew all about the supernova.

> "What do you want to do?"
> "I want to go to Chile--on Saturday"

Making a NOVA is not unlike conducting an experiment. Nominally, the process lasts six months: two months for research, to write the script and arrange the filming schedule--framing hypotheses, designing protocols; a month for filming--collecting data; and three months for editing and finishing--analysis and writing a paper. (One significant difference is that while negative results may be unpopular in science, in television they are entirely unacceptable. Having let you go out and make a film, they show it to 20 million people whether you are pleased with the result or not.) Clearly the supernova NOVA was not to follow the usual pattern. Like the astronomers whose footsteps we dogged, much of our work was carried out at a trot. It gave us a ready answer for colleagues who inquired how we had gotten the project moving so fast: For once the schedule was dictated not by WGBH, but by events in the universe.

Harold Urey was once asked the reason for his success. "I work only on important problems," he replied. I cannot claim only to make films about important subjects, although sometimes I have. But always I have chosen topics which, for a while at least, have fascinated me. In this my superiors at NOVA have been willing accomplices, mostly. For while the series must strive for balance, it is producers who make programs and there is no substitute for a producer's enthusiasm. Supernova 1987A was evidently important and there was the additional intrigue that I had no idea how the story, or rather the film, would end. With hindsight, one further factor played a part. It needs some explaining, since it involves Hollywood, Aristotle, and a graph.

Film is a funny medium to work in. Vivid, supple, it is capable of conveying surface reality in a way nothing else can. By nature it is a performing art, or first cousin to one. The crucial difference is that film lacks the feedback from the audience which is possible in a live performance. The film director sets his decisions in celluloid and hopes the audience will follow along. Pacing becomes perhaps the most delicate aspect of film craft.

Action is its key ingredient, if you will allow me to stretch the term to include talking as well as walking, and its metier is telling stories. That is to say film works on the emotions. It is not good at conveying information, or not very densely. The printed page is better at that. Film's history proves the point: it first flourished in Hollywood, and still feature films are the most glamorous, the wealthiest, and the most prestigious end of the business. What do they do?--they tell stories. In features, as in all fiction, information tends to be incidental. Just enough is given to enable the audience to follow the plot.

The maker of science films carries a heavy burden, for inevitably his (or her) stories contain arcane facts and obscure concepts. Explain too little--the easy option--and you cheat the audience; explain too much and you forfeit drama. Furthermore, science dwells in the realm of ideas, from which action and drama are not easily secured. Yet if I am correct, the basic rules of story-telling, with their essential peaks and troughs, are ignored at peril. Which brings me to Aristotle.

Stories, explains Aristotle[1], must have "a beginning, a middle, and an end." Order is important: "A beginning is that which does not necessarily come after something else, although something else comes about or exists after it. An end, on the contrary, is...not itself followed by anything." From which it is evident that "A middle is that which follows something else, and is itself followed by something" (Aristotle et al. 1982). It is easy to poke fun at Aristotle's solemn formulation, but someone had to do it first and the Greeks were rather good at this kind of thing. Translated into graphical terms, the results are shown in Figure 1.

The graph shows the ebb and flow of drama in a NOVA (or in any story whatsoever). The horizontal axis represents time; the vertical, units of drama. For a film, they are measured in DeMilles--with divisions of mega-, micro-, and milledemilles. With feature films the scale is logarithmic, but for documentaries, with their less dramatic content, a linear scale generally will do. The initial bump is the prologue, introduction, or tease, designed to capture the audience's attention and tell them what the film will be about. Then there is a period of exposition and development rising to the climax, followed by a sharp fall.

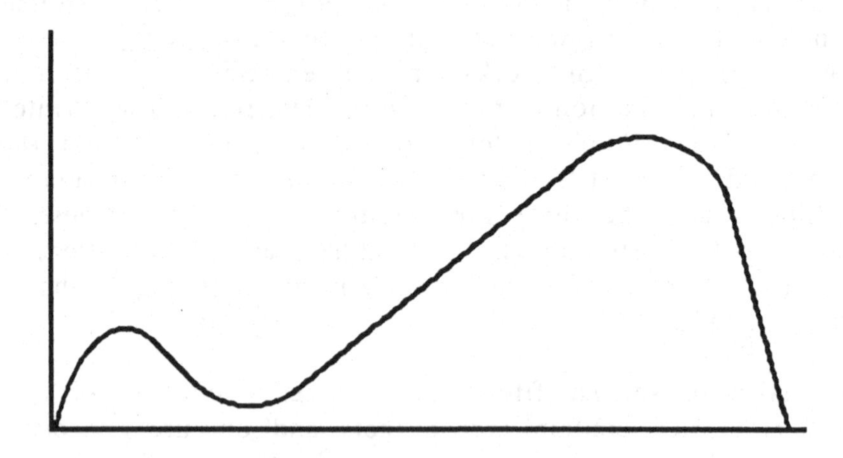

Figure 1: Dramatic curve of a NOVA

1 To be accurate, Aristotle is talking of tragedies, but his argument is perfectly general.

The curve, incidentally, is fractal in nature. Take any section--representing a sequence--enlarge it, and a similar pattern is revealed.

What is the relevance of this to the supernova? Compare the graph to the supernova's light curve (there are dozens scattered throughout the book). The fit is remarkable. A film about the supernova was, as it were, a marriage made in heaven!

One last point on this, to illustrate theory in practice. I said earlier that I did know where the film would end; in fact, I did. On my first trip to Chile, at the very start of filming, I shot not only the opening--the discovery--but the ending too. Setting Ian Shelton against a sunset and Oscar Duhalde against the stars, I asked for their reflections on what discovering the supernova had meant. Wherever the supernova led, I figured it would be pertinent to leave the last word to them.

Memories, then. The film is the official record, but memory captures other incidents for whatever meaning they contain. Let me start with an admission: even with a story like the supernova, which unraveled as we went along, only rarely were we present when critical events took place. Only once, as a matter of fact--the night at Cerro Tololo, in Chile, when Costas Papaliolios and Peter Nisenson discovered the second object, an unexpected companion of the supernova, two light-weeks distant. (Actually, we were also present two months later when Peter, returning for more data, lost the object again). Even then, at the time, we did not grasp the significance of the moment. We had it on film, and a few weeks later when they announced their results, we felt pretty smug. At least we would have done, but for a mishap. On the way back from Chile, the airline lost our sound tapes.

In the end we fared better than the scientists: the tapes finally turned up. Flown back to Chile by mistake, they were sold by customs to a Santiago television company before Juan Lazzerini thoughtfully returned them to us. Meanwhile, at the time of writing anyway, the mysterious additional object remains disappeared.

Chile is a magical place from which to see the stars. In the Andes, at dusk, as snow-tipped peaks fade from pink to blue, you can stand on any modest mountain-top and the splendor of the southern skies takes your breath away. The Milky Way dominates, a million glistening points of light spray-painted across the heavens. To one side lie two faint patches--the Magellanic Clouds, our closest galactic neighbors, seen by and named for Magellan on his epic circumnavigation of the planet. Now home, of course, to the supernova. From each telescope dome music emanates, classical, heavy metal, or jazz, as if, still astonished, they are singing back at the night.

The observatories--Las Campanas and Cerro Tololo American-run, La Silla or E.S.O, the European Southern Observatory, self-evidently not--squat on Andean foothills along the southern fringes of the Atacama Desert. Devoid of people, the land is like California before the gold rush; condors, on the other hand, are common. (No mystery there.) And one early morning, at Campanas, we saw a fox. It watched us for a while, then bored with filming, scampered off light-footed towards the rising sun.

Each observatory possesses a personality. Las Campanas has an intimate, small-town air, where everyone's business is known to everybody. "Campanas" means "bells", and Oscar Duhalde can take you to a place where, until you tire, you can toss rocks down the hillside and make the mountain ring. Cerro Tololo, larger, is more formal. Suburban, perhaps, but lacking the full downtown rigidity of Kitt Peak, its States-side counterpart. And E.S.O.? Well, E.S.O is an armed camp. When we arrived at La Silla, our reception was distinctly chilly. We were refused permission to film and even my best lines came to nothing.

> "Society supports science and so is entitled to know what you do."
> "This observatory is supported by Europe; your program is for Americans."
> "But science belongs to everybody!"

Later I complained to a young astronomer who befriended us: "This place is a dictatorship." He shook his head. No, its a _theocracy_, he said.

New Zealand was more accommodating, if equally short of people. We had gone there to fly at 41,000 feet on NASA's Kuiper Airborne Observatory and observe the supernova in the infrared. The pre-flight briefing began in business-like fashion: "The Dodgers beat the Padres 4 to 2. Clemens finally won one for the Bos-sox..." The aircraft is a Lockheed C141, a heavy military transport divided into three: the cockpit, the cavernous main cabin where the astronomers and Mission Director operate, and, in between, sealed-off and floating on air-bearings, the telescope. Our flight would last five and one half hours and would head east before arching back over the island on the supernova run. Here is an excerpt from the log:

7: 8: 59	BLC UP STARTING TO OPEN THE DOME
7: 14: 23	WE ARE IN THIN CLOUDS
7: 29: 33	WE MAY HAVE A PROBLEM WITH THE COMPRESSOR DUMP SIGNAL!
7: 29: 50	HAVE A SIGNAL IN THE INFRA RED
7: 42: 18	HAVE SHUT DOWN COMPRESSORS WILL SWITCH TO #1 AND #2
7: 42: 28	RUNNING AGAIN
7: 44: 42	INTO CLOUDS AGAIN
7: 47: 23	BAD CLOUDS
7: 48: 18	********Begin SN 1987A (LMC SUPERNOVA)

That was the pattern of the trip. Problems with the compressors--which power the air-bearings and had to be bled by hand--and problems with the weather. As we re-approached Christchurch on the supernova leg, the astronomers innocently suggested aborting the flight. The Mission Director, who had been with the K.A.O. from the beginning, was too wily to be caught. He explained: if the flight was abandoned for technical reasons, NASA would provide another flight. But weather--NASA never guaranteed that. Besides, intermittently, data were coming in.

On our single day off, we--Kathy White, my associate producer, and I--planned to visit Arthur's Pass, an area of mists and waterfalls which marks the center of the Southern Alps. At the airport shop, Kathy tried to buy a New Zealand map. "No, I want one with <u>all</u> the roads." "Those are all the roads," she was told.

Australia came next, but I want to get to Japan. I'll allow myself a couple of lines before I move on. I had first visited Australia two years earlier to film the primordial stromatolites at Shark Bay, and, south of Port Headland, the earliest life-forms, fossil bacteria, preserved in black chert. What is Australia like, I had asked a friend. Oh, it's wonderful, he grinned; like California, only without California's deep sense of history. Wonderful it was, and the remark is unfair. Although to whom, I'm not sure.

This time, as we left the Anglo-Australian Observatory at Siding Spring, a dozen clamoring cockatoos passed overhead. In the newspaper a few days earlier--this is true, I swear--a visiting curator from Karachi Zoo had expressed surprise at seeing a similar flock flying wild. She had not realized they came from Australia, she explained. Back home the birds spoke only Pakistani.

As billed, the Bullet Train took 5 hours 40 minutes to reach Hiroshima. It was my last stop in Japan, impromptu, and by far the most telling. The filming was over: Ginga, at the time the sole operational x-ray satellite, and Kamiokande II, the Japanese-American experiment buried deep in a zinc mine, where Yoji Totsuka and Al Mann were still jubilant after their detection of neutrinos from the supernova. Billions had passed through the Earth, from the bottom up; Kamiokande had grabbed a handful as they flashed by. The burst had lasted 13 seconds and two minutes later the device turned off for routine self-calibration. Little wonder the experimenters were delighted.

The visit to Hiroshima was a pilgrimage, I suppose; over the years my interested in the Manhattan Project had grown. I had met several of the bomb-builders and read about many more; my friend Tom, professor of physics at Harvard and a novelist of distinction, had made it the backdrop of his first book; I knew Enrico Fermi's grand-daughter through some strange chance; and just a month earlier I had visited Stirling Colgate, the father of supernova neutrino-theory, in Los Alamos. Now I would see the other end of things.

In Tokyo I gathered a bunch of books by survivors to read on the train. As the single westerner in carriage filled with Japanese, I read furtively, disturbed by an obscure sense of shame. I knew the arguments that had led to the decision, but there was no avoiding the fact: we--my parents' generation--had dropped it on them. The survivors accounts were simply written. They told of the flash and the shock and, in the aftermath, of haunted figures, their clothes burned off, their skin hanging in black flaps. Other tales--childrens' stories--told of mothers or siblings trapped in the wreckage, to be burned where they lay in the fires that followed. On the train I found myself fighting tears. There was no trick; the experiences of those who lived through it had been unspeakable.

I arrived in Hiroshima after dark. It seemed a thriving modern city, entirely rebuilt. Later that evening, when I made my way to the island near the center which houses the Peace Park, I found a museum, an eternal flame, and monuments to peace and to those who had perished. Yet they were not what I had come to see. I wanted some more direct connection to the past. Eventually I reached the Aioi Bridge, distinctively T-shaped in plan, which connects the point of the island to the banks on either side. With a start I remembered that this was the formal target of the bomb. Across the river rose the ruins of the A-bomb Dome, a familiar sight at last! Preserved, too, as a monument, it remains the sole visual link to those 40-year old, rubble-filled photographs.

In the sunshine next morning the park was transformed. Families on Sunday outings picnicked on the grass, and a group of schoolgirls in sailor-suits waved flirtatiously at me. Outside the museum the fire department drew up a collection of fire trucks and, with varying skill but uniform enthusiasm, children painted pictures of them. Inside the museum drawings by survivors again turned back the clock. Untutored and child-like, they were as moving as the written accounts. One showed "Professor Takenaka," a solitary figure surrounded by fire. In translation, the caption read:

> "Where could that rice ball have come from which he held in his hand? The professor, naked, standing before the flames, holding a rice ball, seemed to me like the symbol of the modest hopes of humanity."

In the early 1950's, when Fred Reines and Clyde Cowan planned their epic detection of the neutrino, they thought first to use the neutrinos produced in an atomic explosion. Later they decided to use the steady stream of the particles emitted by a nuclear reactor. Enrico Fermi declared he liked the new arrangement better: they would not have to set off a bomb every time they wanted to check the results. Neutrinos, neutrinos! Neatly they encapsulate the central dilemma of science. And for me they forged another link between Hiroshima's past and its present.

Two last supernova anecdotes, reflecting differing facets of the scientific endeavor. The first concerns Bob Kirshner, who, as is now notorious, confused us all early on by announcing that Sanduleak - 69: 202, the presumed progenitor of the supernova, was still there. After re-scrutinizing his data and admitting the error Bob characteristically found a one-liner with which to sum up: "The supernova was a bit dim to begin with--like some of the observers."

The second concerns Oscar Duhalde and may be my most cherished memory of the whole event. One evening at Las Campanas, the first part of the 40-inch telescope's schedule had been given over to the supernova. When the observations had been collected, the astronomers began reconfiguring the instrument for their own work. Alone in one corner Oscar sat hunched over a computer console. Bright-eyed as he watched the supernova spectra appear, he was muttering to himself "Fantastic! Fantastic! Fantastic!"

I end with a quotation. Unashamedly, it is from me. It was written after my first trip to Chile, when I was trying to make sense of the event:

> To the right of the Milky Way is a patch of faint light, twice the apparent diameter of the moon. In the top left-hand corner, barely visible to my unaccustomed eyes, is a pinpoint of light--the supernova. Through my binoculars I can see the Tarantula Nebula, a tiny smear, to its north. For a while I stare hard, fixing the image in my mind, trying to imagine the tumultuous events which are generating the supernova's light. I <u>know</u> it is different, the experts keep telling me so. But as I stand on that mountain top half a world from home, I admit to myself that it looks indistinguishable from any other star. The significance is not in the sight itself, but in its meaning--a gift of the men and women throughout time who have turned their eyes to the sky, marveled, and struggled to explain what they saw.

I had a wonderful time. Thank you for letting me come along.

Acknowledgements

To give what I have written the appearance of a real paper, I have allowed myself one footnote, one reference, and one graph. It is now a pleasure to add an acknowledgement, or two. My gratitude and thanks, first to Kathy White, associate producer, who contributed to every aspect of Death of a Star; to Bill Lattanzi (film editor), John Hazzard (cameraman), John Lomberg & Dan Krech (animation), and Peter Melnick (music); to Paula Apsell and Bill Grant and the rest of the NOVA staff. PBS stations, Johnson & Johnson, and Allied Signal support NOVA, as did on this occasion, the NSF. Under grant #MDR 8751407 I achieved Principle Investigator status at last!

References

Aristotle, Horace, Longinius: Classical Literary Criticism. Penguin Classics, 1982

Index